UNIDIRECTIONAL WAVE MOTIONS

NORTH-HOLLAND SERIES IN

APPLIED MATHEMATICS AND MECHANICS

VOLUME 23

NORTH-HOLLAND PUBLISHING COMPANY
AMSTERDAM · NEW YORK · OXFORD

UNIDIRECTIONAL WAVE MOTIONS

H. LEVINE

Department of Mathematics,
Stanford University

1978

NORTH-HOLLAND PUBLISHING COMPANY
AMSTERDAM · NEW YORK · OXFORD

North-Holland ISBN: 0 444 85043 0

Published by:

North-Holland Publishing Company
Amsterdam · New York · Oxford

Sole distributors for the U.S.A. and Canada:

Elsevier North-Holland, Inc.
52 Vanderbilt Avenue
New York, NY 10017

Library of Congress Cataloging in Publication Data

Levine, Harold, 1922–
Unidirectional wave motions.

(North-Holland series in applied mathematics and mechanics; v. 23)
Bibliography: p. 499
Includes index.
1. Wave-motion, Theory of. I. Title.
QA927.L44 531'.113'0151 77-15565
ISBN 0 444 85043 0

Printed in The Netherlands

To my wife, Barbara,
who inspires in all ways

Preface

Wave phenomena, with their multi-faceted aspects and diverse physical associations, sustain a considerable and continuing interest. It is well nigh impossible to survey, let alone assimilate, the accumulated findings over the centuries; and hence the acquisition of both perspective and procedural facility in this important sphere becomes ever more difficult. Students are conventionally introduced to wave theory as it bears on the particular subject matter of courses in mathematics, physics, and engineering, while the benefits of systematization in concept and technique remain unexplored for the most part. Books about wave theory in general are not so numerous, though some recent titles which manifest authoritative content and broad scope facilitate progress towards greater appreciation and technical mastery of the subject. This entrant to the published ranks aims to describe, in a comprehensive manner, the formulations and their consequent elaborations which have found demonstrable value in wave analysis; the deliberate focus on unidirectional waves permits a relatively simple mathematical development, without leaving significant gaps in methodology and capability. Since it is the full resources of mathematics which underlie successful dealings with all manner of wave problems, there are sufficient grounds for a detailed account of methods per se, encompassing both the active or contemporary representatives as well as those previously – and possibly again to be – judged useful. Sections of comparable size constitute the organizational framework of the book, marking a departure from the more typical arrangement of subject matter in chapters, and it is hoped that direct access to individual sections may enhance the utility of the work; problems are included, some taken (with permission) from examination papers set in The University of Cambridge, to display the nature of analytical solutions and further the ends of technical competence.

An author's indebtedness, in connection with an undertaking such as this, to information and benefit received from many sources cannot be adequately documented; special thanks are due, however, to T. T. Wu and F. G. Leppington for stimulating inputs, and to J. Schwinger for initially spotlighting the elegance and versatility of applied mathematics. I am also grateful to Priscilla R. Feigen for an abundance of practical support and to the Office of Naval Research for prior material assistance.

H. Levine

Contents

Part III

Introduction

Wave motions figure prominently, now as heretofore, in numerous and diverse physical phenomena, and their understanding is mirrored in the evolution of the related theories. An impressive roster of technological developments pertains to instances of wave propagation in materials or acoustical/electrical settings; and the less common though significant natural disturbances of seismic or tidal origin direct attention to both the practical and theoretical aspects of wave motions on a terrestrial scale.

Since the ramifications of wave motion have long permeated the physical sciences, a vast corpus of general and special theory, techniques of approximation and descriptive patterns exists and is continually being augmented. On the purely mathematical side, investigations have contributed to the fund of knowledge concerning the nature of continuous and discontinuous solutions to the equations of wave motion, and to the conditions for the existence and uniqueness of solutions when various initial and boundary conditions are laid down, as befit the excitation and subsequent interaction of the waves with spatial inhomogeneities. The approximations devised by Fresnel (before the differential equations governing the propagation of light in vacuo were known), and later by Kirchhoff, in connection with the theory of optical diffraction at an aperture in an opaque screen, and the rules of geometrical optics, are examples of efficient and economical procedures for some restricted purposes; such improvisations do not rest on firm mathematical foundations or provide quantitatively accurate details for all circumstances, but they testify nevertheless to the utility of sound physical insight in dealing with complex situations. In this book we propose to follow a middle course between the extremes of a formal mathematics of wave motion on the one hand, and of a collection of physically motivated approximations on the other; the object will be to describe and apply a sufficiently general mathematical basis which is capable of furnishing both the gross and fine details of varied wave phenomena, and which brings to the subject the benefits of integration and systematization.

The subject of wave motions may be broadly (though not exclusively) subdivided under the headings: Description, Excitation, and Interaction. To the first belongs the particulars of individual wave equations, the geometry or kinematics of wave forms, and general matters bearing on the transport of energy and momentum, as well as dispersion or frequency sensitivity. Since an ever growing number of wave problems find their origins rooted in anisotropic or random media (including elastic substances, plasmas, and electrically conducting fluids), geometrical and analytical techniques of considerable scope are essential to achieving a full account of the manifold and distinctive phenomena possible. The second heading embraces aspects of wave generation by localized and extended sources, respectively; and it is noteworthy, in this context, that plane wave forms, attributable to infinitely remote sources, shed no light on the (non-dissipative) geometrical attenuation of wave amplitudes which reflects physical reality. Point source, or so-called Green's, functions for all types of wave motions enjoy a prominent status; that is, both differential and integral representations of said functions, and their asymptotic form when the source and observation points are widely separated, find ready use in the formulation and analysis of specific problems. Sources with finite extension, such as acoustical oscillators and electrical antennas, have a practical importance, prompting studies of directional characteristics (or radiation patterns) and the influence thereupon of changes in the ratio between the emitted wave length and the source dimension. Apart from such actual or primary sources, it is advantageous to obtain representations of the wave function (via the Green's functions) in terms of equivalent sources, whose strengths are related to assigned or unknown values of the former on particular lines or surfaces; these representations are especially useful as regards the third and final heading of our subject matter, wherein the effects of interaction between specified primary waves and medium irregularities (e.g., obstacles, inclusions, or local variations in the material parameters). Information about these irregularities or scatterers, ranging from microscopic to terrestrial scales, may be gleaned through the attendant phenomena of reflection, refraction, and diffraction, which are fundamental to wave theory. There is an impressive roster of exact and approximate results for scattering in different configurations, whose existence rests on utilizing the full resources of mathematical analysis, and the application of complex function theory, integral transforms, asymptotic expansions, and variational formulations in particular; point source functions are recognizable throughout the analysis, being either the very object of calculation in composite (rather than uniform) media, or acting as a kernel for the representation of secondary waves that symbolize an interaction of the primary wave with the inhomogeneity.

Topics allocated to different subdivisions of the subject matter are interwoven, of course, in particular studies and their exposition need not therefore be rigidly circumscribed by the original ordering. Since it is manifestly impossible to detail even a fraction of the innumerable wave problems and their resolutions within the confines of a single volume, we shall focus our attention on the class of problems which involve wave propagation along a fixed direction; this affords a relative simplicity in procedure and expression, without incurring any significant loss of perspective as would impede analogous studies of multi-directional/dimensional wave phenomena.

PART I

§1. *Flexible String Movements*

The vibrations of strings have long enjoyed an especial pride of place for introducing (and demonstrating) elementary concepts of wave motion inasmuch as linear mass distributions afford a simple prototype of continuous systems and permit the realization of unidirectional wave motions. Despite the restricted scope inherent in uni- rather than multi-directional wave propagation, analytical techniques whose capabilities are fully brought out in the latter circumstances can already be utilized, with the least technical complications, if a lone position coordinate or specific direction of travel is involved. Thus, we may prepare the way for an extensive use of source or Green's functions in wave theory by describing their application to string vibration problems, and we may also introduce general procedures relating to the estimation of Fourier integrals and the asymptotic solution of differential equations; it is likewise appropriate to illustrate, in a one-dimensional setting, the effective manner whereby scattering matrix and variational formulations contribute to wave analysis.

The simplest attributes of an ideal string include perfect uniformity, or distribution of mass along its length, and flexibility, which signifies that the tension or resultant force acting at any (small) normal section is directed along the tangent at the corresponding point of the string. Two types of vibrations, namely longitudinal and transverse, may be distinguished; in the former, the string retains a straight profile and the displacements of its material elements are collinear; in the latter, the elements move within a plane that is perpendicular to the line of the string. If the amplitude of such motions be sufficiently small, these types have an independent character and theoretical description. Large scale motions of the ideal string are governed by non-linear partial differential equations (which involve both longitudinal and transverse displacements), and their analysis is formidable. At the outset, therefore, we consider the features of a linear theory suitable for restricted (infinitesimal) vibrations; and withhold attention from the relationship between linear and non-linear formulations as well as the consequences of non-uniformity and imperfect flexibility of the string.

Let us refer to a string which is located on the x-axis in its equilibrium configuration and envisage planar vibrations specified by time varying displacements, $y(x, t)$, therefrom. We overlook for the present all save the tensile force and suppose that the latter has an invariable magnitude, P. The dynamical (i.e., Newtonian) equation for transverse motion of a differential element, obtained after resolving and combining the tensions at the respective endpoints, indicates a proportionality between local acceleration and curvature of the string; and if the square of the slope, $(\partial y/\partial x)^2$, be neglected in comparison with unity, as befits small and smooth displacements, their connection is given by the so-called linear wave equation

$$\frac{\partial^2 y}{\partial x^2} = \frac{1}{c^2}\frac{\partial^2 y}{\partial t^2}. \tag{1.1}$$

Here the coefficient factor c has the dimensions of a velocity and a constant magnitude

$$c = \sqrt{(P/\rho)} \tag{1.2}$$

if the linear density of the string, ρ, is uniform. The representative (1.1) of second order partial differential equations is atypical insofar as its general solution can be exhibited, namely

$$y(x, t) = f(x - ct) + g(x + ct) \tag{1.3}$$

where f, g designate any twice differentiable functions of one variable; this conclusion is directly foreshadowed by the form which the differential equation assumes on introducing the independent variables $\xi = x - ct$, $\eta = x + ct$, viz.,

$$\frac{\partial^2 y}{\partial \xi \partial \eta} = 0. \tag{1.4}$$

The functional character of the term $f(x - ct)$ in (1.3) is such as to imply a progressive motion or wave form with invariable aspect, inasmuch as its envelope (or numerical spectrum for all values of x) at the instant t is a replica of that at an earlier time, say $t = 0$, in the sense of equality between the values of this term at the pairs of locations x, $x + ct$; evidently the constant c defines the speed with which the wave form advances. The opposite signs of $\partial f/\partial x$ and $\partial f/\partial t$ manifest by the relation

$$\frac{\partial f}{\partial x} = \frac{-1}{c}\frac{\partial f}{\partial t} \tag{1.5}$$

are in keeping with a sense of progression towards the right (that is, increasing values of x) and the accompanying sketch (see Fig. 1) of two successive profiles makes clear the anti-correlation in sign between the local slope and transverse velocity. Likewise, we may verify that the second term in (1.3), $g(x + ct)$, represents a progressive motion which travels towards the left (or negative x-direction) with the same rate of advance c and satisfies the first order differential equation

$$\frac{\partial g}{\partial x} = \frac{1}{c}\frac{\partial g}{\partial t}. \tag{1.6}$$

A fixed value for the magnitude of these oppositely directed motions obtains when the locus of coordinate and time variables is a straight line, viz.:

$$x \mp ct = \text{constant}, \tag{1.7}$$

respectively; the distinct families of parallel lines generated by assigning

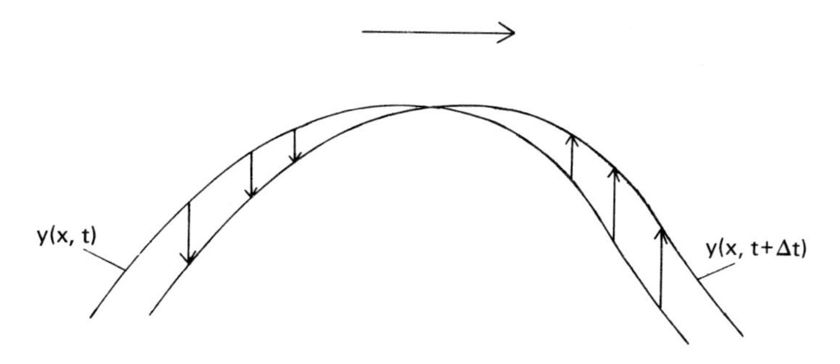

Fig. 1. A moving profile.

different values to the constant in (1.7) are the so-called characteristics of the wave equation (1.1) and find joint specification through the differential relation

$$dx^2 - c^2\, dt^2 = 0 \tag{1.8}$$

whose solutions are other than straight lines in the circumstance that c is a function of x.

Both terms of (1.3) are involved when the displacement is sought in the aftermath ($t > 0$) of arbitrary given initial conditions (at $t = 0$) relating to the displacement and velocity, say

$$y(x, 0) = s_0(x), \qquad \frac{\partial}{\partial t} y(x, 0) = v_0(x); \tag{1.9}$$

and, in the case of a string with unlimited span, the general solution (1.3) then takes the explicit form attributable to d'Alembert (1747)

$$y(x, t) = \frac{1}{2}\left[s_0(x - ct) + s_0(x + ct) + \frac{1}{c}\int_{x-ct}^{x+ct} v_0(\xi)\, d\xi \right], \qquad -\infty < x < \infty, \quad t > 0 \tag{1.10}$$

The latter representation may also be derived by employing Riemann's (1860) method for integration of the wave equation with auxiliary data (here, of the initial variety), in which the characteristics (1.7) are prominently featured. Thus, in order to obtain the solution $y(X, T)$ when $X, T > 0$ are assigned arbitrarily, we single out the characteristics

$$C_1 : x - ct = X - ct$$

$$C_2 : x + ct = X + ct$$

which have only the point $P_1(X, T)$ in common and then consider the triangle whose other vertices P_2, P_3 lie at the points of intersection of these characteristics with the x-axis (see Fig. 2). The version of Gauss' formula,

$$\int\left(\frac{\partial^2 y}{\partial x^2}-\frac{1}{c^2}\frac{\partial^2 y}{\partial t^2}\right)\mathrm{d}x\,\mathrm{d}t=\int\left(\frac{\partial y}{\partial x}\mathrm{d}t+\frac{1}{c^2}\frac{\partial y}{\partial t}\mathrm{d}x\right), \tag{1.11}$$

which equates integrals extended over a domain and its boundary curve shows that solutions of the homogeneous wave equation (1.1) have a vanishing integral

$$\int\left(\frac{\partial y}{\partial x}\mathrm{d}t+\frac{1}{c^2}\frac{\partial y}{\partial t}\mathrm{d}x\right)=0 \tag{1.12}$$

relative to any closed contour; for the triangular shape, in particular, with sides oriented along the characteristic directions and the initial value or x-axis, the corresponding portions of the contour integral take the form

$$\int_{P_1}^{P_2}\frac{1}{c}\frac{\partial y}{\partial s}\mathrm{d}s+\int_{P_2}^{P_3}\frac{1}{c^2}\frac{\partial y}{\partial t}\mathrm{d}x-\int_{P_3}^{P_1}\frac{1}{c}\frac{\partial y}{\partial s}\mathrm{d}s=0$$

where s is a variable along the appropriate characteristic.
Accordingly,

$$2y(X,T)=y(X-cT,0)+y(X+cT,0)+\frac{1}{c}\int_{X-cT}^{X+cT}\frac{\partial y(x,0)}{\partial t}\mathrm{d}x, \tag{1.13}$$

which expresses $y(X, T)$ in terms of the initial values for y and $\partial y/\partial t$ at the endpoints and along the breadth, respectively, of a segment $X-cT\leq x\leq X+cT$ of the x-axis that is cut out by the two characteristics through (X, T). If the initial data are given all along the x-axis, then (1.12) evidently characterizes $y(X, T)$ for all X and any positive T in the d'Alembert fashion (1.10).

Since the differential equation (1.1) is linear, each of the terms in (1.3) is a

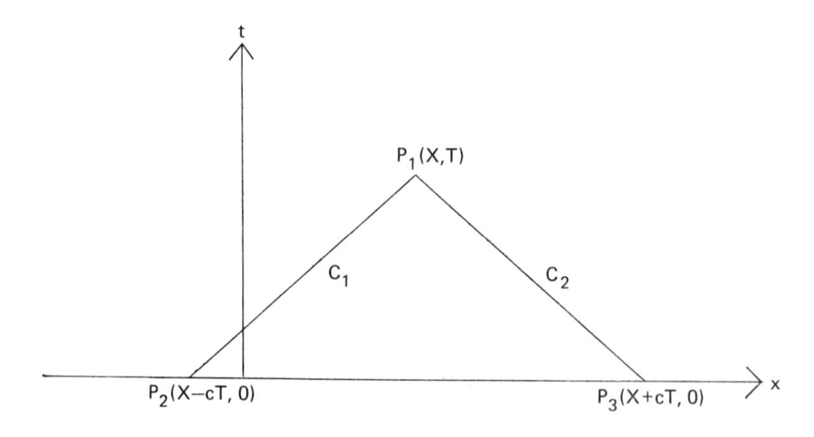

Fig. 2. Characteristic triangle for an initial value problem.

solution in its own right, and it may therefore be inferred that, under suitable conditions of excitation, disturbances of any shape can travel with a specific velocity in either direction along the string while retaining their profile intact. For instance, if wave making is due to a prescribed transverse displacement at the origin, say,

$$y(0, t) = F(t), \qquad t > 0 \tag{1.14}$$

(and $F(t) \equiv 0$, $t < 0$), then the concomitant wave functions take the forms

$$y(x, t) = \begin{cases} F\left(t - \dfrac{x}{c}\right), & x > 0, \\ F\left(t + \dfrac{x}{c}\right), & x < 0, \end{cases} \qquad t > 0 \tag{1.15}$$

on the respective subintervals to the right and left thereof. Both forms are subsumed in the compact representation

$$y^{\text{ret}}(x, t) = F\left(t - \frac{|x|}{c}\right), \qquad x \neq 0, \quad t > 0 \tag{1.16}$$

which evidently constitutes a regular solution of the wave equation everywhere save for the 'source' point $x = 0$ (at which the x-derivative of y^{ret} is discontinuous) and is descriptive of motion travelling away therefrom in symmetric fashion; this is termed a retarded wave or causal solution inasmuch as the delay between the duplication of values for the displacements at and away from the origin equals the travel time between the respective points at the wave speed c. An opposite behavior is manifest by the corresponding advanced wave solution

$$y^{\text{adv}}(x, t) = G\left(t + \frac{|x|}{c}\right), \qquad x \neq 0, \quad t > 0 \tag{1.17}$$

whose representative forms, in $x \gtrless 0$, refer to motions converging on the singular point $x = 0$, and is thus less suited to the embodiment of practical modes of wave stimulation in the finite domain.

§2. *Discontinuous Solutions of the Wave Equation*

Solutions of the homogeneous partial differential equation (1.1), which represent one-dimensional wave motions, may be distinguished according as they, together with successive derivatives, are continuous functions at points (x, t) of space-time, or not; in the latter circumstance the wave function and/or specific derivatives exhibit a discontinuous nature at loci of a fixed or moving type. An instance of discontinuous behavior in first order $(x-)$ derivative is furnished by the retarded wave solution (1.16) with reference to the fixed or immobile source point $x=0$; when the analogous behavior pertains to a moving point and thus typifies a travelling discontinuity, the straight lines or characteristics (1.7), whose role in determining a solution of the initial value problem for the wave equation is already established, again take on a special significance. To detail the further significance of characteristics in explicit fashion, let us consider that solution of the wave equation (1.1) which has null initial values (for all x)

$$y(x,0)=s_0(x)=0, \qquad \frac{\partial}{\partial t}y(x,0)=v_0(x)=0 \tag{2.1}$$

and satisfies the relation

$$y(0,t)=F(t), \qquad t>0 \tag{2.2}$$

at the particular location $x=0$; the stipulations (2.1), (2.2) are evidently compatible if the given function $F(t)$ has the twin assignments expressed by

$$F(0)=0, \qquad F'(0)=0 \tag{2.3}$$

at the initial instant of time.

There is an even symmetry about $x=0$ manifest by the excitation thus contemplated, and it suffices accordingly to focus attention on the range $x>0$ ($t>0$). On employing the general integral (1.3) of the wave equation, the initial conditions are accomodated when $f'(x)=g'(x)=0$ and the functions $f(x)$, $g(x)$ with positive arguments reduce to constants whose sum vanishes, i.e.,

$$f(x)=A, \qquad g(x)=-A, \qquad x>0.$$

The remaining condition

$$f(-ct)+g(ct)=F(t), \quad t>0$$

can be displayed in the form

$$f(-ct)=F(t)+A$$

and it thus appears that

$$f(x-ct)=\begin{cases} A, & x-ct>0 \\ F\left(t-\dfrac{x}{c}\right)+A, & x-ct<0 \end{cases}$$

$$g(x+ct)=-A, \qquad x>0, \quad t>0$$

constitute the requisite determinations of the components of the total wave function, $y(x,t)=f(x-ct)+g(x+ct)$.

The solution to the problem posed is therefore given by the complementary formulas

$$y(x, t) = \begin{cases} y_-(x, t) \equiv 0, & t < x/c \\ y_+(x, t) = F\left(t - \dfrac{x}{c}\right), & t > x/c \end{cases} \tag{2.4}$$

and represents an excitation or disturbance moving away from the origin with the velocity c [compare (1.15)]. At the front of the disturbance, whose instantaneous position coordinate is $x = ct$, the limiting values of the wave functions $y_\mp(x, t)$ and their derivatives up to the second order are, respectively,

$$\begin{gathered} y_- = 0; \qquad y_+ = F(0) = 0 \\ \frac{\partial y_-}{\partial x} = \frac{\partial y_-}{\partial t} = 0; \qquad \frac{\partial y_+}{\partial x} = -\frac{1}{c} F'(0) = 0, \qquad \frac{\partial y_+}{\partial t} = F'(0) = 0 \\ \frac{\partial^2 y_-}{\partial x^2} = \frac{\partial^2 y_-}{\partial x \partial t} = \frac{\partial^2 y_-}{\partial t^2} = 0; \\ \frac{\partial^2 y_+}{\partial x^2} = \frac{1}{c^2} F''(0) \neq 0, \qquad \frac{\partial^2 y_+}{\partial x \partial t} = -\frac{1}{c} F''(0) \neq 0, \qquad \frac{\partial^2 y_+}{\partial t^2} = F''(0) \neq 0. \end{gathered} \tag{2.5}$$

Hence, a discontinuous behavior for the second order derivatives of the wave function obtains at all points on the straight line or characteristic C,

$$\Psi(x, t) \overset{\text{def}}{=} \frac{x}{c} - t = 0$$

that issues from the origin in the first quadrant of the x, t-plane. The sectoral domains of existence and regularity for the wave functions $y_\mp(x, t)$ are evidently located on opposite sides of C (see Fig. 3) where the individual change or jump in second order derivatives of the wave function proves to be

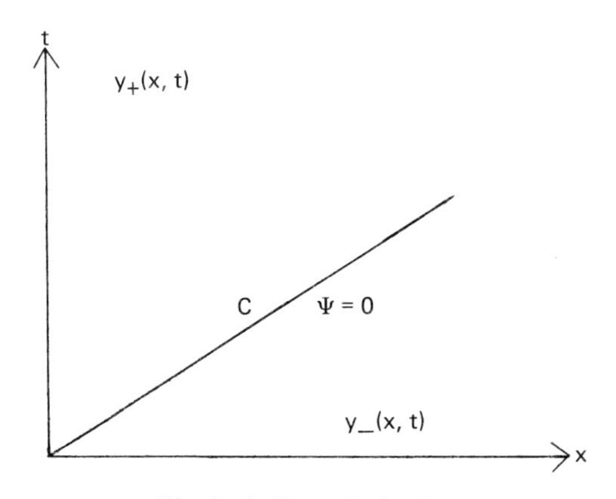

Fig. 3. A discontinuity locus.

$$\left[\frac{\partial^2 y}{\partial x^2}\right] = \frac{\partial^2 y_+}{\partial x^2} - \frac{\partial^2 y_-}{\partial x^2} = \frac{1}{c^2} F''(0),$$
$$\left[\frac{\partial^2 y}{\partial x \partial t}\right] = \frac{\partial^2 y_+}{\partial x \partial t} - \frac{\partial^2 y_-}{\partial x \partial t} = -\frac{1}{c} F''(0), \tag{2.6}$$
$$\left[\frac{\partial^2 y}{\partial t^2}\right] = \frac{\partial^2 y_+}{\partial t^2} - \frac{\partial^2 y_-}{\partial t^2} = F''(0).$$

It follows from (2.6) that

$$\frac{\left[\frac{\partial^2 y}{\partial x^2}\right]}{1/c^2} = \frac{\left[\frac{\partial^2 y}{\partial x \partial t}\right]}{-1/c} = \frac{\left[\frac{\partial^2 y}{\partial t^2}\right]}{1} = F''(0) \tag{2.7}$$

and thus the respective jumps in second order derivatives can be correlated according to the scheme

$$\frac{\left[\frac{\partial^2 y}{\partial x^2}\right]}{\left(\frac{\partial \Psi}{\partial x}\right)^2} = \frac{\left[\frac{\partial^2 y}{\partial x \partial t}\right]}{\frac{\partial \Psi}{\partial x} \cdot \frac{\partial \Psi}{\partial t}} = \frac{\left[\frac{\partial^2 y}{\partial t^2}\right]}{\left(\frac{\partial \Psi}{\partial t}\right)^2} \tag{2.8}$$

where

$$\Psi(x, t) = \frac{x}{c} - t. \tag{2.9}$$

If we express the compatibility relations (2.7) in the form

$$\left[\frac{\partial^2 y}{\partial x^2}\right] = \mu \left(\frac{\partial \Psi}{\partial x}\right)^2, \quad \left[\frac{\partial^2 y}{\partial x \partial t}\right] = \mu \frac{\partial \Psi}{\partial x} \cdot \frac{\partial \Psi}{\partial t}, \quad \left[\frac{\partial^2 y}{\partial t^2}\right] = \mu \left(\frac{\partial \Psi}{\partial t}\right)^2 \tag{2.10}$$

with a discontinuity parameter

$$\mu = F''(0) \neq 0$$

and substitute therefrom in the connection that holds on C, viz.,

$$\left[\frac{\partial^2 y}{\partial x^2}\right] - \frac{1}{c^2}\left[\frac{\partial^2 y}{\partial t^2}\right] = 0, \tag{2.11}$$

which is a ready consequence of the wave equation, the outcome

$$\left\{\left(\frac{\partial \Psi}{\partial x}\right)^2 - \frac{1}{c^2}\left(\frac{\partial \Psi}{\partial t}\right)^2\right\} \mu = 0$$

implies that

$$\left(\frac{\partial \Psi}{\partial x}\right)^2 - \frac{1}{c^2}\left(\frac{\partial \Psi}{\partial t}\right)^2 = 0 \tag{2.12}$$

on C, or $\Psi(x, t) = 0$. Inasmuch as Ψ does not directly enter into (2.12), its particular value (0) appropriate to the characteristic line through the origin is without any special importance and (2.12) constitutes a proper partial differential equation of the first order. The latter affords a complete determination of the characteristics, now established as the loci which mark discontinuous behavior for the wave function (or its derivatives), and evidently the point that corresponds to the position of any such locus in a one-dimensional coordinate plot advances at the speed c.

The compatibility relations (2.10) can be derived without resort to the representations (2.4) of $y_{\mp}(x, t)$ that are indicative of a particular manner of wave excitation. Let $\Psi(x, t) = \text{constant}$ designate a curve C in the x, t-plane where a function $y(x, t)$ evidences continuous variation though, in the first instance, its x and t partial derivatives manifest jumps. If the subscripts + and − refer to the limiting values of y and its derivatives as a point on the curve is reached from opposite sides thereof, the expressions of continuous behavior for y at neighboring points P, Q on the curve

$$y_-(P) = y_+(P), \qquad y_-(Q) = y_+(Q)$$

imply that, when Q approaches P,

$$\mathrm{d}y_-(P) = \mathrm{d}y_+(P)$$

or

$$\frac{\partial y_-}{\partial x}\,\mathrm{d}x + \frac{\partial y_-}{\partial t}\,\mathrm{d}t = \frac{\partial y_+}{\partial x}\,\mathrm{d}x + \frac{\partial y_+}{\partial t}\,\mathrm{d}t \tag{2.13}$$

where

$$\frac{\partial \Psi}{\partial x}\,\mathrm{d}x + \frac{\partial \Psi}{\partial t}\,\mathrm{d}t = 0. \tag{2.14}$$

Rewriting (2.13) in terms of the jumps or differences between limiting values of the first order derivatives,

$$\left[\frac{\partial y}{\partial x}\right]\mathrm{d}x + \left[\frac{\partial y}{\partial t}\right]\mathrm{d}t = 0 \tag{2.15}$$

and assigning a Lagrange multiplier λ to the equation of constraint (2.14) that reflects the given disposition of P and Q, we obtain

$$\left\{\left[\frac{\partial y}{\partial x}\right] - \lambda\frac{\partial \Psi}{\partial x}\right\}\mathrm{d}x + \left\{\left[\frac{\partial y}{\partial t}\right] - \lambda\frac{\partial \Psi}{\partial t}\right\}\mathrm{d}t = 0; \tag{2.16}$$

further to the assumption $\partial\Psi/\partial t \neq 0$, say, the choice

$$\lambda = [\partial y/\partial t]/(\partial \Psi/\partial t)$$

eliminates the second part of (2.16) and, since dx is arbitrary, there follows

$$\left[\frac{\partial y}{\partial x}\right] = \lambda \frac{\partial \Psi}{\partial x}.$$

We may consequently assert that the jumps in derivative of y are proportional to the respective derivatives of Ψ, viz.,

$$\left[\frac{\partial y}{\partial x}\right] = \lambda \frac{\partial \Psi}{\partial x}, \qquad \left[\frac{\partial y}{\partial t}\right] = \lambda \frac{\partial \Psi}{\partial t} \tag{2.17}$$

with a parameter λ of indeterminate a priori status.

When both y and its first order derivatives have a continuous behavior at C, and the second order derivatives undergo jumps, similar reasoning yields

$$\left[\frac{\partial^2 y}{\partial x^2}\right] = \lambda_1 \frac{\partial \Psi}{\partial x}, \qquad \left[\frac{\partial^2 y}{\partial x \partial t}\right] = \lambda_2 \frac{\partial \Psi}{\partial x} = \lambda_1 \frac{\partial \Psi}{\partial t}, \qquad \left[\frac{\partial^2 y}{\partial t^2}\right] = \lambda_2 \frac{\partial \Psi}{\partial t} \tag{2.18}$$

on utilizing separate multipliers for an expression of the discontinuities that obtain after differentiation of $\partial y/\partial x$ and $\partial y/\partial t$. Hence

$$\frac{\lambda_1}{\dfrac{\partial \Psi}{\partial x}} = \frac{\lambda_2}{\dfrac{\partial \Psi}{\partial t}} = \mu$$

and the relations (2.18) take the form

$$\left[\frac{\partial^2 y}{\partial x^2}\right] = \mu \left(\frac{\partial \Psi}{\partial x}\right)^2, \qquad \left[\frac{\partial^2 y}{\partial x \partial t}\right] = \mu \frac{\partial \Psi}{\partial x}\frac{\partial \Psi}{\partial t}, \qquad \left[\frac{\partial^2 y}{\partial t^2}\right] = \mu \left(\frac{\partial \Psi}{\partial t}\right)^2$$

previously given.

§3. *String Profiles with a Moving Vertex*

It is comparatively simple (i.e., without invoking a general theory) to depict a string motion that incorporates a mobile discontinuity in slope (or vertex), if we have regard for the equality between discontinuity and wave speeds. The construction proceeds from an evident fact that the homogeneous wave equation for string displacements,

$$\frac{\partial^2 y}{\partial x^2} = \frac{1}{c^2}\frac{\partial^2 y}{\partial t^2}, \tag{3.1}$$

admits particular and separable solutions which are linearly dependent on both x and t; and which, therefore, describe instantaneous profiles made up of straight sections. If only two of the latter are envisaged and the ends ($x = 0, l$) of the string are permanently fixed, a suitable pair of displacement and transverse velocity expressions take the forms

$$\begin{aligned} y_-(x,t) &= Ax(\tau - t), & v_-(x,t) &= -Ax, & 0 \le x < \xi \\ y_+(x,t) &= B(l-x)t, & v_+(x,t) &= B(l-x), & \xi < x \le l \end{aligned} \tag{3.2}$$

wherein A and B specify the magnitude of the angular velocity with which the respective sections rotate about the fixed points. When the continuity of the displacement at the vertex $x = \xi$ is enforced and

$$y_-(\xi, t) = y_+(\xi, t) = \hat{y}(\xi, t), \text{ say,} \tag{3.3}$$

we find

$$\xi = \frac{Blt}{A\tau + (B-A)t}$$

and thus the requirement

$$\frac{d\xi}{dt} = \pm c \tag{3.4}$$

furnishes the twin conditions

$$B = A$$

and

$$l/\tau = \pm c. \tag{3.5}$$

The string displacements which derive from the individual sign choices in (3.4), (3.5) are

$$y_-(x,t) = Ax\left(\frac{l}{c} - t\right), \qquad y_+(x,t) = A(l-x)t,$$

$$y_-(x,t) = Ax\left(-\frac{l}{c} - t\right), \qquad y_+(x,t) = A(l-x)t,$$

respectively; and these imply equal and opposite vertex displacements, viz.

$$\hat{y}\left(\xi, \pm\frac{\xi}{c}\right) = \pm\frac{A\xi}{c}(l-\xi) = \pm\frac{4\hat{Y}}{l^2}\xi(l-\xi) \tag{3.6}$$

if

$$\hat{Y} = (l/2)^2\frac{A}{c}.$$

Since (3.6) makes plain that the vertex traces a parabolic locus on the interval $0 < \xi < l$, with maximal displacement $\hat{Y}$, we may conceive of a continuing string motion (see Fig. 4) wherein the profile is realized by two straight portions (P_1P', $P'P_2$ or P_1P'', $P''P_2$) and features a vertex (P' or P'') which

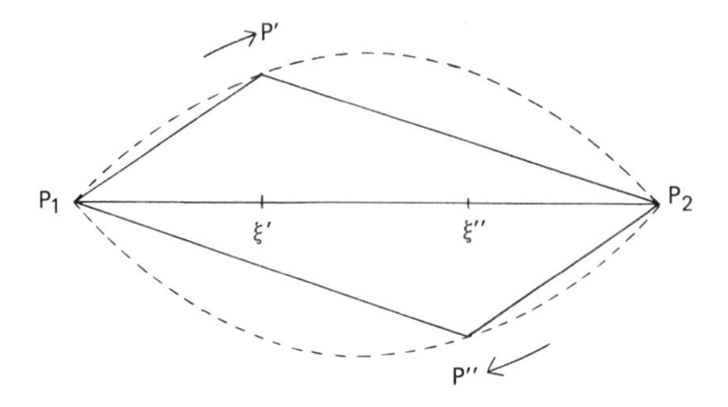

Fig. 4. Vertex motions on a string.

alternately describes parabolic arcs on opposite sides of the single line P_1P_2, as the point (ξ' or ξ'') corresponding to the abscissa of the vertex travels forwards and backwards along P_1P_2 with constant speed c. A quantitative rendering of the overall string profile in such motions, whose period $T = 2l/c$, is provided by the trigonometric development

$$y(x, t) = \frac{8\hat{Y}}{\pi^2} \sum_{n=1}^{\infty} \frac{1}{n^2} \sin \frac{n\pi x}{l} \sin \frac{2n\pi}{T}(t - t_0) \tag{3.7}$$

that predicts a rectilinear (or equilibrium) configuration P_1P_2 at the epochs

$$t - t_0 = m\frac{T}{2}, \quad m = 0, 1, \ldots .$$

By way of supporting this contention, we refer to the trigonometric (Fourier) development,

$$y = \frac{2\hat{y}l^2}{\pi^2\xi(l-\xi)} \sum_{n=1}^{\infty} \frac{1}{n^2} \sin \frac{n\pi x}{l} \sin \frac{n\pi\xi}{l} \tag{3.8}$$

for a static broken line profile whose (single) vertex coordinates are ξ, $\hat{y}$; and next observe that the relationships

$$\frac{2\hat{y}l^2}{\pi^2\xi(l-\xi)} = \pm\frac{8\hat{Y}}{\pi^2}, \tag{3.9}$$

$$\sin \frac{n\pi\xi}{l} = \pm \sin \frac{2n\pi}{T}(t - t_0) \tag{3.10}$$

secure a complete match between the developments (3.7), (3.8). Evidently (3.9) characterizes the parabolic arcs (3.6), while (3.10) regulates variations $\xi(t)$ in vertex location that are fully consistent with a prior description thereof. It is fitting to note, moreover, that Helmholtz (1898) linked a motion of the type (3.7) with the response of a bowed violin string.

An alternative representation of said motion, also found by Helmholtz, is expressible in terms of a pair of oppositely moving progressive wave forms; which means, specifically, the determination of functions f, g such that

$$y(x, t) = f(x - ct) + g(x + ct). \tag{3.11}$$

The fixed end-point boundary conditions $y(0, t) = y(l, t) = 0$ imply that f and g possess a periodic nature, with unchanged value after the whole argument increases or decreases by the amount $2l$. Individual determinations of f and g may be achieved through the relations (where primes symbolize an argument derivative)

$$\begin{aligned} 2cf' &= c\frac{\partial y}{\partial x} - \frac{\partial y}{\partial t}, \\ 2cg' &= c\frac{\partial y}{\partial x} + \frac{\partial y}{\partial t} \end{aligned} \tag{3.12}$$

once the right-hand members are known over a periodicity range.

An arbitrary point of the string moves transversely to the equilibrium line P_1P_2, in the circumstances contemplated, with uniform though distinct speeds in two separate and recurring phases which change over at the instants of time when the chosen point marks a vertex of the overall profile. The duration of one phase, wherein the point (x) crosses P_1P_2 from, say, the lower to the upper vertex position, is equal to $2t_1 = 2x/(2l/T) = xT/l$, since a vertex travels at the speed $c = 2l/T$; and that of the other phase, in which a reversed sense of motion figures, is equal to $T - 2t_1 = T(1 - (x/l))$. Fixing the origin of time so that $y \uparrow 0$, upward motions occur during the intervals

$$\ldots(-t_1, t_1), (T - t_1, T + t_1), (2T - t_1, 2T + t_1), \ldots$$

corresponding to one set of phases, while those of the other set are executed in the intervals

$$\ldots(t_1, T - t_1), (T + t_1, 2T - t_1), (2T + t_1, 3T - t_1), \ldots.$$

If we utilize the vertex displacement formula (3.7.), the expressions of steady motion for a consecutive pair of phases,

$$\begin{aligned} y &= \frac{8\hat{Y}}{l}\frac{l - x}{T}t, \quad -t_1 < x < t_1, \\ y &= \frac{8\hat{Y}}{l}\frac{x}{T}\left(\frac{T}{2} - t\right), \quad t_1 < t < T - t_1 \end{aligned} \tag{3.13}$$

follow, with direct adaptation to other intervals of the same phase on effecting the replacements $t \to t - T, t - 2T, \ldots$. After calculating, by means of the latter expressions, the partial derivatives of y which enter into (3.12), it turns out that

$$2cf' = \frac{8\hat{Y}}{T} + \frac{8\hat{Y}}{lT}(x - ct), \qquad t_1 < t < T + t_1,$$

$$2cg' = \frac{8\hat{Y}}{T} - \frac{8\hat{Y}}{lT}(x + ct), \quad -t_1 < t < T - t_1 \tag{3.14}$$

with values of $x \mp ct$ that vary between 0 and $\mp 2l$ on the stipulated time intervals; accordingly, integration of (3.14) characterizes the functions $f(\zeta)$, $g(\zeta)$ over the ranges $-2l < \zeta < 0$ and $0 < \zeta < 2l$, respectively, which encompass an entire period. The resultant determinations,

$$f(\zeta) = \frac{\hat{Y}}{l^2}\zeta(2l + \zeta), \quad -2l < \zeta < 0,$$

$$g(\zeta) = \frac{\hat{Y}}{l^2}\zeta(2l - \zeta), \qquad 0 < \zeta < 2l \tag{3.15}$$

are depicted (by two parabolic arcs) in Fig. 5, along with the periodic continuations which enable us to specify f, g for all other arguments; and it is easy to confirm that the aforesaid determinations jointly guarantee the requisite boundary values at the fixed ends of the string, viz. $y(x, t) = f(x - ct) + g(x + ct) = 0, x = 0, l$.

To generate changing string profiles, or values of $y(x, t)$, $0 < x < l$, as the time varies, we merely translate, in the sense conveyed by arrows and at a speed c, the sequences of parabolic arcs representative of f, g and add (algebraically) their respective ordinates; Fig. 6 brings out that a vertex is correlated with a point of closest approach between two parabolas.

Although the slope $\partial y/\partial x$ and transverse velocity $\partial y/\partial t$ of the string are each discontinuous at a vertex in its profile, there exists a particular linear combination of the two quantities which has a continuous overall nature. This is the so-called convected time derivative, namely

$$\frac{D}{Dt}y \overset{\text{def}}{=} \frac{\partial}{\partial t}y + U\frac{\partial}{\partial x}y$$

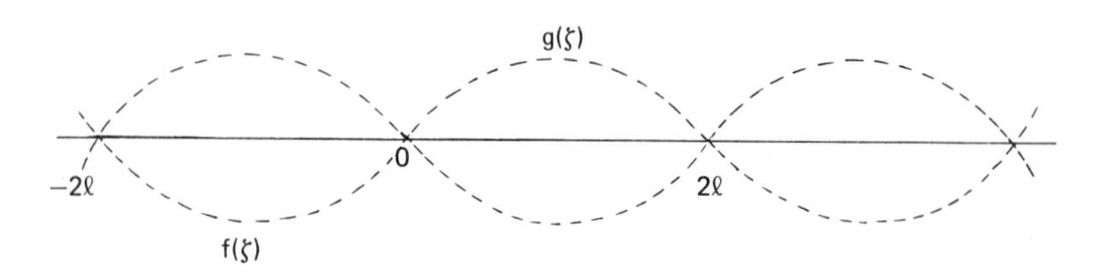

Fig. 5. Parabolic arcs.

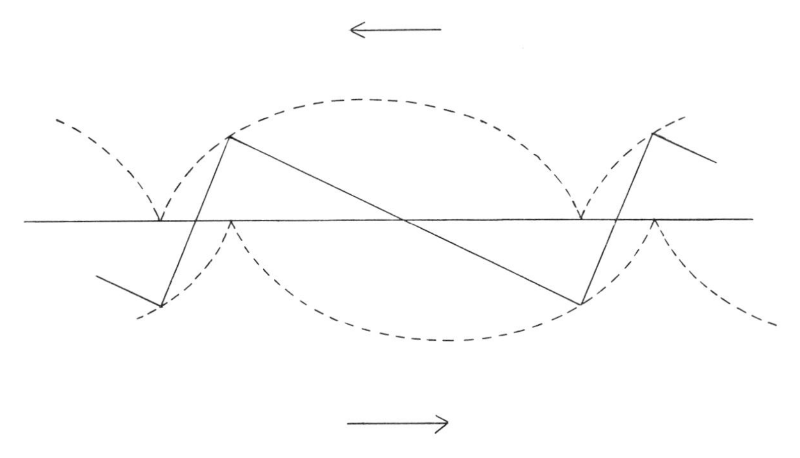

Fig. 6. An evolutionary string profile.

if U represents any (uniform) vertex speed; for verification of same, we utilize the analogues of (3.2)

$$y_{-}(x,t) = A\frac{x}{U}(l - Ut), \qquad y_{+}(x,t) = A(l - x)t$$

and deduce therefrom the equality

$$\frac{\partial y_{-}}{\partial t} + U\frac{\partial y_{-}}{\partial x} = \frac{\partial y_{+}}{\partial t} + U\frac{\partial y_{+}}{\partial x} = A(l - 2Ut).$$

To establish, in the present context, that U and c have a common magnitude, let us designate by

$$\Delta = \frac{\partial y}{\partial x}\bigg|_{-}^{+}, \qquad \Delta' = \frac{\partial y}{\partial t}\bigg|_{-}^{+}$$

the appertaining discontinuities and note the kinematic relation

$$U\Delta + \Delta' = 0 \tag{3.16}$$

which expresses the continuity of the derivative Dy/Dt. Next, we apply a dynamical relation,

$$P\Delta\,\mathrm{d}t = -\rho\Delta'\,\mathrm{d}\xi, \tag{3.17}$$

that connects the momentum change for a section $\mathrm{d}\xi$ encompassing the vertex with the activity of the tension. When the expressions $P/\rho = c^2$, $\mathrm{d}\xi/\mathrm{d}t = U$ are introduced into (3.17) there follows

$$c^2\Delta + U\Delta' = 0 \tag{3.18}$$

and thus, if the measures of discontinuity do not jointly vanish, compatibility of (3.16), (3.18) implies that

$$U^2 = c^2$$

or $U = \pm c$.

§4. *Periodic Wave Functions*

Wave forms of a periodic type play an important role in elaborating the theory of linear partial differential equations, since they offer the advantages of a specific nature and ready superposition for the purposes of generating broad classes of solution. The representative, for the one-dimensional wave equation (1.1), of a simply periodic wave motion that proceeds in the direction of increasing x can be written as

$$y(x, t) = A \cos [k(x - ct) + \delta] \tag{4.1}$$

or

$$y(x, t) = A \cos [kx - \omega t + \delta] \tag{4.2}$$

if A, k, and δ are constants, and moreover

$$\omega = kc. \tag{4.3}$$

At any given point, x, a string displacement of the preceding variety undergoes an oscillation, of period $T = 2\pi/\omega$ and within the amplitude limits $\pm A$, whose phase depends on the magnitude of δ. Furthermore, it appears that the quantity $\lambda = 2\pi/k$ defines the spatial periodicity, or the so-called wave length, which is revealed by the overall displacement profile at any instant of time.

To gauge the forward progress of this periodic wave motion, let us suppose that reference is made to a definite phase of the displacement, such that $kx - \omega t + \delta =$ constant; evidently, in order to keep pace with any individual value of the phase (and thus retain an invariable aspect of the displacement) a unique rate of advance along the direction of wave propagation must be maintained, namely

$$v = \frac{dx}{dt} = \frac{\omega}{k} = c, \tag{4.4}$$

which may be termed the phase velocity. It is also apparent that the ratio $c/\lambda = \omega/2\pi = 1/T = \nu$, or radian frequency, can be identified with the number of wave crests (i.e., displacement maxima) which pass any point in a unit time interval; and that $\nu/c = 1/\lambda = k/2\pi$ designates the number of waves per unit length, though k itself will henceforth be called the wave number.

Inasmuch as actual conditions of excitation preclude the possibility of realizing an exactly periodic wave form (or regular train that can be continued without substantive change to the respective limits of space and time) the latter must be considered a mathematical abstraction. There is, nevertheless, a considerable scope for the manipulation of such wave forms, principally by superposition, in the closer simulation of actual states of vibration. The combination of any number of wave trains (4.1) with varying amplitudes and phases, but fixed wave length, is again a form of the same type. However, if we merely superpose a pair of trains of equal amplitude A and vanishing phase, δ, though with distinct angular frequencies ω_1, ω_2 and corresponding wave numbers k_1, k_2, the result

$$\begin{aligned} y(x, t) &= A \cos (k_1 x - \omega_1 t) + A \cos (k_2 x - \omega_2 t) \\ &= 2A \cos \left(\frac{k_1 - k_2}{2} x - \frac{\omega_1 - \omega_2}{2} t\right) \cos \left(\frac{k_1 + k_2}{2} x - \frac{\omega_1 + \omega_2}{2} t\right) \end{aligned} \tag{4.5}$$

is a form whose respective trigonometric factors characterize progressive motions at two new sets of values for these parameters. When the primary values of the frequency and wave number are nearly equal, $\omega_1 \approx \omega_2$ and $k_1 \approx k_2$, the first factor varies much more slowly in time and position than does the second; and it may be viewed as superposing gradual changes on the latter, whose outcome amounts to the conversion of a regular train into a series of uniformly spaced wave packets that are comprised between adjacent nulls in the complete amplitude function. This aspect is depicted in Fig. 7,

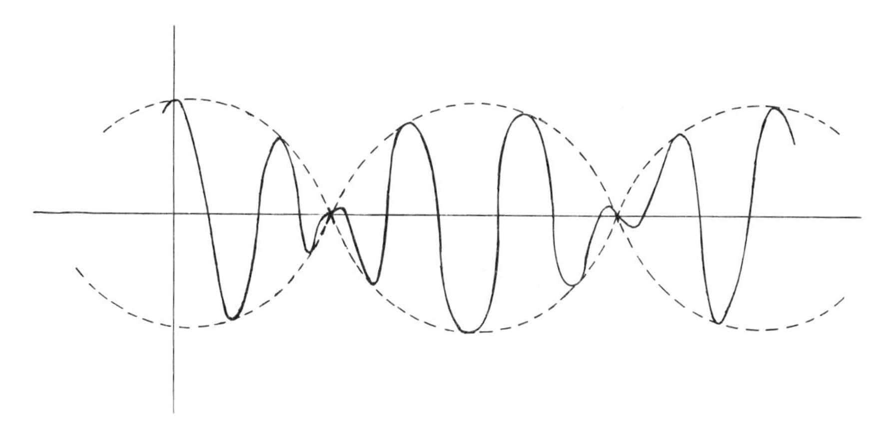

Fig. 7. Wave packets.

wherein the outer (or dashed) profile reflects the spatial variation of the first factor at given instants, and contrasts with the smaller scale appropriate to changes of the inner (or solid) profile. If the phase velocities of the component waves are equal, that is $\omega_1/k_1 = \omega_2/k_2 = c$, their combination, or the system of wave packets, advances without change of form at the same velocity, for both factors in (4.5) are synchronous in their rate of advance, viz.:

$$\frac{\omega_1 - \omega_2}{k_1 - k_2} = \frac{\omega_1 + \omega_2}{k_1 + k_2} = c.$$

A quite different state of affairs obtains, on the other hand, if the phase velocities for the component waves are distinct, a possibility which is ruled out in the case of solutions to the wave equation (1.1) though not for other equations; as then the two factors in (4.5) possess separate rates of forward progress and an unsteady or aperiodic wave pattern results. Furthermore, isolated disturbances or wave packets of limited spatial extension can be fashioned at any instant of time by a sufficiently general combination of periodic wave forms with different wave lengths and phase velocities, though these are found to undergo continued broadening or dispersion in the course of time.

There is a procedural advantage which accrues, in the context of linear wave equations, to the representation of an inherently real and time-periodic wave function in terms of a complex valued form of solution. Since real and imaginary parts of the complex exponential function

$$e^{i(kx - \omega t)}$$

are individual solutions of the homogeneous wave equation (1.1) if (4.3) holds, we may, in particular, express the progressive wave form (4.2) by

$$y(x, t) = \mathrm{Re}\{A\, e^{i(kx - \omega t)}\}, \tag{4.6}$$

where the amplitude A is allowed a complex nature, thus obviating the need for a separate phase factor $\exp(i\delta)$. The merit of such a representation, or its analogue fashioned with the imaginary part, is that the operations of addition differentiation and multiplication can be performed expeditiously with the exponential functions, leaving the real or imaginary part to be extracted afterwards. In the subsequent presentation this final step is usually omitted, so that the equations ordinarily feature the exponential function themselves.

§5. *Variable Wave Patterns*

A departure from the permanence of wave profiles progressing in either direction along the string is generally consequent to the inclusion in the equation of motion of a term which bespeaks an interaction with the surroundings (such as air friction); or to the retention of the second degree term in the local slope of the string, which implies a correspondingly non-linear form of the wave equation. If, in particular, the elements of the string are subject to an external force that is directed towards the equilibrium position and has a magnitude proportional to the distance therefrom, specifically Ph^2y, the equation of small amplitude motions becomes

$$\frac{\partial^2 y}{\partial x^2} = \frac{1}{c^2}\frac{\partial^2 y}{\partial t^2} + h^2 y. \tag{5.1}$$

To obtain the formal solution of an initial value problem for the preceding equation, which generalizes the d'Alembert result (1.10) and makes clear the unsteadying role of the parameter h, we may employ a method of synthesis that is based on periodic constituents. Let the initial displacement and velocity distributions (1.7) have the Fourier integral representations

$$s_0(x) = \int_{-\infty}^{\infty} a(k)\,\mathrm{e}^{\mathrm{i}kx}\,\mathrm{d}k, \qquad v_0(x) = \int_{-\infty}^{\infty} b(k)\,\mathrm{e}^{\mathrm{i}kx}\,\mathrm{d}k \tag{5.2}$$

and observe that the time-periodic progressive wave functions $\exp\{\mathrm{i}(kx \mp \omega t)\}$ are compatible with (5.1) provided that

$$\omega(k) = c\surd(k^2 + h^2) \tag{5.3}$$

and the affiliated phase velocity

$$v = c\surd(1 + (h/k)^2) \tag{5.4}$$

manifests a dependence on the wave number. A solution of (5.1) can thus be expressed in the form

$$y(x, t) = \int_{-\infty}^{\infty} A(k)\,\mathrm{e}^{\mathrm{i}(kx - \omega(k)t)}\,\mathrm{d}k + \int_{-\infty}^{\infty} B(k)\,\mathrm{e}^{\mathrm{i}(kx + \omega(k)t)}\,\mathrm{d}k \tag{5.5}$$

and imposition of the conditions (5.2) yields the relations

$$A(k) = \frac{1}{2}\left[a(k) + \frac{\mathrm{i}}{\omega(k)}b(k)\right], \qquad B(k) = \frac{1}{2}\left[a(k) - \frac{\mathrm{i}}{\omega(k)}b(k)\right]$$

whence

$$y(x, t) = \int_{-\infty}^{\infty} a(k)\,\mathrm{e}^{\mathrm{i}kx} \cos(\surd(k^2 + h^2)ct)\,\mathrm{d}k + \frac{1}{c}\int_{-\infty}^{\infty} b(k)\,\mathrm{e}^{\mathrm{i}kx}\frac{\sin(\surd(k^2 + h^2)ct)}{\surd(k^2 + h^2)}\,\mathrm{d}k. \tag{5.6}$$

If the expressions for the Fourier transforms $a(k)$, $b(k)$ of the initial displacement and velocity, namely

$$a(k) = \frac{1}{2\pi}\int_{-\infty}^{\infty} s_0(\xi)\,\mathrm{e}^{-\mathrm{i}k\xi}\,\mathrm{d}\xi,$$

$$b(k) = \frac{1}{2\pi}\int_{-\infty}^{\infty} v_0(\xi)\,\mathrm{e}^{-\mathrm{i}k\xi}\,\mathrm{d}\xi,$$

are introduced in (5.6) we obtain

$$y(x,t)=\frac{1}{\pi}\int_{-\infty}^{\infty} s_0(\xi)\,\mathrm{d}\xi\int_0^{\infty}\cos k(x-\xi)\cos(\sqrt{(k^2+h^2)}ct)\,\mathrm{d}k$$
$$+\frac{1}{\pi c}\int_{-\infty}^{\infty} v_0(\xi)\,\mathrm{d}\xi\int_0^{\infty}\cos k(x-\xi)\frac{\sin(\sqrt{(k^2+h^2)}ct)}{\sqrt{(k^2+h^2)}}\,\mathrm{d}k,$$

or

$$y(x,t)=\frac{1}{\pi c}\frac{\partial}{\partial t}\int_{-\infty}^{\infty}\mathfrak{G}(x-\xi,ct;h)s_0(\xi)\,\mathrm{d}\xi+\frac{1}{\pi c}\int_{-\infty}^{\infty}\mathfrak{G}(x-\xi,ct;h)v_0(\xi)\,\mathrm{d}\xi \tag{5.7}$$

where

$$\mathfrak{G}(x,\tau;h)=\int_0^{\infty}\cos\zeta x\frac{\sin(\sqrt{(\zeta^2+h^2)}\tau)}{\sqrt{(\zeta^2+h^2)}}\,\mathrm{d}\zeta \tag{5.8}$$

The latter function, whose appearance in (5.7) signifies that it represents a particular solution of the original differential equation (5.1), has a discontinuous behavior in relation to the argument variables x,τ. For details bearing on this aspect, which is connected with the finite speed of propagation of disturbances governed by the equation (5.1), let us first rewrite (5.8) in the sequential versions

$$\mathfrak{G}(x,\tau;h)=\frac{1}{2}\int_{-\infty}^{\infty}\mathrm{e}^{\mathrm{i}\zeta x}\frac{\sin(\sqrt{(\zeta^2+h^2)}\tau)}{\sqrt{(\zeta^2+h^2)}}\,\mathrm{d}\zeta$$
$$=\frac{1}{2}\int_{-\infty}^{\infty}\mathrm{e}^{\mathrm{i}hx\sinh\varphi}\sin(h\tau\cosh\varphi)\,\mathrm{d}\varphi \tag{5.9}$$

where a transformation of variables, $\zeta=h\sinh\varphi$, figures. On next resolving the sine factor into complex exponentials and setting

$$\alpha=hx=\mu\sinh\bar\varphi,\qquad \beta=h\tau=\mu\cosh\bar\varphi$$

so that (5.10)

$$\beta^2-\alpha^2=h^2(\tau^2-x^2)=\mu^2,$$

we obtain

$$\mathfrak{G}=\frac{1}{4\mathrm{i}}\int_{-\infty}^{\infty}[\exp(\mathrm{i}\,\mu\cosh(\varphi+\bar\varphi))-\exp(-\mathrm{i}\,\mu\cosh(\varphi-\bar\varphi))]\,\mathrm{d}\varphi$$
$$=\frac{1}{4\mathrm{i}}\cdot\pi\,\mathrm{i}[H_0^{(1)}(\mu)+H_0^{(2)}(\mu)]=\frac{\pi}{2}J_0(\mu),\quad \mu>0 \tag{5.11}$$

with the customary notations for Hankel functions of the first and second kinds, $H_0^{(1)}$, $H_0^{(2)}$ and the (zero order) Bessel function, $J_0(\mu)=\frac{1}{2}[H_0^{(1)}(\mu)+H_0^{(2)}(\mu)]$. When the hyperbolic functions in (5.10) are exchanged, or

$$\alpha = \nu \cosh \bar{\varphi}, \qquad \beta = \nu \sinh \bar{\varphi},$$
$$\alpha^2 - \beta^2 = h^2(x^2 - \tau^2) = \nu^2 = -\mu^2 \tag{5.12}$$

it appears that

$$\begin{aligned} \mathfrak{G} &= \frac{1}{4\mathrm{i}} \int_{-\infty}^{\infty} [\exp(\mathrm{i}\nu \sinh(\varphi + \bar{\varphi})) - \exp(\mathrm{i}\nu \sinh(\varphi - \bar{\varphi}))] \, \mathrm{d}\varphi \\ &= 0, \quad \nu > 0 \end{aligned} \tag{5.13}$$

since the infinite limits permit the removal of $\pm \bar{\varphi}$ from the respective terms, on suitable shifts of the integration variable, and thus imply their mutual cancellation.

In accordance with (5.7) and the representation which follows from (5.10)–(5.13). viz.

$$\mathfrak{G}(x, \tau; h) = \begin{cases} \frac{\pi}{2} J_0(h\sqrt{(\tau^2 - x^2)}), & \tau^2 > x^2 \\ 0, & \tau^2 < x^2, \end{cases} \qquad h > 0 \tag{5.14}$$

the solution of an initial value problem for the differential equation (5.1) achieves the explicit characterizations

$$\begin{aligned} y(x, t) &= \frac{1}{2c} \frac{\partial}{\partial t} \int_{x-ct}^{x+ct} J_0[h\sqrt{(c^2t^2 - (x - \xi)^2)}] s_0(\xi) \, \mathrm{d}\xi \\ &\quad + \frac{1}{2c} \int_{x-ct}^{x+ct} J_0[h\sqrt{(c^2t^2 - (x - \xi)^2)}] v_0(\xi) \, \mathrm{d}\xi \\ &= \frac{1}{2}[s_0(x - ct) + s_0(x + ct)] - \frac{1}{2} hct \int_{x-ct}^{x+ct} \frac{J_1[h\sqrt{(c^2t^2 - (x - \xi)^2)}]}{\sqrt{(c^2t^2 - (x - \xi)^2)}} s_0(\xi) \, \mathrm{d}\xi \\ &\quad + \frac{1}{2c} \int_{x-ct}^{x+ct} J_0[\sqrt{(c^2t^2 - (x - \xi)^2)}] v_0(\xi) \, \mathrm{d}\xi \end{aligned} \tag{5.15}$$

where particulars of the second are linked with the specific properties $J_0(0) = 1$ and $\mathrm{d}J_0(\mu)/\mathrm{d}\mu = -J_1(\mu)$ of the Bessel functions.

Two features stand out in this generalization of d'Alembert's formula (1.10) which is, fittingly, recovered if the parameter h vanishes; the first reflects a (previously assumed) role for c as the speed involved in transmitting the influence of (initial) data to other locations at later times, while the second is incorporated in the fact that (5.15) contains (integral) terms which are other than functions of $x \mp ct$. It is a consequence of the latter feature that the evolution of string displacements subject to (5.1) cannot be realized through the direct superposition of oppositely moving profiles which retain a permanent form.

Another instance of variable displacement patterns obtains when the elements of the string experience a damping force proportional to their velocity, as implied by the equation of motion

$$\frac{\partial^2 y}{\partial x^2} = \frac{1}{c^2}\left[\frac{\partial^2 y}{\partial t^2} + 2\varepsilon\frac{\partial y}{\partial t}\right] \tag{5.16}$$

where the damping coefficient $\varepsilon\,(>0)$ has the dimension of a frequency. The complex exponential

$$y(x, t) = A\,\mathrm{e}^{\mathrm{i}(kx-\omega t)} \tag{5.17}$$

recommends itself as a particular solution of (5.16), since any number of x or t differentiations leave its functional character unchanged, though here the requisite condition

$$\omega^2 = k^2c^2 - 2\,\mathrm{i}\varepsilon\omega \tag{5.18}$$

reveals that ω and k cannot jointly be real. If the wave number k is presumed to be real and the damping ever so slight, the complex form of the frequency which obtains from (5.18),

$$\omega \doteq kc - \mathrm{i}\varepsilon, \quad \varepsilon \ll |\omega| \tag{5.19}$$

points to a slow subsidence in time of the solution (5.17); at this stage of approximation, moreover, the attenuation factor $\exp(-\varepsilon t)$ represents the full extent of the modification to the simply periodic wave function in the absence of damping. The same factor can therefore be applied to a general combination of unperturbed travelling wave functions, say (1.3), and the resulting expression

$$y(x, t) = \mathrm{e}^{-\varepsilon t}[f(x-ct) + g(x+ct)] \tag{5.20}$$

constitutes an approximate solution of (5.16) so long as the predominant wave number components in the Fourier resolution of the localized wave forms $f(x)$ and $g(x)$ satisfy the inequality

$$k \gg \frac{\varepsilon}{c}.$$

Alternatively, when ω is regarded as a real magnitude and the measure of damping remains small, the requisite solution of (5.18) furnishes a complex-valued wave number

$$k \doteq \frac{\omega}{c} + \mathrm{i}\frac{\varepsilon}{c}, \quad \frac{\varepsilon}{c} \ll |k| \tag{5.21}$$

that is indicative of a spatial attenuation for the time-periodic wave function (5.17) on the interval $x > 0$. After making the decompositions

$$f(x-ct) = \mathrm{e}^{(\varepsilon/c)(ct-x)}F\left(t - \frac{x}{c}\right)$$

and

$$g(x+ct)=\mathrm{e}^{(\varepsilon/c)(ct+x)}G\left(t+\frac{x}{c}\right)$$

the expression (5.20) is converted to the form

$$y(x,t)=\mathrm{e}^{-(\varepsilon/c)x}F\left(t-\frac{x}{c}\right)+\mathrm{e}^{(\varepsilon/c)x}G\left(t+\frac{x}{c}\right) \tag{5.22}$$

in which the respective terms represent spatially attenuated versions of the waves (1.15) launched towards the right and left by the prescribed displacements $F(t)$, $G(t)$ at the origin. The versatility of the primitive exponential form (5.17) is thus given fresh emphasis by its role in uncovering various manifestations of temporal or spatial attenuation in wave excitations.

§6. *Causality and the Superposition of Exponential Wave Functions*

The manner of selecting and combining elementary solutions of constant coefficient partial differential equations, in order to achieve representations for other types of solution, has features that warrant closer scrutiny; problems wherein signals or inputs at a specific location are given lend themselves to describing such features and, by way of a simple example at the outset, suppose that it is desired to solve the wave equation

$$\frac{\partial^2 y}{\partial x^2}=\frac{1}{c^2}\frac{\partial^2 y}{\partial t^2} \tag{6.1}$$

on the interval $x>0$, assuming a periodic input function

$$y(0,t)=\mathrm{e}^{-\mathrm{i}\omega_0 t}, \quad t>0 \tag{6.2}$$

at the endpoint $x=0$. Elementary solutions of the wave equation (6.1) have the form $\exp\{-\mathrm{i}\omega(\mp(x/c))\}$ and the combinations thereof,

$$y_{\mp}(x,t)=\int_{-\infty}^{\infty}A(\omega)\exp\left\{-\mathrm{i}\omega\left(t\mp\frac{x}{c}\right)\right\}\mathrm{d}\omega \tag{6.3}$$

satisfy (6.2) if

$$A(\omega)=\frac{1}{2\pi}\int_0^\infty \mathrm{e}^{\mathrm{i}(\omega-\omega_0)\tau}\,\mathrm{d}\tau=\mathrm{Lim}_{\varepsilon\to 0+}\frac{\mathrm{i}}{\omega-\omega_0+\mathrm{i}\epsilon}. \tag{6.4}$$

Thus

$$y_{\mp}(x,t)=\frac{\mathrm{i}}{2\pi}\int_{-\infty}^{\infty}\frac{\exp\left\{\mathrm{i}\omega\left(t\mp\frac{x}{c}\right)\right\}}{\omega-\omega_0}\mathrm{d}\omega: \qquad \overset{\frown}{\underset{\omega_0}{\longrightarrow}} \tag{6.5}$$

constitute a pair of wave functions in accord with (6.2), the integral being made precise, insofar as the singular point $\omega = \omega_0$ is concerned, by the relative disposition of the contour as shown.

An elementary residue calculation in the complex ω-plane yields

$$y_-(x,t) = \begin{cases} 0, & t < x/c \\ e^{-i\omega_0(t-(x/c))}, & t > x/c \end{cases};$$

$$y_+(x,t) = e^{-i\omega_0(t+(x/c))}, \quad x, t > 0 \tag{6.6}$$

from which it is apparent that $y_+(x,t)$ lacks the causal nature consonant with an initially null distribution, $y(x,0)=0$, $x>0$, since the magnitude of the former differs from zero on the entire positive coordinate range for all and, in particular, arbitrarily small values of t. If the integration contour in (6.5) proceeds below the point $\omega = \omega_0$, then $y_+(x,t)=0$, $x, t>0$ and we may therefore rule out any combination of exponential wave functions depending on $t+(x/c)$ for the problem at hand.

A more intricate example is encountered after replacing (6.1) with the differential equation

$$\frac{\partial^2 y}{\partial x^2} = \frac{1}{c^2}\frac{\partial^2 y}{\partial t^2} + h^2 y \tag{6.7}$$

and the periodic input with a general variation,

$$y(0,t) = F(t), \quad t > 0. \tag{6.8}$$

Elementary solutions of the modified wave equation (6.7) can be displayed in the form

$$\exp\left\{i\left(\pm\left(\frac{\omega^2}{c^2} - h^2\right)^{1/2} x - \omega t\right)\right\}$$

and, guided by earlier considerations, we propose an assemblage

$$y(x,t) = \int_{-\infty}^{\infty} A(\omega)\exp\left\{i\left(\left(\frac{\omega^2}{c^2} - h^2\right)^{1/2} x - \omega t\right)\right\} d\omega$$

$$= \frac{1}{i}\frac{\partial}{\partial x}\int_{-\infty}^{\infty} A(\omega)\exp\left\{i\left(\left(\frac{\omega^2}{c^2} - h^2\right)^{1/2} x - \omega t\right)\right\}\frac{d\omega}{\left(\frac{\omega^2}{c^2} - h^2\right)^{1/2}}$$

whose weighting factor, $A(\omega)$, is given by

$$A(\omega) = \frac{1}{2\pi}\int_0^{\infty} F(\tau)\, e^{-i\omega\tau}\, d\tau \tag{6.10}$$

in keeping with the stipulated input. To ensure that components of the integrand (6.9) for which ω assumes positive values exceeding hc shall represent disturbances travelling outwards from the input location, the choice of argument,

$$\arg\left(\frac{\omega^2}{c^2}-h^2\right)^{1/2}=0, \quad \omega>hc$$

is appropriate; then $\arg((\omega^2/c^2)-h^2)^{1/2}=\pi/2$ on the upper side of a branch cut that extends along the interval $(-hc, hc)$ of the real axis in the ω-plane and the corresponding argument equals π on the range $\omega<-hc$ thereof. An integral of the form (6.9) taken along a contour (Γ) which parallels the real axis and passes just above the branch cut thus contains a spectrum of exponential wave functions that propagate ($|\omega|>hc$) or attenuate ($|\omega|<hc$) in the positive x-direction, as befits the manner of their excitation (see Fig. 8).

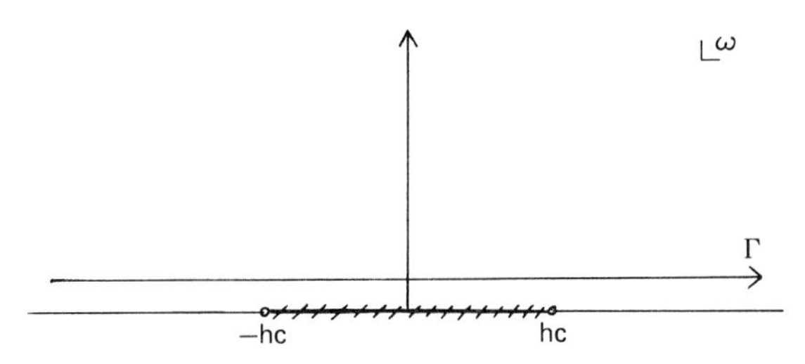

Fig. 8. An integration contour.

On substituting the representation (6.10) for $A(\omega)$ into (6.9) we obtain

$$y(x,t)=\frac{1}{2\pi\,\mathrm{i}}\frac{\partial}{\partial x}\int_0^\infty G(x,t-\tau;h)F(\tau)\,\mathrm{d}\tau \tag{6.11}$$

where

$$G(x,t-\tau;h)=\int_\Gamma \exp\left\{-\mathrm{i}\omega(t-\tau)+\mathrm{i}\left(\frac{\omega^2}{c^2}-h^2\right)^{1/2}x\right\}\frac{\mathrm{d}\omega}{\left(\frac{\omega^2}{c^2}-h^2\right)^{1/2}} \tag{6.12}$$

and, since the absolute magnitude of the ω-integrand along a semi-circular arc in the upper half of the ω-plane tends to zero with increasing radius if $\tau>t-(x/c)$, it follows that

$$y(x,t)=\frac{1}{2\pi\,\mathrm{i}}\frac{\partial}{\partial x}\int_0^{t-(x/c)} G(x,t-\tau;h)F(\tau)\,\mathrm{d}\tau, \quad t>\frac{x}{c}$$

and (6.13)

$$y(x,t)=0, \quad t<\frac{x}{c}.$$

When $0<\tau<t-(x/c)$ the contour Γ in (6.12) may be replaced with one that proceeds (in the clockwise sense) around the branch cut, thereby yielding an integral representation for G which has both finite and fixed limits; the latter is expressible, after use of the determination $\arg\sqrt{((\omega^2/c^2)-h^2)}=-\pi/2$ at the lower side of the cut, in the different forms

$$
\begin{aligned}
G(x,t-\tau;h) &= \frac{2}{\mathrm{i}}\int_{-hc}^{hc} \exp\{-\mathrm{i}\omega(t-\tau)\}\frac{\cosh\left(\left(h^2-\dfrac{\omega^2}{c^2}\right)^{1/2}x\right)}{\left(h^2-\dfrac{\omega^2}{c^2}\right)^{1/2}}\,\mathrm{d}\omega \\
&= \frac{2c}{\mathrm{i}}\int_{-1}^{1} \cos\{hc(t-\tau)\mu\}\frac{\cosh(hx\sqrt{(1-\mu^2)})}{\sqrt{(1-\mu^2)}}\,\mathrm{d}\mu \qquad (6.14)\\
&= \frac{2c}{\mathrm{i}}\int_0^{\pi} \cos\{hc(t-\tau)\cos\vartheta\}\cosh\{hx\sin\vartheta\}\,\mathrm{d}\vartheta.
\end{aligned}
$$

Having regard, collectively, for the characterizations

$$
\begin{aligned}
I &= \int_0^{\pi} \cos(\alpha\cos\vartheta)\cosh(\beta\sin\vartheta)\,\mathrm{d}\vartheta \\
&= \frac{1}{4}\int_0^{\pi} \{\exp(\mathrm{i}\alpha\cos\vartheta+\beta\sin\vartheta)+\exp(-\mathrm{i}\alpha\cos\vartheta-\beta\sin\vartheta) \\
&\quad + \exp(\mathrm{i}\alpha\cos\vartheta-\beta\sin\vartheta)+\exp(-\mathrm{i}\alpha\cos\vartheta+\beta\sin\vartheta)\}\,\mathrm{d}\vartheta \\
&= \frac{1}{2}\int_0^{\pi} [\cos(\mu\cos(\vartheta-\varphi))+\cos(\mu\cos(\vartheta+\varphi))]\,\mathrm{d}\vartheta,
\end{aligned}
$$

where

$$
\alpha=\mu\cos\varphi,\qquad \beta=\mathrm{i}\mu\sin\varphi
$$
$$
\alpha^2-\beta^2=\mu^2,
$$

and also the Bessel function integral

$$
\frac{1}{\pi}\int_0^{\pi} \cos(\gamma\cos(\vartheta+\delta))\,\mathrm{d}\vartheta = J_0(\gamma)
$$

it appears that

$$
I=\pi J_0\sqrt{(\alpha^2-\beta^2)},\quad \alpha^2>\beta^2.
$$

Thus

$$
G(x,t-\tau;h)=\frac{2\pi c}{\mathrm{i}}J_0[h\sqrt{(c^2(t-\tau)^2-x^2)}],\quad c(t-\tau)>x
$$

or (6.15)

$$
G(x,t-\tau;h)=\frac{4c}{\mathrm{i}}\mathfrak{G}(x,t-\tau;h)
$$

in the notation of (5.8). Finally, therefore, we reach a quadrature formula expressing the causal solution of the signalling problem,

$$y(x,t) = -\frac{2c}{\pi}\frac{\partial}{\partial x}\int_0^{t-(x/c)} \mathfrak{G}(x, t-\tau; h)F(\tau)\,d\tau \qquad t > x/c$$
$$= -c\frac{\partial}{\partial x}\int_0^{t-(x/c)} J_0[h\sqrt{(c^2(t-\tau)^2 - x^2)}]F(\tau)\,d\tau, \tag{6.16}$$

that evidently reduces to (2.4) in the limit $h \to 0$.

§7. *Conditions for Permanence and Exponential Decay of Progressive Wave Profiles*

The linear partial differential equations (1.1), (5.1), and (5.16) each possess the same pair of second order terms and thus belong to a common type, known as hyperbolic; their solutions evidence a considerable diversity nonetheless and only the first, for example, is rigorously satisfied by progressive wave forms that have both an arbitrary and permanent profile. It becomes a matter of some interest, accordingly, to ascertain the circumstances wherein profiles with the latter attributes are admissible solutions of the general (and homogeneous) representative of constant coefficient hyperbolic equations in two independent variables x, t, viz.

$$\alpha\frac{\partial^2 y}{\partial x^2} + 2\beta\frac{\partial^2 y}{\partial x \partial t} + \gamma\frac{\partial^2 y}{\partial t^2} + \delta\frac{\partial y}{\partial x} + \mu\frac{\partial y}{\partial t} + \nu y = 0; \tag{7.1}$$

more precisely, we seek those conditions on the ensemble of coefficients $\alpha, \ldots, \nu$ which assure the existence of solutions

$$y_1 = f(x - c_1 t), \qquad y_2 = f(x + c_2 t), \quad c_1, c_2 > 0 \tag{7.2}$$

that characterize oppositely directed progressive waves.

On substituting the expressions for y_1, y_2 into (7.1) we find

$$\begin{aligned}(\alpha - 2\beta c_1 + \gamma c_1^2)f''(x - c_1 t) + (\delta - \mu c_1)f'(x - c_1 t) + \nu f(x - c_1 t) = 0,\\ (\alpha + 2\beta c_2 + \gamma c_2^2)f''(x + c_2 t) + (\delta + \mu c_2)f'(x + c_2 t) + \nu f(x + c_2 t) = 0\end{aligned} \tag{7.3}$$

where primes signify argument derivatives. If the individual equations (7.3) are valid, irrespective of any commitment as to the particulars of f (other than differentiability), it follows that the coefficients of f, f', and f'' vanish separately; hence

$$\alpha - 2\beta c_1 + \gamma c_1^2 = 0, \qquad \delta - \mu c_1 = 0, \quad \nu = 0, \tag{7.4}$$

$$\alpha + 2\beta c_2 + \gamma c_2^2 = 0, \qquad \delta + \mu c_2 = 0, \quad \nu = 0. \tag{7.5}$$

Null values for both δ and μ, which carry over a like assignment for ν, are implicit in the proviso that c_1 and c_2 have positive magnitudes; and the condition

$$\beta^2 - \alpha\gamma > 0, \tag{7.6}$$

linking (7.1) with the hyperbolic class of differential equations, guarantees the reality of solutions

$$c_1 = \frac{\beta \pm \sqrt{(\beta^2 - \alpha\gamma)}}{\gamma}, \qquad c_2 = \frac{-\beta \pm \sqrt{(\beta^2 - \alpha\gamma)}}{\gamma} \tag{7.7}$$

to the first of the relations comprised in (7.4) and (7.5). When $\sqrt{(\beta^2 - \alpha\gamma)} > \beta$, the acceptable determinations

$$c_1 = \frac{\beta + \sqrt{(\beta^2 - \alpha\gamma)}}{\gamma}, \qquad c_2 = \frac{-\beta + \sqrt{(\beta^2 - \alpha\gamma)}}{\gamma}, \qquad \begin{matrix} \alpha < 0 \\ \gamma > 0 \end{matrix} \tag{7.8}$$

dictate unequal wave speeds along the $\pm x$ directions, a feature that is evidently contingent upon the presence of a mixed second-order derivative, $2\beta(\partial^2 y/\partial x \partial t)$, in the equation of motion. Given the inequalities $0 < \sqrt{(\beta^2 - \alpha\gamma)} < \beta$ on the other hand, which necessitate a common algebraic sign for α and γ, we arrive at the requisite determinations

$$c_1 = \frac{\beta \pm \sqrt{(\beta^2 - \alpha\gamma)}}{\gamma} \quad \text{if} \quad \gamma > 0, \qquad c_2 = \frac{-\beta \pm \sqrt{(\beta^2 - \alpha\gamma)}}{\gamma} \quad \text{if} \quad \gamma < 0, \tag{7.9}$$

reflecting a further degree of asymmetry in the nature of progressive wave motions.

The hyperbolic differential equation containing only second-order terms,

$$\alpha \frac{\partial^2 y}{\partial x^2} + 2\beta \frac{\partial^2 y}{\partial x \partial t} + \gamma \frac{\partial^2 y}{\partial t^2} = 0, \tag{7.10}$$

singled out as the prototype for travelling waves, may be recast in the canonical form (1.1) relative to a coordinate system moving along the x-direction. Thus, if we choose new independent variables $\xi = x - Vt$, $\eta = t$, of which the former defines a position coordinate measured from an origin that progresses at uniform speed V in the x-direction, the appertaining version of (7.10) becomes

$$(\alpha - 2\beta V + \gamma V^2) \frac{\partial^2 y}{\partial \xi^2} + 2(\beta - \gamma V) \frac{\partial^2 y}{\partial \xi \partial \eta} + \gamma \frac{\partial^2 y}{\partial \eta^2} = 0. \tag{7.11}$$

With the selection

$$V = \frac{\beta}{\gamma}, \quad \gamma \neq 0 \tag{7.12}$$

that eliminates the mixed partial derivative term from (7.11), there obtains

$$\frac{\partial^2 y}{\partial \xi^2} = \frac{1}{v^2}\frac{\partial^2 y}{\partial \eta^2}, \qquad v^2 = \frac{\beta^2 - \alpha\gamma}{\gamma^2}, \tag{7.13}$$

namely, the simplest version of a wave equation in a stationary reference frame; invoking the specifications for V and v, the values of c_1 in (7.9) are expressible as

$$c_1 = V \pm v$$

and thus determine, in the present context, the unequal (absolute) speeds of disturbances headed in opposite directions with regard to the moving coordinate frame.

If the coefficients α, β, and γ in (7.1) or (7.10) satisfy the inequality $\beta^2 - \alpha\gamma < 0$ and the differential equations belong to the so-called elliptic class, then progressive waves characterized by real speeds are ruled out. The remaining or parabolic class is bound up with a coefficient equality, viz. $\beta^2 - \alpha\gamma = 0$; fixing our attention on the particular solution $y_1 = f(x - c_1 t)$ and taking account of the currently applicable representations [which stem from (7.3), (7.4)].

$$\beta = \sqrt{(\alpha\gamma)}, \qquad c_1 = \frac{\beta}{\gamma} = \sqrt{(\alpha/\gamma)}, \qquad \delta = \mu c_1, \quad \nu = 0$$

$$\delta\frac{\partial y}{\partial x} + \mu\frac{\partial y}{\partial t} = \frac{\mu}{\sqrt{(\gamma)}}\left(\sqrt{(\alpha)}\frac{\partial y}{\partial x} + \sqrt{(\gamma)}\frac{\partial y}{\partial t}\right), \quad \alpha, \gamma > 0$$

it appears that (7.1) admits conversion to the form

$$\left(\sqrt{(\alpha)}\frac{\partial}{\partial x} + \sqrt{(\gamma)}\frac{\partial}{\partial t}\right)\left(\sqrt{(\alpha)}\frac{\partial}{\partial x} + \sqrt{(\gamma)}\frac{\partial}{\partial t}\right)y + \frac{\mu}{\sqrt{(\gamma)}}\left(\sqrt{(\alpha)}\frac{\partial}{\partial x} + \sqrt{(\gamma)}\frac{\partial}{\partial t}\right)y = 0 \tag{7.14}$$

wherein a specific differential operator, $L = \sqrt{(\alpha)}(\partial/\partial x) + \sqrt{(\gamma)}(\partial/\partial t)$, recurs. Once x and t are replaced by another pair of independent variables ζ, τ according to the scheme

$$x = \sqrt{(\alpha)}\zeta, \quad t = \sqrt{(\gamma)}\zeta + \tau$$

(7.14) acquires the status of an ordinary differential equation

$$\frac{d^2 y}{d\zeta^2} + \frac{\mu}{\sqrt{(\gamma)}}\frac{dy}{d\zeta} = 0$$

and there follows readily the consequence that an arbitrary function of the variable

$$\tau = t - \sqrt{(\gamma)}\zeta = t - \sqrt{(\gamma/\alpha)}x = t - \frac{x}{c_1}$$

constitutes a progressive wave integral of the original equation, (7.1).

We may examine, in a similar manner, the compatibility of the latter equation with a temporally modulated wave-type solution, namely

$$y(x, t) = A(t)f(x - ct); \tag{7.15}$$

thus, substitution of (7.15) into (7.1) yields

$$(\alpha - 2\beta c + \gamma c^2)A(t)f''(x - ct) + [(\delta - \mu c)A(t) + 2(\beta - \gamma c)A'(t)]f'(x - ct) \\ + [\nu A(t) + \mu A'(t) + \gamma A''(t)]f(x - ct) = 0$$

and, bearing in mind once again the arbitrariness of f, the above equation is satisfied when

$$\alpha - 2\beta c + \gamma c^2 = 0$$
$$(\delta - \mu c)A(t) + 2(\beta - \gamma c)A'(t) = 0$$
$$\nu A(t) + \mu A'(t) + \gamma A''(t) = 0.$$

The first of these relations yields the same pair of characteristic velocities encountered before, while the second provides a first-order differential equation for $A(t)$, with the exponential solution

$$A(t) = e^{-\varepsilon t}$$

if ε satisfies

$$(\delta - \mu c) - 2\varepsilon(\beta - \gamma c) = 0;$$

and then the third becomes

$$\gamma\varepsilon^2 - \mu\varepsilon + \nu = 0;$$

that is, a specific interrelation between the coefficients of the original differential equation sufficient for the validity of the assumed form of solution. Making use of these results, it follows that the equation of motion of a string which is subject to the combined action of external forces proportional to the displacement and velocity,

$$\frac{\partial^2 y}{\partial x^2} = \frac{1}{c^2}\frac{\partial^2 y}{\partial t^2} + 2\frac{\varepsilon}{c^2}\frac{\partial y}{\partial t} + h^2 y,$$

admits the damped wave solution

$$y(x, t) = e^{-\varepsilon t}f(x - ct)$$

provided that $\varepsilon = ch$.

§8. *Movements of a Heavy Chain and a Compressible Medium*

A consideration of the dynamical systems cited in the above heading enables us to gather additional details bearing on wave-like phenomena, since their excitations are regulated by hyperbolic partial differential equations. We have in mind that the uniform chain executes small amplitude planar movements about a vertical and equilibrium shape, with its free end situated lowermost. Assume that the equilibrium line of the chain coincides with the (positive) x-axis, the free end being located at the origin thereof; then, if $y(x, t)$ specifies the transverse displacement at any point, the dynamical equation of motion takes the form

$$\rho\frac{\partial^2 y}{\partial t^2}=\frac{\partial}{\partial x}\left(P(x)\frac{\partial y}{\partial x}\right) \tag{8.1}$$

where $P(x)$ and ρ represent the (variable) tension and (constant) density. On equating the local magnitude of the tension to the weight of the chain underneath, viz. $P(x)=\rho g x$, g designating the gravitational constant, we obtain in place of (8.1) a variable coefficient equation,

$$\frac{\partial^2 y}{\partial t^2}=c^2(x)\frac{\partial^2 y}{\partial x^2}+g\frac{\partial y}{\partial x}, \qquad c^2(x)=P(x)/\rho=gx, \tag{8.2}$$

which is, in fact, of the hyperbolic type. After replacing x by another variable,

$$\tau=\int_0^x\frac{\mathrm{d}x}{c(x)}=\int_0^x\frac{\mathrm{d}x}{\sqrt{(gx)}}=2\left(\frac{x}{g}\right)^{1/2}, \tag{8.3}$$

the equation (8.2) becomes

$$\frac{\partial^2 y}{\partial t^2}=\frac{\partial^2 y}{\partial \tau^2}+\frac{1}{\tau}\frac{\partial y}{\partial \tau}, \qquad \tau\geq 0, \quad t>0. \tag{8.4}$$

It is appropriate, for the resolution of an initial value problem in which the preceding equation figures, to construct a one-parameter family of particular solutions that possess a separate dependence on the variables t and τ; thus, selecting a periodic time factor and writing

$$y(\tau, t)=F(\tau)\cdot\begin{matrix}\cos\\ \sin\end{matrix}\ \omega t \tag{8.5}$$

we connect $F(\tau)$ with a cylinder function (or Bessel) differential equation,

$$\frac{\mathrm{d}^2F}{\mathrm{d}\tau^2}+\frac{1}{\tau}\frac{\mathrm{d}F}{\mathrm{d}\tau}+\omega^2F=0,$$

whose bounded (and arbitrarily scaled) solution on the range $\tau \geq 0$ is

$$F(\tau) = J_0(\omega\tau). \tag{8.6}$$

By superposition of the solutions (8.5), (8.6) in the fashion

$$y(\tau, t) = \int_0^\infty A(\omega)J_0(\omega\tau) \cos \omega t\, \omega\, d\omega + \int_0^\infty B(\omega)J_0(\omega\tau) \sin \omega t\, d\omega, \tag{8.7}$$

and with reference to the Fourier–Bessel reciprocal formulas

$$\mathfrak{F}(\tau) = \int_0^\infty f(\omega)J_0(\omega\tau)\omega\, d\omega, \qquad f(\omega) = \int_0^\infty \mathfrak{F}(\tau)J_0(\omega\tau)\tau\, d\tau \tag{8.8}$$

the functions $A(\omega)$, $B(\omega)$ acquire their determinations

$$\begin{aligned} A(\omega) &= \int_0^\infty s(\tau)J_0(\omega\tau)\tau\, d\tau, \\ B(\omega) &= \int_0^\infty v(\tau)J_0(\omega\tau)\tau\, d\tau \end{aligned} \tag{8.9}$$

in terms of the initial assignments

$$y(\tau, 0) = s(\tau), \qquad \frac{\partial}{\partial t} y(\tau, 0) = v(\tau), \quad \tau > 0. \tag{8.10}$$

For the special albeit idealized circumstance wherein a chain, originally at rest, experiences an impulse in the single point $x = x_1$ corresponding to a definite value $\tau = \tau_1 = 2\sqrt{(x_1/g)}$ of the affiliated variable, the characterisations

$$A(\omega) = 0, \qquad B(\omega) = J_0(\omega\tau_1)$$

are obtained from (8.9) if

$$\int_0^\infty v(\tau)\tau\, d\tau = 1, \qquad v(\tau) = 0, \quad \tau \neq \tau_1;$$

and the resultant deflection of the chain has the (non-dimensional) expression

$$y(\tau, t) = \int_0^\infty J_0(\omega\tau)J_0(\omega\tau_1) \sin \omega t\, d\omega. \tag{8.11}$$

The integral of (8.11), containing three parameters τ, τ_1, and t, is unaffected by interchange of a pair, $\tau = 2\sqrt{(x/g)}$ and $\tau_1 = 2\sqrt{(x_1/g)}$, which locate the points of observation and impulse, respectively; such a reciprocal behavior of excitation levels at source and observer locations embraces a large and diverse number of systems describable in terms of a linear theory.

At the free end of the chain, where $\tau = 0$ and $J_0(\omega\tau) = 1$, the representation (8.11) yields

$$y(0, t) = \begin{cases} 0, & t < \tau_1 \\ (t^2 - \tau_1^2)^{-1/2}, & t > \tau_1 \end{cases}, \tag{8.12}$$

in accordance with the discontinuous nature of the integral

$$\int_0^\infty J_0(\alpha x) \sin \beta x \, dx = \begin{cases} 0, & 0 < \beta < \alpha \\ (\beta^2 - \alpha^2)^{-1/2}, & 0 < \alpha < \beta \end{cases}.$$

Passing over the fact, attributable to our schematic mode of impulsive action, that the displacement (8.12) commences (at $t = \tau_1$) with a magnitude which prompts a question about the original hypothesis of small scale chain movements everywhere, three observations regarding same are in order; one advertises the time delay, τ_1, before a consequence of the impulse becomes manifest at the free end, another identifies the magnitude of τ_1 as the time for propagation with speed $c(x)$ over the range $(0, x_1)$, and the last calls attention to the perpetuity of the end movement, although the considered state of motion has a local character originally, in contrast with the subsidence of analogous excitations for three-dimensional and extended physical systems.

The infinite integral of (8.11) may be converted, without any restriction on the values of τ, τ_1, and t, to a complete elliptic integral whose limits are finite; it suffices here to record some qualitative aspects stemming therefrom, namely, that a null value of the integral and hence of the displacement obtains when $t < |\tau - \tau_1|$, and also that distinct versions of the displacement apply, according as $\tau + \tau_1 > t > |\tau - \tau_1|$ or $t > \tau + \tau_1$, respectively. Since τ defines the travel time for a disturbance to reach the point x after leaving the free end, the inequalities $\tau + \tau_1 > t > |\tau - \tau_1|$ jointly select the interval between the arrival at x of a primary stimulus (directly communicated by the impulse) and a secondary one (which has been reflected at the free end).

Let us next envisage a medium wherein the pressure p and density ρ depend on a single coordinate x, measured positively in the vertical direction, say, with the interrelation

$$\partial p_0 / \partial x = -g\rho_0 \tag{8.13}$$

between their static or equilibrium values that involves the gravitational constant g. For small vertical or planar movements in this 'atmosphere,' specified by the local displacement $\xi(x, t)$, there is a dynamical equation of motion,

$$\rho_0 \frac{\partial^2 \xi}{\partial t^2} = -\frac{\partial p}{\partial x} - g\rho_0 = -\frac{\partial}{\partial x}(p - p_0), \tag{8.14}$$

and a mass conservation or continuity relation

$$\rho\left(1 + \frac{\partial \xi}{\partial x}\right) = \rho_0. \tag{8.15}$$

If the pressure/density interrelation or so-called equation of state has the form

$$p/p_0 = (\rho/\rho_0)^\gamma, \quad \gamma > 0 \tag{8.16}$$

changes in pressure connected with relative motion in the medium are given by

$$p - p_0 = p_0\left\{\left(1 + \frac{\partial \xi}{\partial x}\right)^{-\gamma} - 1\right\} \doteq -\gamma p_0 \frac{\partial \xi}{\partial x}$$

to a first order of approximation, and thus (8.14) can be rewritten as a linear equation for the displacement exclusively,

$$\frac{\partial^2 \xi}{\partial t^2} = c^2 \frac{\partial^2 \xi}{\partial x^2} - \gamma g \frac{\partial \xi}{\partial x}, \tag{8.17}$$

whose hyperbolic nature is secured by the coefficient inequality

$$c^2 = \gamma \frac{p_0}{\rho_0} > 0. \tag{8.18}$$

Evidently c describes a propagation speed for the disturbances which evolve in accordance with (8.17) and, though it is reasonable to anticipate a non-uniform magnitude for same in large scale (compressible) atmospheric movements (or sound waves), we shall here assume that

$$c^2 = \gamma g l = \text{constant}; \tag{8.19}$$

then the relations (8.13), (8.16) imply an exponential variation, of effective range l, for the equilibrium density (and pressure), viz.

$$\rho_0(x) = A e^{-x/l}. \tag{8.20}$$

Adopting units of length and time equal to $2l$ and $\sqrt{(4l/\gamma g)} = 2l/c$, respectively, or dimensionless variables η, τ given by

$$x = 2l\eta, \qquad t = \frac{2l}{c}\tau \tag{8.21}$$

the differential equation (8.17) becomes

$$\frac{\partial^2 \xi}{\partial \tau^2} = \frac{\partial^2 \xi}{\partial \eta^2} - 2\frac{\partial \xi}{\partial \eta} \tag{8.22}$$

whence, after the substitution

$$\xi = \psi e^\eta \tag{8.23}$$

therein, it turns out that ψ satisfies an equation resembling (5.1),

$$\frac{\partial^2 \psi}{\partial \tau^2} = \frac{\partial^2 \psi}{\partial \eta^2} - \psi. \tag{8.24}$$

An integral of (8.24) fashioned in terms of particular solutions having a separated variable form,

$$\psi(\eta, \tau) = \int_{-\infty}^{\infty} A(k) \cos(\surd(k^2+1)\tau)\, \mathrm{e}^{ik\eta}\, dk + \int_{-\infty}^{\infty} B(k) \frac{\sin(\surd(k^2+1)\tau)}{\surd(k^2+1)}\, \mathrm{e}^{ik\eta}\, dk,$$

accords with the initial assignments (8.25)

$$\psi(\eta, 0) = s(\eta), \qquad \frac{\partial\psi(\eta, 0)}{\partial\tau} = v(\eta) \tag{8.26}$$

if

$$A(k) = \frac{1}{2\pi}\int_{-\infty}^{\infty} s(\eta)\, \mathrm{e}^{-ik\eta}\, \mathrm{d}\eta, \qquad B(k) = \frac{1}{2\pi}\int_{-\infty}^{\infty} v(\eta)\, \mathrm{e}^{-ik\eta}\, \mathrm{d}\eta. \tag{8.27}$$

Both $s(\eta)$ and $A(k)$ vanish identically when there is no initial displacement and $B(k) = 1/2\pi$ for an extreme localization of $v(\eta)$, relative to $\eta = 0$, which is implicit in the attributes

$$\int_{-\infty}^{\infty} v(\eta)\, \mathrm{d}\eta = 1, \quad v(\eta) = 0, \quad \eta \neq 0;$$

on the basis of such (hypothetical) particulars we obtain from (8.25)

$$\begin{aligned} \psi(\eta, \tau) &= \frac{1}{\pi}\mathfrak{G}(\eta, \tau; 1) \quad \text{[cf. (5.8)]} \\ &= \begin{cases} 0, & \tau^2 < \eta^2 \\ \tfrac{1}{2}J_0(\surd(\tau^2 - \eta^2)), & \tau^2 > \eta^2 \end{cases} \quad \text{[cf. (5.14)]} \end{aligned} \tag{8.28}$$

and a resulting displacement

$$\xi(x, t) = \frac{1}{2}(\rho_0(0)\rho_0(x))^{-1/2} J_0\left(\frac{\surd(c^2t^2 - x^2)}{2l}\right), \quad ct > x \tag{8.29}$$

expressed in terms of the original variables x, t. The latter may be identified as the response brought about by an input of momentum at the time $t = 0$, in the amount $\rho_0(0)c$ per unit area of the plane $x = 0$; and the symmetrical occurrence of density factors referring to source and observer locations is consistent with the reciprocity principle. Conservation of momentum afterwards ($t > 0$) implies that

$$\begin{aligned} \rho_0(0)c &= \frac{\mathrm{d}}{\mathrm{d}t}\int_{-ct}^{ct} \rho_0(x)\xi(x, t)\, \mathrm{d}x \\ &= \frac{1}{2}\rho_0(0)\frac{\mathrm{d}}{\mathrm{d}t}\int_{-ct}^{ct} \mathrm{e}^{-x/2l} J_0\left(\frac{\surd(c^2t^2 - x^2)}{2l}\right) \mathrm{d}x \\ &= \rho_0(0)\frac{\mathrm{d}}{\mathrm{d}t}\int_{0}^{ct} \cosh\frac{x}{2l} J_0\left(\frac{\surd(c^2t^2 - x^2)}{2l}\right) \mathrm{d}x \\ &= \rho_0(0)c\frac{\mathrm{d}}{\mathrm{d}\tau}\int_{0}^{\tau} \cosh\eta J_0(\surd(\tau^2 - \eta^2))\, \mathrm{d}\eta, \end{aligned}$$

or

$$\int_0^\tau \cosh \eta J_0(\surd(\tau^2-\eta^2))\,\mathrm{d}\eta = \tau; \tag{8.30}$$

if we apply the transformation $\eta = \tau \sin \varphi$, $0 < \varphi < \pi/2$, in (8.30) the corresponding statement is that

$$\begin{aligned} I(\tau) &= \int_0^{\pi/2} \cosh(\tau \sin \varphi) J_0(\tau \cos \varphi) \cos \varphi \, \mathrm{d}\varphi \\ &= 1, \end{aligned} \tag{8.31}$$

independently of τ. Now

$$\begin{aligned} \frac{\mathrm{d}I}{\mathrm{d}\tau} = \int_0^{\pi/2} &\{\sinh(\tau \sin \varphi) J_0(\tau \cos \varphi) \sin \varphi \cos \varphi \\ &- \cosh(\tau \sin \varphi) J_1(\tau \cos \varphi) \cos^2 \varphi\}\,\mathrm{d}\varphi, \end{aligned} \tag{8.32}$$

and

$$\begin{aligned} &\int_0^{\pi/2} \cosh(\tau \sin \varphi) J_1(\tau \cos \varphi) \cos^2 \varphi \, \mathrm{d}\varphi \\ &= \frac{1}{\tau} \int_0^{\pi/2} J_1(\tau \cos \varphi) \cos \varphi \, \mathrm{d}\{\sinh(\tau \sin \varphi)\} \\ &= -\frac{1}{\tau} \int_0^{\pi/2} \sinh(\tau \sin \varphi) \frac{\mathrm{d}}{\mathrm{d}\varphi}\{J_1(\tau \cos \varphi) \cos \varphi\}\,\mathrm{d}\varphi \\ &= \int_0^{\pi/2} \sinh(\tau \sin \varphi) J_0(\tau \cos \varphi) \sin \varphi \cos \varphi \, \mathrm{d}\varphi, \end{aligned} \tag{8.33}$$

where the final expression incorporates a usage of the formula

$$\frac{\mathrm{d}}{\mathrm{d}\varphi} J_1(\tau \cos \varphi) = -\tau \sin \varphi \{J_0(\tau \cos \varphi) - \frac{1}{\tau \cos \varphi} J_1(\tau \cos \varphi)\}.$$

It follows from (8.32) and (8.33) that

$$\mathrm{d}I/\mathrm{d}\tau = 0$$

and, inasmuch as $I(0) = 1$, the validity of (8.31) is upheld.

§9. *A Third Order (Viscoelastic) Equation of Motion*

To further our appreciation of factors affecting the nature and evolution of progressive disturbances, we now consider a differential equation of motion

$$\left(1 + \frac{2}{\omega_0} \frac{\partial}{\partial t}\right) \frac{\partial^2 z}{\partial x^2} = \frac{1}{c^2} \frac{\partial^2 z}{\partial t^2} \tag{9.1}$$

wherein the third (and highest) order term may be correlated with a force of viscous origin; this equation has relevance, in particular, to small amplitude longitudinal displacements, $z(x, t)$, along a taut straight wire and, as written, features a positive coefficient ω_0, with the dimensions of a frequency, in the viscous term. If the coordinates x and t are replaced by a dimensionless pair ξ, τ, respectively, such that

$$x = \frac{c\xi}{\omega_0}, \qquad t = \frac{\tau}{\omega_0} \tag{9.2}$$

the corresponding equation for $z(\xi, t)$ becomes

$$2\frac{\partial^3 z}{\partial \xi^2 \partial \tau} + \frac{\partial^2 z}{\partial \xi^2} - \frac{\partial^2 z}{\partial \tau^2} = 0. \tag{9.3}$$

We assume that a full range of values, $-\infty < \xi < \infty$, applies to the position variable and inquire after the solution of (9.3) in $\tau > 0$ which is determined by the twin initial assignments

$$z(\xi, 0) = s_0(\xi), \qquad \frac{\partial z(\xi, 0)}{\partial \tau} = v_0(\xi) \tag{9.4}$$

for displacement and velocity. On representing the initial distributions in terms of Fourier integrals,

$$s_0(\xi) = \int_{-\infty}^{\infty} a(\kappa)\, e^{i\kappa\xi}\, d\kappa, \qquad v_0(\xi) = \int_{-\infty}^{\infty} b(\kappa)\, e^{i\kappa\xi}\, d\kappa, \tag{9.5}$$

the concomitant version of $z(\xi, \tau)$,

$$z(\xi, \tau) = \int_{-\infty}^{\infty} [a(\kappa) A_1(\tau, \kappa) + b(\kappa) A_2(\tau, \kappa)]\, e^{i\kappa\xi}\, d\kappa \tag{9.6}$$

provides the desired integral of (9.3), if $A_{1,2}$ are solutions of the common equation

$$\left(\frac{\partial^2}{\partial \tau^2} + 2\kappa^2 \frac{\partial}{\partial \tau} + \kappa^2\right) A_{1,2}(\tau, \kappa) = 0 \tag{9.7}$$

which meet the specifications

$$\begin{matrix} A_1 = 1 & A_2 = 0 \\ \partial A_1/\partial \tau = 0 & \partial A_2/\partial \tau = 1 \end{matrix}, \quad \tau = 0. \tag{9.8}$$

The explicit forms of $A_{1,2}$ can be readily derived and, following their substitution in (9.6), we obtain

$$\begin{aligned} z(\xi, \tau) = \int_{-\infty}^{\infty} \Bigg\{ a(\kappa) \left[\frac{\kappa \sinh(\kappa\sqrt{(\kappa^2 - 1)}\tau)}{\sqrt{(\kappa^2 - 1)}} + \cosh(\kappa\sqrt{(\kappa^2 - 1)}\tau) \right] \\ + b(\kappa) \frac{\sinh(\kappa\sqrt{(\kappa^2 - 1)}\tau)}{\kappa\sqrt{(\kappa^2 - 1)}} \Bigg\} e^{-\kappa^2\tau + i\kappa\xi}\, d\kappa; \end{aligned} \tag{9.9}$$

to establish conditions sufficient for a properly defined integral it is first noted that the products of hyperbolic and exponential functions have a finite value in the limit $|\kappa|\to\infty$, so long as $\tau\geq 0$, whence the integral assuredly converges if $\kappa^2 a(\kappa)$ and $b(\kappa)$ are of bounded magnitude everywhere. The integrand of (9.9), say $f(\tau, \kappa)$, characterizes a regular function of the complex variable $\tau = \mu + i\nu$ and, inasmuch as

$$\begin{aligned}|\sinh(\kappa\sqrt{(\kappa^2-1)}\tau)| &\leq \cosh(\kappa\sqrt{(\kappa^2-1)}\mu)\\ |\cosh(\kappa\sqrt{(\kappa^2-1)}\tau)| &\leq \cosh(\kappa\sqrt{(\kappa^2-1)}\mu)\end{aligned}, \quad \kappa > 1$$

the property expressed by

$$\int_{\kappa(>1)}^{\infty} |f(\tau, \kappa)|\, d\kappa \leq \int_{\kappa(>1)}^{\infty} \cosh(\kappa\sqrt{(\kappa^2-1)}\mu)\, e^{-\kappa^2\mu} \times\left[|a(\kappa)|\left(\frac{\kappa}{\sqrt{(\kappa^2-1)}}+1\right)+|b(\kappa)|\frac{1}{\kappa\sqrt{(\kappa^2-1)}}\right] d\kappa < \infty$$

leads to the conclusion that $z(\xi, \tau)$ is a regular function of its second argument in the half-plane $\operatorname{Re}\tau > 0$. On the imaginary axis, $\operatorname{Re}\tau = 0$, $z(\xi, \tau)$ has a continuous (instead of regular) behavior if the approach to any point thereupon is made along a path lying to the right of this axis; it is the fact of uniform convergence for the integral (9.9) (and its τ-derivative) with regard to the variable τ in the domain $\operatorname{Re}\tau \geq 0$ that supports the foregoing contention [Roy (1914)].

The properties cited above enable us, in particular, to characterize the function $z(\xi, \tau)$ defined by (9.9) as an analytic (or infinitely differentiable) function of the real variable τ, when $\tau > 0$, with continuous variation at $\tau = 0$ for both z and $\partial z/\partial\tau$. Consequently, in the limit $\tau \to 0+$, the representation (9.9) accords with the given initial values of displacement and velocity and the solution of the viscoelastic differential equation (9.1) thus shares, insofar as its dependence on the temporal argument is concerned, the corresponding attributes of the solution

$$z(\xi, \tau) = \frac{1}{2\sqrt{(\pi\tau)}} \int_{-\infty}^{\infty} u(\xi') \exp\left\{-\frac{(\xi-\xi')^2}{4\tau}\right\} d\xi' \tag{9.10}$$

to an equation of the simple diffusion type,

$$\frac{\partial z}{\partial\tau} = \frac{\partial^2 z}{\partial\xi^2} \tag{9.11}$$

with the initial specification

$$z(\xi, 0) = u(\xi), \quad -\infty < \xi < \infty. \tag{9.12}$$

The noteworthy aspects of the manner in which the solution to the viscoelastic equation depends on the real argument ξ are twofold, namely that

$z(\xi, \tau)$ is a continuous function of ξ, for any value of τ with positive real part and that analyticity of $z(\xi, \tau)$ is contingent upon a like property of the initial distributions $s_0(\xi)$, $v_0(\xi)$. In the latter correlation we may perceive a fundamental difference between the evolution of disturbances affected by viscosity and of temperature profiles governed by the equation (9.11) of heat conduction; for (9.10) implies that, whatever the nature of the initial temperature distribution $u(\xi)$, the subsequent temperature possesses an analytic character in regard to the coordinate variable.

The circumstance that irregular features in the initial data are imparted to the solution also prevails in the case of the second order wave equation; if, for instance, either the initial displacement or velocity ceases to be analytic at the point $\xi = \xi_0$, the corresponding solution of

$$\frac{\partial^2 z}{\partial \xi^2} = \frac{\partial^2 z}{\partial \tau^2}$$

has the same aspect at the locations $\xi_0 \pm \tau$, whose traces in the (ξ, τ) plane are evidently the characteristic lines emanating from $(\xi_0, 0)$. By contrast, non-analyticity of the solution to the third order equation (9.3) occurs at the fixed point $\xi = \xi_0$, for all values $\tau > 0$.

§10. *Influence of Viscoelasticity on Signal Transmission*

The striking consequences, as regards the solution of initial value problems, that stem from a viscoelastic term in the equation of motion are rivalled by those manifest in signalling problems; for the latter we envisage a half-range of the position coordinate, say $x > 0$, and the two point specifications

$$z(0, t) = F(t), \qquad z(\infty, t) = 0 \tag{10.1}$$

along with the null initial values

$$z(x, 0) = \frac{\partial z(x, 0)}{\partial t} = 0, \quad x > 0. \tag{10.2}$$

To prepare for the accomodation of such conditions when the basic equation of motion departs from the simplest wave prototype [and general integrals are unavailable], we introduce a method of analysis based on Laplace transformation and first rederive the familiar solution that befits distortionless, finite speed, signal transmission in the absence of viscoelasticity. Let

$$\bar{z}(x, p) = \int_0^\infty e^{-pt} z(x, t)\, dt, \quad \operatorname{Re} p > 0 \tag{10.3}$$

define the temporal Laplace transform of the displacement function and observe that if the latter satisfies the equation

$$\frac{\partial^2 z}{\partial x^2} = \frac{1}{c^2}\frac{\partial^2 z}{\partial t^2},$$

then, taking account of (10.2),

$$\frac{\mathrm{d}^2 \bar{z}}{\mathrm{d}x^2} - \frac{p^2}{c^2}\bar{z} = 0. \tag{10.4}$$

The integral of (10.4) that complies with (10.1) is given by

$$\bar{z}(x, p) = \bar{F}(p)\,\mathrm{e}^{-px/c} \tag{10.5}$$

where

$$\bar{F}(p) = \int_0^\infty \mathrm{e}^{-pt} F(t)\,\mathrm{d}t \tag{10.6}$$

and hence, applying the inverse transform representation,

$$z(x, t) = \frac{1}{2\pi\,\mathrm{i}}\int_\Gamma \mathrm{e}^{p[t-(x/c)]}\bar{F}(p)\,\mathrm{d}p = \begin{cases} F\left(t - \frac{x}{c}\right), & t > \frac{x}{c} \\ 0, & t < \frac{x}{c} \end{cases} \tag{10.7}$$

inasmuch as all the singularities of $\bar{F}(p)$ are located to the left of the infinite contour Γ which passes from $\alpha - \mathrm{i}\infty$ to $\alpha + \mathrm{i}\infty$ along the line $\mathrm{Re}\,p = \alpha =$ constant.

If we adopt the viscoelastic equation of motion

$$\frac{\partial^3 z}{\partial x^2 \partial t} - \frac{1}{\mu_0}\left(\frac{\partial^2 z}{\partial t^2} - c^2\frac{\partial^2 z}{\partial x^2}\right) = 0, \tag{10.8}$$

with the relationship

$$\mu_0 = \frac{2c^2}{\omega_0} \tag{10.9}$$

between the parameters of the current and prior versions, there obtains

$$\bar{z}(x, p) = \frac{z_0}{p}\exp\left\{-\frac{px}{\surd(c^2 + \mu_0 p)}\right\} \tag{10.10}$$

in place of (10.5), if

$$F(t) = z_0, \quad \text{a constant.} \tag{10.11}$$

The problem is thus resolved with an explicit integral represenation

$$z(x, t) = \frac{z_0}{2\pi\,\mathrm{i}}\int_\Gamma \exp\left\{pt - \frac{px}{\surd(c^2 + \mu_0 p)}\right\}\frac{\mathrm{d}p}{p} \tag{10.12}$$

which, notwithstanding the simple assignment of a fixed signal strength, cannot be given in terms of elementary functions. Estimates for $z(x, t)$ at small and large values of t are derivable from (10.12), however, and these advance the current objectives by an exposure of viscoelastic effects in different epochs.

Immediately following the onset of signal activity, an approximation to the exact transform $\bar{z}(x, p)$ for large p is apt, viz.

$$\bar{z}(x, p) = \frac{z_0}{p} \exp\left(-\sqrt{\left(\frac{p}{\mu_0}\right)} x\right)\left\{1 + \frac{1}{2}\frac{c^2 x}{\mu_0^{3/2} p^{1/2}} + \cdots\right\},$$

and there obtains from the concomitant version of (10.12), after replacing the contour Γ with a loop around the origin that has its endpoints at $-\infty$, the characterization

$$z(x, t)/z_0 = \operatorname{erfc}\left(\frac{x}{2\sqrt{(\mu_0 t)}}\right) + \frac{c^2 x \sqrt{t}}{\mu_0^{3/2}} \int_{x/2\sqrt{(\mu_0 t)}}^{\infty} \operatorname{erfc} \xi \, d\xi + \cdots, \quad t \to 0 \tag{10.13}$$

where

$$\operatorname{erfc} \zeta = \frac{2}{\sqrt{\pi}} \int_{\zeta}^{\infty} e^{-\alpha^2} d\alpha = 1 - \operatorname{erf} \zeta \tag{10.14}$$

designates the complementary error function and the (primary) error function or probability integral

$$\operatorname{erf} \zeta = \frac{2}{\sqrt{\pi}} \int_0^{\zeta} e^{-\alpha^2} d\alpha, \quad \operatorname{erf} \infty = 1 \tag{10.15}$$

possesses an antisymmetrical behavior, i.e.,

$$\operatorname{erf}(-\zeta) = -\operatorname{erf} \zeta. \tag{10.16}$$

The first term in (10.13) furnishes the complete solution of a heat conduction problem in one-dimension; namely, it specifies the temperature $z(x, t)$ along a semi-infinite rod, or solution of the equation

$$\frac{\partial^2 z}{\partial x^2} = \frac{1}{\mu_0}\frac{\partial z}{\partial t},$$

corresponding to a zero initial level in $x > 0$ and a constant value z_0 at $x = 0$ thereafter, if the parameter μ_0 is suitably interpreted. Thus, the response of a dynamical system governed by the differential equation (10.8) has a diffusive (rather than propagating) nature in its early phases, a feature linked with the greater importance of the first two terms in that equation.

The approximation (10.13) relinquishes its accuracy once the argument of the complementary error function in the leading term, $x/2\sqrt{(\mu_0 t)}$, assumes values less than unity; and from the condition

$$\frac{c^2 x}{\mu_0^{3/2}} \cdot \frac{x}{2\mu_0^{1/2}} = 1,$$

which reflects a comparable magnitude of the second term, we may infer a change in character of the signal when the effective range

$$x = 0\left(\frac{\mu_0}{c}\right) = 0\left(\frac{c}{\omega_0}\right)$$

is surpassed.

The selection of an appropriate path is instrumental for estimating the integral (10.12) after an extended period of signal activity, and the guiding principle (which underlies the method of steepest descent) is to secure a localized contribution thereupon. In (10.12) the singularities of the integrand that are situated to the left of the path, Γ, comprise a simple pole at the origin and a branch point at $p = -c^2/\mu_0$ on the negative real axis; if we employ the relationship

$$c^2 + \mu_0 p = \zeta^2 c^2$$

for a change of variables, the consequent representation

$$z(x, t)/z_0 = \frac{1}{\pi \mathrm{i}} \int_{\bar{\Gamma}} \exp\left\{\frac{c^2}{\mu_0}(\zeta^2 - 1)\left(t - \frac{x}{c\zeta}\right)\right\} \frac{\zeta \, \mathrm{d}\zeta}{\zeta^2 - 1}, \tag{10.17}$$

with simple pole singularities of the integrand at $\zeta = \pm 1$ and an essential singularity at $\zeta = 0$, involves a path $\bar{\Gamma}$ which approaches the origin in the direction $-\pi/4$, maintains a finite separation (>1) therefrom, and recedes along the direction $\pi/4$. We consider the argument of the exponential function,

$$f(\zeta) = \frac{c^2}{\mu_0} t(\zeta^2 - 1)\left(1 - \frac{x}{ct\zeta}\right), \tag{10.18}$$

in order to locate the saddle points, namely zeros of $f'(\zeta)$, and deduce that they are specified by roots of the third degree equation

$$2\zeta^3 - \frac{x}{ct}\zeta^2 - \frac{x}{ct} = 0.$$

The existence of a real root $\zeta \doteq \zeta^*$, which lies in one or the other of the intervals $(1, (x/ct))$, $((x/ct), 1)$ according as x/ct is greater or smaller than unity, can be readily confirmed, along with the fact that both complex roots have negative real parts.

If $t \gg 0$ and $x/ct = 1$ the pole at $\zeta = 1$ of the integrand in (10.17) is in close proximity to the saddle point ζ^* through which a path of steepest descent passes; the latter, situated in the half-plane $\operatorname{Re} \zeta > 0$, with the lines $(x/2ct) \pm \mathrm{i}\infty$ as asymptotes, provides a useful alternative to the integration contour $\bar{\Gamma}$ since the steady decrease in (real) values of $f(\zeta)$ thereupon with increasing separation from the saddle point implies that contributions to the integral from the immediate neighborhood of $\zeta = 1$ are alone of significance. An estimate that takes advantage of this feature is readily forthcoming if we make the substitution

$$\zeta = 1 + \eta$$

in (10.17) and approximate to the integrand for small η, obtaining

$$z(x, t)/z_0 \doteq \frac{1}{2\pi \mathrm{i}} \int_{-\mathrm{i}\infty}^{\mathrm{i}\infty} \exp\left\{\frac{2c^2}{\mu_0}\eta\left(t - \frac{x}{c} + \frac{x\eta}{c}\right)\right\}\frac{\mathrm{d}\eta}{\eta}$$

with a contour that passes to the right of the origin. On replacing the variable η by another, namely

$$\xi = \frac{2c}{\mu_0}\eta^2,$$

it follows that

$$z(x, t)/z_0 = \frac{1}{4\pi \mathrm{i}} \int \exp\left\{\xi x + (ct - x)\sqrt{\left(\frac{2c\xi}{\mu_0}\right)}\right\}\frac{\mathrm{d}\xi}{\xi}$$

where the contour loops round the origin and has its ends at $\xi = -\infty$; hence (recalling (10.14), (10.16))

$$z(x, t)/z_0 = \frac{1}{2}\left[1 + \operatorname{erf}\left(\frac{ct - x}{\sqrt{(2\mu_0 t)}}\right)\right] = \frac{1}{2}\operatorname{erfc}\left(\frac{x - ct}{\sqrt{(2\mu_0 t)}}\right), \qquad t \gg 0, \quad \frac{x}{ct} \doteq 1 \tag{10.19}$$

furnishes the desired estimate and reveals the existence of a precursor to the anticipated response at a location x after the direct travel time $t = x/c$, followed by a rising amplitude which tends to that of the end displacement or input level. According to (10.19) the excitation, in its later stages, possesses a wave or propagating nature though the profile is steadily broadened or dispersed on account of a viscous presence in the system.

The incompatibility of sharp fronted signal propagation with a viscoelastic equation of motion can be established in a general manner by considerations relative to the non-uniqueness of its solutions. We are better served, in this regard, by giving attention to the system of equations

$$\begin{aligned} \frac{\partial v}{\partial t} &= c^2\frac{\partial^2 z}{\partial x^2} + \mu_0\frac{\partial^2 v}{\partial x^2}, \\ \frac{\partial z}{\partial t} &= v \end{aligned} \tag{10.20}$$

which replace the single equation (10.8); and remark, first of all, on the fact that isolation of the first order time derivatives in (10.20) underlies the uniqueness of solutions for v and z determined by their given values at $t = 0$. Let us next effect a transformation of variables $(x, t) \to (\xi, \eta)$ and examine the consequent system with a view to representing particular derivatives of z and v that pertain to one of the new variables, say η. The interrelations

$$\frac{\partial}{\partial x} = p_0\frac{\partial}{\partial \eta} + p_1\frac{\partial}{\partial \xi}, \qquad \frac{\partial}{\partial t} = q_0\frac{\partial}{\partial \eta} + q_1\frac{\partial}{\partial \xi} \tag{10.21}$$

where

$$p_0 = \frac{\partial \eta}{\partial x}, \qquad p_1 = \frac{\partial \xi}{\partial x}, \qquad q_0 = \frac{\partial \eta}{\partial t}, \qquad q_1 = \frac{\partial \xi}{\partial t}$$

are available for recasting the equations (10.20), and it appears from the last of the pair that

$$\frac{\partial z}{\partial \eta} = \frac{1}{q_0}\left(v - q_1 \frac{\partial z}{\partial \xi}\right) \tag{10.22}$$

provided $q_0 \neq 0$; after transformation of the companion equation and reliance on (10.22) for eliminating the derivatives $\partial z/\partial \eta$ and $\partial^2 z/\partial \eta^2$ therein we find that, provided $p_0 \neq 0$, $\partial^2 v/\partial \eta^2$ is expressible in term of first and second order derivatives of both v and z, those of second order involving the other independent variable, ξ. Such a representation for the system of differential equations implies the unicity of a solution compatible with arbitrarily specified values of v, $\partial v/\partial \eta$, z and $\partial z/\partial \eta$ on a curve $\eta(x, t) = \text{constant}$ in the x, t plane; the indicated resolution for the highest order derivatives with respect to η and the subsequent uniqueness property no longer obtains if

$$q_0 = p_0 = 0 \tag{10.23}$$

and the curve $\eta(x, t) = \text{constant}$ then corresponds to a front or discontinuity locus. From the evolutionary relation

$$\eta(x + \mathrm{d}x, t + \mathrm{d}t) - \eta(x, t) = p_0\, \mathrm{d}x + q_0\, \mathrm{d}t = 0,$$

which applies to an individual locus and defines its rate of progress, namely

$$\frac{\mathrm{d}x}{\mathrm{d}t} = -\, q_0/p_0,$$

we deduce that the conditions (10.23) rule out the possibility of signal propagation with sharp fronts.

§11. *String Vibrations and the Doppler Effect*

A variant of the problem considered earlier (§1), namely that of ascertaining the excitation which results from a prescribed displacement $y(0, t) = F(t)$ at one point, $x = 0$, on an infinite string, enables us to again employ the characteristics of the wave equation in a constructive role and to infer some aspects of general significance. We shall suppose, as before, a stationary rectilinear configuration of the string prior to the initial instant of time, $t = 0$, and imagine that a vibrating attachment thereafter imparts a local displacement $F(t)$, the current novelty being that the point of application, $x = \beta c t$, is assumed to move with the uniform speed βc, $\beta < 1$.

To derive the appertaining form of the string displacement we utilize the connection (cf. 1.11))

$$y(P_1) - y(P_2) + y(P_3) - y(P_4) = 0 \tag{11.1}$$

between solutions of the homogeneous wave equation at the four consecutive vertices P_1, P_2, P_3, and P_4 of a parallelogram in the x, t plane whose sides are fashioned from segments of the characteristic lines $x \mp ct = \text{constant}$ (see Fig. 9). Let us represent the pair of characteristics $x = \pm ct$ for positive values of t and also the attachment trajectory $x = \beta ct$; if desired, values of the function $F(t)$ may be associated with ordinates of corresponding magnitude normal to the line $x = \beta ct$. When $x > ct > 0$ a parallelogram of the preceding type, with vertices at $P^*(x, t)$, $P_2(x + ct, 0)$, $P_3(x, -t)$, and $P_4(x - ct, 0)$, does not contact the trajectory $x = \beta ct$ that 'carries' an assigned displacement and it follows from (11.1) that $y(P^*) = y(x, t) = 0$; a similar conclusion is reached when $x < -ct$ and, accordingly, for any $t > 0$ those portions of the string at a distance from the origin exceeding ct remain undisturbed. The continuous dependence of a wave function of displacement $y(x, t)$ on its argument variable x, for any stipulated value of t, permits us to infer that y vanishes along both of the characteristics $x = \pm ct$, inasmuch as $y(x, t) = 0$, $|x| > ct$.

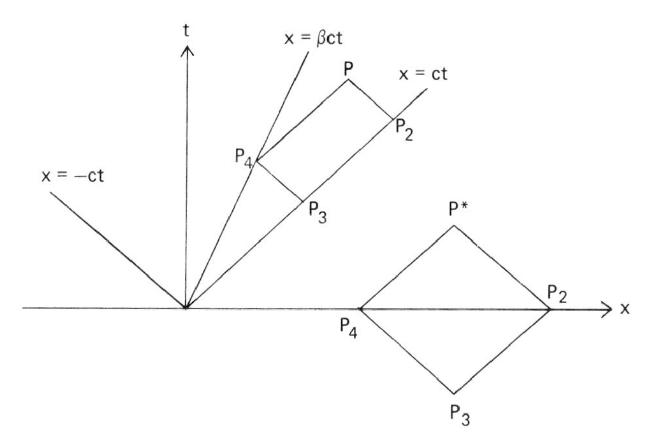

Fig. 9. A moving attachment trajectory and characteristics.

If we next consider a point $P(x, t)$ for which $\beta ct < x < ct$ and construct the parallelogram $PP_2P_3P_4$ with a single vertex

$$P_4\left(\beta\frac{ct - x}{1 - \beta}, \frac{t - \dfrac{x}{c}}{1 - \beta}\right)$$

on the line $x = \beta ct$, and the remaining pair P_2, P_3 on the characteristic $x = ct$, the relation (11.1) simplifies to

$$y(P) = y(P_4) = F(t(P_4)) \tag{11.2}$$

since $y(P_2) = y(P_3) = 0$. Thus

$$y(x, t) = F\left(\frac{t - \dfrac{x}{c}}{1 - \beta}\right), \quad \beta ct \leq x \leq ct \tag{11.3}$$

specifies the displacement of the string ahead of the moving attachment or source, and the corresponding version

$$y(x, t) = F\left(\frac{t + \dfrac{x}{c}}{1 + \beta}\right), \quad -ct \leq x \leq \beta ct \tag{11.4}$$

applies behind the source. A particular case of interest is that wherein the displacement function $F(t)$ has the frequency ω_0, for the adaptations thereto of the general expressions (11.3), (11.4) reveal the distinct frequencies of string vibrations

$$\begin{aligned} \omega_+ &= \frac{\omega_0}{1-\beta}, \quad \beta ct < x < ct, \\ \omega_- &= \frac{\omega_0}{1+\beta}, \quad -ct < x < \beta ct \end{aligned} \tag{11.5}$$

on either side of the moving attachment. A frequency ω_+ in excess of the source frequency ω_0 obtains on that part of the string towards which the attachment is proceeding and a lesser frequency ω_- pertains to the opposite part of the string. These are manifestations of the so-called Doppler effect that occurs in the event of relative motion between source and observation points.

An alternative derivation of the Doppler shift in frequency, which relies upon elementary notions of signalling and kinematics, can be simply elaborated with reference to a localized source that traverses a rectilinear path and a given point of observation or stationary observer located at a minimum distance d therefrom. If c denotes the speed with which signals (i.e., waves) propagate away from any instantaneous position of the source and l specifies its corresponding distance from the observer, the difference between the times of emission and reception, t and τ, respectively, is

$$\tau - t = \frac{l}{c}. \tag{11.6}$$

On introducing the source coordinate x along its trajectory it follows that

$$l = \sqrt{(d^2 + x^2)}$$

and thus, after fixing the origin of time, $t = 0$, when $x = 0$ and the separation l is least, we have $x = \beta ct$ and consequently

$$\tau - t = \frac{1}{c}\sqrt{(d^2 + (\beta ct)^2)}. \tag{11.7}$$

Two signals emitted from the source with a time delay Δt are separated in their arrival at the point of observation by a time interval $\Delta\tau$ whose magnitude follows from (11.7); if Δt and $\omega_0 = 2\pi/\Delta t$ characterize the period and proper frequency of the source the concomitant frequency $\omega = 2\pi/\Delta\tau$ is registered at the observation point. Hence, for small magnitudes of Δt and $\Delta\tau$,

$$\frac{\omega_0}{\omega} = \frac{\Delta\tau}{\Delta t} \doteq \frac{\mathrm{d}\tau}{\mathrm{d}t} = 1 + \frac{\beta^2 ct}{\sqrt{(d^2 + (\beta ct)^2)}} \tag{11.8}$$

which makes it clear that $\omega \gtrless \omega_0$ according as $t \lessgtr 0$, the relevant phases of the motion being those wherein the source approaches or recedes from the observer, respectively. A simpler version of (11.8) obtains when the observer is located on the source trajectory ($d = 0$), namely

$$\omega_0/\omega = 1 \pm \beta, \quad t \gtrless 0 \tag{11.9}$$

and the latter predictions are in accord with (11.5).

The frequency ratio (11.8) acquires its dependence on the time at emission, t, through the explicit representation (11.7) for τ (and thence for $\mathrm{d}\tau/\mathrm{d}t$) in terms of t; to display the analogous Doppler formula with the local time at the observation point, τ, as a parameter we merely invert (11.7), yielding

$$t = \frac{1}{1-\beta^2}\{\tau - \sqrt{(\beta^2\tau^2 + (1-\beta^2)d^2/c^2)}\} \tag{11.10}$$

and thereafter deduce that

$$\frac{\omega}{\omega_0} = \frac{\mathrm{d}t}{\mathrm{d}\tau} = \frac{1}{1-\beta^2}\left\{1 - \frac{\beta^2\tau}{\sqrt{(\beta^2\tau^2 + (1-\beta^2)d^2/c^2)}}\right\}. \tag{11.11}$$

Evidently, the prior limiting form (11.9) can be recovered from (11.11) when $d = 0$ and $\tau \lessgtr 0$.

§12. *Excitation of a String by a Fixed or Moving Local Force*

The two inputs which d'Alembert's solution (1.10) of the Cauchy problem relies upon, namely initial displacement and velocity, impart distinctive aspects to the consequent motion. For a strictly limited initial displacement $s_0(x)$, in particular, the appertaining response elsewhere has a well defined onset and termination though both features do not carry over when the initial velocity has a similar nature. If $s_0(x)$ vanishes identically, the wave function

$$y(x, t) = Y(x + ct) - Y(x - ct) \tag{12.1}$$

where

$$Y(x) = \frac{1}{2c}\int_{-\infty}^{x} v_0(\xi)\,\mathrm{d}\xi \tag{12.2}$$

applies for every choice of initial velocity and assumes the explicit form

$$y(x, t) = \frac{V}{4c}[|x + ct + \varepsilon| - |x + ct - \varepsilon| + |x - ct - \varepsilon| - |x - ct + \varepsilon|] \tag{12.3}$$

given a uniform distribution

$$v_0(x) = \begin{pmatrix} V, & |x| < \varepsilon \\ 0, & |x| > \varepsilon \end{pmatrix} = \frac{V}{\pi} \int_{-\infty}^{\infty} e^{ikx} \frac{\sin k\varepsilon}{k} \, dk \tag{12.4}$$

whose range is 2ε. An evident symmetry of the representation (12.3) with regard to x enables us to focus on the domain $x > 0$, say; and it can be readily deduced from (12.3) that, relative to points for which $x < \varepsilon$, the displacement is given by

$$y(x, t) = \begin{cases} Vt, & 0 \leq ct \leq \varepsilon - x \\ \dfrac{V}{2}\left(t - \dfrac{x}{c} + \dfrac{\varepsilon}{c}\right), & \varepsilon - x \leq ct \leq \varepsilon + x \\ \dfrac{V\varepsilon}{c}, & ct \geq \varepsilon + x \end{cases} \tag{12.5}$$

whereas the corresponding versions on the interval $x > \varepsilon$ take the form

$$y(x, t) = \begin{cases} 0, & 0 \leq ct \leq x - \varepsilon \\ \dfrac{V}{2}\left(t - \dfrac{x}{c} + \dfrac{\varepsilon}{c}\right), & x - \varepsilon \leq ct \leq x + \varepsilon \\ \dfrac{V\varepsilon}{c}, & ct \geq x + \varepsilon. \end{cases} \tag{12.6}$$

These expressions reveal the persistence of a steady deflection both within and outside the region that marks the extent of a truncated initial velocity distribution.

It is appropriate to conceive of establishing the initial velocity through the action of an impulse or force applied over the segment $|x| < \varepsilon$ during an infinitesimal period of time; if an impulse I with the magnitude

$$I = \mathrm{Lim}_{\varepsilon \to 0} \, 2\rho\varepsilon V$$

is delivered at the point $x = \xi$ just prior to the time $t = \tau$ the residual displacement, according to (12.6), has the uniform value

$$y(x, t) = \frac{I}{2\rho c} \quad \text{when} \quad |x - \xi| < c(t - \tau). \tag{12.7}$$

Let us next suppose that a steady and transversely oriented point force $\mathfrak{F}$ commences its action on a string at $t = 0$, proceeding thereafter from the origin $x = 0$ in the positive direction at the constant speed βc. The string thus receives an impulse of amount $\mathfrak{F}\, dt$ in each time interval dt and, as a result of this stimulus, oppositely directed wave motions are excited thereupon. We may anticipate that distinct forms of excitation are realized according as $\beta < 1$ or > 1; and, in preparation for the analysis which suits the respective cases, we depict in Fig. 10 the future portions ($t > 0$) of characteristic lines $x = \pm ct$ issuing from the origin of the x, t plane along with the trajectories $x = \beta ct$ of the moving force, through the points I, I' when $\beta < 1, > 1$.

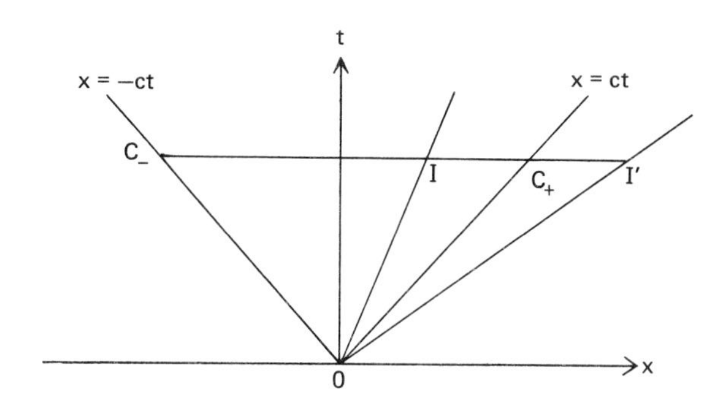

Fig. 10. Characteristics and a moving force locus.

If $\beta < 1$ we infer that, corresponding to the time $t(>0)$, forward and backward wave motions are confined, respectively, to the ranges IC_+, IC_- between the impulse point I and the aforementioned characteristics. Focussing exclusively on the excitation due to the applied force, the wave function has a representation

$$y(x, t) = f(x - ct) + g(x + ct) \tag{12.8}$$

where the first component is different from zero on IC_+ and the second is different from zero on IC_-. To determine the latter explicitly, we utilize a trio of conditions

$$f(0) + g(0) = 0, \tag{12.9}$$

$$f((\beta - 1)ct) = g((\beta + 1)ct), \tag{12.10}$$

$$\int_{-ct}^{\beta ct} \rho c g'(x + ct)\,\mathrm{d}x - \int_{\beta ct}^{ct} \rho c f'(x - ct)\,\mathrm{d}x = \mathfrak{F} t \tag{12.11}$$

which, in turn, signify a null displacement at $t = 0$, assure continuity of the displacement at the impulse point and account for the momentum transfer to the string during the interval of force activity. Integration of (12.11), where primes signify an argument derivative, yields

$$g((\beta + 1)ct) + f((\beta - 1)ct) - g(0) - f(0) = \mathfrak{F} t/\rho c$$

and thus, referring to (12.9), (12.10), it follows that

$$f((\beta - 1)ct) = g((\beta + 1)ct) = \mathfrak{F} t/2\rho c,$$

whence

$$f(x-ct)=\frac{\mathfrak{F}}{2\rho c}\frac{t-\dfrac{x}{c}}{1-\beta}, \quad \beta ct \leq x \leq ct,$$

and (12.12)

$$g(x+ct)=\frac{\mathfrak{F}}{2\rho c}\frac{t+\dfrac{x}{c}}{1+\beta}, \quad -ct \leq x \leq \beta ct$$

furnish the desired string displacement.

If $\beta > 1$ the full range of disturbance on the string at time t corresponds to $I'C_-$, with a backward moving wave $g(x+ct)$ on its entirety and a forward moving wave $f(x-ct)$ on the portion $I'C_+$ that lies between the characteristic $x = ct$ and the trajectory $x = \beta ct$. The conditions which are appropriate to the determination of these functions include

$$f(0)=0, \qquad g(0)=0, \tag{12.13}$$

indicative of an undisturbed state at $t = 0$, and

$$f((\beta-1)ct)+g((\beta+1)ct)=0, \quad t>0 \tag{12.14}$$

that expresses a null displacement at the instantaneous location of the moving force. Continuity of the displacement at the characteristic $x = ct$ requires the equality of the respective limits, $g(2ct)$ and $f(0)+g(2ct)$, and this is assured by (12.13). The dynamical equation, involving $\mathfrak{F}$, now takes the form

$$\int_{-ct}^{ct} \rho c g'(x+ct)\,\mathrm{d}x + \int_{ct}^{\beta ct} \rho c[g'(x+ct)-f'(x-ct)]\,\mathrm{d}x = \mathfrak{F}t \tag{12.15}$$

or, on simplification,

$$-g(0)+g((\beta+1)ct)-f((\beta-1)ct)+f(0)=\frac{\mathfrak{F}t}{\rho c};$$

accordingly,

$$f(x-ct)=\frac{\mathfrak{F}}{2\rho c}\frac{t-\dfrac{x}{c}}{\beta-1}, \quad ct \leq x \leq \beta ct$$

(12.16)

$$g(x+ct)=\frac{\mathfrak{F}}{2\rho c}\frac{t+\dfrac{x}{c}}{\beta+1}, \quad -ct \leq x \leq \beta ct$$

and the requisite string displacements are given by

$$y(x, t) = \begin{cases} g(x+ct) = \dfrac{\mathfrak{F}}{2\rho c} \dfrac{t + \dfrac{x}{c}}{\beta + 1}, & -ct \leq x \leq ct \\ f(x-ct) + g(x+ct) = \dfrac{\mathfrak{F}\beta}{\rho c} \dfrac{t - \dfrac{x}{\beta c}}{\beta^2 - 1}, & ct \leq x \leq \beta ct. \end{cases} \tag{12.17}$$

An independent derivation of the preceding formulae, which is based on solution of a differential equation for the string displacement that incorporates a moving force term, appears in §23.

§13. *Boundary Conditions and Normal Modes of Vibration*

If a taut string of finite length is secured at its endpoints, say $x = 0, L$, a study of the displacement profiles centers on the differential equation of motion and the boundary conditions

$$y(0, t) = y(L, t) = 0 \tag{13.1}$$

when we refrain from inquiry into their manner of excitation. Having regard for the particular equation (1.1) and imposing the second boundary condition on its general solution,

$$y(x, t) = f\left(t - \frac{x}{c}\right) + g\left(t + \frac{x}{c}\right),$$

it follows that

$$f\left(t - \frac{L}{c}\right) + g\left(t + \frac{L}{c}\right) = 0$$

whence the representation

$$y(x, t) = f\left(t - \frac{x}{c}\right) - f\left(t - \frac{2L - x}{c}\right) \tag{13.2}$$

applies to an arbitrary time interval; the combination of terms in (13.2) shows that compliance with the boundary condition at $x = L$ is tantamount to full reflection (and reversal in amplitude) of the 'incident' profile $f(t - (x/c))$. A similar feature is manifest at the other end, $x = 0$, and the deduction from (13.1), (13.2),

$$f(t) = f\left(t - \frac{2L}{c}\right) \tag{13.3}$$

dictates a periodic behavior of the function $f(t)$. The indicated periodicity range, $2L/c$, for $f(t)$ implies that time periodic states of vibration, wherein

$$f(t) = A\,\mathrm{e}^{-\mathrm{i}\omega t}, \tag{13.4}$$

are feasible only if the frequency ω satisfies the relation

$$\mathrm{e}^{-2\,\mathrm{i}\omega L/c} = 1$$

which furnishes an explicit sequence of characteristic values,

$$\omega_n = n\pi c/L, \quad n = 1, 2, \ldots . \tag{13.5}$$

In accordance with (13.2), (13.4) and (13.5) we obtain a family of displacement functions

$$y_n(x, t) = 2\,\mathrm{i}A \sin\frac{n\pi x}{L}\,\mathrm{e}^{-\mathrm{i}\omega_n t}, \quad 0 \le x \le L$$

whose real parts are proportional to

$$\sin\frac{n\pi x}{L} \cdot \begin{matrix}\sin\\ \cos\end{matrix}\left(\frac{n\pi ct}{L}\right) \tag{13.6}$$

when A is real or purely imaginary, and otherwise feature a linear combination of the trigonometric time factors; these may aptly be termed standing wave patterns inasmuch as the above derivation suggests their establishment by reflection of travelling wave patterns at the ends of the string. If the proper (or eigen-) functions $\sin(n\pi x/L)$, $n = 1, 2, \ldots$, which vanish at the endpoints of the interval $(0, L)$, are referred to a coordinate origin at the midpoint thereof, the subgroups

$$\sin\frac{2p\pi\xi}{L}, \qquad \cos\frac{(2q-1)\pi\xi}{L}, \qquad -\frac{L}{2} \le \xi \le \frac{L}{2}, \quad p, q = 1, 2, \ldots$$

possess odd and even symmetry in ξ or $(x - (L/2))$. The so-called normal mode functions (13.6), that satisfy both the wave equation and the boundary conditions (13.1), can be combined with unique weight factors so as to provide a series expansion of the solution to any initial value problem for the string with fixed ends.

The separation of variables reflected in (13.6) suggests that a resolution

$$y(x, t) = \varphi(x)\psi(t) \tag{13.7}$$

is suitable, a priori, for determining the normal mode functions. A direct relationship between the proper coordinate factors and the specific boundary conditions is apparent, though the parallel importance of the equation of motion needs also to be acknowledged. Let us, by way of bringing out the influential role of the latter equation, suppose that the string has a finite electrical conductivity or resistance and is thus capable of transmitting current along its length; if the string executes motions, with displacement $y(x, t)$, in a plane which is transverse to the direction of a uniform magnetic field, H, a time-varying induced current, J, proportional to the product of H and the velocity integral, $\int_{-L/2}^{L/2} (\partial y/\partial t)\,\mathrm{d}x$, results in a local (drag) force of electromagnetic nature whose magnitude is given in terms of the product HJ. Accordingly, the equation of small amplitude motions takes the form

$$\frac{\partial^2 y}{\partial x^2} - \frac{1}{c^2}\frac{\partial^2 y}{\partial t^2} = a^2 \int_{-L/2}^{L/2} \frac{\partial y}{\partial t}\,\mathrm{d}x + F(t), \tag{13.8}$$

where $F(t)$ accounts for any impressed potential difference between the ends and the positive coefficient of the integral depends on (among other parameters) both H and the resistance of the string. In contrast with the corresponding equation for motions affected by a damping force that is proportional to the local velocity, the preceding involves an electromagnetic force which depends on the velocity of all elements of the string and gives (13.8) the character of an integro-differential equation.

To investigate the normal motions we set $F(t) = 0$ and substitute the product representation (13.7) into the homogeneous version of (13.8), thus obtaining an equation

$$\frac{\mathrm{d}^2\varphi}{\mathrm{d}x^2} \cdot \psi - \frac{1}{c^2}\varphi \cdot \frac{\mathrm{d}^2\psi}{\mathrm{d}t^2} = a^2 \frac{\mathrm{d}\psi}{\mathrm{d}t} \int_{-L/2}^{L/2} \varphi(x)\,\mathrm{d}x \tag{13.9}$$

which links the single variable functions of x and t. Evidently, a proper mode status is retained by those field independent vibrations with antisymmetry about the midpoint of the string and the eigenfunctions $\sin(2p\pi x/L)$, $p = 1, \ldots$ are appropriate when the endpoints remain fixed. For motions of a symmetric type whose scale conforms with the normalization

$$\int_{-L/2}^{L/2} \varphi(x)\,\mathrm{d}x = 1, \tag{13.10}$$

we reduce the non-separable relation (13.9) to an ordinary differential equation describing $\varphi(x)$ on the assumption that

$$\psi(t) = \mathrm{e}^{-\mu t} \tag{13.11}$$

and readily establish

$$\varphi(x) = \frac{b}{\mu}\left[1 - \frac{\cosh \dfrac{\mu x}{c}}{\cosh \dfrac{\mu L}{2c}}\right], \quad b = (ac)^2 \tag{13.12}$$

as the companion factor which satisfies the boundary conditions $\varphi(\pm L/2) = 0$. The scale requirement (13.10) enables us to obtain an equation which determines μ, namely

$$\mathrm{e}^{\mu L/c}[\mu^2 - Lb\mu + 2bc] + [\mu^2 - Lb\mu - 2bc] = 0, \tag{13.13}$$

and once the roots of the latter are known, the functions φ, ψ and their product (or normal mode function) acquire definite characterizations.

It is a simple matter to confirm that the left-hand member of (13.13) vanishes at $\mu = 0$ and does not vanish for any other real value of μ, whence the roots are necessarily complex-valued; in fact, equations of the form

$$p_1(z)\,e^z + p_2(z) = 0, \tag{13.14}$$

where $p_1(z)$ and $p_2(z)$ are polynomials in z, have infinitely many roots. A proof of this contention (devised by Herglotz (1903)) commences with the representation that holds if their number is finite, viz.,

$$p_1(z)\,e^z + p_2(z) = e^{f(z)} p_3(z) \tag{13.15}$$

wherein $p_3(z)$ is a polynomial and $f(z)$ is a regular function for all finite values of z. From the resulting identity

$$e^{f(z)} = r_1(z)\,e^z + r_2(z),$$

involving a pair of rational functions $r_1(z)$, $r_2(z)$, we deduce that

$$\begin{aligned} e^{\operatorname{Re} f(z)} &< |r_1(z)|\,e^{|z|} + |r_2(z)| \\ &< M|z|^n[e^{|z|} + 1] \end{aligned}$$

by utilizing the common estimate

$$|r_1(z)| < M|z|^n, \qquad |r_2(z)| < M|z|^n, \quad |z| > R$$

which is contingent on appropriate choice of M, n and R. Thus,

$$\operatorname{Re} f(z) < \log[M|z|^n(e^{|z|} + 1)]$$

and since the right-hand member, divided by $|z|$, tends to unity as $|z| \to \infty$, the consequent estimate

$$\operatorname{Re} f(z) < A|z|, \quad |z| > R$$

implies that $f(z)$ is a linear function of z. We have, accordingly,

$$p_1(z)\,e^z + p_2(z) = e^{Cz} p_3(z), \tag{13.16}$$

or, following a sufficient number of differentiations and the elimination of $p_2(z)$, an identity

$$P_1(z)\,e^z = P_3(z)\,e^{Cz}, \qquad P_{1,3}(z) \not\equiv 0 \tag{13.17}$$

which calls for the assignment $C = 1$; the prior version (13.16) thus requires, as a matter of consistency, the specification $p_2 \equiv 0$, contrary to the original hypothesis, and the stated property of roots for (13.14) is confirmed.

An approximate determination of roots with large absolute magnitude is simply arrived at if we recast (13.13) in the form

$$e^{\mu L/c} = -\frac{1 - \dfrac{Lb}{\mu} - \dfrac{2bc}{\mu^2}}{1 - \dfrac{Lb}{\mu} + \dfrac{2bc}{\mu^2}}$$

and expand the right side in powers of $1/\mu$; retaining only the initial pair of terms,

$$e^{\mu L/c} \doteq -\left(1 - \frac{4bc}{\mu^2}\right),$$

from which it appears that

$$\mu_n = \frac{4(acL)^2}{(2n+1)^2\pi^2} + i\frac{\pi c}{L}(2n+1), \quad n \gg 1 \tag{13.18}$$

characterizes a first approximation to the real and imaginary parts of the large magnitude roots. The positive values for Re μ_n imply, in conjunction with (13.11), a temporal subsidence of the string motions.

§14. *Solution of an Inhomogeneous Wave Equation on a Finite Coordinate Interval*

The resolution of a combined initial and boundary value problem for the inhomogeneous (linear) wave equation,

$$\frac{\partial^2 y}{\partial x^2} - \frac{1}{c^2}\frac{\partial^2 y}{\partial t^2} = -q(x, t), \tag{14.1}$$

is achieved through the superposition of components that satisfy the given and related homogeneous equations, respectively. Let us consider, at the outset, any point $P(x, t)$ within the triangular domain I of the x, t plane in Fig. 11 which is delineated by a segment OL $(0 \le x \le L, t = 0)$ of the x-axis and portions of the two characteristics $x = ct$, $x = L\text{-}ct$ that intersect at $Q(L/2, L/2c)$; and evaluate the integrals appearing in the relation (1.11), when $\partial y/\partial t$ and $\partial y/\partial x$ are ascribed a continuous behavior relative to t and x, respectively, and the triangular domain with vertices at $P(x, t)$, $P_1(x - ct, 0)$, $P_2(x + ct, 0)$ is utilized. There follows a sum representation for solutions of (14.1), namely

$$y(x, t) = Y_1(x, t) + Y_2(x, t), \tag{14.2}$$

with a first term

$$Y_1(x, t) = \frac{1}{2}\left[s_0(P_1) + s_0(P_2) + \frac{1}{c}\int_{P_1P_2} v_0(\xi)\, d\xi\right] \tag{14.3}$$

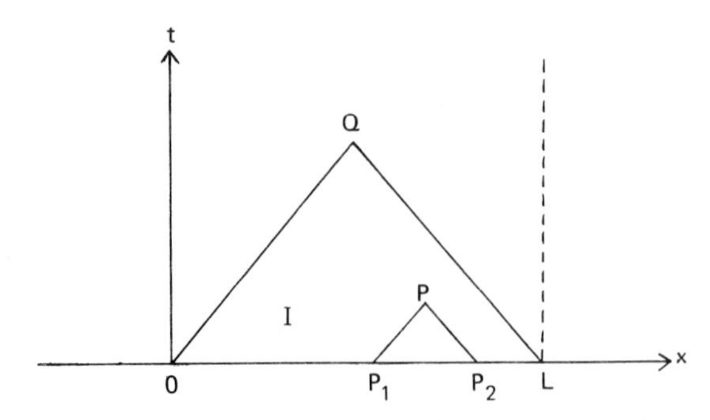

Fig. 11. Characteristic triangles for a bounded string.

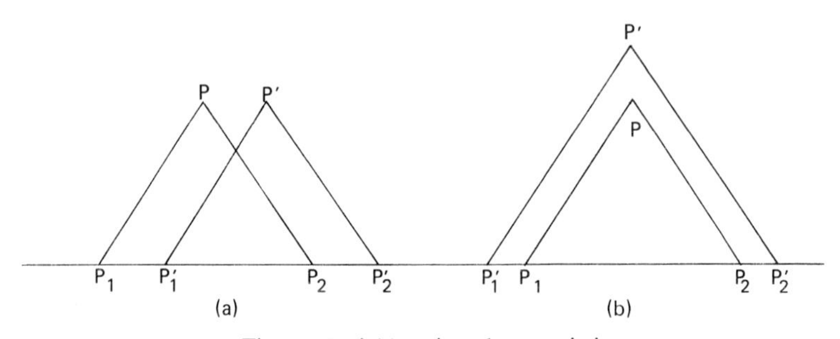

Fig. 12. Neighbouring characteristics.

that involves the arbitrarily chosen functions $s_0(x)$, $v_0(x)$, and the second term

$$Y_2(x, t) = \frac{c}{2} \int_{PP_1P_2} q(x', t')\, \mathrm{d}x'\, \mathrm{d}t' \tag{14.4}$$

which acquires its functional dependence through the analytical description of the triangle PP_1P_2. The former, $Y_1(x, t)$, featuring a d'Alembert combination of terms, satisfies the homogeneous differential equation

$$\frac{\partial^2 Y_1}{\partial x^2} - \frac{1}{c^2} \frac{\partial^2 Y_1}{\partial t^2} = 0 \tag{14.5}$$

and assumes the values

$$Y_1(x, 0) = s_0(x), \qquad \frac{\partial Y_1(x, 0)}{\partial t} = v_0(x) \tag{14.6}$$

at $t = 0$, whereas the latter constitutes a particular solution of (14.1) that vanishes in the limit $t \to 0$; jointly, these functions make up a representation, (14.2), which demonstrates the uniqueness of a solution to (14.1) meeting the initial specifications furnished by any independent choices for $s_0(x)$, $v_0(x)$.

That a solution of such form exists is confirmed, when the functions s_0, $\mathrm{d}s_0/\mathrm{d}x$ and v_0 have continuous behavior, on calculating the partial derivatives of Y_1 and Y_2; we are assisted, in the latter regard, by noting the relative disposition of neighboring domains PP_1P_2 and $P'P_1'P_2'$ (see Fig. 12) pertinent to coordinate (a) and time (b) differentiation, and eventually arrive at the individual expressions

$$\begin{aligned}
\frac{\partial Y_1}{\partial x} &= \frac{1}{2}\left\{\left(\frac{\partial s_0}{\partial x}\right)_{P_1} + \left(\frac{\partial s_0}{\partial x}\right)_{P_2} + \frac{1}{c}(v_0)_{P_2} - \frac{1}{c}(v_0)_{P_1}\right\} \\
\frac{\partial^2 Y_1}{\partial x^2} &= \frac{1}{2}\left\{\left(\frac{\partial^2 s_0}{\partial x^2}\right)_{P_1} + \left(\frac{\partial^2 s_0}{\partial x^2}\right)_{P_2} + \frac{1}{c}\left(\frac{\partial v_0}{\partial x}\right)_{P_2} - \frac{1}{c}\left(\frac{\partial v_0}{\partial x}\right)_{P_1}\right\} \\
\frac{\partial Y_1}{\partial t} &= \frac{1}{2}\left\{-c\left(\frac{\partial s_0}{\partial x}\right)_{P_1} + c\left(\frac{\partial s_0}{\partial x}\right)_{P_2} + (v_0)_{P_2} + (v_0)_{P_1}\right\} \\
\frac{\partial^2 Y_1}{\partial t^2} &= \frac{1}{2}\left\{c^2\left(\frac{\partial^2 s_0}{\partial x^2}\right)_{P_1} + c^2\left(\frac{\partial^2 s_0}{\partial x^2}\right)_{P_2} + c\left(\frac{\partial v_0}{\partial x}\right)_{P_2} - c\left(\frac{\partial v_0}{\partial x}\right)_{P_1}\right\}
\end{aligned} \tag{14.7}$$

together with

$$\begin{aligned}
\frac{\partial Y_2}{\partial x} &= \frac{c}{2}\int_{P_2}^{P} q(x',t')\,\mathrm{d}t' - \frac{c}{2}\int_{P_1}^{P} q(x',t')\,\mathrm{d}t' \\
\frac{\partial^2 Y_2}{\partial x^2} &= \frac{c}{2}\int_{P_2}^{P} \frac{\partial}{\partial x'}q(x',t')\,\mathrm{d}t' - \frac{c}{2}\int_{P_1}^{P} \frac{\partial}{\partial x'}q(x',t')\,\mathrm{d}t' \\
\frac{\partial Y_2}{\partial t} &= \frac{c^2}{2}\int_{P_2}^{P} q(x',t')\,\mathrm{d}t' + \frac{c^2}{2}\int_{P_1}^{P} q(x',t')\,\mathrm{d}t' \\
\frac{\partial^2 Y_2}{\partial t^2} &= \frac{c^2}{2}\int_{P_2}^{P} \frac{\partial}{\partial t'}q(x',t')\,\mathrm{d}t' + \frac{c^2}{2}\int_{P_1}^{P} \frac{\partial}{\partial t'}q(x',t')\,\mathrm{d}t' + \frac{1}{2}c^2(q)_{P_1} + \frac{1}{2}c^2(q)_{P_2}
\end{aligned} \tag{14.8}$$

where the integrals are conducted along the characteristics

$$P_1P: x'-x = c(t'-t), \qquad P_2P: x'-x = -c(t'-t), \quad 0 \le t' \le t.$$

On the basis of the above determinations, we find

$$\begin{aligned}
&\left(\frac{\partial^2}{\partial x^2} - \frac{1}{c^2}\frac{\partial^2}{\partial t^2}\right)(Y_1+Y_2) \\
&= -\frac{1}{2}\int_{P_1}^{P}\left[\frac{\partial q}{\partial t'}\mathrm{d}t' + \frac{\partial q}{\partial x'}\mathrm{d}x'\right] - \frac{1}{2}\int_{P_2}^{P}\left[\frac{\partial q}{\partial t'}\mathrm{d}t' + \frac{\partial q}{\partial x'}\mathrm{d}x'\right] - \frac{1}{2}(q)_{P_1} - \frac{1}{2}(q)_{P_2} \\
&= -\tfrac{1}{2}[2(q)_P - (q)_{P_1} - (q)_{P_2}] - \tfrac{1}{2}(q)_{P_1} - \tfrac{1}{2}(q)_{P_2} = -(q)_P,
\end{aligned}$$

thereby establishing the fact that (14.2) is a solution of the inhomogeneous equation (14.1); moreover, (14.2), (14.7), and (14.8) imply the initial relationship

$$y(x,0) = s_0(x), \qquad \frac{\partial}{\partial t}y(x,0) = v_0(x)$$

and also reveal the continuous behavior expected of $\partial y/\partial t$, $\partial y/\partial x$ if the functions $s_0(x)$, $v_0(x)$ belong to the stipulated class.

With the description of a procedure for and the results of integrating (14.1) in the domain $I(0 < t < L/2c)$ set out, our subsequent attention is directed to achieving a solution of the initial value problem valid everywhere in the semi-infinite strip domain $\hat{I}: 0 < x < L$, $t > 0$ (see Fig. 13). It becomes necessary to impose a pair of boundary conditions, such as

$$y(0, t) = 0, \qquad y(L, t) = 0, \quad t > 0$$

and only these are considered in the present section. We adopt a method that involves a continuation of the desired function $y(x, t)$ throughout the half-plane $t > 0$ and regard the latter as partitioned among the congruent strips $nL < x < (n+1)L$, $n = 0, \pm 1, \ldots, t > 0$. Let the values of similarly extended functions $s_0(x)$, $v_0(x)$ and $q(x, t)$ at points symmetrically placed with respect to the boundary lines of the strips differ in their algebraic sign; that is, each function assumes a common (albeit individual) magnitude at the representative points P, while the reverse magnitude obtains at the points P^*. In particular, the continuations of $s_0(x)$, $v_0(x)$ from $0 < x < L$ to the whole coordinate range are given by $\pm s_0(x), \pm v_0(x)$, with the $\pm$ sign according as x lies in a strip than can be superposed on the original one, $\hat{I}$, after a displacement which is an even or odd multiple of L. On the basis of the indicated characterizations for the extended trio of functions, the formulas (14.2)–(14.4) are capable of specifying a (unique) function $y(x, t)$, with continuous nature in the half-plane $t > 0$, which fulfills the necessary requirements for a solution of the inhomogeneous wave equation in the domain $\hat{I}$; namely, said function vanishes at $x = 0, L$ and, in conjunction with its first order time derivative, assumes the given values $s_0(x)$, $v_0(x)$, respectively, on $0 < x < L$.

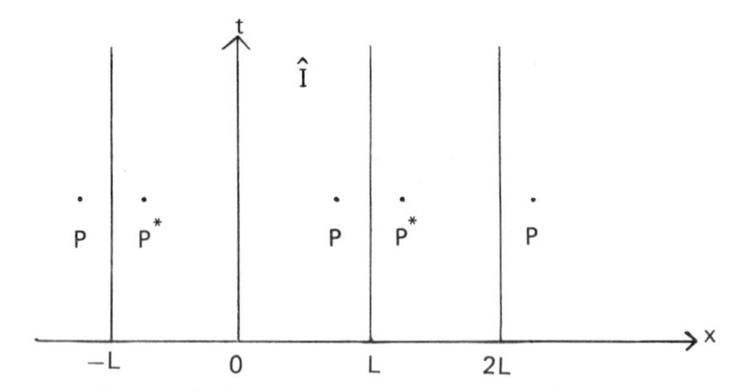

Fig. 13. Reflection across boundaries of a strip.

§15. *Inhomogeneous Boundary Conditions*

An extension of the data, collectively defined by initial values and the inhomogeneous member of the wave equation on a finite coordinate interval, is employed in our previous handling of a problem with homogeneous boundary conditions. We may proceed otherwise and refer exclusively to the domain $\hat{I}, 0 < x < L, t > 0$, in which the solution is desired, applying Riemann's method of integration with contours that do not cross the boundary lines thereof, namely $\Gamma_1: 0 \leq x \leq L, t = 0, \Gamma_2: x = 0, t \geq 0$ and $\Gamma_3: x = L, t \geq 0$ as in Fig. 14.

Let us introduce designations for the wave function and its normal derivative on the respective portions of said (non-characteristic) lines, i.e.,

$$\begin{aligned} y(x,0) &= s_0(x), & \frac{\partial}{\partial t} y(x,0) &= v_0(x) \quad \text{on } \Gamma_1 \\ y(0,t) &= F_1(t), & \frac{\partial}{\partial x} y(0,t) &= G_1(t) \quad \text{on } \Gamma_2 \\ y(L,t) &= F_2(t), & \frac{\partial}{\partial x} y(L,t) &= G_2(t) \quad \text{on } \Gamma_3 \end{aligned} \tag{15.1}$$

and note the inhomogeneous nature conferred on the boundary conditions by functions $F_{1,2}(t)$ and $G_{1,2}(t)$ which are different from zero. A continuous variation of the functions

$$s_0(x), \quad s_0'(x), \quad v_0(x), \quad F_1(t), \quad F_1'(t), \quad G_1(t), \quad F_2(t), \quad F_2'(t), \quad G_2(t)$$

is assumed, along with the conditions

$$\begin{aligned} s_0(0) &= F_1(0), & s_0'(0) &= G_1(0), & s_0'(L) &= G_2(0) \\ s_0(L) &= F_2(0), & v_0(0) &= F_1'(0), & v_0(L) &= F_2'(0) \end{aligned} \tag{15.2}$$

which secure a smooth behavior at the corners $(0,0)$ and $(L,0)$ of the domain $\hat{I}$. The representations furnished by Riemann's integral for solutions of the

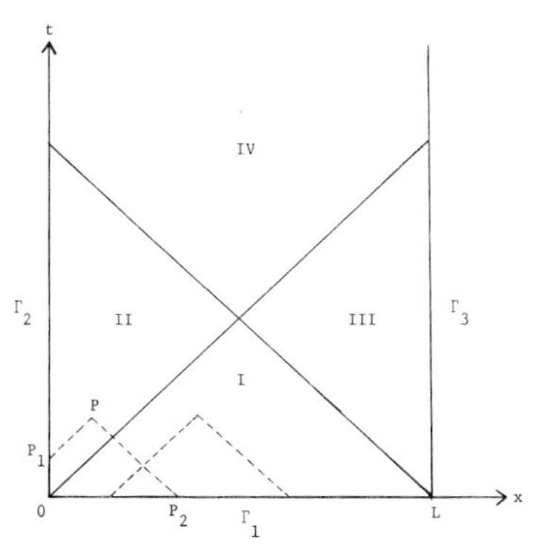

Fig. 14. Domains of dependence.

wave equation,

$$2y_P = y_{P_1} + y_{P_2} + \frac{1}{c}\int_{P_1}^{P_2}\left[\frac{\partial y}{\partial t}\,\mathrm{d}x + c^2\frac{\partial y}{\partial x}\,\mathrm{d}t\right] \tag{15.3}$$

have a different form in the four subdomains I, II, III, IV of the semi-infinite strip $\hat{I}$; thus, when the point P lies in the first and the integral of (15.3) extends along the x-axis, we obtain, with recourse to (15.1),

$$2y_P(x, t) = s_0(x - ct) + s_0(x + ct) + \frac{1}{c}\int_{x-ct}^{x+ct} v_0(\xi)\,\mathrm{d}\xi, \quad ct < x, \quad x + ct < L \tag{15.4}$$

whereas for points P in the second and a contour of integration P_1P_2 formed by sections P_1O and OP_2 along the boundary lines Γ_2, Γ_1, respectively, we obtain

$$\begin{aligned} 2y_P(x, t) = {} & F_1\left(t - \frac{x}{c}\right) + s_0(x + ct) - c\int_0^{t-(x/c)} G_1(\tau)\,\mathrm{d}\tau \\ & + \frac{1}{c}\int_0^{x+ct} v_0(\xi)\,\mathrm{d}\xi, \quad ct > x, \quad x + ct < L. \end{aligned} \tag{15.5}$$

For the other subdomains III, IV the corresponding representations are

$$\begin{aligned} 2y_P(x, t) = {} & s_0(x - ct) + F_2\left(t - \frac{L-x}{c}\right) \\ & + \frac{1}{c}\int_{x-ct}^{L} v_0(\xi)\,\mathrm{d}\xi + c\int_0^{t-(L-x)/c} G_2(\tau)\,\mathrm{d}\tau, \quad ct < X, \quad x + ct > L \end{aligned} \tag{15.6}$$

and

$$\begin{aligned} 2y_P(x, t) = {} & F_1\left(t - \frac{x}{c}\right) + F_2\left(t - \frac{L-x}{c}\right) - c\int_0^{t-(x/c)} G_2(\tau)\,\mathrm{d}\tau \\ & + \frac{1}{c}\int_0^{L} v_0(\xi)\,\mathrm{d}\xi + c\int_0^{t-(L-x)/c} G_2(\tau)\,\mathrm{d}\tau \qquad ct > x, \quad x + ct > L \end{aligned} \tag{15.7}$$

where the latter stems from a three-part integration contour composed of sections P_1O, OL, and LP_2.

A solution of the wave equation is uniquely determined in the domain $\hat{I}$ following the (independent) assignment of values for both y and $\partial y/\partial t$ on Γ_1 and for either y or $\partial y/\partial x$ on Γ_2, Γ_3; if y (that is, the functions $F_1(t)$, $F_2(t)$) be given on Γ_2, Γ_3, for instance, a separate calculation of the functions $G_1(t)$, $G_2(t)$ becomes necessary prior to utilizing the formulas (15.4)–(15.7). The relationships between x and t partial derivatives of a wave function at points located on the same characteristic line [cf. Prob. I(a), 7] are available for such purpose, with different versions according as $t < L/c$ or $t > L/c$. For $t < L/c$ the values of G_1, F_1 and G_2, F_2 at points on Γ_2, Γ_3, respectively, are linked with those of s_0, v_0 at specific points on Γ_1, and the relevant connections take the forms

$$\left.\begin{aligned} G_1(t) + \frac{1}{c}F_1'(t) &= \frac{\mathrm{d}}{\mathrm{d}\xi}s_0(\xi)|_{\xi=ct} + \frac{1}{c}v_0(ct) \\ G_2(t) - \frac{1}{c}F_2'(t) &= \frac{\mathrm{d}}{\mathrm{d}\xi}s_0(\xi)|_{\xi=L-ct} - \frac{1}{c}v_0(L - ct). \end{aligned}\right\} \quad t < \frac{L}{c} \tag{15.8}$$

When $t > L/c$, on the other hand, the initial values no longer enter in connections of such nature, since the appertaining points are situated on Γ_2 and Γ_3, and we find that

$$\left.\begin{aligned} G_1(t)+\frac{1}{c}F_1'(t) &= G_2\left(t-\frac{L}{c}\right)+\frac{1}{c}F_2'\left(t-\frac{L}{c}\right) \\ G_2(t)-\frac{1}{c}F_2'(t) &= G_1\left(t-\frac{L}{c}\right)-\frac{1}{c}F_1'\left(t-\frac{L}{c}\right). \end{aligned}\right\} \quad t>\frac{L}{c} \tag{15.9}$$

The relations (15.8), (15.9) thus enable us to determine $G_1(t)$ and $G_2(t)$ for all positive values of t [with a smooth fit at $t = L/c$ assured by the conditions (15.2)] and thence, on application of (15.4)–(15.7), to obtain a solution of the combined initial-boundary value problem in $\hat{I}$.

Let us next simplify the system of equations involved in calculating G_1, G_2, on the assumption that $F_2 \equiv 0$, and proceed to establish individual difference equation characterizations for these functions. After the indicated simplification is achieved in (15.8) and the consequent representations for $G_1(t-(L/c))$ and $G_2(t-(L/c))$ are substituted in (15.9), it follows that (dropping the subscript from the remaining function $F_1(t)$ of boundary type)

$$\left.\begin{aligned} G_1(t) &= -\frac{1}{c}F'(t)+\frac{\mathrm{d}}{\mathrm{d}\xi}s_0(\xi)|_{\xi=2L-ct}-\frac{1}{c}v_0(2L-ct) \\ G_2(t) &= -\frac{2}{c}F'\left(t-\frac{L}{c}\right)+\frac{\mathrm{d}}{\mathrm{d}\xi}s_0(\xi)|_{\xi=ct-L}+\frac{1}{c}v_0(ct-L), \end{aligned}\right\} \quad L<ct<2L \tag{15.10}$$

(with "and" between the two equations)

which expressions, together with (15.8), directly furnish the values of $G_1(t)$ and $G_2(t)$ throughout the interval $0<ct<2L$.

If $ct>2L$ we can arrange for and readily effect the elimination of G_1 or G_2 from the system (15.9), by first displaying the latter in the alternative versions

$$G_1(t)=G_2\left(t-\frac{L}{c}\right)-\frac{1}{c}F'(t) \qquad G_1\left(t-\frac{L}{c}\right)=G_2\left(t-\frac{2L}{c}\right)-\frac{1}{c}F'\left(t-\frac{L}{c}\right)$$

$$G_2\left(t-\frac{L}{c}\right)=G_1\left(t-\frac{2L}{c}\right)-\frac{1}{c}F'\left(t-\frac{2L}{c}\right) \qquad G_2(t)=G_1\left(t-\frac{L}{c}\right)-\frac{1}{c}F'\left(t-\frac{L}{c}\right)$$

and then adding the component equations of the respective pairings; this yields, subsequent to a shift of the argument variables,

$$\left.\begin{aligned} G_1\left(t+\frac{2L}{c}\right) &= G_1(t)-\frac{1}{c}\left[F'\left(t+\frac{2L}{c}\right)+F'(t)\right], \\ G_2\left(t+\frac{2L}{c}\right) &= G_2(t)-\frac{2}{c}F'\left(t+\frac{L}{c}\right). \end{aligned}\right\} \quad t>0 \tag{15.11}$$

The outcome of introducing a sequence of values

$$t+\frac{2L}{c},\quad t+\frac{4L}{c},\cdots,t+\frac{2(N-1)L}{c}$$

into (15.11) and summing up the resultant equations is an independent pair of difference relations

$$G_1\left(t+\frac{2NL}{c}\right)-G_1(t)=-\frac{1}{c}\left[F'\left(t+\frac{2NL}{c}\right)\right.$$
$$\left.+2F'\left(t+\frac{(2N-2)L}{c}\right)+\cdots+2F'\left(t+\frac{2L}{c}\right)+F'(t)\right],$$
$$G_2\left(t+\frac{2NL}{c}\right)-G_2(t)=-\frac{2}{c}\left[F'\left(t+\frac{(2N-1)L}{c}\right)\right. \tag{15.12}$$
$$\left.+F'\left(t+\frac{(2N-3)L}{c}+\cdots+F'\left(t+\frac{L}{c}\right)\right]$$

that hold for any integer $N\geq 1$.

When the function $F(t)$ has a simply periodic nature, say $F'(t)=\mathrm{e}^{-\mathrm{i}\omega t}$, we find

$$G_1\left(t+\frac{2NL}{c}\right)-G_1(t)=-\frac{2}{c}\,\mathrm{e}^{-\mathrm{i}\omega(t+(NL/c))}\sin\frac{N\omega L}{c}\cot\frac{\omega L}{c}$$
$$=-\frac{2}{c}N\,\mathrm{e}^{-\mathrm{i}n\pi ct/L},\quad \omega=\frac{n\pi c}{L},\quad n=1,2,\ldots, \tag{15.13}$$
$$G_2\left(t+\frac{2NL}{c}\right)-G_2(t)=-\frac{2}{c}\,\mathrm{e}^{-\mathrm{i}\omega(t+(NL/c))}\frac{\sin\dfrac{N\omega L}{c}}{\sin\dfrac{\omega L}{c}}$$
$$=-\frac{2}{c}N(-1)^n\,\mathrm{e}^{-\mathrm{i}n\pi ct/L},\quad \omega=\frac{n\pi c}{L}$$

with steadily larger magnitudes for G_1, G_2 as t increases (and a corresponding rise in amplitude of the wave function (15.7)) if the 'applied' frequency ω equals any one of the eigenfrequencies $n\pi c/L$.

§16. *Wave Motions on a String with a Point Load*

To the circumstances, already enumerated, which preclude form invariant wave propagation along a string we may append the occurrence of abrupt or gradual changes in its composition or density. An alteration of the mass density for a string results from loading, and a noteworthy albeit hypothetical instance obtains when the additional mass is concentrated at a given point of an otherwise uniform string.

A progressive wave will be partly reflected and partly transmitted after arrival at the load point and the component string profiles (associated with primary and secondary waves) are generally distinct from each other. Let us suppose that the added mass M is located at $x=0$ (barring its freedom to slide along the string) and that the incident wave form is $A(t-(x/c))$; then we postulate for the resulting displacements to the left and right of the load point

$$y(x,t)=\begin{cases} A\left(t-\dfrac{x}{c}\right)+B\left(t+\dfrac{x}{c}\right), & x<0 \\ C\left(t-\dfrac{x}{c}\right), & x>0 \end{cases} \tag{16.1}$$

where the functions B and C specify the reflected and transmitted waves, respectively. To determine the latter in terms of A we call upon a pair of conditions jointly involving the displacements (16.1); these are, firstly, that the displacement is continuous at $x=0$, yielding the relationship

$$y(0,t)\equiv Y(t)=A(t)+B(t)=C(t), \tag{16.2}$$

and, secondly, that the slope of the string is discontinuous there, in accordance with the equation for the small transverse motions of the mass M,

$$M\left(\frac{\partial^2 y}{\partial t^2}\right)_{x=0}=P\left\{\left(\frac{\partial y}{\partial x}\right)_{x\to 0+}-\left(\frac{\partial y}{\partial x}\right)_{x\to 0-}\right\} \tag{16.3}$$

which furnishes the relationship

$$M\frac{\mathrm{d}^2 Y}{\mathrm{d}t^2}=\frac{P}{c}\left\{-\frac{\mathrm{d}C}{\mathrm{d}t}+\frac{\mathrm{d}A}{\mathrm{d}t}-\frac{\mathrm{d}B}{\mathrm{d}t}\right\}. \tag{16.4}$$

Eliminating $\mathrm{d}B/\mathrm{d}t$ from (16.4) with the help of (16.2), and noting the identity of the functions Y and C, it follows on integrating once that, apart from an additive constant α, say,

$$\tau\frac{\mathrm{d}Y}{\mathrm{d}t}+Y(t)=A(t), \tag{16.5}$$

wherein

$$\tau=\frac{Mc}{2P}=\frac{M}{2\rho c}. \tag{16.6}$$

The general solution of the inhomogeneous first order equation (16.5) comprises an arbitrary constant multiple, β, of the exponential function $\exp(-t/\tau)$, which is a solution of the associated homogeneous equation, and a particular integral of the full equation, namely

$$Y(t)=\mathrm{e}^{-t/\tau}\int_{-\infty}^{t}\mathrm{e}^{t''/\tau}A(t'')\frac{\mathrm{d}t''}{\tau}$$

if $Y(-\infty)=0$. After appropriate changes of variable in the latter expression, we find

$$Y(t)=C(t)=\int_0^{\infty}\mathrm{e}^{-t'/\tau}A(t-t')\frac{\mathrm{d}t'}{\tau} \tag{16.7}$$

$$=\int_0^{\infty}\mathrm{e}^{-\zeta}A(t-\tau\zeta)\,\mathrm{d}\zeta. \tag{16.8}$$

If the load is situated at the position $x = \xi$ rather than $x = 0$, and the displacement amplitude there is denoted by $y(\xi, t) = Y_\xi(t)$, then

$$Y_\xi(t) = C\left(t - \frac{\xi}{c}\right) = \int_0^\infty e^{-t'/\tau} A\left(t - \frac{\xi}{c} - t'\right) \frac{dt'}{\tau}. \tag{16.9}$$

From the condition $A(t'' - (x/c)) = 0$, $t'' < t$, $x \geq 0$, which expresses the fact that the (localized) incident wave has yet to reach the point x at the time t, it is clear that

$$\int_0^\infty e^{-t'/\tau} A\left(t - \frac{x}{c} - t'\right) \frac{dt'}{\tau} = 0, \quad x \geq 0$$

and thus the requirement of causality (or quiescence before the arrival of the primary stimulus) is ensured for motions on the positive half of the string ($x \geq 0$) if both of the integration constants α, β vanish. Accordingly, the amplitude of vibration of the load and the transmitted wave amplitude are fully specified by the above integrals, which reveal that these quantities are not synchronous with the primary wave amplitude, but depend instead on values of the latter at the given as well as earlier times; in effect, the weighting factor $\exp(-t/\tau)$ singles out the values of A during the time interval $(t - \tau, t)$, so that τ acquires the role of a correlation time between the incident and transmitted waves.

The representation (16.8) is convenient for small magnitudes of τ and furnishes the development

$$Y(t) = A(t) - \tau \frac{dA}{dt} + \tau^2 \frac{d^2A}{dt^2} - \cdots, \tag{16.10}$$

whereas (16.7) leads to an expansion in reciprocal powers of τ,

$$Y(t) = \frac{1}{\tau} \int_0^\infty A(t - t')\, dt' - \frac{1}{\tau^2} \int_0^\infty t' A(t - t')\, dt' + \cdots, \tag{16.11}$$

suitable for large values of τ.

On integrating by parts in (16.7), we find

$$\begin{aligned} C(t) &= \int_0^\infty d(-e^{-t'/\tau}) A(t - t') \\ &= [-e^{-t'/\tau} A(t - t')]_{t'=0}^{t'=\infty} + \int_0^\infty e^{-t'/\tau} \frac{d}{dt'} A(t - t')\, dt' \\ &= A(t) - \frac{d}{dt} \int_0^\infty e^{-t'/\tau} A(t - t')\, dt', \end{aligned}$$

whence

$$\begin{aligned} B(t) &= -\frac{d}{dt}\int_0^\infty e^{-t'/\tau} A(t-t')\, dt' \\ &= -\tau \frac{d}{dt}\int_0^\infty e^{-\zeta} A(t-\tau\zeta)\, d\zeta, \end{aligned} \tag{16.12}$$

in consequence of (16.2).

It is in keeping with intuitive expectations that the reflected wave amplitude vanishes if M or $\tau \to 0$, while the incident and transmitted wave amplitudes are then indistinguishable; moreover, if M or $\tau \to \infty$, the incident wave undergoes complete reflection with $B(t) = -A(t)$.

Let us now specialize the incident wave in the foregoing problem by assigning to it a simply periodic dependence on the time. Omitting an explicit indication that the real parts thereof, say, are eventually of significance, the displacements on the string are then assumed to have the forms

$$y(x,t) = \begin{cases} A\, e^{i(kx-\omega t)} + B\, e^{i(-kx-\omega t)}, & x<0 \\ C\, e^{i(kx-\omega t)}, & x>0 \end{cases} \tag{16.13}$$

to the left and right of the load at $x = 0$, with A, B, and C designating the incident, reflected and transmitted wave amplitudes, respectively. In these forms recognition is given to the fact that the displacement has a common angular frequency or period anywhere on the string, and that the wave number is likewise unique, since the wave velocity is of fixed numerical value for any progressive motions along the string. Thus, given the primary amplitude, A, only the two magnitudes B, C await determination, for which purpose we call upon the same pair of joining conditions at the load point as were employed previously for the corresponding aperiodic vibrations. The requisite continuity of the displacement at all times yields the relation

$$A + B = C \tag{16.14}$$

and from the discontinuity in the slope of the string implied by (16.3) it follows that $-M\omega^2 C = i\rho k[C - A + B]$ or

$$A - B = \left(1 - i\frac{Mk}{\rho}\right) C. \tag{16.15}$$

Combining (16.14) and (16.15) we obtain

$$\frac{B}{iMk} = \frac{C}{2\rho} = \frac{A}{2\rho - iMk}, \tag{16.16}$$

and B, C are thereby available in terms of the primary amplitude factor A; evidently the reflected and transmitted waves differ both in amplitude and phase relative to the incident wave, and their limiting amplitudes are 0 and A, respectively, when $M \to 0$, as would be anticipated.

The information contained in (16.16) is also provided by the solutions (16.8), (16.12) of the same problem that allow of a more general time dependence for the incident wave. Thus, if $A(t)$ be given the particular form

$$A(t) = \mathrm{e}^{-\mathrm{i}\omega t},$$

we deduce from (16.8) that

$$C(t) = \mathrm{e}^{-\mathrm{i}\omega t} \int_0^\infty \mathrm{e}^{-(1-\mathrm{i}\omega\tau)\zeta}\, \mathrm{d}\zeta = \frac{A(t)}{1-\mathrm{i}\omega\tau}$$

and the substance of the second equality in (16.16) is recovered, inasmuch as

$$\omega\tau = kc \cdot \frac{M}{2\rho c} = \frac{Mk}{2\rho}.$$

Likewise, if the same representation for $A(t)$ is introduced into (16.12), it follows that

$$B(t) = \mathrm{i}\omega\tau\, \mathrm{e}^{-\mathrm{i}\omega t} \int_0^\infty \mathrm{e}^{-(1-\mathrm{i}\omega\tau)\zeta}\, \mathrm{d}\zeta = \frac{\mathrm{i}\omega\tau}{1-\mathrm{i}\omega\tau} A(t),$$

which is fully compatible with the first equality in (16.16).

It may be observed that the controlling and dimensionless magnitude $\omega\tau$ in the amplitude connections expresses a ratio of two characteristic times, namely the correlation time for the load, $\tau = M/2\rho c$, and the period $T = 2\pi/\omega$ of the string oscillations.

§17. *A Green's Function Approach*

Instead of employing distinct solutions of the wave equation appropriate to the periodic string displacements on either side of the load and then subjecting them to the requisite interrelations at their point of confluence, we may base the calculation on a single uniformly valid representation of the displacement which is obtained with the help of an auxiliary quantity, or Green's function.

If the displacement is expressed as (the real part of)

$$y(x, t) = s(x)\, \mathrm{e}^{-\mathrm{i}\omega t} \tag{17.1}$$

its (complex) coordinate factor, $s(x)$, is to be sought via the explicit time-independent conditions

$$\left(\frac{\mathrm{d}^2}{\mathrm{d}x^2} + k^2\right) s(x) = 0, \tag{17.2}$$

$$s(0+) = s(0-), \tag{17.3}$$

and

$$\left(\frac{\mathrm{d}s}{\mathrm{d}x}\right)_{0+} - \left(\frac{\mathrm{d}s}{\mathrm{d}x}\right)_{0-} = -\frac{M\omega^2}{P} s(0), \tag{17.4}$$

of which the first holds at all points on the string, while the latter pair are operative at the position of the load. These relations need to be supplemented by stipulations which make the problem unique, namely that $A\exp(ikx)$ is a part of the solution of (17.2) in $x<0$, as befits the given primary wave form, and that the remainder of the solution in $x \lessgtr 0$ consists of outgoing (or secondary) wave forms $\exp(ik|x|)$ with respect to their source point $x=0$.

A function which evidences the aforesaid outgoing wave character relative to an arbitrary source location, say $x=x'$, may be associated with the inhomogeneous version of (17.2),

$$\left(\frac{d^2}{dx^2}+k^2\right)G(x,x') = -\delta(x-x') \qquad -\infty<x,x'<\infty \tag{17.5}$$

wherein the right-hand member contains a symbolic, or Dirac delta, function; the latter is defined by the statements

$$\int_{x_0}^{x_1}\delta(x-x')\,dx = \begin{cases} 1, & x_1-x'>0>x_0-x' \\ 0, & x_1-x'>x_0-x'>0 \quad \text{or} \quad 0<x_0-x'<x_1-x' \end{cases} \tag{17.6}$$

which imply that the integral vanishes unless the domain of integration encompasses the point $x = x'$, and has the value unity in this circumstance. It is a consequence of the above definition that $\delta(x-x')$ vanishes for all x save x' and possesses a singularity at $x=x'$ which is sufficient to render an integrated value thereof equal to unity; furthermore, an arbitrary continuous function of the coordinate x can be represented by a linear superposition of δ-functions

$$f(x) = \int_{-\infty}^{\infty}\delta(x-x')f(x')\,dx' \tag{17.7}$$

for the entire contribution to the integral derives from the point $x'=x$. No ordinary function can claim the properties (17.6), in fact, though particular examples of functions exhibiting these attributes in the limit, and thus approximating to the delta function with arbitrary precision, include

$$\delta(x-x') = \operatorname*{Lim}_{\varepsilon\to 0}\frac{1}{\pi}\frac{\varepsilon}{(x-x')^2+\varepsilon^2}, \tag{17.8}$$

$$\delta(x-x') = \operatorname*{Lim}_{\varepsilon\to 0}\frac{1}{\varepsilon}\,e^{-\pi(x-x')^2/\varepsilon^2}. \tag{17.9}$$

Although the proper mathematical basis for bolstering up these notions is supplied by the concept of generalized functions or distributions, we shall here regard the δ-function as an ordinary, differentiable function.

In accordance with the properties thus ascribed to the inhomogeneous member of (17.5), we may regard $G(x,x')$ as an everywhere continuous function of x, for any parametric value of x', which satisfies the homogeneous equation (17.2) on the intervals $x \lesseqgtr x'$ and has a fixed discontinuity in its first derivative at x', viz.:

$$\left.\frac{\mathrm{d}G}{\mathrm{d}x}\right]_{x'+0} - \left.\frac{\mathrm{d}G}{\mathrm{d}x}\right]_{x'-0} = -1. \tag{17.10}$$

The regular function

$$G(x, x') = \frac{1}{2k}\,\mathrm{e}^{\mathrm{i}k|x-x'|} = G(x', x) \tag{17.11}$$

complies with all the foregoing requirements and may be termed an outgoing wave Green's function for time-periodic and one-dimensional motions on the understanding that the time dependence is supplied by the factor $\exp(-\mathrm{i}\omega t)$; if the time factor be chosen in the form $\exp(\mathrm{i}\omega t)$ the analogous outgoing wave Green's function is merely a complex conjugate of our previous one. It will be appreciated that (17.11) is but a particular solution of (17.5), and the reason for excluding both linearly independent solutions of the corresponding homogeneous equation, $\exp(\pm \mathrm{i}kx)$, is that they pertain to wave motions which simply travel from one end to the other of the infinite x-interval.

The integral representation for $\delta(x)$ which obtains from the formula

$$\frac{1}{\pi}\frac{\varepsilon}{x^2+\varepsilon^2} = \frac{1}{2\pi}\int_{-\infty}^{\infty} \mathrm{e}^{\mathrm{i}\zeta x}\,\mathrm{e}^{-\varepsilon|\zeta|}\,\mathrm{d}\zeta,$$

namely

$$\delta(x) = \frac{1}{2\pi}\int_{-\infty}^{\infty} \mathrm{e}^{\pm \mathrm{i}\zeta x}\,\mathrm{d}\zeta \tag{17.12}$$

suggests that a particular solution of (17.5) takes the form

$$G(x, x') = \frac{1}{2\pi}\int_{-\infty}^{\infty} \frac{\mathrm{e}^{\mathrm{i}\zeta(x-x')}}{\zeta^2 - k^2}\,\mathrm{d}\zeta \tag{17.13}$$

though an essential complication is manifest by the singularities of the integrand at $\zeta = \pm k$. This difficulty can be resolved in a purely formal manner by supposing at the outset that k has an infinitesimal positive imaginary part, whence the aforesaid singularities are displaced from the real ζ-axis or path of integration. Then a residue method of evaluation of the integral for $x - x' \gtrless 0$, linked with the first order pole singularities at $\pm k$ in the upper and lower halves of the ζ-plane, respectively, confirms the equivalence of the Green's function representations (17.11), (17.13). It is evident from (17.11) that a positive imaginary component of k implies an exponential diminution of the Green's function which depends proportionately on the size of the interval between the source and observation points x', x; and hence, if this component has an infinitesimal magnitude the corresponding attrition of the Green's function is revealed only at very large separation $|x - x'|$ between the latter points. If a negative imaginary part of the wave number is envisioned for the interpretation of (17.13), it will be found that the complex conjugate of the Green's function (17.11) is thereby characterized, and the previous implications regarding exponential decay remain unaltered. The same characterization can be deduced wihout the prior attribution of complex values for the wave number by regarding the integral in (17.13) as conducted along a path in the complex ζ-plane which deviates from the real axis in the neighborhood of the points $\pm k$, passing below $+k$ and above $-k$ in order to realize the particular form (17.11).

From its manner of specification we may infer that the Green's function is inherently suited to representing the secondary waves which arise in the problem of scattering by a concentrated load on a string. To see how this is accomplished, we proceed from the differential equations (17.2), (17.5) for the spatial part of the string displacement and the Green's function, respectively, first multiplying the former by $G(x, x')$ and the latter by $s(x)$, then subtracting one from the other and next integrating with respect to x. Utilizing the 'sifting' property of the delta function expressed by (17.7), there results

$$s(x') = \int_{-X_1}^{X_2} \left[G(x, x') \frac{\mathrm{d}^2 s(x)}{\mathrm{d}x^2} - s(x) \frac{\mathrm{d}^2 G(x, x')}{\mathrm{d}x^2} \right] \mathrm{d}x;$$

and since the integrand may be converted to a total differential,

$$s(x') = \left[G(x, x') \frac{\mathrm{d}s}{\mathrm{d}x} - s(x) \frac{\mathrm{d}G}{\mathrm{d}x} \right]_{x=-X_1}^{x=X_2} + \left[G(x, x') \frac{\mathrm{d}s}{\mathrm{d}x} - s(x) \frac{\mathrm{d}G}{\mathrm{d}x} \right]_{x=0+}^{x=0-} \tag{17.14}$$

where the second pair of terms appears in consequence of deleting an infinitesimal portion of the integration range centered at the position of the load, at which point $\mathrm{d}s/\mathrm{d}x$ is discontinuous. (It is not necessary to make a similar deletion at the point $x = x'$ where G has a discontinuous first derivative, since this feature is already accounted for by the inhomogeneous member of the Green's function differential equation.) The first part of the expression (17.14), which seemingly depends on the arbitrary locations $x = -X_1, X_2$ chosen, has in fact the value $A \exp(\mathrm{i}kx')$ as a direct calculation with the explicit form (17.11) and the requisite behaviors

$$s(x) = \begin{cases} A\,\mathrm{e}^{\mathrm{i}kx} + B\,\mathrm{e}^{-\mathrm{i}kx}, & x < 0 \\ C\,\mathrm{e}^{\mathrm{i}kx}, & x > 0 \end{cases} \tag{17.15}$$

shows. So long as $x' \neq 0$, it is only the discontinuity in $\mathrm{d}s/\mathrm{d}x$ at $x = 0$, which is given by (17.4), that provides a non-vanishing contribution to the second part, and we obtain consequently

$$s(x') = A\,\mathrm{e}^{\mathrm{i}kx'} + G(0, x') \frac{M\omega^2}{P} s(0)$$

where the restriction $x' \neq 0$ can now be lifted.

Dropping the prime attached to the coordinate variable and invoking the symmetry of the Green's function (17.11) under an interchange of its arguments x, x' the compact representation

$$y(x,t) = A\,\mathrm{e}^{\mathrm{i}(kx-\omega t)} + \frac{M\omega^2}{P} G(x,0)y(0,t), \qquad -\infty < x < \infty \tag{17.16}$$

follows. In the latter we observe that the scale of the secondary disturbance is proportional to the displacement at the load, $y(0,t)$, and that its variable strength is furnished by the Green's function whose source coordinate lies at the point of loading. The displacement $y(0,t)$ is deduced after placing $x=0$ in (17.16), with the result that

$$y(0,t) = \frac{A\,\mathrm{e}^{-\mathrm{i}\omega t}}{1 - \dfrac{M\omega^2}{P} G(0,0)} = \frac{A\,\mathrm{e}^{-\mathrm{i}\omega t}}{1 - \mathrm{i}\dfrac{Mk}{2\rho}}, \tag{17.17}$$

and on substituting this value in (17.16) we arrive finally at the explicit form of the string displacement everywhere, viz.:

$$y(x,t) = A\,\mathrm{e}^{\mathrm{i}(kx-\omega t)} + \frac{\mathrm{i}\dfrac{Mk}{2\rho}}{1-\mathrm{i}\dfrac{Mk}{2\rho}} A\,\mathrm{e}^{\mathrm{i}k|x| - \mathrm{i}\omega t}, \qquad -\infty < x < \infty. \tag{17.18}$$

It is readily confirmed from the preceding expression that the complex amplitudes of the reflected and transmitted waves in the domains $x<0, x>0$ are in agreement with (16.16); when they are divided by the incident wave amplitude we obtain the so-called reflection and transmission coefficients, namely

$$r = \frac{B}{A} = \frac{\mathrm{i}Mk/2\rho}{1 - \mathrm{i}\dfrac{Mk}{2\rho}} \tag{17.19}$$

and

$$t = \frac{C}{A} = \frac{1}{1-\mathrm{i}\dfrac{Mk}{2\rho}} = 1 + r. \tag{17.20}$$

We may observe that if the ensemble of conditions (17.2)–(17.4) is condensed into the single relation

$$\left(\frac{\mathrm{d}^2}{\mathrm{d}x^2} + k^2\right) s(x) = -\frac{M\omega^2}{P} s(0)\delta(x), \tag{17.21}$$

the close relationship between $s(x)$ and the Green's function becomes evident; and it appears that any differences in their respective behavior is accounted for by solutions of the related homogeneous equation. Thus the appropriate integral of (17.21) for the above scattering problem differs from a multiple, $M\omega^2 s(0)/P$, of the function

$$G(x,0) = \frac{\mathrm{i}}{2k}\,\mathrm{e}^{\mathrm{i}k|x|}$$

by the term $A\,\mathrm{e}^{\mathrm{i}kx}$ which puts in evidence the given form of excitation or primary wave.

§18. *Excitation of a String by the Impulsive Stimulus of an Attached Load*

In the problem just discussed the existence of secondary waves on the string is attributable to the inertial reaction of an attached load in response to the displacement which an incident or primary wave forces upon it; the slope of the string evidences a discontinuity at the load point and the otherwise passive load is capable of acting as a source of outgoing or scattered waves. Let us now conceive of a situation in which there are no incoming waves and, instead, an impulse delivered to the load (at $x = 0$) gives rise to displacements on the string; we shall suppose, further, that the load experiences an additional (linear) restoring force and may thus execute (simply periodic) oscillations should the string become slack. If the impulsive force is applied transversely to the undisplaced string configuration, the resulting motions are evidently symmetrical about the load point, and we can therefore confine our attention, for purposes of analysis, on one-half of the string (where $x > 0$, say). In accordance with this symmetry, the equation of motion for a load of mass M takes the form

$$M\left(\frac{\partial^2 y}{\partial t^2} + \omega_0^2 y\right)_{x=0} = 2P\left(\frac{\partial y}{\partial x}\right)_{x\to 0+} \tag{18.1}$$

where ω_0 designates its natural frequency of oscillation when dynamically uncoupled from the string (through a reduction of the tension, P, to a negligible magnitude). We express the string displacement by a progressive wave form

$$y(x,t) = f(ct - x), \quad x > 0 \tag{18.2}$$

as befits a disturbance moving away from the site of impulse or excitation, and then the relation (18.1) is turned into an ordinary homogeneous differential equation for the (real) profile function f, viz.,

$$\frac{\mathrm{d}^2 f(\xi)}{\mathrm{d}\xi^2} + 2\frac{\rho}{M}\frac{\mathrm{d}f(\xi)}{\mathrm{d}\xi} + \frac{\omega_0^2}{c^2} f(\xi) = 0, \quad \xi > 0. \tag{18.3}$$

The solution of (18.3) possesses different representations according as ω_0/c

is greater or smaller than ρ/M and, corresponding to the first of these alternatives, we obtain

$$f(\xi) = C\,\mathrm{e}^{-\rho\xi/M} \sin \kappa(\xi + \delta), \quad \xi > 0 \tag{18.4}$$

where

$$\kappa = \sqrt{\left(\frac{\omega_0^2}{c^2} - \frac{\rho^2}{M^2}\right)} \tag{18.5}$$

and C, δ are constants. An equilibrium state for the string and load before the moment of impulse ($t = 0$) implies that $f(\xi) = 0$, $\xi \le 0$, and thus $\delta = 0$. Accordingly,

$$y(x, t) = \begin{cases} C \exp\left[-(ct - |x|)\dfrac{\rho}{M}\right] \sin \kappa(ct - |x|), & |x| < ct \\ 0, & |x| > ct \end{cases} \tag{18.6}$$

on making explicit the even symmetry with regard to x; and the remaining constant C is fixed in terms of the magnitude of the impulse or the initial velocity $\dot{Y}(0) = (\partial/\partial t)y(0, t)|_{t=0}$.

When the second alternative applies, the relevant solution of (18.3) assumes the form

$$y(x, t) = \begin{cases} C\{\exp[\kappa_1(ct - |x|)] - \exp[\kappa_2(ct - |x|)]\}, & |x| < ct \\ 0, & |x| > ct \end{cases} \tag{18.7}$$

where

$$\binom{\kappa_1}{\kappa_2} = -\frac{\rho}{M} \pm \left(\frac{\rho^2}{M^2} - \frac{\omega_0^2}{c^2}\right)^{1/2} = -\frac{\rho}{M} \pm \hat{\kappa}. \tag{18.8}$$

Since $(\omega_0/c)/(\rho/M) = 2\omega_0\tau$, where $\tau = M/2\rho c$ is the load correlation time introduced in §16, the foregoing alternatives pertain to periods of natural vibration for the mass which are smaller or greater than $4\pi\tau$, respectively. The load displacement, $Y(t) = y(0, t)$, is compactly rendered by the expressions

$$Y(t) = \dot{Y}(0)\,\mathrm{e}^{-\rho ct/M} \frac{\sin \kappa ct}{\kappa c} \tag{18.9}$$

or

$$Y(t) = \dot{Y}(0)\,\mathrm{e}^{-\rho ct/M} \frac{\sinh \hat{\kappa} ct}{\hat{\kappa} c} \tag{18.10}$$

according as $\omega_0 \gtrless \tau/2$, and it appears from the first version, (18.9), that $\mathrm{d}Y/\mathrm{d}t$ vanishes at the unending sequence of times which satisfy the relation

$$\tan \kappa ct = M\kappa/\rho,$$

whereas the second implies a null value for $\mathrm{d}Y/\mathrm{d}t$ at but one instant of time, specified by the equation

$$\tanh \hat{\kappa} ct = M\hat{\kappa}/\rho.$$

It is the magnitude of $\mathrm{d}Y/\mathrm{d}t$, in fact, which determines the rate of energy loss from the load, for we establish, on multiplying the equation (18.1) by $\mathrm{d}Y/\mathrm{d}t$ and utilizing the relation $(\partial y/\partial x)_{x\to 0+} = -(1/c)\,\mathrm{d}Y/\mathrm{d}t$, that

$$-\frac{\mathrm{d}}{\mathrm{d}t}\left\{\frac{1}{2}M\left(\frac{\mathrm{d}Y}{\mathrm{d}t}\right)^2+\frac{1}{2}M\omega_0^2Y^2\right\}=\frac{2P}{c}\left(\frac{\mathrm{d}Y}{\mathrm{d}t}\right)^2 \tag{18.11}$$

where the left-hand member evidently involves a sum of the kinetic and potential energies for the elastically controlled load. An integration of (18.11) for $t \geq 0$ yields, when the particular initial value $Y(0) = 0$ is chosen, the energy transfer relationship

$$\frac{1}{2}M(\dot{Y}(0))^2=\frac{2P}{c}\int_0^\infty\left(\frac{\mathrm{d}Y}{\mathrm{d}t}\right)^2\mathrm{d}t=\frac{2P}{c}\omega_0^2\int_0^\infty Y^2(t)\,\mathrm{d}t \tag{18.12}$$

wherein the declared equality of integrals may be confirmed through integration by parts in the former and use of the load equation

$$\frac{\mathrm{d}^2Y}{\mathrm{d}t^2}+\omega_0^2Y=-\frac{2P}{Mc}\frac{\mathrm{d}Y}{\mathrm{d}t}. \tag{18.13}$$

The individual representations (18.9), (18.10) of the load displacement function $Y(t)$ are fully consistent with (18.12) and enable us, furthermore, to calculate the amounts of mechanical energy, ΔE, which pass from the load to the string during the finite time interval $(0, \tau)$, viz.:

$$\Delta E=\frac{2P}{c}\int_0^\tau\left(\frac{\mathrm{d}Y}{\mathrm{d}t}\right)^2\mathrm{d}t=\frac{1}{2}M(\dot{Y}(0))^2\{1-\varepsilon(\tau)\} \tag{18.14}$$

$$\begin{aligned}\varepsilon(\tau)&=\mathrm{e}^{-2\rho c\tau/M}\left[1+\frac{1}{2}\left(\frac{\sin\kappa c\tau}{\kappa c\tau}\right)^2-\frac{\sin 2\kappa c\tau}{2\kappa c\tau}\right]\\&=\mathrm{e}^{-2\rho c\tau/M}\left[1+\frac{1}{2}\left(\frac{\sinh\hat{\kappa} c\tau}{\hat{\kappa} c\tau}\right)^2-\frac{\sin 2\hat{\kappa} c\tau}{2\hat{\kappa} c\tau}\right]\end{aligned} \tag{18.15}$$

in the respective cases $2\omega_0\tau > 1$, $2\tau\hat{\kappa}c < 1$.

When the aperiodic wave form (18.6) is given a complex representation,

$$y(x,t)=C\,\mathrm{Im}\exp\left[\mathrm{i}\left(\kappa+\mathrm{i}\frac{\rho}{M}\right)(ct-|x|)\right],\quad |x|<ct \tag{18.16}$$

the subsidence of the string excitation in time can be linked with a complex-valued wave number, namely

$$k = \kappa + \mathrm{i}\frac{\rho}{M}. \tag{18.17}$$

§19. *Characteristic Functions and Complex Eigenfrequencies*

The standing wave and time-periodic displacement profiles that vanish at the endpoints $x = 0, L$ of a vibrating string, namely

$$\begin{aligned} &\sin\frac{\omega_n}{c}x(a_n \cos \omega_n t + b_n \sin \omega_n t) \\ &= \frac{\mathrm{i}}{4}a_n[y(x, t, \omega_n) - y(x, t, -\omega_n)] - \frac{1}{4}b_n[y(x, t, \omega_n) + y(x, t, -\omega_n)] \end{aligned} \tag{19.1}$$

where

$$y(x, t, \omega_n) = \mathrm{e}^{-\mathrm{i}\omega_n(t+x/c)} - \mathrm{e}^{-\mathrm{i}\omega_n(t-x/c)}, \tag{19.2}$$

correspond to a discrete spectrum of characteristic (or eigen) frequencies

$$\omega_n = \frac{n\pi c}{L}, \quad n = 1, 2, \ldots \tag{19.3}$$

and form an orthogonal system on this interval, since

$$\int_0^L y(x, t, \pm\omega_m)y(x, t, \pm\omega_n)\,\mathrm{d}x = 0, \quad m \neq n; \tag{19.4}$$

completeness of the system implies that a linear combination of the characteristic functions (19.1) can be made (by suitable and individual determination of the coefficients a_n, b_n) to assume arbitrarily specified initial values for the wave function and its time derivative along the string.

As the length of the string increases the frequency spectrum (19.3) becomes ever more dense and, in the limit $L \to \infty$, the counterpart of (19.2), viz.

$$y(x, t, \omega) = \mathrm{e}^{-\mathrm{i}\omega(t+x/c)} - \mathrm{e}^{-\mathrm{i}\omega(t-x/c)}, \tag{19.5}$$

represents a particular wave function which, for any real value of ω, vanishes at $x = 0$ and is bounded for all $x, t > 0$; evidently, the latter pair of terms define a standing wave pattern attributable to oppositely directed travelling wave profiles that have equal amplitude and a relative phase difference equal to π at $x = 0$.

To solve an initial value problem on the range $0 < x < \infty$, given the condition of a vanishing wave function at $x=0$, we may superpose the characteristic solutions (19.5) in the fashion

$$y(x, t) = \int_0^\infty [A(\omega)\{y(x, t, \omega) - y(x, t, -\omega)\} + B(\omega)\{y(x, t, \omega) + y(x, t, -\omega)\}] \, d\omega \tag{19.6}$$

and thereby obtain the individul relationships

$$\begin{aligned} y(x, 0) = s_0(x) &= -4\,i \int_0^\infty A(\omega) \sin \frac{\omega x}{c} \, d\omega \\ \frac{\partial y(x, 0)}{\partial t} = v_0(x) &= -4 \int_0^\infty B(\omega) \sin \frac{\omega x}{c} \, d\omega \end{aligned} \qquad x > 0 \tag{19.7}$$

between the original displacement/velocity distributions and the frequency dependent amplitude factors $A(\omega)$, $B(\omega)$ of the resulting wave function. On inverting the Fourier sine integrals of (19.7) there obtains

$$\begin{aligned} A(\omega) &= \frac{i}{2\pi c} \int_0^\infty s_0(x) \sin \frac{\omega x}{c} \, dx \\ B(\omega) &= -\frac{1}{2\pi \omega c} \int_0^\infty v_0(x) \sin \frac{\omega x}{c} \, dx \end{aligned} \tag{19.8}$$

whence, after substitution in (19.6) and evaluation of the ω-integrals, it follows that

$$y(x, t) = \frac{1}{2}[s_0(x + ct) + \operatorname{sgn}(x - ct) s_0(|x - ct|)] + \frac{1}{2c} \int_{|x-ct|}^{x+ct} v_0(\xi) \, d\xi \tag{19.9}$$

using the customary nomenclature for the sign function

$$\operatorname{sgn} \alpha = \pm 1, \quad \alpha \gtrless 0.$$

The above solution is identical, if $t < x/c$, with that for the initial value problem on an infinite string and the boundary condition at $x = 0$ plays no role therein; an alternative derivation of (19.9), relying directly on the d'Alembert representation (1.10) and the continuations

$$y(x, 0) = \begin{cases} s_0(x), & x > 0, \\ -s_0(-x), & x < 0, \end{cases} \qquad \frac{\partial}{\partial t} y(x, 0) = \begin{cases} v_0(x), & x > 0, \\ -v_0(-x), & x < 0 \end{cases}$$

of the initial data, facilitates the interpretation of the solution at subsequent times and brings out the influence of the boundary condition. Specifically, if we consider a resolution of the initial profiles into components that have opposite senses of propagation, it is only the incoming part which undergoes reflection at $x = 0$ with a consequent sign reversal of its amplitude.

Let us next suppose that a harmonic oscillator of mass M is attached to the end $x = 0$ of the string and is capable of transverse oscillations with the natural frequency ω_0 if the string is without tension. The displacement $y(0, t)$ of the oscillator satisfies the equation of motion

$$\left(\frac{d^2}{dt^2} + \omega_0^2\right) y(0, t) = \frac{P}{M} \frac{\partial y(0, t)}{\partial x} \tag{19.10}$$

which also serves as a boundary condition for the string displacement. A solution of the wave equation in $x > 0$,

$$y(x, t, \omega) = e^{-i\omega(t+x/c)} - S(\omega)\, e^{-i\omega(t-x/c)} \tag{19.11}$$

is consonant with (19.10) provided that

$$S(\omega) = \frac{\omega^2 - \omega_0^2 - 2\, i\omega\gamma}{\omega^2 - \omega_0^2 + 2\, i\omega\gamma} = \frac{(\omega + \omega_1)(\omega + \omega_2)}{(\omega - \omega_1)(\omega - \omega_2)} \tag{19.12}$$

where

$$\gamma = \frac{P}{2Mc} = \frac{\rho c}{2M}, \tag{19.13}$$

and

$$\omega_{1,2} = \pm \sqrt{(\omega_2^2 - \gamma^2)} - i\gamma. \tag{19.14}$$

We encounter in (19.11) a wave function that describes the total reflection of an incoming time-periodic wave train with (real) frequency ω, since the complex amplitude $S(\omega)$ of the outgoing train has unit absolute value; the corresponding oscillator amplitude, say $A(\omega)$, which follows from (19.11) and the representation

$$y(0, t, \omega) = A(\omega)\, e^{-i\omega t}, \tag{19.15}$$

proves to be

$$A(\omega) = 1 - S(\omega) = \frac{4\, i\omega\gamma}{\omega^2 - \omega_0^2 + 2\, i\omega\gamma} \tag{19.16}$$

and acquires its largest magnitude, namely 2, when $\omega = \omega_0$ (as befits the appertaining boundary condition $\partial y/\partial x = 0, x = 0$).

Evidently the particular values $\omega = \omega_{1,2}$ locate simple poles of the rational function $S(\omega)$ and these may be termed complex eigenfrequencies of the composite dynamical system (string and oscillator); inasmuch as $\text{Im}\, \omega_{1,2} < 0$ the concomitant exponential functions $e^{-i\omega_{1,2}t}$ diminish with increasing values of t and are thus appropriate for representing transient states of vibration. An orthogonal property of the wave functions (19.11), realized when the oscillator is absent and $S(\omega) = 1$, does not obtain; nevertheless, these functions may be superposed to form solutions of the wave equation that comply with the boundary condition (19.10) and the particular aggregate

$$y(x,t) = A\,\mathrm{Re}\,\mathrm{i}\int_{-\infty}^{\infty} \mathrm{e}^{-\mathrm{i}\omega t}(\mathrm{e}^{-\mathrm{i}\omega x/c} - S(\omega)\,\mathrm{e}^{\mathrm{i}\omega x/c})\frac{\mathrm{d}\omega}{\omega}$$

$$= 2A\,\mathrm{Re}\int_{-\infty}^{\infty} \mathrm{e}^{-\mathrm{i}\omega t}\sin\frac{\omega x}{c}\frac{\mathrm{d}\omega}{\omega} + A\,\mathrm{Re}\,\mathrm{i}\int_{-\infty}^{\infty} \mathrm{e}^{-\mathrm{i}\omega(t-x/c)}(1-S(\omega))\frac{\mathrm{d}\omega}{\omega},$$

where the integration contour passes above the poles of $S(\omega)$, defines a real-valued profile

$$y(x,t) = 4\pi A\frac{\omega_1+\omega_2}{\omega_1-\omega_2}[\mathrm{e}^{-\mathrm{i}\omega_2(t-x/c)} - \mathrm{e}^{-\mathrm{i}\omega_1(t-x/c)}]$$

$$= 8\pi A\frac{\gamma}{\sqrt{(\omega_0^2-\gamma^2)}}\,\mathrm{e}^{-\gamma(t-x/c)}\sin\sqrt{(\omega_0^2-\gamma^2)}(t-x/c) \qquad t > x/c$$

that is suited to the transient motion behind the front of a disturbance on the string resulting from an impulse delivered to the oscillator [compare (18.6)].

The wave function (19.11), whose incoming component has a monochromatic or single frequency character, is adapted to the circumstance of general time variation through multiplication by a function $\mathfrak{A}(\omega)$, which defines the temporal Fourier transform of the primary component, and subsequent integration relative to ω; thus, the oscillator amplitude takes the form

$$Y(t) = \frac{1}{2\pi}\int_{-\infty}^{\infty} \mathfrak{A}(\omega)(1-S(\omega))\,\mathrm{e}^{-\mathrm{i}\omega t}\,\mathrm{d}\omega \tag{19.17}$$

if the incident wave amplitude at the position thereof is represented by

$$\mathfrak{A}(t) = \frac{1}{2\pi}\int_{-\infty}^{\infty} \mathfrak{A}(\omega)\,\mathrm{e}^{-\mathrm{i}\omega t}\,\mathrm{d}\omega. \tag{19.18}$$

With the particular choice

$$\mathfrak{A}(t) = \begin{cases} 0, & t < -T \\ \mathrm{e}^{-\mathrm{i}\omega_0 t}, & -T < t < T \\ 0, & t > T \end{cases} \tag{19.19}$$

for a limited duration stimulus at the natural frequency of the oscillator, we find

$$\mathfrak{A}(\omega) = 2\frac{\sin(\omega-\omega_0)T}{\omega-\omega_0} \tag{19.20}$$

and consequently (19.17) becomes

$$Y(t,T) = \frac{1}{\pi}\int_{-\infty}^{\infty} \frac{\sin(\omega-\omega_0)T}{\omega-\omega_0}\cdot\frac{4\,\mathrm{i}\gamma\omega}{(\omega-\omega_1)(\omega-\omega_2)}\,\mathrm{e}^{-\mathrm{i}\omega t}\,\mathrm{d}\omega. \tag{19.21}$$

Since

$$\lim_{T\to\infty} \mathfrak{A}(\omega) = 2\pi\delta(\omega-\omega_0)$$

the corresponding version of (19.21), viz.

$$Y(t,\infty)=\frac{4\,\mathrm{i}\gamma\omega_0}{(\omega_0-\omega_1)(\omega_0-\omega_2)}\,\mathrm{e}^{-\mathrm{i}\omega_0 t}=2\,\mathrm{e}^{-\mathrm{i}\omega_0 t}$$

evidently reproduces a prior result for the steady state response of the oscillator at the condition of resonance; after the differentiation of $Y(t, T)$ with respect to T, which implies that

$$\frac{\partial Y}{\partial T}=\frac{4\,\mathrm{i}\gamma}{\pi}\int_{-\infty}^{\infty}\cos(\omega-\omega_0)T\frac{\omega\,\mathrm{e}^{-\mathrm{i}\omega t}}{(\omega-\omega_1)(\omega-\omega_2)}\,\mathrm{d}\omega,$$

and a residue evaluation of the latter integral, yielding

$$\frac{\partial Y}{\partial T}=\frac{\gamma}{\sqrt{(\omega_0^2-\gamma^2)}}\{\omega_1\,\mathrm{e}^{-\mathrm{i}\omega_1 t-\mathrm{i}(\omega_1-\omega_0)T}-\omega_2\,\mathrm{e}^{-\mathrm{i}\omega_2 t-\mathrm{i}(\omega_2-\omega_0)T}\},\quad 0<t<T$$

we arrive at the characterization

$$\begin{aligned}Y(t,T)&=2\,\mathrm{e}^{-\mathrm{i}\omega_0 t}-\int_T^{\infty}\frac{\partial Y(t,\tau)}{\partial\tau}\,\mathrm{d}\tau\\&=2\,\mathrm{e}^{-\mathrm{i}\omega_0 t}+\frac{\mathrm{i}\gamma}{\sqrt{(\omega_0^2-\gamma^2)}}\,\mathrm{e}^{\mathrm{i}\omega_0 T}\left\{\frac{\omega_2}{\omega_0-\omega_2}\,\mathrm{e}^{-\mathrm{i}\omega_2(t+T)}-\frac{\omega_1}{\omega_0-\omega_1}\,\mathrm{e}^{-\mathrm{i}\omega_1(t+T)}\right\},\\&\qquad 0<t<T\end{aligned}$$

whose form indicates a small departure from the steady state amplitude once the duration of stimulus, T, exceeds the decay time, $1/\gamma$, stemming from the (common) imaginary part of the complex eigenfrequencies ω_1, ω_2. When $t>T$ the oscillator response, $Y(t, T)$, has no component with the primary time variation, $\mathrm{e}^{-\mathrm{i}\omega_0 t}$, and its attenuation is bound up with the complex nature of ω_1, ω_2.

§20. *Resonant Reflection and Forced Motion*

Reflection of an incident wave with the frequency ω, from a mass oscillator with the natural frequency ω_0 at one end of a string, is accounted for in the wave function (19.11) by a factor

$$S(\omega)=\frac{f(-\omega)}{f(\omega)}\tag{20.1}$$

where $f(\omega)$ is an entire function that has two (simple) zeros in the lower half of the complex ω-plane; moreover, the local version of (20.1),

$$S(\omega)\doteq\frac{\omega-\omega_0-\mathrm{i}\delta}{\omega-\omega_0+\mathrm{i}\delta},\quad\omega\to\omega_0,\quad\delta=2\gamma\tag{20.2}$$

indicates a resonance peak of width δ for the oscillator amplitude, $1-S(\omega)$, when the incident wave frequency has a magnitude close to that of its natural frequency. It may be presumed that the behavior manifest in (20.2) is typical for each natural frequency $\omega_0, \omega_1, \ldots$ of a scattering arrangement; and for confirmation in a particular case, let us consider a semi-infinite string with a fixed end and a passive load M at a finite distance therefrom. The characteristic frequency equation for a finite and isolated length l of string that is fixed at one end and attached to a load M at the other, viz.

$$\cot \frac{\omega l}{c} = \left(\frac{M}{\rho l}\right) \frac{\omega l}{c} \tag{20.3}$$

has infinitely many (positive and negative) roots; if

$$\varepsilon = \frac{\rho l}{M} \tag{20.4}$$

is small compared with unity, these are specified approximately by

$$\frac{\omega_0 l}{c} = \sqrt{\varepsilon}, \qquad \frac{\omega_n l}{c} = n\pi + \frac{\varepsilon}{n\pi}, \qquad n = \pm 1, \pm 2, \ldots, \qquad \varepsilon \to 0. \tag{20.5}$$

When the segment $(-l, 0)$ is joined to an infinite section $(0, \infty)$ of string with the same density, and there is an incoming time-periodic wave in the latter, we adopt the representation

$$s(x) = \exp\left(-\mathrm{i}\frac{\omega}{c}x\right) - \exp\left(\mathrm{i}\frac{\omega}{c}(x+2l)\right) + s_1(x), \qquad x \geq -l \tag{20.6}$$

after leaving out the complex factor $\exp(-\mathrm{i}\omega t)$. The component function $s_1(x)$ in (20.6), which satisfies the differential equation

$$\left(\frac{\mathrm{d}^2}{\mathrm{d}x^2} + \frac{\omega^2}{c^2}\right) s_1(x) = -\frac{M\omega^2}{P}(1 - \mathrm{e}^{2\mathrm{i}kl} + s_1(0))\delta(x) \tag{20.7}$$

and the supplementary conditions

$$s_1(-l) = 0, \qquad s_1(x) \sim \exp\left(\mathrm{i}\frac{\omega x}{c}\right), \quad x > 0, \tag{20.8}$$

may be expressed in the form

$$s_1(x) = \frac{M\omega^2}{P}(1 - \mathrm{e}^{2\mathrm{i}kl} + s_1(0))G(x, 0) \tag{20.9}$$

where

$$G(x, x') = \frac{c}{\omega} \sin \frac{\omega}{c}(x_< + l) \exp\left(\mathrm{i}\frac{\omega}{c}(x_> + l)\right), \qquad -l < x, \quad x' < \infty \tag{20.10}$$

($x_<, x_>$ the smaller/larger of x, x' respectively) is such that

$$\left(\frac{d^2}{dx^2}+\frac{\omega^2}{c^2}\right)G(x,x') = -\,\delta(x-x')$$
$$G(-l,x')=0, \qquad G(x,x')\sim \exp\left(i\frac{\omega x}{c}\right), \quad x>x'. \tag{20.11}$$

With the help of the determination that follows directly from (20.9),

$$s_1(0) = (1-e^{2ikl})\frac{(M\omega^2/P)G(0,0)}{1-(M\omega^2/P)G(0,0)},$$

we obtain, through (20.5), (20.8), an explicit representation for the overall wave function,

$$s(x) = \exp\left(-i\frac{\omega}{c}x\right) - \exp\left(i\frac{\omega}{c}(x+2l)\right) + (1-e^{2ikl})\frac{(M\omega^2/P)G(x,0)}{1-(M\omega^2/P)G(0,0)},$$
$$-l\le x<\infty. \tag{20.12}$$

On the range $x>0$, namely in the exterior of the segment $(-l,0)$ that may be viewed as a reflecting unit with the discrete frequency spectrum (20.5), we find

$$s(x) = \exp\left(-i\frac{\omega}{c}x\right) - S(\omega)\exp\left(i\frac{\omega}{c}(x+2l)\right), \quad x\ge 0 \tag{20.13}$$

where the coefficient factor $S(\omega)$ has the standard form

$$S(\omega) = \frac{f(-\omega)}{f(\omega)} \tag{20.1}$$

and is rendered explicit by the determination

$$f(\omega) = 1-\frac{\omega l}{\varepsilon c}\sin\frac{\omega l}{c}\exp\left(i\frac{\omega l}{c}\right). \tag{20.14}$$

Total reflection of the incoming wave at any (real) frequency ω is a consequence of the property

$$S(\omega)S^*(\omega) = |S(\omega)|^2 = 1 \tag{20.15}$$

and, inasmuch as $f(\omega)$ is an entire function of the complex variable ω with infinitely many zeros, $S(\omega)$ constitutes a meromorphic function. Estimates for the zeros of $f(\omega)$, i.e.,

$$\hat{\omega}_0 = \omega_0 - i\frac{\varepsilon c}{2l}, \quad \hat{\omega}_n = \omega_n - i\left(\frac{\varepsilon}{n\pi}\right)^2\frac{c}{l}, \quad n=\pm 1, \pm 2, \ldots, \quad \varepsilon\to 0, \tag{20.16}$$

associate their real parts with the proper frequencies ω_0, ω_n defined earlier and reveal that the imaginary parts are all negative; the corresponding residues of $S(\omega)$ turn out to be

$$R_0 = \operatorname{Res} S(\omega)|_{\omega=\hat{\omega}_0} = \quad -\mathrm{i}\varepsilon\frac{c}{l},$$

and (20.17)

$$R_n = \operatorname{Res} S(\omega)|_{\omega=\hat{\omega}_n} = -2\,\mathrm{i}\left(\frac{\varepsilon}{n\pi}\right)^2\frac{c}{l}, \quad n=\pm 1, \ldots .$$

Having regard for an arbitrarily large closed contour C_∞ upon which S remains bounded and equating the independent evaluations of an integral attached thereto, namely

$$I = \frac{1}{2\pi\,\mathrm{i}}\int_{C_\infty}\frac{S(\omega')}{\omega'-\omega}\,\mathrm{d}\omega' = S(\omega) - \sum_{n=-\infty}^{\infty}\frac{R_n}{\omega-\hat{\omega}_n}$$

and

$$I = \frac{1}{2\pi\,\mathrm{i}}\int_{C_\infty} S(\omega')\left[\frac{1}{\omega'}+\frac{\omega}{(\omega')^2}+\cdots\right]\mathrm{d}\omega' = S(0) + \sum_{n=-\infty}^{\infty}\frac{R_n}{\hat{\omega}_n},$$

there follows an exact representation of the Mittag–Leffler type,

$$S(\omega) = 1 + \sum_{n=-\infty}^{\infty} R_n\left[\frac{1}{\omega-\hat{\omega}_n}+\frac{1}{\hat{\omega}_n}\right] \tag{20.18}$$

and its approximate version

$$S(\omega) = 1 + \sum_{n=-\infty}^{\infty}\frac{R_n}{\omega-\hat{\omega}_n}, \quad \varepsilon\to 0. \tag{20.19}$$

The direct consequences of (20.19), namely that

$$S(\omega) \doteq \frac{\omega-\omega_0-\mathrm{i}\dfrac{\varepsilon c}{2l}}{\omega-\omega_0+\mathrm{i}\dfrac{\varepsilon c}{2l}}, \qquad \omega\to\omega_0$$

and (20.20)

$$S(\omega) \doteq \frac{\omega-\omega_n-\mathrm{i}\left(\dfrac{\varepsilon}{n\pi}\right)^2\dfrac{c}{l}}{\omega-\omega_n+\mathrm{i}\left(\dfrac{\varepsilon}{n\pi}\right)^2\dfrac{c}{l}}, \qquad \omega\to\omega_n$$

make evident a uniformity in the reflected wave features associated with the individual resonant frequencies ω_0, ω_n

Choosing an incident wave form which is sharply limited in extent, rather than frequency, viz.,

$$\begin{aligned} s^{\mathrm{inc}}(x,t) &= x_0\delta(x-x_0+ct), \quad x_0>0 \\ &= \frac{x_0}{2\pi c}\int_{-\infty}^{\infty}\exp\left\{-\mathrm{i}\omega\left(t+\frac{x-x_0}{c}\right)\right\}\mathrm{d}\omega, \end{aligned} \tag{20.21}$$

we obtain, by use of the latter synthesis and the reflection formula (20.13), the outgoing wave representation in $x>0$,

$$\begin{aligned} s^{\text{ref}}(x,t) &= -\frac{x_0}{2\pi c}\int_{-\infty}^{\infty} S(\omega)\exp\left\{-\mathrm{i}\omega\left(t-\frac{x+x_0+2l}{c}\right)\right\}\mathrm{d}\omega \\ &= -x_0\delta(x+x_0+2l-ct) \\ &\quad +\mathrm{i}\frac{x_0}{c}H(ct-x-x_0-2l)\sum_{n=-\infty}^{\infty} R_n\exp\left\{-\mathrm{i}\hat{\omega}_n\left(t-\frac{x+x_0+2l}{c}\right)\right\} \end{aligned} \tag{20.22}$$

after reliance on the expression (20.19) for $S(\omega)$. The first term in (20.22) corresponds to a direct reflection of the incident profile while the sum, containing the Heaviside factor

$$H(\xi)=\begin{cases}1, & \xi>0\\ 0, & \xi<0\end{cases}, \tag{20.23}$$

comprise terms that are linked with excitations on the section $(-l,0)$.

In the case of a monochromatic incident wave train, it follows from (20.12) that the response within the aforesaid section,

$$s(x)=A\sin\frac{\omega}{c}(x+l), \quad -l\le x\le 0 \tag{20.24}$$

has an amplitude factor

$$A=\frac{2\operatorname{cosec}\dfrac{\omega l}{c}}{1+\mathrm{i}\left(\cot\dfrac{\omega l}{c}-\dfrac{\omega l}{\varepsilon c}\right)} \tag{20.25}$$

which assumes large values at the frequencies given by the characteristic equation (20.3); the specific estimates for $\varepsilon\to 0$,

$$A\doteq 2/\sqrt{\varepsilon},\quad \omega=\omega_0;\quad A\doteq 2(-1)^n n\pi/\varepsilon,\quad \omega=\omega_n,$$

possess magnitudes considerably greater than that of the mass displacement at the resonant frequencies,

$$s(0)=2, \qquad \omega=\omega_0,\,\omega_n$$

which is merely twice that of the incident wave.

If the forced motion of the loaded string is attributable to a given displacement $\bar{A}\,\mathrm{e}^{-\mathrm{i}\omega t}$ at the extremity $x=-l$, the concomitant representation

$$s(x)=\bar{A}\exp\left\{\mathrm{i}\frac{\omega}{c}(x+l)\right\}+\frac{\bar{A}\,\mathrm{e}^{\mathrm{i}\omega l/c}(M\omega^2/P)G(x,0)}{1-(M\omega^2/P)G(0,0)}, \quad -l\le x<\infty \tag{20.26}$$

replaces a prior one, (20.12), both incorporating the same source function $G(x,x')$ defined in (20.11). The consequence of (20.26)

$$s(x)=\frac{\bar{A}\operatorname{cosec}\dfrac{\omega l}{c}\,e^{i\omega x/c}}{\cot\dfrac{\omega l}{c}-\dfrac{\omega l}{\varepsilon c}-i},\qquad x\geq 0 \tag{20.27}$$

leads, in turn, to the estimates

$$s(x)\doteq\frac{\bar{A}}{\sqrt{\varepsilon}}\frac{\exp(i\omega x/c)}{\dfrac{2l}{\varepsilon c}(\omega-\omega_0)+i},\qquad s(x)\doteq\bar{A}\frac{n\pi}{\varepsilon}(-1)^{n+1}\frac{\exp(i\omega x/c)}{\left(\dfrac{n\pi}{\varepsilon}\right)^2\dfrac{l}{c}(\omega-\omega_n)+i}$$

$$\omega\to\omega_0\qquad\qquad\qquad \omega\to\omega_n$$

which indicate a resonant amplification for the outgoing wave motions when the frequency ω approaches equality with any representatives of the characteristic set $\omega_0, \omega_1, \ldots$. A further deduction from (20.26),

$$s(x)=\bar{A}\operatorname{cosec}\frac{\omega l}{c}\,\frac{\exp(i\omega x/c)\dfrac{\omega l}{\varepsilon c}\sin\dfrac{\omega x}{c}}{\cot\dfrac{\omega l}{c}-\dfrac{\omega l}{\varepsilon c}-i},\qquad -l\leq x\leq 0 \tag{20.28}$$

can be recast in the form

$$s(x)=\bar{A}\,e^{i\omega l/c}\frac{\exp\left(i\omega\dfrac{x}{c}\right)-\gamma\exp\left(-i\omega\dfrac{x}{c}\right)}{1-\gamma\exp\left(2\,i\omega\dfrac{l}{c}\right)},\qquad \gamma=\frac{1}{1+2\,i\dfrac{\varepsilon c}{\omega l}} \tag{20.29}$$

thus suggesting a development

$$\begin{aligned}s(x)=\bar{A}\Big[&\exp\left\{i\frac{\omega}{c}(x+l)\right\}+\gamma\exp\left\{i\frac{\omega}{c}(x+3l)\right\}+\gamma^2\exp\left\{i\frac{\omega}{c}(x+5l)\right\}+\cdots\\&-\gamma\exp\left\{-i\frac{\omega}{c}(x-l)\right\}-\gamma^2\exp\left\{-i\frac{\omega}{c}(x-3l)\right\}-\cdots\Big],\qquad x\leq 0\end{aligned} \tag{20.30}$$

whose individual terms are linked to multiply reflected waves on the section $(-l, 0)$; transfer of energy (or 'leakage') into the sector $x>0$ accounts for an amplitude decrease of such waves that is more pronounced in the higher orders.

Problems I(a)

1. (a) Discuss the Cauchy problem for the one-dimensional wave equation,

$$\frac{\partial^2 y}{\partial x^2} = \frac{1}{c^2}\frac{\partial^2 y}{\partial t^2},$$

if the data, namely the values of y and of its normal derivative, are given on a characteristic $x - ct = \text{const.}$

(b) Find the wave function $y(x, t)$ in $t > 0$ if its values on the respective characteristics $x \mp ct = 0$ are specified by $\varphi(x)$, $x > 0$; $\psi(x)$, $x < 0$, with the interrelation $\varphi(0) = \psi(0)$.

2. When the particular assignments

$$y = f(x), \qquad \frac{\partial y}{\partial n} = 0$$

relative to a function $y(x, t)$ and its normal derivative on the line

$$ct = \alpha x, \quad 0 \leq \alpha < 1$$

are given, deduce that

$$y(x, t) = \frac{1}{2(1+\alpha^2)}\left\{(1-\alpha)^2 f\left(\frac{x-ct}{1-\alpha}\right) + (1+\alpha)^2 f\left(\frac{x+ct}{1+\alpha}\right)\right\}$$

furnishes the solution of a Cauchy problem for the homogeneous wave equation.

3. Let it be required to determine a solution of the homogeneous wave equation within a sector of the x, t plane that has the boundary lines

$$t = 0, \qquad x > 0,$$

$$t - \frac{x}{v} = 0, \qquad v < c,$$

with the respective stipulations

$$y(x, 0) = f(x), \qquad \frac{\partial}{\partial t} y(x, 0) = g(x), \qquad x \geq 0,$$

$$y|_{t=x/v} = h(x), \qquad x \geq 0$$

thereupon; assume that $f(0) = h(0)$ and obtain the condition for a smooth (i.e., continuously differentiable) solution throughout the sector which, evidently, contains the characteristic line $x - ct = 0$.

4. Inasmuch as the variables x, ct enters the wave equation $\partial^2 y/\partial x^2 =$

$\partial^2 y/\partial(ct)^2$ symmetrically, it may be concluded from the d'Alembert solution thereof that

$$y(x, t) = \frac{1}{2}[f(ct - x) + f(ct + x)] + \frac{1}{2c}\int_{ct-x}^{ct+x} F(\xi)\, d\xi \qquad (*)$$

constitutes a solution for all time t on the range $0 < x < \infty$, with the boundary values

$$y(0, t) = f(ct), \qquad \frac{\partial}{\partial x} y(0, t) = \frac{1}{c} F(ct), \qquad -\infty < t < \infty.$$

Does the above solution have an evident physical nature? What relationship between the functions f and F imparts to (*) an outgoing wave character?

5. Obtain a solution of the homogeneous wave equation which has null initial values

$$y(x, 0) = \frac{\partial}{\partial t} y(x, 0) = 0, \quad x > 0$$

on the considered interval $x > 0$ and a specified derivative

$$\frac{\partial}{\partial x} y(0, t) = F(t), \quad t > 0$$

at the endpoint $x = 0$. Find the corresponding solution if the latter condition is replaced by

$$\frac{\partial}{\partial x} y(0, t) - \gamma y(0, t) = F(t), \quad t > 0.$$

6. Let it be required to demonstrate the uniqueness of the solution to a Cauchy problem for the one-dimensional wave equation which is posed in terms of assigned values for y and $\partial y/\partial t$ on an arc Γ with continuous tangent dt/dx of magnitude less than $1/c$. Show that, when both y and $\partial y/\partial t$ have null values on Γ, the appertaining wave function $y(x, t)$ (with continuous second order partial derivatives) vanishes identically in the triangle formed by Γ and two characteristics that intersect at a point whose time coordinate exceeds

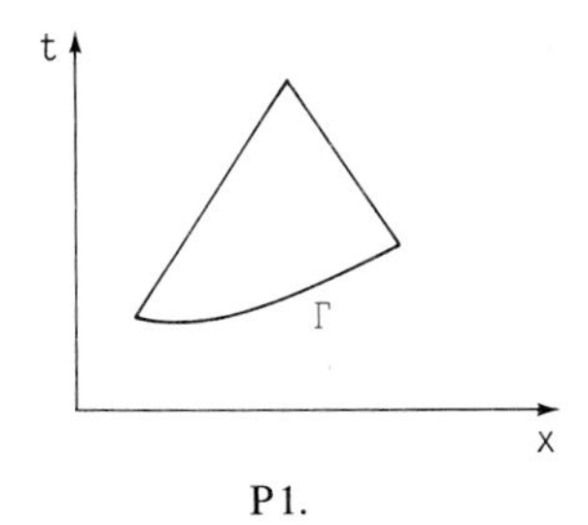

P1.

any of those on Γ (see Fig. P1). [Hint: Consider the curvilinear integral

$$\int\left[\left\{\frac{1}{2}\left(\frac{\partial y}{\partial ct}\right)^2+\frac{1}{2}\left(\frac{\partial y}{\partial x}\right)^2\right\}dx+2\frac{\partial y}{\partial t}\frac{\partial y}{\partial x}dt\right]$$

extended along the aforesaid triangle.]

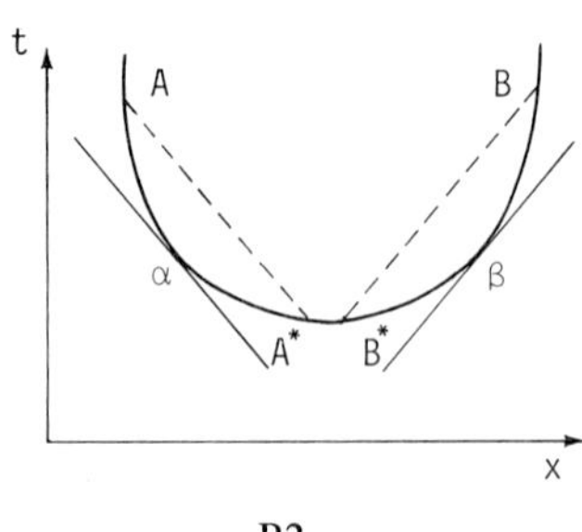

P2.

7. Consider a Cauchy problem wherein characteristic lines are tangent to the arc joining A, B at the points α, β (see Fig. P2); show that the first partial derivatives of a wave function satisfy the interrelations

$$\left(\frac{\partial y}{\partial x}\right)_A+\frac{1}{c}\left(\frac{\partial y}{\partial t}\right)_A=\left(\frac{\partial y}{\partial x}\right)_{A^*}+\frac{1}{c}\left(\frac{\partial y}{\partial t}\right)_{A^*}$$

$$\left(\frac{\partial y}{\partial x}\right)_B-\frac{1}{c}\left(\frac{\partial y}{\partial t}\right)_B=\left(\frac{\partial y}{\partial x}\right)_{B^*}-\frac{1}{c}\left(\frac{\partial y}{\partial t}\right)_{B^*}$$

and cannot therefore be assigned arbitrarily. If the points A^*, B^* are coincident, justify the consequent representations

$$2\left(\frac{\partial y}{\partial x}\right)_{A^*}=\left(\frac{\partial y}{\partial x}\right)_A+\left(\frac{\partial y}{\partial x}\right)_B+\frac{1}{c}\left[\left(\frac{\partial y}{\partial t}\right)_A-\left(\frac{\partial y}{\partial t}\right)_B\right],$$

$$2\left(\frac{\partial y}{\partial t}\right)_{A^*}=c\left[\left(\frac{\partial y}{\partial x}\right)_A-\left(\frac{\partial y}{\partial x}\right)_B\right]+\left(\frac{\partial y}{\partial t}\right)_A+\left(\frac{\partial y}{\partial t}\right)_B$$

on the basis of Riemann's solution for a wave function in the domain A^*BA.

8. If a localized impulse $I=\mathfrak{F}\Delta\tau$ is delivered to a section $\Delta\xi$ of string during the infinitesimal interval of time $\Delta\tau$, the resultant change of velocity

$$\left(\frac{\partial y(\xi,t)}{\partial t}\right)_{\tau+0}-\left(\frac{\partial y(\xi,t)}{\partial t}\right)_{\tau-0}=\frac{I}{\rho\Delta\xi}=\frac{\mathfrak{F}}{\rho}\frac{\Delta\tau}{\Delta\xi} \tag{*}$$

signifies a discontinuous behavior for $\partial y/\partial t$ at points of the locus $\Gamma'\{\xi=\xi(\tau)\}$ which corresponds to a particular motion of the impulse along the string. A like behavior of the coordinate derivative, $\partial y/\partial x$, is thus implied by the force equation

$$\mathfrak{F} = -P\left[\left(\frac{\partial y(x,\tau)}{\partial x}\right)_{\xi-0} - \left(\frac{\partial y(x,\tau)}{\partial x}\right)_{\xi+0}\right]$$

which, together with (*), establishes the relation

$$\left[\left(\frac{\partial y(\xi,t)}{\partial t}\right)_{\tau+0} - \left(\frac{\partial y(\xi,t)}{\partial t}\right)_{\tau-0}\right] = -\frac{c^2}{\mathrm{d}\xi/\mathrm{d}\tau}\left[\left(\frac{\partial y(x,\tau)}{\partial x}\right)_{\xi-0} - \left(\frac{\partial y(x,\tau)}{\partial x}\right)_{\xi+0}\right]$$

on Γ'. Having regard for these discontinuities in both first order partial derivatives of y, discuss the solution of the Cauchy problem at a point P, given that PA, PB are portions of characteristic lines and that the values of $\partial y/\partial x$ and $\partial y/\partial t$ are assigned on the arc Γ between A and B (see Fig. P3).

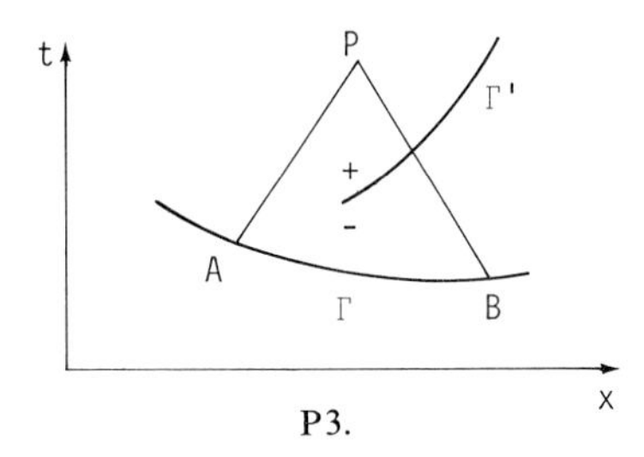

P3.

9. Discuss the mixed, initial and boundary value, problem for the modified wave equation

$$\frac{\partial^2 y}{\partial x^2} - \frac{1}{c^2}\frac{\partial^2 y}{\partial t^2} - h^2 y = 0,$$

assuming that $y(x,0)$ and $\partial y(x,0)/\partial t$ vanish identically on the range of positive x, while $\partial y(0,t)/\partial x = F(t), t > 0$ is a prescribed function of time.

10. Let the solution to an initial value problem for the preceding differential equation be displayed in the form

$$\begin{aligned} y(x,t) &= \frac{1}{2}\int_{-\infty}^{\infty}\left[a(k) - \frac{b(k)}{\mathrm{i}c\surd(k^2+h^2)}\right]\exp\{\mathrm{i}(kx - \surd(k^2+h^2)ct)\}\,\mathrm{d}k \\ &\quad + \frac{1}{2}\int_{-\infty}^{\infty}\left[a(k) + \frac{b(k)}{\mathrm{i}c\surd(k^2+h^2)}\right]\exp\{\mathrm{i}(kx + \surd(k^2+h^2)ct)\}\,\mathrm{d}k \\ &= y_-(x,t) + y_+(x,t), \end{aligned}$$

say, where

$$2\pi(k) = \int_{-\infty}^{\infty} y(x,0)\,\mathrm{e}^{-ikx}\,\mathrm{d}x, \qquad 2\pi b(k) = \int_{-\infty}^{\infty}\frac{\partial y(x,0)}{\partial t}\,\mathrm{e}^{-ikx}\,\mathrm{d}x$$

and the integration contour in the k-plane passes above a branch cut joining $k = -\mathrm{i}h$ and $k = \mathrm{i}h$. Supposing the square integrable functions $y(x,0)$, $\partial y(x,0)/\partial t$ to be identically zero in $x > 0$, show that $y_-(x,t)$ and $y_+(x,t)$ are null functions, respectively, when $x > ct$ and $x > -ct$; what interpretation of these results can be given?

11. Transform the damped wave equation (5.10) in accordance with the substitution

$$y(x, t) = e^{-\varepsilon t} s(x, t)$$

and deduce that the solution for $x, t \geq 0$ which satisfies the conditions

$$y = \frac{\partial y}{\partial t} = 0, \quad t = 0; \qquad y = F(t), \quad x = 0$$

is expressed by

$$y(x, t) = F\left(t - \frac{x}{c}\right) e^{-\varepsilon x/c} + i\varepsilon x \int_0^{t - x/c} \frac{J_0'[i\varepsilon \sqrt{((t-\tau)^2 - x^2/c^2)}]}{\sqrt{(c^2(t-\tau)^2 - x^2)}} F(\tau)\, e^{-\varepsilon(t-\tau)}\, d\tau$$

when $x < ct$ and vanishes otherwise, the prime being meant to indicate an argument derivative.

12. Verify that a solution of the damped wave equation

$$c^2 \frac{\partial^2 y}{\partial x^2} = \frac{\partial^2 y}{\partial t^2} + 2\kappa c \frac{\partial y}{\partial t} \qquad (*)$$

which complies with the localized initial conditions

$$y(x, 0) = A\delta(x), \qquad \frac{\partial y(x, 0)}{\partial t} = 0, \quad -\infty < x < \infty$$

has non-vanishing values if $ct > |x|$, and takes the form

$$y(x, t) = \frac{A}{2} e^{-\kappa ct} \Big[\delta(|x| - ct) + \kappa I_0(\kappa \sqrt{(c^2 t^2 - x^2)}) + \frac{\kappa ct}{\sqrt{(c^2 t^2 - x^2)}} I_1(\kappa \sqrt{(c^2 t^2 - x^2)}) \Big] \qquad (**)$$

where I_0, I_1 designate modified Bessel functions, viz.

$$J_0(iz) = I_0(z), \qquad J_1(iz) = iI_1(z).$$

Utilize a large argument or asymptotic approximation of these functions and establish that the consequent version of (**),

$$y(x, t) \sim A \left(\frac{\kappa}{2\pi ct} \right)^{1/2} e^{-(\kappa x^2/2ct)}, \quad ct \gg x$$

represents a solution of the diffusion equation obtained from (*) by omitting the first term on the right-hand side thereof.

13. Show that a solution of the inhomogeneous wave equation

$$\frac{\partial^2 y}{\partial t^2} - c^2 \frac{\partial^2 y}{\partial x^2} = q(x, t)$$

which vanishes at $x = 0$ and $x = 1$ for all positive values of the time may be characterized through the integro-differential equation

$$y(x, t) + \frac{1}{c^2}\frac{\partial^2}{\partial t^2}\int_0^1 K(x, \xi)y(\xi, t)\,\mathrm{d}\xi = Q(x, t)$$

where

$$K(x, \xi) = \tfrac{1}{2}(x + \xi) - \tfrac{1}{2}|x - \xi| - x\xi,$$

$$Q(x, t) = \frac{1}{c^2}\int_0^1 K(x, \xi)q(\xi, t)\,\mathrm{d}\xi.$$

14. Suppose that a homogeneous string under considerable tension slides at constant velocity v over a pair of smooth rollers which are set a distance L apart. Establish a differential equation for the small amplitude transverse motions of the string between the rollers and examine, in particular, the modes of harmonic time variation with a view to characterizing the longest period motion.

15. Consider the differential equation

$$\frac{\partial^2 y}{\partial t^2} + 2\varepsilon\frac{\partial y}{\partial t} - c^2\frac{\partial^2 y}{\partial x^2} = F(t), \quad -\frac{l}{2} < x < \frac{l}{2}, \quad t > 0 \tag{*}$$

relevant to damped vibrations of a string in a homogeneous field of force, and suppose that the boundary conditions $y(\pm l/2, t) = 0$ prevail. Having regard for the function

$$S(x) = \frac{4}{\pi}\sum_{n=1,3,\ldots}^{\infty}\frac{(-1)^{(n+3)/2}}{n}\cos\frac{n\pi x}{l},$$

whose value equals unity on the interval $-(l/2) < x < (l/2)$, obtain the details of a solution to (*),

$$y(x, t) = \sum_{n=1,3,\ldots}^{\infty} f_n(t)\cos\frac{n\pi x}{l},$$

which, together with its first order time derivative, vanishes when $t = 0$. If $F(t) = F$ (a constant), verify the representation

$$y(0, t) \equiv Y(t) = \frac{4F}{\pi}\sum_{n=1,3,\ldots}^{\infty}\frac{(-1)^{(n+3)/2}}{n\left(\left(\frac{n\pi c}{l}\right)^2 - \varepsilon^2\right)^{1/2}}\int_0^t \mathrm{e}^{-\varepsilon\tau}\sin\left(\left(\left(\frac{n\pi c}{l}\right)^2 - \varepsilon^2\right)^{1/2}\tau\right)\mathrm{d}\tau$$

for the displacement at the center of the string and show that

$$\begin{aligned}\frac{\mathrm{d}^2 Y}{\mathrm{d}t^2} = \frac{4F}{\pi}\mathrm{e}^{-\varepsilon t}\Bigg\{ &-\varepsilon\int_0^t \mathrm{d}\tau \sum_{n=1,3,\ldots}^{\infty}\frac{(-1)^{(n+3)/2}}{n}\cos\left(\left(\left(\frac{n\pi c}{l}\right)^2 - \varepsilon^2\right)^{1/2}\tau\right)\\ &+ \sum_{n=1,3,\ldots}^{\infty}\frac{(-1)^{(n+3)/2}}{n}\left[\cos\left(\left(\left(\frac{n\pi c}{l}\right)^2 - \varepsilon^2\right)^{1/2}t\right) - \cos\frac{n\pi ct}{l}\right]\\ &+ \sum_{n=1,3,\ldots}^{\infty}\frac{(-1)^{(n+3)/2}}{n}\cos\frac{n\pi ct}{l}\Bigg\}\end{aligned}$$

wherein the first two terms characterize continuous functions of t and the last term implies a discontinuous behavior for $\mathrm{d}^2Y/\mathrm{d}t^2$ at $t = t^* = l/2c$, viz.

$$\frac{\mathrm{d}^2 Y}{\mathrm{d}t^2}\Bigg|_{t=t^*-0}^{t=t^*+0} = -2F\,\mathrm{e}^{-\varepsilon t^*}.$$

Prove that

$$Y(t) = \frac{F}{4\varepsilon^2}(\mathrm{e}^{-2\varepsilon t} - 1 + 2\varepsilon t), \quad 0 < t < t^*$$

and interpret the consequent results

$$\frac{\mathrm{d}^2 Y}{\mathrm{d}t^2} = F\,\mathrm{e}^{-2\varepsilon t^*} > 0, \quad t = t^* - 0$$

$$\frac{\mathrm{d}^2 Y}{\mathrm{d}t^2} = F\,\mathrm{e}^{-2\varepsilon t^*} - 2F\,\mathrm{e}^{-\varepsilon t^*}$$

$$= F\,\mathrm{e}^{-2\varepsilon t^*}(1 - 2\,\mathrm{e}^{\varepsilon t^*}) < 0, \qquad t = t^* + 0$$

with reference to a curve which describes the motion of the center of the string.

16. Let us suppose a taut string of mass density ρ, located in a magnetic field H and carrying a current I, to form part of a circuit with resistance R and self-inductance L; and adopt the coupled dynamical and circuit equations

$$\frac{\partial^2 y}{\partial t^2} + 2\varepsilon\frac{\partial y}{\partial t} - c^2\frac{\partial^2 y}{\partial x^2} = \frac{HI}{\rho}, \quad c^2 = P/\rho$$

$$E = RI + L\frac{\mathrm{d}I}{\mathrm{d}t} + H\int_0^l \frac{\partial y}{\partial t}\,\mathrm{d}x \tag{*}$$

where y, ε denote the displacement and damping constant of the string and $E(t)$ accounts for an external electromotive force impressed on the circuit. Giving attention to free vibrations ($E(t) = 0$) and utilizing the representations

$$y(x, t) = s(x)\,\mathrm{e}^{-\mathrm{i}\omega t}, \quad I(t) = J\,\mathrm{e}^{-\mathrm{i}\omega t}$$

derive, for a string with fixed endpoints ($x = 0, l$) the complex frequency equation

$$\tilde{\omega}^2 \sin\frac{\tilde{\omega}l}{c} + \mu\left\{l \sin\frac{\tilde{\omega}l}{c} - \frac{2c}{\tilde{\omega}}\left(1 - \cos\frac{\tilde{\omega}l}{c}\right)\right\} = 0 \tag{**}$$

wherein

$$\tilde{\omega}^2 = \omega^2 + 2\,\mathrm{i}\varepsilon\omega, \qquad \mu = \frac{\mathrm{i}\omega H^2}{\rho(R - \mathrm{i}\omega L)}.$$

Confirm that (**) admits a sequence of proper values for $\tilde{\omega}$, namely

$$\tilde{\omega}^2 = \omega^2 + 2\,\mathrm{i}\varepsilon\omega = \left(\frac{2n\pi c}{l}\right)^2, \quad n = 1, 2, \ldots,$$

which are independent of μ (or H) and characterize such modes of vibration.
Assuming R to be of great magnitude, establish the approximate determinations compatible with (**),

$$\tilde{\omega} \doteq (2m+1)\frac{\pi c}{l} - 4\,\mathrm{i}\frac{l^2 H^2}{\pi^3 \rho c R}\frac{\omega_m}{(2m+1)^3},$$

$$\text{where } \left[\omega_m^2 + 2\,\mathrm{i}\varepsilon\omega_m = \left(\frac{2m+1}{l}\pi c\right)^2\right], \quad m = 0, 1, \ldots,$$

$$= (2m+1)\frac{\pi c}{l} - 4\,\mathrm{i}\frac{H^2 l}{\pi^2 \rho R}\frac{1}{(2m+1)^2}, \quad \varepsilon = 0, \tag{***}$$

and explain why the concomitant damping factor of the string motions,

$$\exp\left(-4\frac{H^2 l t}{\pi^2 \rho R (2m+1)^2}\right)$$

is less significant when m takes large values. Show that, if the resistance is very slight,

$$\tilde{\omega} \doteq \omega_n - 2\,\mathrm{i}\frac{\rho c R}{l^2 H^2},$$

where ω_n designates a root of the transcendental equation

$$\frac{2c}{\omega l}\tan\frac{\omega l}{2c} = 1$$

and remark on an aspect of the damping factor which contrasts with the prior estimate.
Employ a series representation for the string displacement,

$$y(x,t) = \sum_{n=1}^{\infty} \varphi_n(t)\sin\frac{n\pi x}{l},$$

whose individual terms satisfy the boundary conditions, and, pursuant to the assumptions

$$\varphi_n(t) = c_n\,\mathrm{e}^{-\mathrm{i}\omega t}, \qquad I(t) = J\,\mathrm{e}^{-\mathrm{i}\omega t},$$

obtain a frequency equation, on the basis of the system (*), in case $\varepsilon = L = 0$; compare the latter with (**).

17. Derive, in connection with the preceding configuration, an expansion of the string displacement, appropriate when R is large and (**) applies, viz.

$y(x, t) =$

$$\sum_{n=1,3,\ldots}^{\infty} \exp\left\{-2\alpha_n \frac{t}{l}\right\}\left[A_n\left\{\frac{\alpha_n}{R}\left(1-\cos\frac{n\pi x}{l}\right)\cos\frac{n\pi ct}{l}+\sin\frac{n\pi x}{l}\sin\frac{n\pi ct}{l}\right\}\right.$$
$$\left.+B_n\left\{\frac{\alpha_n}{R}\left(1-\cos\frac{n\pi x}{l}\right)\sin\frac{n\pi ct}{l}-\sin\frac{n\pi x}{l}\cos\frac{n\pi ct}{l}\right\}\right]$$

where $\alpha_n = 2H^2l^2/\rho Rn^2\pi^2$; and discuss the matter of specifying the coefficients A_n, B_n through assignments for y and its time derivative at $t = 0$.

18. If the wave equation

$$\frac{\partial^2\psi}{\partial x^2} - \frac{1}{c^2}\frac{\partial^2\psi}{\partial t^2} = 0$$

holds in any coordinate system, disturbances governed thereby propagate with a common speed c. Such is not the case when the systems are in relative motion at the speed v and the Galilean transformations apply, viz., $x' = x - vt$, $t' = t$, inasmuch as

$$\frac{\partial^2\psi}{\partial x^2} - \frac{1}{c^2}\frac{\partial^2\psi}{\partial t^2} = \frac{\partial^2\psi}{\partial x'^2} - \frac{1}{c^2}\left(\frac{\partial}{\partial t'} - v\frac{\partial}{\partial x'}\right)^2\psi$$

and thus a convected wave equation obtains in the system (x', t'). Let it be intended to characterize transformations between coordinate systems $S:(x, t)$ and $S':(x', t')$ which preserve form invariance of the wave equation. Assume that the origins of S and S' coincide when $t = t' = 0$, and that the origin of S' is located at $x = vt$, $t > 0$, relative to S; thus

$$x' = f(x - vt), \qquad f(0) = 0. \tag{*}$$

Combining (*) and the reciprocal formula

$$x = f(x' + vt)$$

involving the same function f (whose inverse is denoted by f^{-1}),

$$t' = \frac{1}{v}\{f^{-1}(x) - f(x - vt')\} = F(x, t) \tag{**}$$

where F and t are identical in the limit $v \to 0$. On the basis of the (formal) coordinate/time interrelations (*), (**) and the postulate that the wave equation manifests the same appearance in terms of (x, t) or (x', t'), deduce the explicit representation

$$f = \left(1 - \frac{v^2}{c^2}\right)^{-1/2}(x - vt)$$

and the Lorentz transformation expressions

$$x' = \left(1 - \frac{v^2}{c^2}\right)^{-1/2} (x - vt), \qquad t' = \left(1 - \frac{v^2}{c^2}\right)^{-1/2} \left(t - \frac{v}{c^2} x\right).$$

19. Provide arguments in support of the coupled equations for small transverse motions of a non-uniform string, namely

$$\rho(x) \frac{\partial \xi}{\partial t} - \frac{\partial \eta}{\partial x} = 0, \qquad \frac{\partial \eta}{\partial t} - P \frac{\partial \xi}{\partial x} = 0,$$

where $\rho(x)$ and P refer to the mass density and (constant) tension, while $\xi(x, t)$, $\eta(x, t)$ are the transverse components of velocity and tension. Find the equation of the characteristic curves and discuss the aspect of uniqueness for the solution to a problem in which ξ, η are prescribed at $t = 0$ $(x \geq 0)$ and ξ is also given at $x = 0$ $(t \geq 0)$.

20. Establish, and express in compact matrix notation, the first order systems of partial differential equations which are substitutes for the individual equations

$$\frac{1}{c^2} \frac{\partial^2 y}{\partial t^2} - \frac{\partial^2 y}{\partial x^2} \pm h^2 y = 0,$$

assigning a unit coefficient to the respective time derivatives.

21. The telegrapher's equations, suitable for low frequency excitations on coaxial or parallel two-wire lines, have the matrix representation

$$\begin{pmatrix} C & 0 \\ 0 & L \end{pmatrix} \begin{pmatrix} \partial V / \partial t \\ \partial I / \partial t \end{pmatrix} + \begin{pmatrix} 0 & 1 \\ 1 & 0 \end{pmatrix} \begin{pmatrix} \partial V / \partial x \\ \partial I / \partial x \end{pmatrix} + \begin{pmatrix} G & 0 \\ 0 & R \end{pmatrix} \begin{pmatrix} V \\ I \end{pmatrix} = 0$$

where $V(x, t), I(x, t)$ designate the (transverse) voltage and (longitudinal) currents, while C, L, R, and G specify the uniform shunt capacitance, self-inductance, resistance and shunt conductance per unit length of the configuration. Recast this coupled system of equations in the normal form

$$\begin{pmatrix} \partial V / \partial t \\ \partial I / \partial t \end{pmatrix} + M_1 \begin{pmatrix} \partial V / \partial x \\ \partial I / \partial x \end{pmatrix} + M_2 \begin{pmatrix} V \\ I \end{pmatrix} = 0,$$

obtaining explicit determinations of the matrices M_1, M_2, and derive the relation

$$\frac{\partial}{\partial t} \left(\frac{1}{2} C V^2 + \frac{1}{2} L I^2 \right) + \frac{\partial}{\partial x} (V I) + G V^2 + R I^2 = 0.$$

22. If the resistance and conductance of the aforesaid lines be negligible, the simpler system

$$L \frac{\partial I}{\partial t} = - \frac{\partial V}{\partial x}, \qquad C \frac{\partial V}{\partial t} = - \frac{\partial I}{\partial x}$$

implies that voltage and current obey the same equation

$$\left(\frac{\partial^2}{\partial x^2} - LC\frac{\partial^2}{\partial t^2}\right)\binom{V}{I} = 0$$

and hence possess the travelling wave representations

$$I = f(x-ct) + g(x+ct), \qquad c = (LC)^{-1/2}$$
$$V = Z(f(x-ct) - g(x+ct)), \qquad Z = (L/C)^{1/2}.$$

Suppose that a shunt inductance $\hat{L}$ is inserted at $x = 0$, with a consequent discontinuity in voltage there which conforms to the relationship $V(0+,t) - V(0-,t) + \hat{L}(\partial I(0,t)/\partial t) = 0$; discuss the reflection of both arbitrary and time-periodic wave profiles on an infinite line loaded in such fashion and afterwards extend the analysis so as to allow for distinct parameters L, C on the sections $x \gtrless 0$.

23. Show that voltage and current individually satisfy a second order differential equation

$$\left(LC\frac{\partial^2}{\partial t^2} + [RC + GL]\frac{\partial}{\partial t} + RG - \frac{\partial^2}{\partial x^2}\right)\binom{V}{I} = 0 \qquad (*)$$

if resistance and conductance of the line be taken into account; and are, moreover, separately derivable from a function Φ, which also satisfies (*), in accordance with the expressions

$$V(x,t) = -\frac{\partial \Phi}{\partial x}, \qquad I(x,t) = C\frac{\partial \Phi}{\partial t} + G\Phi.$$

Adopting the product representation

$$\Phi(x,t) = e^{-\alpha t/2}\varphi(x,t), \qquad \alpha = \frac{R}{L} + \frac{G}{C}$$

establish the equation

$$\frac{\partial^2 \varphi}{\partial t^2} - c^2\frac{\partial^2 \varphi}{\partial x^2} - h^2\varphi = 0, \qquad h^2 = \frac{1}{4}\left(\frac{R}{L} - \frac{G}{C}\right)^2$$

and the existence of solutions periodic in x, which describe travelling waves if $h < kc$, where k is the wave number. Study an initial value problem for φ, wherein

$$\varphi(x,0) = f(x), \qquad \frac{\partial \varphi(x,0)}{\partial t} = g(x)$$

and determine (i) the behavior of $\varphi(x,t)$, $\Phi(x,t)$ in the limit $t \to \infty$, (ii) the range of influence $[\varphi(x,t) \neq 0]$ in case of initial data with a limited span, i.e., $f, g \neq 0$, $|x| < l$.

24. Derive the Fourier integral solution to an initial value problem of the telegrapher's equation for the voltage, namely

$$V(x,t)=\frac{1}{2\pi}\int_{-\infty}^{\infty}\frac{-\omega_2 V(k)+\mathrm{i}\dot{V}(k)}{\omega_1-\omega_2}\exp\{\mathrm{i}(kx-\omega_1(k)t)\}\,\mathrm{d}k$$
$$+\frac{1}{2\pi}\int_{-\infty}^{\infty}\frac{\omega_1 V(k)-\mathrm{i}\dot{V}(k)}{\omega_1-\omega_2}\exp\{\mathrm{i}(kx-\omega_2(k)t)\}\,\mathrm{d}k,$$

where

$$V(k)=\int_{-\infty}^{\infty}V(x,0)\,\mathrm{e}^{-\mathrm{i}kx}\,\mathrm{d}x,\qquad \dot{V}(k)=\int_{-\infty}^{\infty}\frac{\partial V(x,0)}{\partial t}\,\mathrm{e}^{-\mathrm{i}kx}\,\mathrm{d}x$$

and ω_1, ω_2 designate the roots of the quadratic equation

$$\omega^2+\mathrm{i}\alpha\omega-k^2c^2-\left(\frac{\alpha^2}{4}-h^2\right)=0,$$

carrying over the prior specifications for α, h^2. What inferences can be drawn, as regards these integrals, on the basis of the asymptotic estimates

$$\omega_{1,2}=\mp kc-\mathrm{i}\frac{\alpha}{2}+0\left(\frac{1}{k}\right),\qquad |k|\to\infty$$

and the particular assignments

$$V(x,0)=\begin{cases}\mathrm{e}^{\varepsilon x}, & x<0\\ 0, & x>0\end{cases};\qquad \frac{\partial V(x,0)}{\partial t}=0,\qquad -\infty<x<\infty?$$

25. Investigate, with the aim of achieving a solution of the differential equation

$$\frac{\partial^2 y}{\partial t^2}+h^2\frac{\partial y}{\partial t}-v^2\frac{\partial^2 y}{\partial x^2}=0$$

that reduces to a given function, $f(t)$, at $x=0$, the superposition of particular solutions, viz.:

$$y(x,t)=\mathrm{e}^{-h^2t/2}\,\mathrm{e}^{\mathrm{i}\omega(t-(x/c))},\qquad c^2=\frac{\omega^2v^2}{\omega^2+\frac{1}{4}h^4}$$

which has the form

$$y(x,t)=\mathrm{e}^{-h^2t/2}\int_{-\infty}^{\infty}A(\omega)\,\mathrm{e}^{\mathrm{i}\omega(t-(x/c))}\,\mathrm{d}\omega.$$

§21. *General Excitation of String and Oscillator*

It is instructive to enlarge upon the prior analysis (§18) of excitations for a string with an attached oscillator by admitting any stipulated initial displacement and velocity of the whole system. If we locate the oscillator (with mass M) at one extremity ($x = 0$) of a semi-infinite string, our objective is that of ascertaining the solution of the wave equation

$$\frac{\partial^2 y}{\partial x^2} - \frac{1}{c^2}\frac{\partial^2 y}{\partial t^2} = 0 \tag{21.1}$$

for $x > 0$ say, which complies with the boundary condition

$$M\left(\frac{\partial^2}{\partial t^2} + \omega_0^2\right) y(0, t) = P\frac{\partial}{\partial x} y(0, t) \tag{21.2}$$

and the initial conditions

$$y(x, 0) = s_0(x), \qquad \frac{\partial y(x, 0)}{\partial t} = v_0(x), \quad x > 0. \tag{21.3}$$

The displacement profile $y(x, t)$ evidently has a continuous nature relative to both of its argument variables and the particular function of time,

$$Y(t) = y(0, t) = \lim_{x \to 0+} y(x, t),$$

with the initial value

$$Y(0) = s_0(0) = s_0$$

specifies the instantaneous position of the oscillator.

Utilizing the integral of (21.1) which constitutes a superposition of oppositely directed wave forms,

$$y(x, t) = f(x - ct) + g(x + ct), \tag{21.4}$$

the twin initial conditions (21.3) yield

$$\left.\begin{aligned} f(x) &= \frac{1}{2} s_0(x) - \frac{1}{2c}\int_0^x v_0(\xi)\, d\xi, \\ g(x) &= \frac{1}{2} s_0(x) + \frac{1}{2c}\int_0^x v_0(\xi)\, d\xi, \end{aligned}\right\} \quad x > 0 \tag{21.5}$$

that is, explicit determination of the functions f, g for positive values of their arguments. The second term in (21.4) is thus specified, and it remains to characterize the function

$$f_-(x) \overset{\text{def}}{=} f(-x), \quad x > 0;$$

for this purpose we impose the boundary condition (21.2) on the representation (21.4), obtaining thereby an inhomogeneous second-order differential equation

$$\left(\frac{d^2}{dx^2}+\frac{P}{Mc^2}\frac{d}{dx}+\frac{\omega_0^2}{c^2}\right)f_-(x)=-\left(\frac{d^2}{dx^2}-\frac{P}{Mc^2}\frac{d}{dx}+\frac{\omega_0^2}{c^2}\right)g(x), \quad x>0 \quad (21.7)$$

with a known right-hand member.

Inasmuch as

$$Y(t)=f_-(ct)+g(ct) \quad (21.8)$$

the elimination of f_- between (21.7), (21.8) provides an equation for the oscillator displacement, viz:

$$\begin{aligned}\left(\frac{d^2}{dt^2}+\frac{P}{Mc}\frac{d}{dt}+\omega_0^2\right)Y(t)&=2\frac{P}{M}g'(ct)\\&=\frac{P}{M}\left[s_0'(ct)+\frac{1}{c}v_0(ct)\right],\end{aligned} \quad (21.9)$$

where the damping and external force terms $(P/Mc)(dY/dt)$ and $2(P/M)g'(ct)$, respectively, convey the effect of the coupling between the oscillator and the string. Once a solution of (21.9) is achieved, subject to the initial conditions $Y(0)=s_0$, $dY(0)/dt=v_0(0)=v_0$, the concomitant string displacement readily obtains; thus, if we employ the decomposition

$$y(x,t)=y_1(x,t)+y_2(x,t)$$

along with the specifications

$$\begin{aligned}&y_1(x,0)=s_0(x), && y_2(x,0)=0, && x>0,\\&\partial y_1(x,0)/\partial t=v_0(x), && \partial y_2(x,0)/\partial t=0, && x>0,\\&y_1(0,t)=0, && y_2(0,t)=Y(t), && t>0,\end{aligned}$$

it follows that the respective wave functions $y_1(x,t)$, $y_2(x,t)$ take the forms

$$y_1(x,t)=\begin{cases}\frac{1}{2}[s_0(x-ct)+s_0(x+ct)]+\frac{1}{2c}\int_{x-ct}^{x+ct}v_0(\xi)\,d\xi, & ct<x\\ \frac{1}{2}[s_0(ct+x)-s_0(ct-x)]+\frac{1}{2c}\int_{ct-x}^{ct+x}v_0(\xi)\,d\xi, & ct>x\end{cases}$$

and

$$y_2(x,t)=\begin{cases}Y\left(t-\frac{x}{c}\right), & ct>x\\ 0, & ct<x.\end{cases}$$

It may be directly verified that the expression

$$
\begin{aligned}
Y(t) = {} & \frac{s_0}{\omega_1 - \omega_2}[\omega_1 \, e^{-i\omega_1 t} - \omega_2 \, e^{-i\omega_2 t}] + \frac{iv_0}{\omega_1 - \omega_2}[e^{-i\omega_1 t} - e^{-i\omega_2 t}] \\
& - \frac{\omega_1 + \omega_2}{\omega_1 - \omega_2} \int_0^{ct} [e^{-i\omega_1 (t - (\xi/c))} - e^{-i\omega_2 (t-(\xi/c))}] \left[s_0'(\xi) + \frac{1}{c} v_0(\xi) \right] d\xi
\end{aligned}
\tag{21.10}
$$

meets the various stipulations laid down for the oscillator displacement, provided the quantities ω_1, ω_2 denote the complex eigenfrequencies characterized in (19.13), (19.14).

By way of applying these results, let us revert to the special circumstances contemplated earlier, wherein the string has no initial displacement or velocity and the excitation arises from an impulse delivered to the oscillator; then (21.10) simplifies to

$$
\begin{aligned}
Y(t) &= \frac{iv_0}{\omega_1 - \omega_2}[e^{-i\omega_1 t} - e^{-i\omega_2 t}] \\
&= \frac{v_0}{\left(\omega_0^2 - \left(\frac{\rho c}{2M}\right)^2\right)^{1/2}} e^{-\rho c t/2M} \sin\left\{ \left(\omega_0^2 - \left(\frac{\rho c}{2M}\right)^2\right)^{1/2} t \right\}
\end{aligned}
$$

and the appertaining displacement

$$
y(x, t) = \frac{iv_0}{\omega_1 - \omega_2}\{e^{-i\omega_1 (t - x/c)} - e^{-i\omega_2 (t-x/c)}\}, \quad ct > x
$$

features a particular combination of the wave functions

$$
e^{-i\omega_{1,2}(t - x/c)},
$$

as foreshadowed in (18.6), (18.7). It will be noted that, since $\text{Im}(\omega_{1,2}) < 0$, the latter functions remain bounded for all values of the time on the stipulated range $t > x/c$, although the coordinate factors $e^{i\omega_{1,2} x/c}$ are unbounded in the limit $x \to \infty$.

If the scale factor C in (18.6) is expressed in terms of the initial velocity v_0, i.e.,

$$
C = \frac{v_0}{\kappa c} = \frac{v_0}{\sqrt{(\omega_0^2 - P^2/M^2 c^2)}}
$$

and P is next reduced in value by 1/2, so that the oscillator experiences only a right-hand tensile force, we arrive directly at the displacement function (21.11).

§22. *A String with Two Attached Oscillators; Matched Inner and Outer Expansions*

Motions of the dynamical system formed with a long tense string and a pair of identical mass oscillators linked thereto by transversely positioned springs (see Fig. 15) are affected by a number of parameters; these include the wave speed along the string, $c = \sqrt{(P/\rho)}$, the common natural frequency, ω_0, and separation, l, of the oscillators as well as the elastic constant, K, of the individual springs which couple the masses to the string. On defining a wave length $\lambda_0 = 2\pi c/\omega_0$ associated with the natural frequency of the uncoupled oscillators, we may introduce the dimensionless ratio

$$\varepsilon = l/\lambda_0 \tag{22.1}$$

and, if K is a multiple, α, of the spring constant $M\omega_0^2$ for each oscillator, another dimensionless ratio

$$\gamma = \frac{Kl}{2P} = \frac{\alpha M\omega_0^2 l}{2\rho c^2} = 2\pi^2\alpha\varepsilon^2 M/\rho l \tag{22.2}$$

serves to compare the relative stiffness of connecting springs and the string. In the interest of simplicity it will be assumed that

$$\begin{gathered} \varepsilon \ll 1, \\ \alpha, \quad \gamma \ll \varepsilon, \end{gathered} \tag{22.3}$$

which signifies a weak coupling between the mechanical oscillator and the

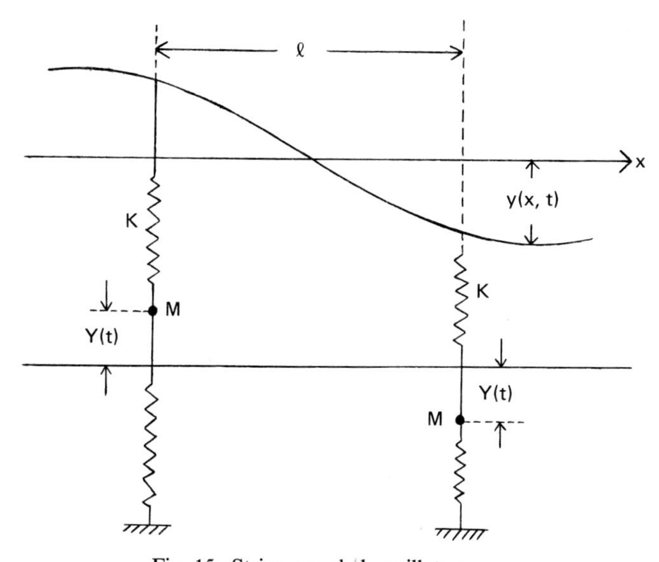

Fig. 15. String coupled oscillators.

string. The latter circumstance enables us to analyze, in approximate though informative fashion, the combined motion of the system by a sequential scheme; thus, we focus attention successively on the string and masses, commencing with a given motion of the latter and a determination of the consequent string displacement before proceeding to form new equations of mass motions which yield a refined determination of the string displacement, and so on. The execution of this program does not require an overall solution of the inhomogeneous, partial differential equation for the string displacement and is achieved, in fact, by matching expansions of the solution that obtain far from or in the vicinity of the masses, the so-called outer and inner expansions. A single oscillator displacement needs to be reckoned with if we limit our considerations to antisymmetrical modes of vibration, wherein the point of the string midway between the masses ($x = 0$) remains permanently fixed. Let $\pm Y(t)$ designate the corresponding displacements (from their equilibrium position) of the masses at $x = \pm l/2$, respectively, and suppose that attention is confined to the right half of the string where $x > 0$ and the displacement $y(x, t)$ satisfies the equation

$$\rho\frac{\partial^2 y}{\partial t^2} - P\frac{\partial^2 y}{\partial x^2} = K\left[Y(t) - y\left(\frac{l}{2}, t\right)\right]\delta\left(x - \frac{l}{2}\right) \tag{22.4}$$

along with the boundary condition

$$y(0, t) = 0, \quad t > 0. \tag{22.5}$$

If the parameters ω_0 and λ_0 or l are utilized for the purpose of defining independent variables with dimensionless nature, then

$$t^* = \omega_0 t \tag{22.6}$$

is an evident choice as regards the time, whereas the alternatives

$$x^* = 2\pi x/\lambda_0 \quad \text{and} \quad X^* = 2\pi x/l = x^*/\varepsilon \tag{22.7}$$

lend themselves to the specification of position coordinates. The latter, X^*, may be termed an inner variable since it is measured relative to a scale set by the oscillator separation, whereas x^* represents an outer variable whose scale is given by a particular wave length of string vibrations. When the wave equation (22.4) is recast in terms of outer variables (x^*, t^*) and we assume that $x^* \gg \varepsilon$ the right hand member disappears since the δ-function argument remains bounded away from zero; on this outer range, accordingly,

$$\frac{\partial^2 y}{\partial t^{*2}} - \frac{\partial^2 y}{\partial x^{*2}} = 0. \tag{22.8}$$

The version of (22.4) that obtains with the use of inner coordinates (X^*, t^*) is

$$\frac{\partial^2 y}{\partial X^{*2}} - \varepsilon^2\frac{\partial^2 y}{\partial t^{*2}} = -\frac{\gamma}{\pi}(Y - \bar{y})\delta(X^* - \pi) \tag{22.9}$$

where

$$\bar{y}(t^*) = y\left(\frac{l}{2}, \frac{t^*}{\omega_0}\right) \tag{22.10}$$

identifies the string displacement at the oscillator position.

On postulating the inner expansion

$$y(X^*, t^*) \sim f_1(X^*, t^*) + \varepsilon f_2(X^*, t^*) + \varepsilon^2 f_3(X^*, t^*) + \cdots \tag{22.11}$$

and collecting the terms of like order in ε which result after substitution in (22.9), we find that the equations

$$\frac{\partial^2 f_1}{\partial X^{*2}} = -\frac{\gamma}{\pi}(Y - \bar{y})\delta(X^* - \pi) \tag{22.12}$$

$$\frac{\partial^2 f_2}{\partial X^{*2}} = 0 \tag{22.13}$$

and

$$\frac{\partial^2 f_3}{\partial X^{*2}} = \frac{\partial^2 f_1}{\partial t^{*2}} \tag{22.14}$$

permit an explicit representation of the individual functions f_1, f_2, f_3. Specifically,

$$f_1 = \begin{cases} \gamma(Y - \bar{y}), & \pi \le X^* \\ \dfrac{\gamma}{\pi}(Y - \bar{y})X^*, & 0 \le X^* \le \pi \end{cases} \tag{22.15}$$

describes the continuous solution of (22.12) which vanishes at $X^* = 0$ and has the requisite discontinuity in derivative

$$\left.\frac{\partial f_1}{\partial X^*}\right]_{\pi-0}^{\pi+0} = -\frac{\gamma}{\pi}(Y - \bar{y}).$$

By virtue of the parameter inequality $\gamma \ll \varepsilon \ll 1$ and the fact that the functions f_1 and $\bar{y}$ have comparable magnitudes, we may replace (22.15) with the simpler form

$$f_1 = \begin{cases} \gamma Y, & \pi \le X^* \\ \dfrac{\gamma}{\pi} Y X^*, & 0 \le X^* \le \pi \end{cases} \tag{22.16}$$

and thereby obtain a first approximation

$$\bar{y}(t^*) \sim \gamma Y(t^*) \tag{22.17}$$

to the string displacement at the oscillator position.

The solution of (22.13) which vanishes at the midpoint of the string is

$$f_2 = A(t^*)X^* \tag{22.18}$$

and for the specification of its coefficient function, $A(t^*)$, the inner expansion

$$\begin{aligned} y(X^*, t^*) &\sim \gamma Y(t^*) + \varepsilon A(t^*)X^* + \cdots, \quad X^* \gg 1 \\ &= \gamma Y(t^*) + A(t^*)x^* + \cdots \end{aligned} \tag{22.19}$$

need merely be matched to, or rendered consistent with, an outer expansion

$$y(x^*, t^*) \sim \hat{f}_1(x^*, t^*) + \varepsilon \hat{f}_2(x^*, t^*) + \cdots \tag{22.20}$$

whose individual terms satisfy the homogeneous wave equation (22.8); if we choose an outgoing wave function on the interval $x^* > 0$ for the first term in (22.20),

$$\hat{f}_1(x^*, t^*) = B(t^* - x^*), \tag{22.21}$$

and subsequently effect the development of same at small values of x^*, the expansion

$$\begin{aligned} y(x^*, t^*) &\sim B(t^*) - x^* \frac{\mathrm{d}B}{\mathrm{d}t^*} + \cdots, \quad x^* \ll 1 \\ &= B(t^*) - \varepsilon \frac{\mathrm{d}B}{\mathrm{d}t^*} X^* + \cdots \end{aligned} \tag{22.22}$$

agrees with (22.19) provided that

$$B(t^*) = \gamma Y(t^*) \tag{22.23}$$

and

$$A(t^*) = -\frac{\mathrm{d}B}{\mathrm{d}t^*} = -\gamma \frac{\mathrm{d}Y(t^*)}{\mathrm{d}t^*}. \tag{22.24}$$

Thus, details of both inner and outer expansions for the solution of the fundamental equation (22.4) emerge when these are matched on a common range of applicability and a composite, albeit approximate, form of the overall solution is achieved. Employing (22.24) it appears from (22.19) that

$$\bar{y}(t^*) \sim \gamma Y(t^*) - \varepsilon \pi \gamma \frac{\mathrm{d}Y(t^*)}{\mathrm{d}t^*} \tag{22.5}$$

constitutes a refinement of the prior determination of the string displacement, and evidently the way is open to improving thereupon by matching a larger number of terms.

The oscillator displacement, $Y(t^*)$, hitherto accorded a given status, must conform to the dynamical equation

$$\frac{\mathrm{d}^2 Y}{\mathrm{d}t^{*2}} + Y + K(Y - \bar{y}) = 0 \tag{22.26}$$

wherein the final term manifests the coupling of oscillator and string. If we avail ourselves of the representation (22.25) so as to eliminate $\bar{y}$ from the preceding, there results

$$\frac{d^2Y}{dt^{*2}}+\varepsilon\pi K\gamma\frac{dY}{dt^*}+(1+K)Y=0 \tag{22.27}$$

with solutions that evidence a gradual decay in time which signifies, of course, a concomitant excitation of the string. Higher order terms make an appearance in the oscillator equation through the extension of (22.25) and some members of the enlarged family of solutions decay rapidly; such solutions, indicative of large perturbations for the free oscillator vibrations, have no significance inasmuch as they contradict the premise $\varepsilon \ll 1$ on which the approximation procedure rests [Burke, 1971].

§23. *Instantaneous and Moving Point Source Functions*

The coordinate dependent Green's function $G(x, x')$ defined in §17 acquires, through the juxtaposition of a periodic time factor, its relationship to wave trains progressing outwards from a "source" point x' at which dG/dx is discontinuous. If the source has an activity that is also localized in time, the corresponding Green's function $G(x, t; x', t')$ receives specification through a partial differential equation, viz.

$$\left(\frac{\partial^2}{\partial x^2}-\frac{1}{c^2}\frac{\partial^2}{\partial t^2}\right)G(x,t;x',t')=-\delta(x-x')\delta(t-t')$$

whose inhomogeneous member singles out the space-time coordinates (x', t') of the source.

The existence of a complex Fourier transform of $G(x, t; x', t')$ relative to the time variable, namely

$$\bar{G}(x-x',\omega)=\int_{-\infty}^{\infty} e^{i\omega(t-t')}G(x,t;x',t')\,d(t-t'), \tag{23.2}$$

is assured in consequence of a null excitation prior to the instant of time t', along with the assumption $\operatorname{Im}\omega>0$ which renders the integral (23.2) convergent at its upper limit; and said transform satisfies an ordinary differential equation

$$\left(\frac{d^2}{dx^2}+\frac{\omega^2}{c^2}\right)\bar{G}(x-x',\omega)=-\delta(x-x'). \tag{23.3}$$

Referring to §17 for the everywhere bounded solution of (23.3),

$$\bar{G} = \frac{\mathrm{i}c}{2\omega} \exp\left\{\mathrm{i}\frac{\omega}{c}|x - x'|\right\},$$

and utilizing the inverse relation to (23.2), it follows that

$$G(x, t; x', t') = \frac{\mathrm{i}c}{4\pi} \int_{-\infty}^{\infty} \exp\left\{\mathrm{i}\frac{\omega}{c}|x - x'| - \mathrm{i}\omega(t - t')\right\}\frac{\mathrm{d}\omega}{\omega}$$

where the integration contour passes above the origin; hence, a residue calculation which takes account of the simple pole at $\omega = 0$ yields the sought for representation

$$G(x, t; x', t') = G(x - x', t - t') = \frac{c}{2} H\left(t - t' - \frac{|x - x'|}{c}\right) \tag{23.4}$$

involving the Heaviside function (20.23).

An alternative procedure for constructing the Green's function is initiated by a Fourier transformation of (23.1) with respect to the coordinate variable, which provides the differential equation

$$\left(\frac{\mathrm{d}^2}{\mathrm{d}t^2} + (\zeta c)^2\right)\hat{G}(\zeta, t - t') = c^2\delta(t - t') \tag{23.5}$$

where

$$\hat{G}(\zeta, t - t') = \int_{-\infty}^{\infty} \mathrm{e}^{\mathrm{i}\zeta(x - x')} G(x, t; x', t')\, \mathrm{d}(x - x'). \tag{23.6}$$

The appropriate solution of (23.5) is

$$\hat{G} = c\frac{\sin \zeta c(t - t')}{\zeta} H(t - t')$$

and, consequently, we obtain

$$\begin{aligned} G(x - x', t - t') &= \frac{c}{2\pi} H(t - t') \int_{-\infty}^{\infty} \mathrm{e}^{-\mathrm{i}\zeta(x - x')} \sin \zeta c(t - t') \frac{\mathrm{d}\zeta}{\zeta} \\ &= \frac{c}{2} H(t - t')[H(x - x' + c(t - t')) - H(x - x' - c(t - t'))] \\ &= \frac{c}{2} H(t - t')[\mathrm{sgn}(x - x' + c(t - t')) - \mathrm{sgn}(x - x' - c(t - t'))] \end{aligned} \tag{23.7}$$

which result is fully equivalent to (23.4).

We deduce from the foregoing expressions that the support of the source function at time t is provided on the interval $[x' - c(t - t'), x' + c(t - t')]$ of the

x-axis; and that

$$y(x,t)=\frac{c}{2}\int H\left(t-t'-\frac{|x-x'|}{c}\right)q(x',t')\,\mathrm{d}x'\,\mathrm{d}t'$$
$$=\frac{c}{2}\int_{\mathfrak{D}} q(x',t')\,\mathrm{d}x'\,\mathrm{d}t' \tag{23.8}$$

represents a particular solution of the inhomogeneous wave equation

$$\frac{\partial^2 y}{\partial x^2}-\frac{1}{c^2}\frac{\partial^2 y}{\partial t^2}=-q(x,t), \quad t>0 \tag{23.9}$$

which has both a null value and time-derivative initially ($t=0$), since the respective vertices of the finite triangular domain $\mathfrak{D}$ are located at (x,t), $(x+ct,0)$, $(x-ct,0)$, and is identical with (14.4).

For a source in uniform with speed U which enters the positive range of the x-axis at $t=0$ the delta function, $\delta(x-x')$, of (23.1) is replaced by $\delta(x-Ut)$; and, further to the assumption that its strength is a periodic function of the time, the requisite wave equation becomes

$$\left(\frac{\partial^2}{\partial x^2}-\frac{1}{c^2}\frac{\partial^2}{\partial t^2}\right)G(x,t)=-\delta(x-Ut)\,\mathrm{e}^{-\mathrm{i}\omega_0 t}. \tag{23.10}$$

On transforming the independent variables x, t in accordance with the relations

$$\xi=x-Ut, \quad \tau=t$$

there obtains, as the subsequent version of (23.10),

$$\left[(1-\beta^2)\frac{\partial^2}{\partial\xi^2}+2\frac{\beta}{c}\frac{\partial^2}{\partial\xi\partial\tau}-\frac{1}{c^2}\frac{\partial^2}{\partial\tau^2}\right]G(\xi,\tau)=-\delta(\xi)\,\mathrm{e}^{-\mathrm{i}\omega_0\tau}, \quad \beta=\frac{U}{c}.$$

Let

$$G(\xi,\tau)=g(\xi)\,\mathrm{e}^{-\mathrm{i}\omega_0\tau} \tag{23.11}$$

so as to isolate the characteristic time factor of the source, and we thus encounter the ordinary differential equation

$$\left[(1-\beta^2)\frac{\mathrm{d}^2}{\mathrm{d}\xi^2}-2\,\mathrm{i}\frac{\omega_0}{c}\frac{\mathrm{d}}{\mathrm{d}\xi}+\frac{\omega_0^2}{c^2}\right]g(\xi)=-\delta(\xi)$$

with the particular solution

$$g(\xi)=\frac{1}{2\pi}\int_{-\infty}^{\infty}\frac{\mathrm{e}^{\mathrm{i}\xi\mu}}{\Omega(\mu)}\,\mathrm{d}\mu, \quad \Omega(\mu)=(1-\beta^2)\mu^2-2\frac{\omega_0}{c}\beta\mu-\frac{\omega_0^2}{c^2}$$

that manifests a discontinuous first derivative at $\xi = 0$. A product resolution of the second order polynomial $\Omega(\mu)$ reveals that

$$g(\xi) = \frac{1}{2\pi} \frac{1}{1-\beta^2} \int_{-\infty}^{\infty} \frac{e^{i\xi\mu}\, d\mu}{\left(\mu - \dfrac{\omega_0/c}{1-\beta}\right)\left(\mu + \dfrac{\omega_0/c}{1+\beta}\right)} \tag{23.12}$$

where the contour passes above and below the points $-((\omega_0/c)/(1+\beta))$, $((\omega_0/c)/(1-\beta))$, $\beta < 1$, respectively, in keeping with the outgoing wave or source behavior for the limiting case $\beta \to 0$. Accordingly,

$$g(\xi) = \begin{cases} \dfrac{ic}{2\omega_0} \exp\left\{ i\xi \dfrac{\omega_0/c}{1-\beta} \right\}, & \xi > 0 \\[2ex] \dfrac{ic}{2\omega_0} \exp\left\{ -i\xi \dfrac{\omega_0/c}{1+\beta} \right\}, & \xi < 0 \end{cases}$$

and the complete wave function

$$G(x, t) = \frac{ic}{2\omega_0} \exp\left\{ \frac{i\omega_0}{1 \mp \beta} \left(\frac{\pm x}{c} - t \right) \right\}, \quad x \gtrless Ut \tag{23.13}$$

exhibits the asymmetry occasioned by relative motion between the source and its surroundings; we perceive in (23.13), for instance, the Doppler shifted frequencies $\omega_\pm = \omega_0(1 \mp \beta)$, ahead of and behind the source, as specified in (11.5).

If $\beta > 1$ the contour in (23.12) proceeds above both poles of the integrand and thus

$$g(\xi) = \begin{cases} 0, & \xi > 0 \\[1ex] \dfrac{ic}{2\omega_0} \left[\exp\left\{ -i\xi \dfrac{\omega_0/c}{\beta+1} \right\} - \exp\left\{ -i\xi \dfrac{\omega_0/c}{\beta-1} \right\} \right], & \xi < 0 \end{cases}$$

which leads to the result

$$G(x, t) = \begin{cases} 0, & x > Ut \\[1ex] \dfrac{ic}{2\omega_0} e^{-i\omega_0 t} \left[\exp\left\{ i(Ut - x) \dfrac{\omega_0/c}{\beta+1} \right\} - \exp\left\{ i(Ut - x) \dfrac{\omega_0/c}{\beta-1} \right\} \right], & x < Ut \end{cases} \tag{23.14}$$

that takes the form

$$G(x, t) = \begin{cases} 0, & x > Ut \\[1ex] \dfrac{Ut - x}{\beta^2 - 1}, & x < Ut \end{cases} \tag{23.15}$$

when the source strength remains fixed ($\omega_0 \to 0$).

The differential equation that pertains to the excitation of a string, with uniform density ρ, by a localized force of fixed magnitude $\mathfrak{F}$ which commences its action at $t=0$ and moves steadily away from the origin at the speed U, is

$$\frac{\partial^2 y}{\partial x^2}-\frac{1}{c^2}\frac{\partial^2 y}{\partial t^2}=-\frac{\mathfrak{F}}{\rho c^2}\delta(x-Ut)H(t). \tag{23.16}$$

Adapting the integral (23.8) of a wave equation with general inhomogeneous member to the present circumstance, we obtain for the induced string displacement

$$y(x,t)=\frac{\mathfrak{F}}{2\rho c}\int_{-\infty}^{\infty}\mathrm{d}t'\int_{-\infty}^{\infty}\mathrm{d}x' H\left(t-t'-\frac{|x-x'|}{c}\right)\delta(x'-Ut')H(t')$$

whence it follows, after reduction, that

$$y(x,t)=\frac{\mathfrak{F}}{2\rho c}\int_{0}^{\infty}H\left(t-t'-\frac{|x-Ut'|}{c}\right)\mathrm{d}t' \tag{23.17}$$

To evaluate (23.17) let us introduce the implicit relation

$$t-\tau=\frac{|x-U\tau|}{c} \tag{23.18}$$

for specifying the instant(s) τ at which a disturbance moving with the speed c, must leave the position of the force (i.e., source point) in order to reach the point of observation at the time t; the solutions of (23.18) are

$$\tau=\frac{ct\pm x}{c\pm U}$$

and, if $U<c$, we may confirm the unique determinations

$$\tau_1=\frac{ct-x}{c-U}, \quad x>Ut$$

$$\tau_2=\frac{ct+x}{c+U}, \quad x<Ut$$

that comply with the causal requirement $\tau_{1,2}<t$. Thus, inasmuch as $\tau_1<0$ for $x>ct$ and $\tau_2<0$ for $x<-ct$, we obtain from (23.17) the representations

$$y(x,t)=\frac{\mathfrak{F}}{2\rho c}\begin{pmatrix}\tau_1\\0\end{pmatrix}, \quad \begin{matrix}Ut<x<ct\\x>ct\end{matrix}$$

and

$$y(x,t)=\frac{\mathfrak{F}}{2\rho c}\begin{pmatrix}\tau_2\\0\end{pmatrix}, \quad \begin{matrix}-ct<x<Ut\\x<-ct\end{matrix}$$

which are in complete agreement with (12.12).

If $U > c$ there are two determinations

$$\tau_1^* = \frac{x - ct}{U - c}, \quad \tau_2^* = \frac{x + ct}{U + c}$$

meeting the requirement $\tau_{1,2}^* < t$ when $x < Ut$ and none when $x > Ut$; since $\tau_{1,2}^* > 0$ and $\tau_1^* < \tau_2^*$ on the range $ct < x < Ut$, it follows that

$$y(x, t) = \frac{\mathfrak{F}}{2\rho c}\begin{pmatrix} \tau_2^* - \tau_1^* \\ 0 \end{pmatrix}, \quad \begin{matrix} ct < x < Ut \\ x > Ut \end{matrix}$$

Furthermore, relying on the properties

$$\tau_1^* < 0, \quad x < ct, \quad \tau_2^* < 0, \quad x < -ct$$

we find that

$$y(x, t) = \frac{\mathfrak{F}}{2\rho c}\begin{pmatrix} \tau_2^* \\ 0 \end{pmatrix}, \quad \begin{matrix} -ct < x < ct \\ x < -ct \end{matrix}$$

and thus the expected agreement with (12.17) results.

§24. *A Densely Loaded String*

Let us now imagine that identical point loads M are attached to an otherwise uniform string in sufficient numbers so that every infinitesimal section dx thereof contains many representatives, and consider some aspects of progressive motion along this arrangement. In response to an incident wave each section becomes the seat of oppositely directed and outgoing secondary waves, and hence we can view the whole excitation on the string in terms of forward and backward moving components. The forward component, on this basis, comprises an incident wave part and a superposition of like directed secondary waves from sections of the string to one side of the observation point, which aggregate may be collectively termed the transmitted wave; and the backward component or reflected wave is made up of the other set of secondary waves from the complementary sections of the string.

If the wave length of the transmitted profile has a large magnitude on the scale of neighboring load separations, we may anticipate that the corresponding progressive wave function closely resembles its counterpart for a continuously loaded string; and that, moreover, the equivalent density of the latter can be linked with an average load density of constant magnitude. To support these contentions and elaborate on the replacement of a discrete substructure by a continuous environment for long wave disturbances, we utilize the prior analysis of scattering by a single load. According to (17.16) the secondary waves that convey the effect of an isolated mass M at the fixed location x' find their expression in the compact form

$$\frac{M\omega^2}{P} G(x, x')y(x', t) = \frac{\mathrm{i}Mk}{2\rho} \,\mathrm{e}^{\mathrm{i}k|x-x'|} y(x', t),$$

where $y(x', t)$ is the total time-periodic wave amplitude at the load point. When the amount of loading is very slight and the disparity between the transmitted and total wave amplitudes is ignored, the transmitted wave which emerges after passage (in the positive sense) through a section of breadth $\mathrm{d}x$ commencing at the origin is

$$Y(\mathrm{d}x, t) = y(\mathrm{d}x, t) + \mathrm{i}\frac{NMk}{2\rho}\,\mathrm{d}x \,\mathrm{e}^{\mathrm{i}k\,\mathrm{d}x} y(0, t) \tag{24.1}$$

approximately, if N denotes the number of mass loads per unit length of the string and their dispersal remains to be accounted for in a subsequent stage of refinement. On substituting the correlated expressions

$$y(\mathrm{d}x, t) = C\,\mathrm{e}^{\mathrm{i}k\,\mathrm{d}x - \mathrm{i}\omega t},$$
$$y(0, t) = C\,\mathrm{e}^{-\mathrm{i}\omega t}$$

in the above representation it follows that

$$\begin{aligned} Y(\mathrm{d}x, t) &= C\,\mathrm{e}^{\mathrm{i}k\,\mathrm{d}x - \mathrm{i}\omega t}\left[1 + \frac{\mathrm{i}NMk}{2\rho}\,\mathrm{d}x\right] \\ &\doteq C \exp\left\{\mathrm{i}k\left(1 + \frac{NM}{2\rho}\right)\mathrm{d}x - \mathrm{i}\omega t\right\}; \end{aligned} \tag{24.2}$$

evidently, the wave number bespoken in this progressive wave form differs from that appropriate to the unloaded string and can be assigned the effective value

$$\kappa = k\left(1 + \frac{NM}{2\rho}\right) = k\left(1 + \frac{1}{2}\frac{\Delta\rho}{\rho}\right) \tag{24.3}$$

where $\Delta\rho = NM$ specifies the small increment of density, viz: $\Delta\rho/\rho \ll 1$. In order to lift the latter restriction we should allow for the inequality of transmitted and total wave amplitudes which is from the point of view adopted a consequence of partial reflection.

There is, in fact, no need to draw such an amplitude distinction and the functional equation

$$y(x, t) = \frac{\mathrm{i}MNk}{2\rho}\int_{-\infty}^{\infty} \mathrm{e}^{\mathrm{i}k|x-x'|} y(x', t)\,\mathrm{d}x', \quad -\infty < x < \infty \tag{24.4}$$

which describes a self-supporting mode of excitation, with oppositely directed component waves subsumed in the integral, serves to characterize travelling disturbances on the string. Thus, solutions of the foregoing homogeneous integral equation take the simple exponential form

$$y(x, t) = \mathrm{e}^{\pm \mathrm{i}\kappa x - \mathrm{i}\omega t} \tag{24.5}$$

provided that the value of κ is determined by the relation

$$1 = \frac{\mathrm{i}NMk}{2\rho} \int_{-\infty}^{\infty} \mathrm{e}^{\mathrm{i}k|x-x'| \mp \mathrm{i}\kappa(x-x')}\, \mathrm{d}x'$$

or

$$1 = \frac{\mathrm{i}MNk}{2\rho} \lim_{\varepsilon \to 0} \int_{-\infty}^{\infty} \mathrm{e}^{-\varepsilon|\xi| + \mathrm{i}k|\xi| \mp \mathrm{i}\kappa\xi}\, \mathrm{d}\xi \tag{24.6}$$

following the temporary reliance on a convergence factor. Evaluation of the integral and limit in (24.6) reveals that

$$\kappa^2 = k^2\left(1 + \frac{\Delta\rho}{\rho}\right) \tag{24.7}$$

in place of the earlier version (24.3) which is, moreover, directly recovered for small values of $\Delta\rho/\rho$. The direct proportionality between the squares of the wave numbers κ^2, k^2 and the respective densities $\rho + \Delta\rho$, ρ of the loaded and unloaded strings is in keeping with the concomitant alteration of wave velocity when the tension remains unchanged.

§25. *A Composite or Sectionally Uniform String*

A point load, whether functioning as a scatterer of incoming waves on the string to which it is attached, or as the seat of outgoing waves in response to an initially local excitation, marks the vertex of a discontinuity in slope of the string. The latter aspect is not encountered, however, relative to motions on an infinite string composed of separate parts having unequal densities, for then the profile of the string evidences a continuous slope throughout, and hence no point can be singled out as the scattering center.

Let us imagine an infinite string to be made up of uniform though distinct portions $x < 0, > 0$, wherein the density has the constant values ρ_1 and ρ_2, respectively. The characteristic wave velocities in these parts of the string are

$$c_1 = \sqrt{(P/\rho_1)}, \qquad c_2 = \sqrt{(P/\rho_2)} \tag{25.1}$$

and when a state of periodic vibrations with a given angular frequency ω prevails, the corresponding wave numbers

$$k_1 = \omega/c_1, \qquad k_2 = \omega/c_2 \tag{25.2}$$

are also distinct, and likewise the wave lengths λ_1, λ_2.

Along with a wave train $A\exp[i(k_1x-\omega t)]$ which is incident on the origin ($x=0$) from the left there will be a reflected wave train $B\exp[-i(k_1x+\omega t)]$ in the same part of the string ($x<0$), and a transmitted motion $C\exp[i(k_2x-\omega t)]$ in the other part ($x>0$). The secondary wave amplitudes B, C are easily related to the primary amplitude A on securing the continuity of both the (total) displacement and its slope at the junction point, and we thus obtain

$$B=\frac{k_1-k_2}{k_1+k_2}A,\qquad C=\frac{2k_1}{k_1+k_2}A. \tag{25.3}$$

Hence the reflected and incident waves are in phase at $x=0$ provided that $k_1>k_2$ (or $\rho_1>\rho_2$), and are otherwise wholly out of phase.

While the above procedure furnishes an explicit solution to stipulated problem, it does not illuminate the nature of the underlying connections between the displacement at different points on the string. In order to develop such features, and to deduce independently the secondary wave amplitudes, we again call upon the concept of a point source or Green's function. Specifically, let us employ an outgoing wave Green's function associated with the wave number k_1,

$$G_1(x,x')=\frac{i}{2k_1}\,e^{ik_1|x-x'|}, \tag{25.4}$$

namely, an integral of the equation

$$\left(\frac{d^2}{dx^2}+k_1^2\right)G_1(x,x')=-\,\delta(x-x'),\quad -\infty<x,x'<\infty \tag{25.5}$$

to arrive at a representation of the spatial part of the string displacement, $s(x)$, that subsumes the complementary differential specifications

$$\left(\frac{d^2}{dx^2}+k_1^2\right)s(x)=0,\quad x<0 \tag{25.6}$$

and

$$\left(\frac{d^2}{dx^2}+k_2^2\right)s(x)=0,\quad x>0 \tag{25.7}$$

together with the asymmetric representations

$$s(x)=\begin{cases}A\,e^{ik_1x}+B\,e^{-ik_1x}, & x<0\\ C\,e^{ik_2x}, & x>0\end{cases} \tag{25.8}$$

which suit an incoming wave on the left.

Thus, if we multiply the differential equation (25.5) by $s(x)$ and integrate over all values of x, it follows that

$$s(x')=-\lim_{X\to\infty}\int_{-X}^{X}s(x)\left[\frac{d^2}{dx^2}+k_1^2\right]G_1(x,x')\,dx,$$

whence, after twice integrating by parts (so as to transfer the second x-derivative from G_1 to s),

$$s(x') = \left[\frac{ds(x)}{dx} G_1(x, x') - s(x)\frac{d}{dx} G_1(x, x')\right]_{x\to-\infty}^{x\to+\infty} - \int_{-\infty}^{\infty} G_1(x, x')\left(\frac{d^2}{dx^2} + k_1^2\right)s(x)\, dx. \quad (25.9)$$

Far to the left of the origin ($x \to -\infty$) the integrated terms make a contribution $A \exp(ikx')$, as a simple calculation with (25.4) and (25.8) shows, and far to the right of the origin ($x \to +\infty$) the same terms vanish if k_2 is (temporarily) endowed with an infinitesimal positive imaginary part, for such a premise carries with it the implication that motions heading away from the origin are ultimately reduced to a negligible scale ($s(x), (ds/dx) \to 0, x \to \infty$). The latter modification of an asymptotic behavior at $x = \infty$, which can be ascribed to the presence of a small damping term (proportional to the local velocity) in the equation of motion, does not conflict with our immediate objectives, namely to determine the amplitudes for the reflected wave and the wave which is launched into the right half of the string and, after this means of excluding an inward directed disturbance on the right, the restoration of a real magnitude for k_2 is permissible.

Noting that the integral in (25.9) vanishes for $x < 0$, as a consequence of the differential equation satisfied there by $s(x)$, and that $d^2s/dx^2 = -k_2^2 s$ for $x > 0$, we obtain finally, on interchanging x and x' and invoking the symmetry of the Green's function,

$$s(x) = A\, e^{ik_1x} + (k_2^2 - k_1^2) \int_0^{\infty} G_1(x, x')s(x')\, dx', \quad -\infty < x < \infty. \quad (25.10)$$

The latter representation of the displacement factor incorporates the primary wave component and confirms that scattering is absent on a uniform string ($k_2 = k_1$); moreover, it defines a function which, together with its first derivative, is continuous everywhere and satisfies the appropriate differential equation (25.6) or (25.7) according as x is negative or positive. To verify the last assertion, we observe that inasmuch as the integration in (25.10) extends over positive values for the second argument, x', of the Green's function, the latter is effectively a solution of the homogeneous version of (25.5) if $x < 0$, and hence the differential operator $((d^2/dx^2) + k_1^2)$ annihilates the right-hand side of (25.10). When $x > 0$, on the other hand, an application of this operator to (25.10) yields

$$\left(\frac{d^2}{dx^2} + k_1^2\right)s(x) = (k_2^2 - k_1^2) \int_0^{\infty} [-\delta(x - x')]s(x')\, dx' = (k_1^2 - k_2^2)s(x),$$

or

$$\left(\frac{d^2}{dx^2} + k_2^2\right)s(x) = 0, \quad x > 0$$

as required.

After introducing the explicit form of $G_1(x, x')$ into (25.10) there obtains

$$s(x) = A\,e^{ik_1x} + \frac{i}{2k_1}(k_2^2 - k_1^2)\int_0^\infty e^{ik_1|x-x'|}s(x')\,dx', \quad -\infty < x < \infty \qquad (25.11)$$

and, in particular,

$$s(x) = A\,e^{ik_1x} + \frac{i}{2k_1}(k_2^2 - k_1^2)\,e^{-ik_1x}\int_0^\infty e^{ik_1x'}s(x')\,dx', \quad x < 0 \qquad (25.12)$$

which permits the identification of the reflected wave amplitude, viz.:

$$B = \frac{i}{2k_1}(k_2^2 - k_1^2)\int_0^\infty e^{ik_1x}s(x)\,dx. \qquad (25.13)$$

According to this representation the reflected wave amplitude depends on the totality of the displacements $s(x)$ at all points in the complementary or right half of the string, and it is the Green's function associated with the propagation characteristics of the left-hand portion which acts as the source function in superposing these displacements.

In the above approach to the scattering problem the function $s(x)$ is as yet unknown, and pending its determination the relation (25.10) has only a formal significance. Actually, the values of $s(x)$ for $0 < x < \infty$ are directly obtainable from (25.10), which constitutes an inhomogeneous integral equation of the second kind on the aforesaid interval, with the Green's function in the role of a symmetric kernel; and the values of $s(x)$ for $x < 0$ may then be deduced by evaluation of the right-hand member thereof. Two procedures for solving the integral equation are available, the simpler one relying on the assumption of a single exponential form of solution, namely

$$s(x) = C\,e^{i\kappa x}, \quad x > 0 \qquad (25.14)$$

where C and κ are unspecified constants. If the range of integration in (25.11) is partitioned at the location $x > 0$,

$$s(x) = \left[A + \frac{i}{2k_1}(k_2^2 - k_1^2)\int_0^x e^{-ik_1x'}s(x')\,dx'\right]e^{ik_1x} + \left[\frac{i}{2k_1}(k_2^2 - k_1^2)\int_x^\infty e^{ik_1x'}s(x')\,dx'\right]e^{-ik_1x}, \qquad (25.15)$$

and thus, after the substitution of (25.14) herein, we find

$$C\,e^{i\kappa x} = \left(A - \frac{C}{2k_1}\frac{k_2^2 - k_1^2}{\kappa - k_1}\right)e^{ik_1x} + C\,\frac{k_2^2 - k_1^2}{\kappa^2 - k_1^2}\,e^{i\kappa x}, \quad x > 0 \qquad (25.16)$$

when the convergence of the infinite integral in (25.15) is brought about through the assumption that $\text{Im}\,\kappa = \varepsilon(>0)$. Evidently, the pair of relations

$$1 = \frac{k_2^2 - k_1^2}{\kappa^2 - k_1^2},$$

$$C = 2k_1 A \frac{\kappa - k_1}{k_2^2 - k_1^2}$$

ensure the self-consistency of (25.16) for all (positive) values of x; the former characterizes $\kappa = k_2$ as the wave number of the transmitted profile and the latter furnishes the amplitude of same,

$$C = \frac{2k_1}{k_1 + k_2} A,$$

in agreement with previous conclusions.

The structure of (25.15) makes clear that responsibility for eliminating the primary wave function, $\exp(ik_1 x)$, falls upon the displacement factor $s(x')$ in $0 < x' < x$; and if we recast (25.15) in the form

$$s(x) = \left[A + \frac{i}{2k_1}(k_2^2 - k_1^2) \int_0^\infty e^{-ik_1 x} s(x)\, dx \right] e^{ik_1 x}$$

$$+ \frac{i}{2k_1}(k_2^2 - k_1^2) \left[e^{-ik_1 x} \int_x^\infty e^{ik_1 x'} s(x')\, dx' - e^{ik_1 x} \int_x^\infty e^{-ik_1 x'} s(x')\, dx' \right]$$

a primary wave extinction condition

$$\int_0^\infty e^{-ik_1 x} s(x)\, dx = \frac{2\, ik_1 A}{k_2^2 - k_1^2} \tag{25.17}$$

is implied by the asymptotic behavior $s(x) \to 0,\ x \to \infty$.

The fact that the kernel of the integral equation

$$s(x) = A\, e^{ik_1 x} + (k_2^2 - k_1^2) \int_0^\infty G_1(x - x') s(x')\, dx', \quad 0 < x < \infty \tag{25.18}$$

is a difference function of its argument variables enables a means of direct solution by Fourier transformation and complex analysis; and this calls for a preliminary extension of the given equation to the remainder of the x-interval, say

$$s_+(x) + s_-(x) = \begin{Bmatrix} A\, e^{ik_1 x}, & x > 0 \\ 0, & x < 0 \end{Bmatrix} + (k_2^2 - k_1^2) \int_{-\infty}^\infty G_1(x - x') s_+(x')\, dx', \tag{25.19}$$

$$-\infty < x < \infty$$

where $s_+(x) \neq 0,\ x > 0$ (and is identical with $s(x)$), whereas $s_-(x) \neq 0,\ x < 0$ (and characterizes the reflected wave factor). If the premise is made that both the wave numbers k_1 and k_2 have a (common) infinitesimal imaginary part $\varepsilon(>0)$, the functions $s_+(x)$ and $s_-(x)$, which correspond to outgoing wave motions on their respective domains $x>,<0$, undergo exponential attenuation when $|x|\to\infty$; furthermore, their Fourier transforms

$$s_+(\zeta)=\int_0^\infty \mathrm{e}^{\mathrm{i}\zeta x}s_+(x)\,\mathrm{d}x, \tag{25.20}$$

$$s_-(\zeta)=\int_{-\infty}^\infty \mathrm{e}^{\mathrm{i}\zeta x}s_-(x)\,\mathrm{d}x \tag{25.21}$$

then define regular functions of a complex variable ζ in the half-planes $\operatorname{Im}\zeta>-\varepsilon,<\varepsilon$, respectively. Taking note of the result

$$\int_{-\infty}^\infty G_1(x-x')\,\mathrm{e}^{\mathrm{i}\zeta(x-x')}\,\mathrm{d}(x-x')=\frac{1}{\zeta^2-k_1^2},\quad |\operatorname{Im}\zeta|<\varepsilon \tag{25.22}$$

which may be established by reference to (17.12) and (17.13), we find that the outcome of multiplication in (25.19) by $\exp(\mathrm{i}\zeta x)$, and subsequent integration over all values of x, is a transform relation

$$s_+(\zeta)+s_-(\zeta)=\frac{\mathrm{i}A}{\zeta+k_1}+\frac{k_2^2-k_1^2}{\zeta^2-k_1^2}s_+(\zeta) \tag{25.23}$$

that is meaningful (only) in the infinite strip $|\operatorname{Im}\zeta|<\varepsilon$, wherein all of its terms evidence joint analyticity with regard to ζ. After rewriting (25.23) in the successive versions

$$\frac{\zeta^2-k_2^2}{\zeta^2-k_1^2}s_+(\zeta)+s_-(\zeta)=\frac{\mathrm{i}A}{\zeta+k_1}, \tag{25.24}$$

and

$$\frac{\zeta+k_2}{\zeta+k_1}s_+(\zeta)+\frac{\zeta-k_1}{\zeta-k_2}s_-(\zeta)=\mathrm{i}A\frac{\zeta-k_1}{(\zeta-k_2)(\zeta+k_1)}$$

$$=\mathrm{i}A\left[\frac{\zeta-k_1}{(\zeta-k_2)(\zeta+k_1)}-\frac{2k_1}{(k_1+k_2)(\zeta+k_1)}\right]\frac{2\,\mathrm{i}Ak_1}{(k_1+k_2)(\zeta+k_1)}$$

we finally express, via the relation

$$\frac{\zeta+k_2}{\zeta+k_1}s_+(\zeta)-\frac{2\,\mathrm{i}Ak_1}{k_1+k_2}\cdot\frac{1}{\zeta+k_1}=-\frac{\zeta-k_1}{\zeta-k_2}s_-(\zeta)+\frac{\mathrm{i}A}{\zeta+k_1}\left[\frac{\zeta-k_1}{\zeta-k_2}-\frac{2k_1}{k_1+k_2}\right], \tag{25.25}$$

an equality within the aforesaid or common strip of left- and right-hand members which are regular in the (overlapping) half-planes $\operatorname{Im}\zeta>-\varepsilon,<\varepsilon$, respectively. Thus the above relation defines an entire or integral function throughout the ζ-plane, and inasmuch as

$$s_{\pm}(\zeta) \sim 0\left(\frac{1}{\zeta}\right), \quad |\zeta| \to \infty, \quad \operatorname{Im} \zeta \gtrless 0, \tag{25.26}$$

owing to the bounded nature of $s_{\pm}(x)$ at and near $x = 0$, we deduce that the entire function $E(\zeta)$ is identically zero, since $E(\zeta) \to 0$, $|\zeta| \to \infty$. Hence, the members of (25.25) vanish separately and explicit representations of the transforms $s_{\pm}(\zeta)$ follow, namely

$$s_{+}(\zeta) = \frac{2\,\mathrm{i}k_1 A}{k_1 + k_2} \cdot \frac{1}{\zeta + k_2}, \tag{25.27}$$

$$s_{-}(\zeta) = -\,\mathrm{i}A \frac{k_1 - k_2}{k_1 + k_2} \cdot \frac{1}{\zeta - k_1}. \tag{25.28}$$

On comparison of the latter expressions with the versions

$$s_{+}(\zeta) = \frac{\mathrm{i}C}{\zeta + k_2}; \quad s_{-}(\zeta) = \frac{-\,\mathrm{i}B}{\zeta - k_1} \tag{25.29}$$

that are derived from the appertaining functions

$$s_{+}(x) = C\,\mathrm{e}^{\mathrm{i}k_2 x}, \quad x > 0; \quad s_{-}(x) = B\,\mathrm{e}^{-\mathrm{i}k_1 x}, \quad x < 0 \tag{25.30}$$

there obtains

$$B = \frac{k_1 - k_2}{k_1 + k_2} A, \qquad C = \frac{2k_1}{k_1 + k_2} A$$

as given in (25.3); moreover, when the extinction condition (25.17) is restated in the form

$$s_{+}(-k_1) = \frac{2\,\mathrm{i}k_1 A}{k_2^2 - k_1^2} \tag{25.31}$$

its accord with (25.27) becomes evident.

An alternative procedure aimed at determining the individual transforms $s_{\pm}(\zeta)$ from the single equation (25.24) utilizes the rearrangement of same,

$$\frac{(\zeta - k_2)(\zeta + k_2)}{\zeta + k_1} s_{+}(\zeta) - \mathrm{i}A \frac{\zeta - k_1}{\zeta + k_1} = -\,(\zeta - k_1) s_{-}(\zeta) \tag{25.32}$$

wherein the left- and right-hand members characterize functions regular in the half-planes $\operatorname{Im} \zeta > -\varepsilon$, $< \varepsilon$. Since the members of (25.32) have a constant limit as $|\zeta| \to \infty$ in the respective half-planes, the entire function which they jointly define is likewise a constant and we deduce that

$$s_{-}(\zeta) = -\frac{\alpha}{\zeta - k_1}, \tag{25.33}$$

$$s_+(\zeta) = \frac{\mathrm{i}A(\zeta - k_1) + \alpha(\zeta + k_1)}{(\zeta - k_2)(\zeta + k_2)}. \tag{25.34}$$

The magnitude of α is uniquely fixed by the relation

$$\alpha(k_1 + k_2) = \mathrm{i}A(k_1 - k_2) \tag{25.35}$$

that ensures a finite value for $s_+(\zeta)$ at the point $\zeta = k_2$ in its half-plane of regularity; and when the specification (25.35) is employed in (25.33), (25.34) we recover the formulas (25.27), (25.28).

§26. *Another Composite String*

If the departure from uniformity of an infinite string is restricted to a finite (central) portion a new characteristic length is thereby introduced, and wave propagation on the string can reveal a marked degree of sensitivity to the ratio between this and the local wave length. Let us suppose that the portion of the string between $x = 0$ and $x = L$ has a density ρ_2 and a wave number k_2 for simply periodic vibrations, while the density and wave number elsewhere $(x < 0, > L)$ have the values ρ_1, k_1, respectively. We concern ourselves with a particular type of excitation on this composite string, characterized (as usual) by a lone incoming wave component, of amplitude A, which advances on the central portion from the far left; then the overall displacement factors have the forms

$$s(x) = \begin{cases} A \exp(\mathrm{i}k_1x) + B \exp(-\mathrm{i}k_1x), & x < 0 \\ D \exp(\mathrm{i}k_2x) + E \exp(-\mathrm{i}k_2x), & 0 < x < L \\ C \exp(\mathrm{i}k_1x), & x > L \end{cases} \tag{26.1}$$

on the indicated ranges, and the asymmetry of excitation is evidenced by the occurrence of oppositely directed wave functions to the left but not the right of the central portion of the string. The reflected and transmitted wave amplitudes B, C, as well as those of the central motion, D, E, may be individually related to the primary amplitude A on solving the four simultaneous linear equations that ensure the continuity of both s and $\mathrm{d}s/\mathrm{d}x$ at the junction points $x = 0$ and $x = L$. Alternatively, we can avail ourselves of the relation

$$s(x) = A\,\mathrm{e}^{\mathrm{i}k_1x} + (k_2^2 - k_1^2)\int_0^L G_1(x, x')s(x')\,\mathrm{d}x', \quad -\infty < x < \infty \tag{26.2}$$

which is the evident counterpart of (25.10); and if, at the outset, attention be focussed on the interval $0 < x < L$ it follows after introducing the appropriate form of $s(x)$ from (26.1) that

$$0 = e^{ik_1x}\left[A - \frac{1}{2k_1}(k_2+k_1)D + \frac{1}{2k_1}(k_2-k_1)E\right] + e^{-ik_1x}\left[\frac{D}{2k_1}(k_2-k_1)\,e^{i(k_2+k_1)L} - \frac{E}{2k_1}(k_2+k_1)\,e^{-i(k_2-k_1)L}\right], \quad 0<x<L$$

whence the amplitudes D, E are given by the pair of linear equations

$$\begin{aligned}(k_2+k_1)D - (k_2-k_1)E &= 2k_1A\\ (k_2-k_1)\,e^{i(k_2+k_1)L}D - (k_2+k_1)\,e^{-i(k_2-k_1)L}E &= 0\end{aligned} \tag{26.3}$$

in terms of A.

Referring next to the particular consequences of (26.2),

$$s(x) = A\,e^{ik_1x} + \frac{i}{2k_1}(k_2^2-k_1^2)\,e^{\mp ik_1x}\int_0^L e^{\pm ik_1x}s(x)\,dx,$$

according as $x<0$ or $x>L$, and the stipulated representations (26.1) on these ranges we arrive at the characterizations

$$\begin{aligned}B &= \frac{i}{2k_1}(k_2^2-k_1^2)\int_0^L e^{ik_1x}s(x)\,dx\\ &= \frac{1}{2k_1}[-(k_2-k_1)D + (k_2+k_1)E],\end{aligned} \tag{26.4}$$

and

$$\begin{aligned}C &= A + \frac{i}{2k_1}(k_2^2-k_1^2)\int_0^L e^{-ik_1x}s(x)\,dx\\ &= \frac{1}{2k_1}[(k_2+k_1)\,e^{i(k_2-k_1)L}D - (k_2-k_1)\,e^{-i(k_2+k_1)L}E]\end{aligned} \tag{26.5}$$

so that a sequential scheme for determining all the secondary wave amplitudes is thereby completed.

The outcome of solving the linear equations (26.3) for D, E is

$$D = \frac{2k_2(k_2+k_1)}{(k_2+k_1)^2-(k_2-k_1)^2\,e^{2ik_2L}}A, \qquad E = \frac{2k_1(k_2-k_1)\,e^{2ik_2L}}{(k_2+k_1)^2-(k_2-k_1)^2\,e^{2ik_2L}}A \tag{26.6}$$

and thus, after introducing the dimensionless wave number ratio

$$n = k_2/k_1, \tag{26.7}$$

the reflection and transmission coefficients which follow from (26.4), (26.5) prove to be

$$r = \frac{B}{A} = \frac{(n^2-1)(e^{2ink_1L}-1)}{(n+1)^2-(n-1)^2\,e^{2ink_1L}} = \frac{(1-n^2)\sin nk_1L}{(1+n^2)\sin nk_1L + 2in\cos nk_1L} \tag{26.8}$$

and

$$t = \frac{C}{A} = \frac{4n\,\mathrm{e}^{ik_1(n-1)L}}{(n+1)^2 - (n-1)^2\,\mathrm{e}^{2\,ink_1L}} = \frac{2\,in\,\mathrm{e}^{-ik_1L}}{(1+n^2)\sin nk_1L + 2\,in\cos nk_1L} \tag{26.9}$$

A null reflection coefficient obtains in the circumstance that

$$k_2L = nk_1L = 2\pi L/\lambda_2 = N\pi \tag{26.10}$$

with N integral, and for this set of discrete frequencies or wave lengths $\lambda_2 = (N/2)L$ the appertaining transmission coefficient

$$t = (-1)^N\,\mathrm{e}^{-i\pi N/n} \tag{26.11}$$

has its maximum absolute value, unity. It is a simple matter to establish the relations

$$|r|^2 = \frac{(1-n^2)^2\sin^2 nk_1L}{(1+n^2)^2\sin^2 nk_1L + 4n^2\cos^2 nk_1L} = 1 - \frac{4n^2}{(1+n^2)^2\sin^2 nk_1L + 4n^2\cos^2 nk_1L}$$
$$= 1 - |t|^2 \tag{26.12}$$

and to infer therefrom the corresponding extrema

$$|r|^2_{\max} = \left(\frac{1-n^2}{1+n^2}\right), \qquad |t|^2_{\min} = \frac{(2n)^2}{(1+n^2)^2} \tag{26.13}$$

when

$$k_2L = nk_1L = \left(N + \frac{1}{2}\right)\pi \tag{26.14}$$

If we write (26.8) in the version

$$r = \frac{k_1 - k_2}{k_1 + k_2}\,\frac{1 - \mathrm{e}^{2\,ik_2L}}{1 - \left(\dfrac{k_1 - k_2}{k_1 + k_2}\right)^2 \mathrm{e}^{2\,ik_2L}} \tag{26.15}$$

and expand the fraction containing $\mathrm{e}^{2\,ik_2L}$ in ascending powers, the consequent series

$$r = \frac{k_1 - k_2}{k_1 + k_2}\left[1 - \frac{4k_1k_2}{(k_1 + k_2)^2}\,\mathrm{e}^{2\,ik_2L} + 0(\mathrm{e}^{4\,ik_2L}) + \cdots\right] \tag{26.16}$$

has individual terms which admit of ready interpretation. Thus, in the first or wholly L-independent term, we can identify the reflection coefficient

$$r_\infty^{(+)} = \frac{k_1 - k_2}{k_1 + k_2} = \frac{c_2 - c_1}{c_2 + c_1} = \frac{1 - n}{1 + n} \tag{26.17}$$

for a wave incident from the left along a semi-infinite string wherein the wave number is k_1, joined at the origin ($x = 0$) to another uniform semi-infinite string wherein the wave number is k_2. The second term, which has an additional phase $2k_2L + \pi$ relative to the first, can be ascribed to a motion communicated through the section $0 < x < L$, reflected at the junction $x = L$ and then transmitted across the junction $x = 0$ towards the left; its amplitude is compounded from the product of two transmission factors (at $x = 0$) for oppositely directed primary waves

$$t_\infty^{(+)} \cdot t_\infty^{(-)} = \frac{2k_1}{k_1 + k_2} \cdot \frac{2k_2}{k_1 + k_2}$$

and a single reflection factor (relative to $x = L$) that is the negative of (26.17).

Succeeding terms of the expansion (26.16) are evidently linked to further internal reflections within the segment $0 < x < L$, before the eventual re-emergence of a wave with suitably depleted amplitude in $x < 0$. The condition (26.10) for a vanishing overall reflection coefficient is simply one of mutual (destructive) interference between the aggregate of such partial contributions thereto; we note, in fact, that if $\delta = \exp(2\,\mathrm{i}k_2L)$, then

$$\frac{r}{r_\infty^{(+)}} = \lim_{\delta \to 1}\{1 - (1 - r_\infty^{(+)2})\delta - r_\infty^{(+)2}(1 - r_\infty^{(+)2})\delta^2 - \cdots\} = 0$$

because of paired cancellations. Apart from the significance claimed inferentially for $r_\infty^{(+)}$ and $t_\infty^{(+)}$ in the preceding, it may be observed that their forms are directly implied by the expressions (26.8), (26.6) for the amplitude ratios B/A and D/A when $L \to \infty$, if it be assumed that $\operatorname{Im} k_2 > 0$ and thus $\exp(2\,\mathrm{i}k_2L) \to 0$.

The Fourier transform technique employed for the direct solution of the integral equation (25.18) on a semi-infinite interval $0 < x < \infty$ can also be adapted to deal with the corresponding equation on an interval of finite extent, and hence affords an independent means of arriving at the results obtained in this section. A full account of the details will be omitted here, though some features not present in the former analysis deserve mention. Thus, if the central portion of the string is disposed symmetrically about the origin, with its boundaries at $x = -L/2, +L/2$, and if $s_-(x)$, $s_+(x)$ designate the reflected and transmitted wave factors on the intervals $x < -L/2$, $x > L/2$, respectively, their corresponding Fourier transforms

$$s_-(\zeta) = \int_{-\infty}^{-L/2} \mathrm{e}^{\mathrm{i}\zeta x} s_-(x)\,\mathrm{d}x, \qquad s_+(\zeta) = \int_{L/2}^{\infty} \mathrm{e}^{\mathrm{i}\zeta x} s_+(x)\,\mathrm{d}x \tag{26.18}$$

turn out to be interconnected by the relations (compare (25.27), (25.28))

$$s_-(\zeta) = \frac{k_1 - k_2}{\zeta - k_1}\mathrm{e}^{\mathrm{i}(k_2 - \zeta)L/2} s_+(k_2) - \mathrm{i}A\frac{k_1 - k_2}{k_1 - k_2}\mathrm{e}^{-\mathrm{i}(k+\zeta)L/2} \cdot \frac{1}{\zeta - k_1}, \tag{26.19}$$

$$s_+(\zeta) = -\frac{k_1 - k_2}{\zeta + k_1}\mathrm{e}^{\mathrm{i}(k_2 + \zeta)L/2} s_-(-k_2) + \frac{\mathrm{i}A}{\zeta + k_1}\mathrm{e}^{-\mathrm{i}(k_1 - \zeta)L/2 + \mathrm{i}k_2 L} \tag{26.20}$$

which put in evidence the particular and as yet unknown, transforms $s_+(k_2), s_-(-k_2)$. To obtain explicit values for the latter, and thence for $s_-(\zeta), s_+(\zeta)$ generally, we substitute $\zeta = -k_2$ in (26.19) and $\zeta = k_2$ in (26.20), thereby securing a pair of linear equations for $s_+(k_2), s_-(-k_2)$. Once in possession of their values, the reflection and transmission coefficients relative to the point $x = 0$ (and not the left-hand boundary of the central portion) follow from

$$R = \frac{k_1 - k_2}{k_1 + k_2} \mathrm{e}^{-\mathrm{i}k_1 L} + \frac{\mathrm{i}}{A}(k_1 - k_2)s_+(k_2)\, \mathrm{e}^{\mathrm{i}(k_2 - k_1)L/2},$$

$$T = \mathrm{e}^{\mathrm{i}(k_2 - k_1)L} + \frac{\mathrm{i}}{A}(k_1 - k_2)s_-(- k_2)\, \mathrm{e}^{\mathrm{i}(k_2 - k_1)L/2}$$

and the results are in agreement with (26.8), (26.9), viz.: $R = \mathrm{e}^{-\mathrm{i}k_1 L} r$, $T = t$. The transform of the central displacement factor $s(x)$, $-L/2 < x < L/2$, can be directly connected with the foregoing transforms on the external ranges, and thus the remaining aspects of the wave complex are revealed.

It needs to be remarked, in concluding this section, that although a Green's function-integral equation approach calls for more effort than does the direct integration of the space factor differential equation in the simple problems discussed, it is nevertheless capable of important formal and practical elaboration. Such an approach lends itself to the analysis of scattering problems in more than one dimension, where the advantages of its compactness stand out in contrast to the greater complexity of direct integration and superposition of particular solutions of the wave equation; moreover, it can be usefully recast into variational schemes for estimation of the secondary wave amplitudes in problems wherein the differential equations are not susceptible to explicit solution [cf. §75].

§27. *A Multi-section String*

Let us next consider an infinite string with a number N of adjacent sections, located on the intervals $(x_1, x_2), (x_2, x_3), \ldots, (x_N, x_{N+1})$ of span $l_1 = x_2 - x_1, l_2 = x_3 - x_2, \ldots, l_N = x_{N+1} - x_N$, wherein the wave number possesses distinct and constant values $k_1, k_2, \ldots, k_N$; a common value k is assumed to hold elsewhere (i.e., $x < x_1$, $x > x_{N+1}$) and serves for contrasting the material parameters of the individual sections by means of the ratios

$$n_1 = \frac{k_1}{k} = \surd(\rho_1/\rho), \ldots, n_N = \frac{k_N}{k} = \surd(\rho_N/\rho). \tag{27.1}$$

If A_0, B_0 denote the incoming and outgoing amplitude factors of the periodic wave complex to the left of the first section and A_1, B_1 are the corresponding factors within the latter, the continuity requirements for the total wave function and its x-derivative at $x = x_1$ imply that

$$A_0\,e^{ikx_1} + B_0\,e^{-ikx_1} = A_1\,e^{ikn_1x_1} + B_1\,e^{-ikn_1x_1},$$
$$A_0\,e^{ikx_1} - B_0\,e^{-ikx_1} = n_1A_1\,e^{ikn_1x_1} - n_1B_1\,e^{-ikn_1x_1}$$

or

$$\begin{pmatrix} e^{ikx_1} & e^{-ikx_1} \\ e^{ikx_1} & -\,e^{-ikx_1} \end{pmatrix}\begin{pmatrix} A_0 \\ B_0 \end{pmatrix} = \begin{pmatrix} e^{ikn_1x_1} & e^{-ikn_1x_1} \\ n_1\,e^{ikn_1x_1} & -\,n_1\,e^{-ikn_1x_1} \end{pmatrix}\begin{pmatrix} A_1 \\ B_1 \end{pmatrix} \tag{27.2}$$

in terms of a matrix notation. Following multiplication on the left with a reciprocal of the lead matrix, we have

$$\begin{pmatrix} A_0 \\ B_0 \end{pmatrix} = \frac{1}{2}\begin{pmatrix} e^{-ikx_1} & e^{-ikx_1} \\ e^{ikx_1} & -\,e^{ikx_1} \end{pmatrix}\begin{pmatrix} e^{ikn_1x_1} & e^{-ikn_1x_1} \\ n_1\,e^{ikn_1x_1} & -\,n_1\,e^{-ikn_1x_1} \end{pmatrix}\begin{pmatrix} A_1 \\ B_1 \end{pmatrix} \tag{27.3}$$

and when the same continuity requirements are imposed at the common point x_2 of the first and second sections, yielding a relationship

$$\begin{pmatrix} A_1 \\ B_1 \end{pmatrix} = \frac{1}{2}\begin{pmatrix} e^{-ikn_1x_2} & \dfrac{1}{n_1}\,e^{-ikn_1x_2} \\ e^{ikn_1x_2} & -\dfrac{1}{n_1}\,e^{ikn_1x_2} \end{pmatrix}\begin{pmatrix} e^{ikn_2x_2} & e^{-ikn_2x_2} \\ n_2\,e^{ikn_2x_2} & -\,n_2\,e^{-ikn_2x_2} \end{pmatrix}\begin{pmatrix} A_2 \\ B_2 \end{pmatrix}. \tag{27.4}$$

where A_2, B_2 represent the amplitude factors of the component waves in the second section, the result of combining (27.3), (27.4) takes the form

$$\begin{pmatrix} A_0 \\ B_0 \end{pmatrix} = \frac{1}{2}\begin{pmatrix} e^{-ikx_1} & e^{-ikx_1} \\ e^{ikx_1} & -\,e^{ikx_1} \end{pmatrix}\begin{pmatrix} \cos kn_1l_1 & -\dfrac{i}{n_1}\sin kn_1l_1 \\ -\,in_1\sin kn_1l_1 & \cos kn_1l_1 \end{pmatrix}.$$
$$\times\begin{pmatrix} e^{ikn_2x_2} & e^{-ikn_2x_2} \\ n_2\,e^{ikn_2x_2} & -\,n_2\,e^{-ikn_2x_2} \end{pmatrix}\begin{pmatrix} A_2 \\ B_2 \end{pmatrix}.$$

A repetition of this procedure, which utilizes the requisite interconnections and permits the elimination of all internal amplitude factors for the N sections, reveals that

$$\begin{pmatrix} A_0 \\ B_0 \end{pmatrix} = \frac{1}{2}\begin{pmatrix} e^{-ikx_1} & e^{-ikx_1} \\ e^{ikx_1} & -\,e^{ikx_1} \end{pmatrix}\prod_{j=1}^{N}\begin{pmatrix} \cos(kn_jl_j) & -\dfrac{i}{n_j}\sin(kn_jl_j) \\ -\,in_j\sin(kn_jl_j) & \cos(kn_jl_j) \end{pmatrix}. \tag{27.5}$$
$$\times\begin{pmatrix} e^{ikx_{N+1}} & e^{-ikx_{N+1}} \\ e^{ikx_{N+1}} & -\,e^{-ikx_{N+1}} \end{pmatrix}\begin{pmatrix} A_{N+1} \\ B_{N+1} \end{pmatrix};$$

when there is but one incident wave, from the left, we have $B_{N+1} = 0$ and the ratios B_0/A_0, A_{N+1}/A_0 provided by (27.5) specify a pair of reflection and transmission coefficients that characterize the overall scattering properties of the N-section unit. In particular, if the wave numbers and breadths of the sections are such that

$$kn_1l_1 = kn_2l_2 = \cdots = kn_Nl_N = p\pi, \quad p \text{ integral}$$

and the special values $x_1 = 0$, $x_{N+1} = L = l_1 + l_2 + \cdots + l_N$ are introduced in (27.5), the consequent form thereof

$$\begin{pmatrix} A_0 \\ B_0 \end{pmatrix} = \frac{1}{2}\begin{pmatrix} 1 & 1 \\ 1 & -1 \end{pmatrix}\begin{pmatrix} e^{ikL} & e^{-ikL} \\ e^{ikL} & -e^{-ikL} \end{pmatrix}\begin{pmatrix} A_{N+1} \\ 0 \end{pmatrix}$$
$$= \begin{pmatrix} e^{ikL} & 0 \\ 0 & e^{-ikL} \end{pmatrix}\begin{pmatrix} A_{N+1} \\ 0 \end{pmatrix}$$

implies a null reflection coefficient,

$$r = \frac{B_0}{A_0} = 0$$

along with a transmission coefficient

$$t = \frac{A_{N+1}}{A_0} = e^{-ikL} = e^{-i\pi p((1/n_1)+(1/n_2)+\cdots+(1/n_N))}$$

of unit absolute magnitude.

§28. *Initial and Boundary Value Problems for a Sectionally Uniform String of Finite Length*

Suppose that the outer parts are removed from the infinite string which came in for attention above, leaving a finite length made up of N uniform sections with endpoints at $x_1 = 0$ and $x_{N+1} = L$, the points of interconnection being located at $x_2, \ldots, x_N$. Let us assign the individual coordinates ξ_i, $i = 1, \ldots, N$ to these sections, with the ranges $0 \le \xi_i \le l_i = x_{i+1} - x_i$, and inquire after solutions of the respective wave equations

$$\frac{\partial^2 y_1}{\partial \xi_i^2} = \frac{1}{c_i^2}\frac{\partial^2 y_i}{\partial t^2}, \quad c_i^2 = P/\rho_i \tag{28.1}$$

that are coupled through the boundary conditions

$$y_i(l_i, t) = y_{i+1}(0, t), \qquad \frac{\partial y_i(l_i, t)}{\partial \xi_i} = \frac{\partial y_{i+1}(0, t)}{\partial \xi_{i+1}} \tag{28.2}$$

and also satisfy the initial conditions

$$y_i(\xi_i, 0) = s_0(\xi_i), \qquad \frac{\partial y_i(\xi_i, 0)}{\partial t} = v_0(\xi_i). \tag{28.3}$$

One procedure employs the separated variable representations

$$y_i(\xi_i, t) = s_i(\xi_i, \omega) \begin{matrix}\cos \\ \sin\end{matrix}\, \omega t,$$

with a common time factor for the displacement in all sections of the string, and takes account of the conditions (28.2) through the particular integrals

$$s_{i+1}(\xi_{i+1}, \omega) = s_i(l_i, \omega) \cos \frac{\omega \xi_{i+1}}{c_{i+1}} + \frac{c_{i+1}}{\omega} \frac{ds_i(l_i, \omega)}{d\xi_i} \sin \frac{\omega \xi_{i+1}}{c_{i+1}} \tag{28.4}$$

of the ordinary differential equations

$$\frac{d^2 s_i}{d\xi_i^2} + \frac{\omega^2}{c_i^2} s_i = 0 \tag{28.5}$$

for the coordinate and frequency dependent factors $s_i(\xi_i, \omega)$. Following a determination of $s_1(\xi_1, \omega)$ in the first section that is consonant with (28.5) and an appropriate boundary condition at $\xi_1 = 0$, i.e.,

$$s_1(\xi_1, \omega) = \sin \frac{\omega}{c_1} \xi_1,$$

if the end is fixed, the corresponding factor $s_N(\xi_N, \omega)$ for the last section can be found through joint utilization of the recurrences (28.4); and when a boundary condition at $\xi_N = l_N$ is imposed, namely

$$s_N(l_N, \omega) = 0 \tag{28.6}$$

if this end is also fixed, we obtain thereby an equation that yields all the natural frequencies of free vibration of the sectioned string. The latter frequencies form a discrete set $\omega_1 < \omega_2 < \ldots$, in general, and the superposition of product solutions

$$y_i(\xi_i, t) = \sum_{j=1}^{\infty} s_i(\xi_i, \omega_j)[A_j \cos \omega_j t + B_j \sin \omega_j t] \tag{28.7}$$

accomodates the initial assignments of displacement and velocity, provided that A_j, B_j are the coefficients suited to the expansions

$$s_0(\xi_i) = \sum_{j=1}^{\infty} A_j s_i(\xi_i, \omega_j)$$

and $\qquad 0 \le \xi_i \le l_i \qquad$ (28.8)

$$v_0(\xi_i) = \sum_{j=1}^{\infty} \omega_j B_j s_i(\xi_i, \omega_j)$$

which characterize the given functions $s_0(\xi_i)$, $v_0(\xi_i)$ in terms of the particular set $\{s_i(\xi_i, \omega_j)\}$.

Inasmuch as

$$(\omega_j^2 - \omega_k^2) \sum_{i=1}^{N} \frac{1}{c_i^2} \int_0^{l_i} s_i(\xi_i, \omega_j) s_i(\xi_i, \omega_k)\, d\xi_i \tag{28.9}$$

$$= \sum_{i=1}^{N} \int_0^{l_i} \left[s_i(\xi_i, \omega_j) \frac{d^2 s_i(\xi_i, \omega_k)}{d\xi_i^2} - s_i(\xi_i, \omega_k) \frac{d^2 s_i(\xi_i, \omega_j)}{d\xi_i^2} \right] d\xi_i,$$

the orthogonality relation

$$\sum_{i=1}^{N} \frac{1}{c_i^2} \int_0^{l_i} s_i(\xi_i, \omega_j) s_i(\xi_i, \omega_k)\, d\xi_i = 0, \qquad \omega_j \neq \omega_k \tag{28.10}$$

obtains if the sum on the right-hand side of (28.9) vanishes; and this proves to be the case, in fact, when the ends of the string are secured, having regard for the boundary conditions (28.2) at the points of interconnection. Accordingly, on the basis of (28.10), the individual coefficient determinations

$$A_j = \frac{\displaystyle\sum_{i=1}^{N} \frac{1}{c_i^2} \int_0^{l_i} s_0(\xi_i) s_i(\xi_i, \omega_j)\, d\xi_i}{\displaystyle\sum_{i=1}^{N} \frac{1}{c_i^2} \int_0^{l_i} s_i^2(\xi_i, \omega_j)\, d\xi_i} \tag{28.11}$$

and

$$\omega_j B_j = \frac{\displaystyle\sum_{i=1}^{N} \frac{1}{c_i^2} \int_0^{l_i} v_0(\xi_i) s_i(\xi_i, \omega_j)\, d\xi_i}{\displaystyle\sum_{i=1}^{N} \frac{1}{c_i^2} \int_0^{l_i} s_i^2(\xi_i, \omega_j)\, d\xi_i} \tag{28.12}$$

serve to complete a specification of the desired solution. The latter is suitable for ascertaining the state of excitation at any time, though changes of the initial data necessitate the recalculation of the coefficients A_j, B_j that enter into the development (28.7).

A different approach to the problem utilizes the progressive wave solutions of the equations (28.1),

$$y_i(\xi_i, t) = f_i\left(t - \frac{\xi_i}{c_i}\right) + g_i\left(t + \frac{\xi_i}{c_i}\right), \quad i = 1, \ldots, N \tag{28.13}$$

along with the coefficients of reflection and transmission

$$r_{i,i+1} = \frac{c_{i+1} - c_i}{c_{i+1} + c_i}, \qquad t_{i,i+1} = \frac{2c_i}{c_{i+1} + c_i} \tag{28.14}$$

appropriate to incidence from the left on the junction (at $\xi_i = 0$) between unlimited sections having the distinct wave velocities $c_i(\xi_i < 0)$ and $c_{i+1}(\xi_i > 0)$. For the opposite direction of incidence the analogous coefficients are

$$r_{i+1,i} = -r_{i,i+1}, \qquad t_{i+1,i} = \frac{2c_{i+1}}{c_{i+1} + c_i} \tag{28.15}$$

and, evidently,

$$t_{i,i+1}t_{i+1,i} - r_{i,i+1}r_{i+1,i} = 1.$$

If the motion commences at $t = 0$ the functions f_i and g_i can be assigned null values when their arguments are less than $\tau_i = l_i/c_i$ and 0, respectively, since the initial data provide no information relating to such ranges. A determination of f_i and g_i for complementary values of the argument, in conformity with the boundary and initial conditions, proceeds from the interconnections at an arbitrary joining point $\xi_i = 0$,

$$f_i(t) = r_{i,i-1}g_i(t) + t_{i-1,i}f_{i-1}(t - \tau_{i-1}), \tag{28.16}$$

$$t > 0$$

$$g_{i-1}(t + \tau_{i-1}) = r_{i-1,i}f_{i-1}(t - \tau_{i-1}) + t_{i,i-1}g_i(t), \tag{28.17}$$

that express the self-consistency of alternative realizations of oppositely directed wave profiles. Both positive and negative values of the argument for the functions f_i, f_{i-1} may occur in these equations and, before they are capable of fully specifying said functions, it is necessary to employ the information provided by the initial conditions. The latter imply that

$$f_i\left(-\frac{\xi_i}{c_i}\right) = \frac{1}{2}s_i(\xi_i) - \frac{1}{2c_i}\int_0^{\xi_i} v_i(x)\,dx,$$

$$g_i\left(\frac{\xi_i}{c_i}\right) = \frac{1}{2}s_i(\xi_i) + \frac{1}{2c_i}\int_0^{\xi_i} v_i(x)\,dx,$$

thus furnishing the values of f_i and g_i when $0 < \xi_i/c_i < \tau_i = l_i/c_i$; if we use the desginations F_i and G_i to make explicit such (partial) determinations and write

$$f_i(t - \tau_i) = F_i(t - \tau_i), \quad 0 < t < \tau_i \tag{28.18}$$

the direct consequence of (28.16),

$$f_i(t - \tau_i) = r_{i,i-1}g_i(t - \tau_i) + t_{i-1,i}f_{i-1}(t - \tau_{i-1} - \tau_i), \quad \tau_i < t \tag{28.19}$$

combines with (28.18) to yield the recurrence relation

$$f_i(t - \tau_i) - F_i(t - \tau_i) = r_{i,i-1}g_i(t - \tau_i) + t_{i-1,i}f_{i-1}(t - \tau_{i-1} - \tau_i), \quad t > 0 \tag{28.20}$$

that incorporates both initial and boundary conditions. From (28.17) and the characterization

$$g_{i-1}(t) = G_{i-1}(t), \quad 0 < t < \tau_{i-1} \tag{28.21}$$

there obtains a further recurrence relation

$$g_{i-1}(t) - G_{i-1}(t) = r_{i-1,i}f_{i-1}(t - 2\tau_{i-1}) + t_{i,i-1}g_i(t - \tau_{i-1}), \quad t > 0 \tag{28.22}$$

which, coupled with (28.20), affords a means of determining the $2n$ functions

f_i, g_i. Thus, let

$$\bar{h}(p) = \int_0^\infty e^{-pt} h(t)\, dt$$

denote the Laplace transform of a function $h(t)$ and observe that the transformed version of the relations (28.20), (28.22), namely

$$\bar{f}_i(p) - \bar{F}_i(p) = r_{i,i-1}\bar{g}_i(p) + t_{i-1,i}\, e^{-\tau_{i-1}p}\bar{f}_{i-1}(p)$$
$$\bar{g}_i(p) - \bar{G}_i(p) = t_{i+1,i}\, e^{-\tau_i p}\bar{g}_{i+1}(p) + r_{i,i+1}\, e^{-2\tau_i p}\bar{f}_i(p),$$

constitute an inhomogeneous linear system of $2n$ equations for $\bar{f}_i(p), \bar{g}_i(p)$. After the latter functions are found inverse Laplace transformation specifies the time-dependent components $f_i(t), g_i(t)$ of the displacement in each section of the string; their synthesis from multiple reflection and transmission of the initial wave profiles, however, limits the practicality of this scheme to the early stages of the motion [Beddoe, 1966].

§29. *Excitation of a String with Variable Length*

The initial value and signalling problems considered heretofore warrant a fresh analysis, of interest in its own right, if the string has a variable length. Let $x = 0$ and $x = \xi(t)$ mark the extremities of the string at a time t, where the latter coincides with the instantaneous position of a close fitting ring that can move in prescribed fashion along the line of the string. Assuming that the ring merely secures a null displacement of the string at the point of contact, the specifications

$$\left(\frac{\partial^2}{\partial x^2} - \frac{1}{c^2}\frac{\partial^2}{\partial t^2}\right) y(x, t) = 0,$$
$$y(0, t) = F(t), \qquad y(\xi(t), t) = 0, \quad t > t_0, \tag{29.1}$$
$$y(x, t_0) = s_0(x), \qquad \frac{\partial}{\partial t} y(x, t_0) = v_0(x), \quad 0 \le x \le \xi(t_0)$$

jointly define a boundary and initial value problem appropriate to string motions on the interval $[0, \xi(t)]$.

There are alternative ways of proceeding with the analysis, depending on whether the system (29.1) is directly utilized or subjected to a preliminary transformation; if the first option is exercised, the general integral of the homogeneous wave equation,

$$y(x, t) = f\left(t - \frac{x}{c}\right) + g\left(t + \frac{x}{c}\right) \tag{29.2}$$

becomes available for imposing the other conditions, while the second option enables the problem to be reformulated on a fixed interval, although the differential equation then assumes a more complicated form. Both options present technical difficulties if a full measure of generality is sought and wide latitude given in the matter of assigned data, and it must therefore suffice here to indicate some of the aspects encountered in developing the theory.

When the displacement at the end $x = 0$ vanishes, or $F(t) \equiv 0$, self-cancellation of the two terms in (29.2) implies that

$$y(x, t) = f\left(t - \frac{x}{c}\right) - f\left(t + \frac{x}{c}\right); \tag{29.3}$$

thus, if

$$\xi = Ut,$$

which signifies a steady rate of change in effective length of the string, the second boundary condition, $y(Ut, t) = 0$, yields the functional equation

$$f\left(\left(1 - \frac{U}{c}\right)t\right) = f\left(\left(1 + \frac{U}{c}\right)t\right), \quad t > 0 \tag{29.4}$$

wherein the arguments of f have a fixed ratio. To solve the latter, a so-called homogeneous difference equation of the geometric type, with the alternative version

$$f(\gamma\tau) = f(\tau), \quad \tau > 0, \qquad \gamma = \frac{1 + U/c}{1 - U/c} \tag{29.5}$$

let

$$\tau = \mathrm{e}^{\eta}, \qquad f(\mathrm{e}^{\eta}) = \bar{f}(\eta) \tag{29.6}$$

and observe that (29.4) acquires the form

$$\bar{f}(\eta + \log \gamma) = \bar{f}(\eta), \quad -\infty < \eta < \infty \tag{29.7}$$

which manifests a periodicity in $\bar{f}(\eta)$ on the scale

$$\log \gamma = 2\pi/\delta, \tag{29.8}$$

say, in the variable η.

Hence, we may characterize the function $f(\tau)$ by means of the expansion

$$f(\tau) = \sum_{n=0}^{\infty} [A_n \sin(n\delta \log \tau) + B_n \cos(n\delta \log \tau)] \tag{29.9}$$

and subsequently obtain from (29.3) the string displacement representation

$$\begin{aligned} y(x, t) = 2 \sum_{n=1}^{\infty} \sin\left(\frac{n\delta}{2} \log \frac{ct - x}{ct + x}\right)\bigg[A_n \cos\left(\frac{n\delta}{2} \log \frac{c^2t^2 - x^2}{c^2t_0^2}\right) \\ - B_n \sin\left(\frac{n\delta}{2} \log \frac{c^2t^2 - x^2}{c^2t_0^2}\right)\bigg]. \end{aligned} \tag{29.10}$$

To satisfy the initial conditions,

$$y(x, t_0) = s_0(x), \qquad \frac{\partial}{\partial t} y(x, t_0) = v_0(x), \quad 0 < x < Ut_0 \tag{29.11}$$

by appropriate choice of the expansion coefficients A_n, B_n, we note, firstly, that the combinations

$$s_0(-x) = -s_0(x), \qquad v_0(-x) = -v_0(x), \tag{29.12}$$

are in keeping with the odd symmetry (in x) of (29.10) and its time derivative, and, secondly, that

$$f\left(t_0 + \frac{x}{c}\right) = -\frac{1}{2}\left[s_0(x) + \frac{1}{c}\int_0^x v_0(\xi)\, d\xi\right] \tag{29.13}$$

is thereby defined on the range $-Ut_0 < x < Ut_0$. Thus, substituting $\tau = t_0 + (x/c)$ in (29.9) and applying the orthogonality relations

$$\int_{-Ut_0}^{Ut_0} \begin{matrix}\sin\\ \cos\end{matrix}\left(m\delta \log \frac{ct_0 + x}{ct_0}\right) \cdot \begin{matrix}\sin\\ \cos\end{matrix}\left(n\delta \log \frac{ct_0 + x}{ct_0}\right) \frac{dx}{ct_0 + x}$$
$$= \int_{\log(c-U)/c}^{\log(c-U)/c + 2\pi/\delta} \begin{matrix}\sin\\ \cos\end{matrix}(m\delta z) \cdot \begin{matrix}\sin\\ \cos\end{matrix}(n\delta z)\, dz = 0, \quad m \neq n$$

it follows that

$$\left.\begin{matrix}A_n\\ B_n\end{matrix}\right\} = \frac{\delta}{\pi}\int_{-Ut_0}^{Ut_0} f\left(t_0 + \frac{x}{c}\right) \begin{matrix}\sin\\ \cos\end{matrix}\left(n\delta \log \frac{ct_0 + x}{ct_0 - x}\right) \frac{dx}{ct_0 + x}$$
$$= -\frac{\delta}{2\pi}\int_{-Ut_0}^{Ut_0}\left[s_0(x) + \frac{1}{c}\int_0^x v_0(\xi)\, d\xi\right] \begin{matrix}\sin\\ \cos\end{matrix}\left(n\delta \log \frac{ct_0 + x}{ct_0 - x}\right) \frac{dx}{ct_0 + x} \tag{29.14}$$

provide the sought for coefficient determinations in terms of the initial data.

For purposes of retrieving from (29.14) the expressions suited to a string of fixed length, let the integration variable be scaled in accordance with the relation

$$x = Ut_0 \frac{x'}{L}, \qquad -L < x' < L;$$

next, availing ourselves of the estimates

$$\delta = \frac{\pi c}{U} + 0\left(\frac{U}{c}\right)$$

and

$$\log\left(1 \pm \frac{x}{ct_0}\right) = \log\left(1 \pm \frac{Ux'}{cL}\right) = \pm \frac{Ux'}{cL} + 0\left(\frac{U}{c}\right)^2,$$

we obtain

$$A_n = -\frac{1}{2L}\int_{-L}^{L} s_0(x')\sin\frac{n\pi x'}{L}\,\mathrm{d}x' \tag{29.15}$$

$$U/c \to 0$$

$$B_n = -\frac{1}{2cL}\int_{-L}^{L}\left(\int_0^{x'} v_0(\xi)\,\mathrm{d}\xi\right)\cos\frac{n\pi x'}{L}\,\mathrm{d}x' \tag{29.16}$$

after the symmetry properties (29.12) are taken into account. The expression (29.15) is directly corroborated when we refer to a limiting form of the representation (29.10), viz.

$$y(x', t_0) = s_0(x') = -2\sum_{n=1}^{\infty} A_n \sin\frac{n\pi x'}{L}, \quad -L < x' < L, \quad \frac{U}{c}\to 0;$$

and (29.16) may likewise be verified.

If an inhomogeneous boundary condition, $y(0, t) = F(t)$, is given at the extremity of the string which remains in the position $x = 0$, the representation

$$y(x, t) = f\left(t - \frac{x}{c}\right) - f\left(t + \frac{x}{c}\right) + F\left(t + \frac{x}{c}\right) \tag{29.17}$$

takes the place of (29.3); and there now follows, by imposition of the prior constraint $y(\xi(t), t) = 0$ at the moving extremity, a functional equation of inhomogeneous nature,

$$f\left(t - \frac{\xi(t)}{c}\right) - f\left(t + \frac{\xi(t)}{c}\right) + F\left(t + \frac{\xi(t)}{c}\right) = 0, \quad t > t_0$$

that determines f. With the specific choices

$$F(t) = \sin\omega t,$$

$$\xi(t) = Ut$$

the latter equation becomes [compare (29.5)]

$$f(\gamma\tau) = f(\tau) + \sin\omega\gamma\tau, \quad \tau > 0. \tag{29.18}$$

A regular solution of (29.18), characterized by a power series expansion in τ, is available once the derivatives of f at zero argument are calculated therefrom, and this proves to be

$$f(\tau) = f(0) + \sum_{n=0}^{\infty} (-1)^n \frac{(\omega\tau)^{2n+1}}{(2n+1)!}\cdot\frac{1}{1-\gamma^{-(2n+1)}}. \tag{29.19}$$

An expansion, in reciprocal powers of $\gamma(>1)$, for the last factor of the sum enables us to infer that

$$f(\tau) = f(0) + \sum_{n=0}^{\infty} \sin\frac{\omega\tau}{\gamma^n} \tag{29.20}$$

and hence

$$y(x,t) = f\left(t - \frac{x}{c}\right) - f\left(\frac{t + (x/c)}{\gamma}\right)$$
$$= \sin \omega\left(t - \frac{x}{c}\right) - 2\sum_{n=1}^{\infty} \sin\frac{\omega x}{c\gamma^n}\cos\frac{\omega t}{\gamma^n}, \quad ct > x \tag{29.21}$$

is a solution of the time-periodic signalling problem; evidently, the first term connotes a faithful transmission of the end displacement or input, while the sum, that tends to zero if $U \to c$ and $\gamma \to \infty$, represents an aggregate of standing and frequency shifted waves which are activated by the moving termination or ring. It is possible to construct another, namely integral form of, solution to the constant coefficient first-order difference equation (29.18) and thereby examine the circumstance wherein the ring passes slowly through any of the locations $\xi_n = n\pi c/\omega$ that characterize resonant lengths of a string fixed at one end and subject to harmonic displacement with frequency ω at the other end [Greenspan, 1963]. The availability of a general solution (29.9) of the related homogeneous difference equation allows, moreover, for the accomodation of initial conditions in a straightforward manner.

When we seek to develop the analysis in a more conventional frame, that is with a fixed range for one of the independent variables, a transformation of the wave equation becomes necessary at the outset. Let

$$\xi(t) = l + Ut \tag{29.22}$$

specify the ring position at any time, so that l represents the length of the string at $t = 0$, and introduce a new pair of (dimensionless) variables

$$\zeta = \frac{Ux/lc}{1 + \dfrac{Ut}{l}}, \qquad \eta = \log\left(1 + \frac{Ut}{l}\right) \tag{29.23}$$

with the ranges

$$0 \le \zeta \le \frac{U}{c}, \quad 0 \le \eta < \infty \tag{29.24}$$

that correspond to those of the original pair,

$$0 \le x \le \xi(t), \quad 0 \le t < \infty.$$

After effecting this change of independent variables in the wave equation for the string displacement y, we obtain a partial differential equation

$$(1-\zeta^2)\frac{\partial^2 y}{\partial \zeta^2} - 2\zeta\frac{\partial y}{\partial \zeta} + 2\zeta\frac{\partial^2 y}{\partial \zeta \partial \eta} - \frac{\partial^2 y}{\partial \eta^2} + \frac{\partial y}{\partial \eta} = 0, \tag{29.25}$$

with infinitely many separable or product solutions that vanish at $\zeta = 0$ and $\zeta = U/c$, viz.

$$y_n(\zeta, \eta) = \exp(in\delta\eta)\varphi_n(\zeta), \qquad n = 1, 2, \ldots \tag{29.26}$$

where

$$\begin{aligned} \varphi_n(\zeta) &= \frac{\Phi_n(\zeta)}{(1-\zeta^2)^{1/2}} \exp(in\delta \log(1-\zeta^2)^{1/2}), \\ \Phi_n(\zeta) &= \left(\frac{2\delta}{\pi}\right)^{1/2} (1-\zeta^2)^{1/2} \sin\left(n\delta \log\left(\frac{1+\zeta}{1-\zeta}\right)^{1/2}\right); \end{aligned} \tag{29.27}$$

the functions $\Phi_n(\zeta)$ constitute, relative to a weight factor $(1-\zeta^2)^{-2}$, an orthonormal set on $(0, (U/c))$, since

$$\begin{aligned} \int_0^{U/c} \Phi_m(\zeta)\Phi_n(\zeta)(1-\zeta^2)^{-2}\,\mathrm{d}\zeta &= \frac{2\delta}{\pi}\int_0^{\pi/\delta} \sin(m\delta z)\sin(n\delta z)\,\mathrm{d}z \\ &= \delta_{mn}, \end{aligned} \tag{29.28}$$

and, moreover, they satisfy the ordinary, homogeneous differential equation

$$\frac{\mathrm{d}^2\Phi_n}{\mathrm{d}\zeta^2} + \frac{1+n^2\delta^2}{(1-\zeta^2)^2}\Phi_n = 0, \quad n = 1, 2, \ldots. \tag{29.29}$$

The boundary and initial conditions of (29.1) acquire, when expressed in terms of ζ, η, the forms

$$\begin{aligned} &y(0, \eta) = F(\eta), \qquad y\left(\frac{U}{c}, \eta\right) = 0, \quad \eta > 0 \\ &y(\zeta, 0) = s_0(\zeta), \qquad \frac{\partial}{\partial \eta} y(\zeta, 0) = \zeta s_0'(\zeta) + \frac{v_0(\zeta)}{U/l}, \end{aligned} \tag{29.30}$$

and, consequent to the utilization of a Laplace transform

$$\bar{y}(\zeta, p) = \int_0^\infty \mathrm{e}^{-p\eta} y(\zeta, \eta)\,\mathrm{d}\eta, \tag{29.31}$$

we deduce from (29.25), (29.30) that

$$\begin{aligned} (1-\zeta^2)\frac{\mathrm{d}^2\bar{y}}{\mathrm{d}\zeta^2} - 2\zeta(1-p)\frac{\mathrm{d}\bar{y}}{\mathrm{d}\zeta} + p(1-p)\bar{y} &= \zeta s_0'(\zeta) + (1-p)s_0(\zeta) - \frac{v_0(\zeta)}{U/l} \\ &= \mathfrak{F}(\zeta, p), \end{aligned} \tag{29.32}$$

say. To eliminate the first order derivative in (29.32) and thereby arrive at a simpler inhomogeneous equation, we replace the dependent variable y by another

$$Y(\zeta, p) = \bar{y}(\zeta, p)(1-\zeta^2)^{(1-p)/2}, \tag{29.33}$$

which has the determining equations

$$\frac{\mathrm{d}^2 Y}{\mathrm{d}\zeta^2} + \frac{1-p^2}{(1-\zeta^2)^2} Y = \mathfrak{F}(\zeta, p)(1-\zeta^2)^{-(1+p)/2}, \quad 0 < \zeta < \frac{U}{c}, \tag{29.34}$$

$$Y(0,p)=\bar{F}(p)=\int_0^\infty e^{-p\eta}F(\eta)\,d\eta$$
$$Y\left(\frac{U}{c},p\right)=0. \tag{29.35}$$

It is easy to verify that the functions

$$Y_\pm(\zeta,p)=(1-\zeta^2)^{1/2}\left(\frac{1+\zeta}{1-\zeta}\right)^{\pm p/2} \tag{29.36}$$

are (separate) solutions of the homogeneous version of (29.34), and that the particular linear combination

$$\hat{Y}(\zeta,p)=\frac{\left(\log\frac{1}{\gamma}\right)^{p/2}Y_+(\zeta,p)-\left(\log\frac{1}{\gamma}\right)^{-p/2}Y_-(\zeta,p)}{2\sinh\left(\frac{p}{2}\log\gamma\right)}$$
$$=(1-\zeta^2)^{1/2}\frac{\sinh\left(\frac{p}{2}\log\left(\frac{1}{\gamma}\frac{1+\zeta}{1-\zeta}\right)\right)}{\sinh\left(\frac{p}{2}\log\gamma\right)} \tag{29.37}$$

assumes the values $\hat{Y}(0,p)=-1$, $\hat{Y}((U/c),p)=0$ at the endpoints of the ζ-interval; thus, for purposes of integrating (29.34), we may define

$$\Psi(\zeta,p)=\hat{Y}(\zeta,p)+\bar{F}(p)\hat{Y}(\zeta,p) \tag{29.38}$$

and, inasmuch as Ψ vanishes at both endpoints, postulate the expansion

$$\Psi(\zeta,p)=\sum_{n=1}^\infty c_n(p)\Phi_n(\zeta). \tag{29.39}$$

On employing both the differential and boundary specifications for the functions Ψ and Φ_n, together with the orthonormality of the latter set, we obtain

$$c_n(p)=\int_0^{U/c}\Psi(\zeta,p)\Phi_n(\zeta)\frac{d\zeta}{(1-\zeta^2)^2}$$
$$=-\frac{1}{1+n^2\delta^2}\int_0^{U/c}\Psi(\zeta,p)\frac{d^2\Phi_n}{d\zeta^2}\,d\zeta=-\frac{1}{1+n^2\delta^2}\int_0^{U/c}\frac{d^2\Psi}{d\zeta^2}\Phi_n(\zeta)\,d\zeta$$
$$=-\frac{1}{1+n^2\delta^2}\int_0^{U/c}\left\{-\frac{1-p^2}{(1-\zeta^2)^2}\Psi(\zeta,p)+\mathfrak{F}(\zeta,p)(1-\zeta^2)^{-(1+p/2)}\right\}\Phi_n(\zeta)\,d\zeta$$

whence

$$c_n(p)=\frac{1}{p^2+n^2\delta^2}\int_0^{U/c}\mathfrak{F}(\zeta,p)(1-\zeta^2)^{-(5+p)/2}\Phi_n(\zeta)\,d\zeta. \tag{29.40}$$

Problems I(b)

1. Consider a string with mass per unit length ρ which is subject to a tension P. Let one end be capable of transverse motions that are resisted by a force μv which is proportional to the transverse velocity v. Show that the reflection of a wave incident on the end is almost total when μ has either a very small or a very large magnitude; and determine a value of μ for which no reflection occurs.

2. If a string of finite length has one end ($x = 0$) fixed and the other ($x = L$) attached to a mass M that is capable of transverse motion, determine the overall displacement $y(x, t)$, $0 \leq x \leq L$, $t > 0$, or solution of the homogeneous wave equation, given the boundary conditions

$$y = 0, \qquad x = 0; \qquad \frac{\partial^2 y}{\partial t^2} + \frac{P}{M}\frac{\partial y}{\partial x} = 0, \qquad x = L$$

and the specific initial conditions

$$y(x, 0) = Yx/L, \qquad \frac{\partial y(x, 0)}{\partial t} = 0, \quad 0 < x < L.$$

3. Consider a uniform tense string with its extremities at $x = 0$ and $x = \infty$, the former being attached to a small ring of mass M that can move freely along a smooth rod in the transverse or y-direction. If a train of waves whose transverse displacement profile has the form

$$y(x, t) = A \cos \frac{\omega}{c}(x + ct)$$

is incident on the ring, show that

$$y(0, t) = 2A \cos \psi \cos(\omega t - \psi)$$

specifies the accompanying ring displacement, where

$$\tan \psi = M\omega/\rho c$$

and ρ is the line density of the string. Discuss the limiting cases that correspond to very large and small ring mass, respectively. Deduce the amplitude and phase of the reflected wave train.

4. An infinite string (in $0 < x < \infty$) is joined to another of finite length L whose distant end (at $x = -L$) remains fixed. If a wave train with frequency $\omega/2\pi$ in the former undergoes reflection at the junction of the strings, verify that the phase difference between reflected and incident waves equals

$$\pi - 2\tan^{-1}\left(\frac{c'}{c}\tan\frac{\omega L}{c'}\right)$$

where c, c' designate the velocities of wave propagation along the infinite and finite strings.

5. Suppose that the arrangement of strings envisaged in the foregoing problem lies along the x-axis at the initial instant of time, $t = 0$, after which the end $x = -L$ undergoes a specified transverse displacement given by $y(-L, t) = F(t)$. Assuming a progressive wave profile $y(x, t) = f(t - (x/c))$ in $x \geq 0$, establish the difference equation

$$(c + c')f\left(t + \frac{L}{c'}\right) + (c - c')f\left(t - \frac{L}{c'}\right) = 2F(t)$$

and, given that $f(\xi) = 0, \xi < L/c'$, confirm the representation

$$f(t) = \frac{2c}{c + c'}\sum_{n=0}^{N}\left(\frac{c' - c}{c' + c}\right)^n F\left(t - \frac{(2N + 1)L}{c'}\right)$$

for the displacement at the junction of the strings when

$$(2N + 1)L < c't < (2N + 3)L.$$

6. Small amplitude longitudinal motions along a uniform rod, described by the local displacement $\xi(x, t)$ from a fixed plane normal to the rod that has the coordinate x when the latter is unstressed, conform to the wave equation

$$\frac{\partial^2 \xi}{\partial x^2} = \frac{1}{c^2}\frac{\partial^2 \xi}{\partial t^2}, \qquad c^2 = E/\rho$$

where E denotes the Young's modulus of the material (or ratio of longitudinal stress, σ, to the strain, $(\partial\xi/\partial x)$) and ρ its density. Suppose that a bullet or particle with mass M is travelling at speed v ($\ll c$) in the line of a semi-infinite rod and strikes the end face ($x = 0$) at time $t = 0$. If it be assumed that the bullet afterwards ($t > 0$) remains in contact with this face (of cross-sectional area A) and that the end-condition (or equation of motion for the bullet)

$$M\left(\frac{\partial^2 \xi}{\partial t^2}\right)_{x=0} = EA\left(\frac{\partial \xi}{\partial x}\right)_{x=0}$$

applies, determine $\xi(x, t)$ for $x, t > 0$ and also the compressive stress $EA(\partial\xi/\partial x)$ in the rod.

7. Consider a rod or bar of unit length end at $x = 0$ is fixed and which is struck at the other end, $x = 1$, by a particle incident along the axial direction.

Let the equation for longitudinal displacements, $\xi(x, t)$, in the rod be given a non-dimensional form

$$\frac{\partial^2 \xi}{\partial x^2} = \frac{\partial^2 \xi}{\partial t^2}$$

and assume that $s(x, t)$ defines the particle coordinate; then the initial conditions at the instant of impact, $t = 0$, are expressed by

$$\xi = \frac{\partial \xi}{\partial t} = 0, \quad 0 < x < 1; \qquad s = 0, \qquad \frac{\partial s}{\partial t} = -v, \qquad x = 1$$

if v designates the appertaining particle velocity, and the end-conditions become

$$\xi = 0, \qquad x = 0; \qquad \frac{\partial^2 \xi}{\partial t^2} = -\alpha \frac{\partial \xi}{\partial x}, \qquad x = 1, \quad t > 0$$

inasmuch as

$$s(1, t) = \xi(1, t), \quad t > 0.$$

Show that the subsequent excitation of the rod has a progressive wave representation

$$\xi(x, t) = f(t - x) - f(t + x),$$

where

$$f(\zeta) = \frac{v}{4\pi \mathrm{i}} \int_{\Gamma} \frac{\mathrm{e}^{\zeta\mu} \, \mathrm{d}\mu}{\mu(\mu \sinh \mu + \alpha \cosh \mu)}$$

and Γ is an infinite contour on which $\operatorname{Re} \mu > 0$, the singularities of the integrand being situated to the left thereof. By development of $(\mu \sinh \mu + \alpha \cosh \mu)^{-1}$ in powers of $\mathrm{e}^{-\mu}$, derive (and interpret) the characterizations

$$\begin{aligned} f(\zeta) &= 0, \quad \zeta < 1 \\ &= \frac{v}{\alpha}\{1 - \exp(-\alpha(\zeta - 1))\}, \quad 1 < \zeta < 3 \\ &= -\frac{v}{\alpha} \exp(-\alpha(\zeta - 1)) + \frac{v}{\alpha}\{1 + 2\alpha(\zeta - 3)\} \exp(-\alpha(\zeta - 3)), \quad 3 < \zeta < 5 \\ &\quad \ldots\ldots \end{aligned}$$

Obtain the uniformly applicable representation for the particle displacement

$$s(1, t) = -2v \sum_n \frac{\sin \nu_n t}{\nu_n(1 + \alpha \operatorname{cosec}^2 \nu_n)}$$

where the sum involves all roots of the equation

$$\nu - \alpha \cot \nu = 0.$$

8. Suppose that a bar of uniform cross-section A is permanently without longitudinal displacement, ξ, at one end ($x = 0$), whereas the other end ($x = L$) is subjected to a load P after the instant of time $t = 0$. For a one-dimensional elastic stress distribution in the bar, the displacement $\xi(x, t)$, strain $\varepsilon(x, t) = \partial\xi/\partial x$ and the stress $\sigma(x, t) = E\varepsilon(x, t)$ [according to the simplest version of Hooke's Law with the Young's modulus E] individually satisfy the wave equation

$$\left(\frac{\partial^2}{\partial x^2} - \frac{1}{c^2}\frac{\partial^2}{\partial t^2}\right)(\xi, \varepsilon, \sigma) = 0$$

and the supplementary conditions

$$\xi = \frac{\partial \xi}{\partial t} = \varepsilon = \sigma = 0, \quad t \le 0$$

$$\xi(0, t) = 0, \qquad \sigma(L, t) = P/A, \quad t > 0;$$

show that

$$\xi(x, t) = \frac{Pc}{AE} \cdot \frac{1}{2\pi \mathrm{i}} \int_\Gamma \frac{\mathrm{e}^{pt} \sinh \dfrac{px}{c}}{p^2 \cosh \dfrac{pL}{c}} \mathrm{d}p \tag{*}$$

where the contour Γ is a straight line parallel to the imaginary axis of the p-plane, with $\operatorname{Re} p > 0$ thereupon. Evaluate the integral in terms of a static (or time-independent) part and undamped wave components.

If the Hooke's Law relating stress and strain be given the form

$$\sigma + \tau_1 \frac{\partial \sigma}{\partial t} = E\left(\varepsilon + \tau_2 \frac{\partial \varepsilon}{\partial t}\right),$$

establish the counterpart of (*),

$$\xi(x, t) = \frac{Pc}{AE} \cdot \frac{1}{2\pi \mathrm{i}} \int_\Gamma \left(\frac{1 + p\tau_1}{1 + p\tau_2}\right)^{1/2} \frac{\sinh\left(\dfrac{px}{c}\left(\dfrac{1 + p\tau_1}{1 + p\tau_2}\right)^{1/2}\right)}{\cosh\left(\dfrac{pL}{c}\left(\dfrac{1 + p\tau_1}{1 + p\tau_2}\right)^{1/2}\right)} \frac{\mathrm{d}p}{p^2} \tag{**}$$

and prove that the propagation velocity for the disturbance due to the loading has the magnitude $c\sqrt{(\tau_2/\tau_1)}$. Confirm that $p = 0$ and individual roots of the cubic equation

$$\tau_1 p^3 + p^2 + \omega_n^2 \tau_2 p + \omega_n^2 = 0, \qquad \omega_n = (2n + 1)\frac{\pi c}{2L}$$

locate the pole type singularities of the integral (**), with the condition for time decay of residue contributions stemming from the latter, viz., Re $p < 0$, satisfied if

$$\frac{1}{\tau_1} > 0, \quad \frac{\omega_n^2}{\tau_1}\left(\frac{\tau_2}{\tau_1} - 1\right) > 0, \quad \frac{\omega_n^2}{\tau_1} > 0$$

or

$$\tau_1 > 0, \quad \tau_2 > \tau_1.$$

9. Suppose that the ends of a uniform string, lying originally along the x-axis, are attached to massless rings which can slide along transverse wires at $x = 0, L$; and that a frictional resistance of magnitude $\alpha(\partial y/\partial t)$ acts on the ring at $x = 0$ while the other has the prescribed displacement $y = y_0 \cos(\omega t + \delta)$. Obtain the forced profile of the string displacement everywhere and calculate the average rate of energy dissipation at $x = 0$.

10. Assume that a uniform string with line density ρ and tension ρc^2 has its extremities at $x = -L$ and $x = +\infty$, and that a particle of mass M is attached at $x = 0$; when a train of transverse waves, characterized by the displacement amplitude $y = A \cos \omega(t + (x/c))$, is incident upon the load and the end $x = -L$ of the string remains fixed, show that

$$y(x, t) = 2A \operatorname{cosec}\frac{\omega L}{c} \cos\delta \sin\left(\frac{\omega x}{c} + \frac{\omega L}{c}\right)\cos(\omega t - \delta), \quad -L < x < 0$$

where

$$\tan\delta = \frac{M\omega}{\rho c} - \cot\frac{\omega L}{c}.$$

Determine the displacement profile of the reflected wave train in $x > 0$.

11. Let the equation

$$c^2\frac{\partial^2 y}{\partial x^2} = \frac{\partial^2 y}{\partial t^2} + 2\kappa c\frac{\partial y}{\partial t}$$

relate to the displacement $y(x, t)$ of a string whose elements are acted on by a resisting force with magnitude $2\kappa c$ times their momentum, and show that if one end $(x = 0)$ be fixed while the other $(x = l)$ has the prescribed displacement $y(l, t) = A \sin \pi ct/l$, then

$$y(x, t) = A \operatorname{cosech} \kappa l\left[\sin\frac{\pi x}{l}\cosh kx \sin\frac{\pi ct}{l} - \cos\frac{\pi x}{l}\sinh \kappa x \cos\frac{\pi ct}{l}\right]$$

specifies the forced oscillation of the string, so long as $2\kappa l$ is small.

12. Consider the problem of §16, wherein a point load M on a uniform string scatters an incident wave that has the general profile $A(t-(x/c))$; and, by employing the Green's function

$$G(x-x',t-t')=\frac{c}{2}H\left(t-t'-\frac{|x-x'|}{c}\right)$$

discussed in §23, construct the appropriate integral of the differential equation

$$\frac{\partial^2 y}{\partial x^2}-\frac{1}{c^2}\frac{\partial^2 y}{\partial t^2}=\frac{M}{P}\frac{\mathrm{d}^2 Y}{\mathrm{d}t^2}\delta(x), \quad Y(t)=y(0,t)$$

and show that

$$\frac{Mc}{2P}\frac{\mathrm{d}Y}{\mathrm{d}t}+Y(t)=A(t) \quad \text{(as in (16.5)).}$$

13. Let the inhomogeneous wave equation

$$\left(\frac{\partial^2}{\partial x^2}-\frac{1}{c^2}\frac{\partial^2}{\partial t^2}\right)G(x,t;T)=-\delta(x)\frac{\exp(-t^2/T^2)}{\sqrt{(\pi)}T}$$

be adopted for the purpose of characterizing a source with fixed location ($x=0$) and an effective "half-life" (T); and note that, inasmuch as

$$\lim_{T\to 0}\frac{\exp(-t^2/T^2)}{\sqrt{(\pi)}T}=\delta(t),$$

the limiting form

$$G(x,t;0)=\frac{c}{2}H\left(t-\frac{|x|}{c}\right)$$

is descriptive of an instantaneous source function. Verify the representations

$$G(x,t;T)=\frac{c}{4}-\frac{c}{2\pi}\int_0^\infty \sin\omega\left(\frac{|x|}{c}-t\right)\mathrm{e}^{-\omega^2T^2/4}\frac{\mathrm{d}\omega}{\omega}$$

and

$$G(x,t;T)=\frac{c}{2\sqrt{(\pi)}T}\int_{-\infty}^{t-(|x|/c)}\mathrm{e}^{-\tau^2/T^2}\,\mathrm{d}\tau,$$

and discuss their respective merits for obtaining estimates of the source function when T is large or small.

14. Suppose that an instantaneous source is located at the point ξ of a semi-infinite interval $0<x<\infty$ where the wave propagation speed equals c_1, with the explicit representation

$$G_s(x,\xi,t)=\frac{c}{2}H\left(t-\frac{|x-\xi|}{c_1}\right), \quad x>0$$

for its direct stimulus. If a different propagation speed c_2 applies on the adjoining interval $-\infty < x < 0$, establish the forms of the secondary (or reflected and transmitted) wave functions

$$G_r(x, \xi, t) = \frac{c_1}{2} \frac{\mu_1/c_1 - \mu_2/c_2}{\mu_1/c_1 + \mu_2/c_2} H\left(t - \frac{x+\xi}{c_1}\right), \quad x > 0$$

$$G_t(x, \xi, t) = \frac{\beta_1}{\alpha_2} \frac{1}{\mu_1/c_1 + \mu_2/c_2} H\left(t - \frac{\xi}{c_1} + \frac{x}{c_2}\right), \quad x < 0$$

when the junction conditions

$$\alpha_1(G_s + G_r) = \alpha_2 G_t \qquad x = 0$$

$$\beta_1 \frac{\partial}{\partial x}(G_s + G_r) = \beta_2 \frac{\partial}{\partial x} G_t$$

hold, and

$$\mu_1 = \beta_1/\alpha_1, \qquad \mu_2 = \beta_2/\alpha_2.$$

15. Given the pair of equations

$$\left(\frac{d^2}{dt^2} + \gamma \frac{d}{dt} + \omega_0^2\right) Y(t) = \omega_0^2 y(0, t)$$

and

$$\rho \frac{\partial^2 y}{\partial t^2} - P \frac{\partial^2 y}{\partial x^2} = M\omega_0^2[y(0, t) - Y(t)]\delta(x)$$

which involve the displacements $Y(t)$ and $y(x, t)$ for a mass M and a string that are transversely coupled by a spring with elastic constant $M\omega_0^2$; determine the reflection and transmission coefficients r, t if a harmonic wave train is incident on the junction point $x = 0$ and describe the behaviors of

$$|r|^2 = \tau^2 \omega_0^4 \frac{\omega^2 + \gamma^2}{(\omega_0^2(1 + \tau\gamma) - \omega^2)^2 + \omega^2(\gamma + \tau\omega_0^2)^2},$$

$$\tau = M/2\rho c$$

$$|t|^2 = \frac{(\omega_0^2 - \omega^2)^2 + (\gamma\omega)^2}{(\omega_0^2(1 + \tau\gamma) - \omega^2)^2 + \omega^2(\gamma + \tau\omega_0^2)^2}$$

as functions of ω.

16. Study the respective functions

$$\left(\frac{\partial^2}{\partial x^2} - \frac{1}{c^2}\frac{\partial^2}{\partial t^2}\right) G_Q(x, t; x', t') = -Q(t)\delta(x - Ut)$$

and

$$\left(\frac{\partial^2}{\partial x^2} - \frac{1}{c^2}\frac{\partial^2}{\partial t^2}\right) G_D(x, t; x', t') = D(t)\frac{\partial}{\partial x}\delta(x - Ut)$$

which are associated with moving sources of variable strengths $Q(t)$, $D(t)$, the latter being termed a dipole and conceived of as a combination of two simple (monopole) sources with equal and opposite strengths and a vanishingly small separation along the x-direction.

17. The inhomogeneous wave equation for string displacements

$$\frac{\partial^2 y}{\partial x^2} - \frac{1}{c^2}\frac{\partial^2 y}{\partial t^2} = -\frac{I}{\rho c^2}\delta(x - x_0)\frac{T}{\pi(t^2 + T^2)}$$

incorporates an excitation at the point $x = x_0$, whose temporal duration is controlled by the magnitude of T and which characterizes an impulse I when T becomes vanishingly small. Having regard for a string with fixed ends at $x = 0, l$, and initially at rest ($t = -\infty$), show that the time dependent coefficients $\phi_n(t)$ of the harmonic expansion

$$y(x, t) = \sum_{n=1}^{\infty} \phi_n(t) \sin \frac{n\pi x}{l}$$

take the form

$$\phi_n(0) = \frac{2I}{\pi\rho\omega_n l} \sin \frac{n\pi x_0}{l} \int_0^{\infty} \frac{\sin \omega_n T\zeta}{\zeta^2 + 1} \,\mathrm{d}\zeta, \qquad \omega_n = \frac{n\pi c}{l}$$

at $t = 0$, and obtain their asymptotic form when $t/T \gg 1$. Describe the related distribution of amplitudes among the harmonics.

18. Imagine a composite bar made up of two distinct parts with lengths L_1, L_2 and cross-sectional areas A_1, A_2; let the displacement function, u, for longitudinal motions therein satisfy the equations

$$\frac{\partial^2 u_1}{\partial t^2} = c_1^2 \frac{\partial^2 u_1}{\partial x^2}, \quad -L_1 < x < 0, \qquad \frac{\partial^2 u_2}{\partial t^2} = c_2^2 \frac{\partial^2 u_2}{\partial x^2}, \quad 0 < x < L_2$$

where $c_1^2 = E_1/\rho_1$, $c_2^2 = E_2/\rho_2$ specify the characteristic wave speeds in terms of the respective elastic moduli, E_1, E_2, and densities ρ_1, ρ_2. Continuity of displacement and stress at the interface $x = 0$ is expressed by the relations

$$\left.\begin{aligned} u_1 &= u_2 \\ E_1 A_1 \frac{\partial u_1}{\partial x} &= E_2 A_2 \frac{\partial u_2}{\partial x} \end{aligned}\right\} \quad x = 0 \qquad (*)$$

and the conditions

$$\frac{\partial u_1}{\partial x} = 0, \quad x = -L_1, \qquad \frac{\partial u_2}{\partial x} = 0, \quad x = L_2$$

imply stress-free outer ends.

Assuming that the motions are simply periodic in time, demonstrate that the second of the interfacial conditions (*) can be replaced with an integral relation,

$$\int_{-L_1}^{0} A_1\rho_1 \frac{\partial u_1}{\partial t}\,\mathrm{d}x + \int_{0}^{L_2} A_2\rho_2 \frac{\partial u_2}{\partial t}\,\mathrm{d}x = 0,$$

which is indicative of a fixed momentum center of gravity.

19. Consider the specification of a Green's function for the damped wave equation,

$$\left(\frac{\partial^2}{\partial t^2} + 2\varepsilon\frac{\partial}{\partial t} - c^2\frac{\partial^2}{\partial x^2}\right)G(x - x', t - t') = \delta(x - x')\delta(t - t'), \quad \varepsilon > 0$$

which possesses a causal nature and makes evident, through the related property

$$G = 0, \quad t - t' < \frac{|x - x'|}{c},$$

that c is the speed for propagation of effects between points with any spatial separation. Show that, pursuant to appropriate definition of the integration contour for ω,

$$\begin{aligned} G &= \frac{1}{(2\pi)^2}\int_{-\infty}^{\infty} \frac{\exp\{\mathrm{i}k(x-x') - \mathrm{i}\omega(t-t')\}}{(kc)^2 - \omega^2 - 2\,\mathrm{i}\varepsilon\omega}\,\mathrm{d}k\,\mathrm{d}\omega \\ &= \frac{1}{2\pi}\mathrm{e}^{-\varepsilon(t-t')}\int_{-\infty}^{\infty} \exp\{\mathrm{i}k(x-x')\}\frac{\sin[\surd((kc)^2 - \varepsilon^2)(t-t')]}{\surd((kc)^2-\varepsilon^2)}\,\mathrm{d}k \\ &= \begin{cases} \dfrac{1}{2c}\mathrm{e}^{-\varepsilon(t-t')}I_0\left[\varepsilon\left((t - t')^2 - \dfrac{(x-x')^2}{c^2}\right)^{1/2}\right], & t - t' > \dfrac{|x - x'|}{c} \\ 0, & t - t' < \dfrac{|x - x'|}{c}\end{cases} \end{aligned}$$

Construct, with reliance on the foregoing Green's function, a particular solution of the inhomogeneous differential equation

$$\left(\frac{\partial^2}{\partial t^2} + 2\varepsilon\frac{\partial}{\partial t} - c^2\frac{\partial^2}{\partial x^2}\right)y(x, t) = F\delta(x)$$

such that

$$y(x, 0) = \frac{\partial}{\partial t}y(x, 0) = 0;$$

verify the result

$$y(0, t) = \frac{F}{4\varepsilon^2}[\mathrm{e}^{-2\varepsilon t} - 1 + 2\varepsilon t]$$

and interpret the latter in conjunction with Problem 15I(a).

20. Imagine two systems, with the individual densities ρ, ρ' and elastic moduli μ, μ', which are coupled by a local interaction proportional to their relative acceleration. Let y, y' identify the respective displacements and assume that

$$\rho\frac{\partial^2 y}{\partial t^2} = \mu\frac{\partial^2 y}{\partial x^2} - \nu\rho\frac{\partial^2}{\partial t^2}(y - y'), \qquad \rho'\frac{\partial^2 y'}{\partial t^2} = \mu'\frac{\partial^2 y'}{\partial x^2} + \nu\rho\frac{\partial^2}{\partial t^2}(y - y')$$

constitute the pertinent equations of motion, ν being a coupling constant. Establish the symmetrical representations

$$y(x, t) = f\left(\frac{x}{v_1} - t\right) - \frac{\mu'}{\nu\rho}\left(\frac{1}{v_2^2} - \frac{\rho' + \nu\rho}{\mu'}\right)g\left(\frac{x}{v_2} - t\right)$$

$$y'(x, t) = g\left(\frac{x}{v_2} - t\right) + \frac{\mu}{\nu\rho}\left(\frac{\rho + \nu\rho}{\mu} - \frac{1}{v_1^2}\right)f\left(\frac{x}{v_1} - t\right)$$

for waves travelling in the positive x-direction, where $v_1, v_2 (v_1 > v_2)$ are the (positive) roots of the characteristic equation

$$\left(\frac{1}{v^2} - \frac{\rho + \nu\rho}{\mu}\right)\left(\frac{1}{v^2} - \frac{\rho' + \nu\rho}{\mu'}\right) = \frac{\nu^2\rho^2}{\mu\mu'}.$$

If there is a considerable difference in density, viz. ρ and $\nu\rho$ are small compared with ρ' and, moreover, $(\rho/\mu \ll \rho'/\mu')$, derive the estimates

$$v_1 = \left(\frac{\mu}{\rho(1 + \nu)}\right)^{1/2}, \qquad v_2 = \left(\frac{\mu'}{\rho'}\right)^{1/2}$$

and the appertaining wave representations

$$y(x, t) = f\left(\frac{x}{v_1} - t\right) - \frac{\nu}{1 + \nu}\left(\frac{v_2}{v_1}\right)^2 g\left(\frac{x}{v_2} - t\right)$$

$$y'(x, t) = g\left(\frac{x}{v_2} - t\right) + \frac{\nu\rho}{\rho'}f\left(\frac{x}{v_1} - t\right).$$

Interpret the latter in physical terms.

21. Employ the Green's function characterized by

$$\left(\frac{\mathrm{d}^2}{\mathrm{d}t^2} + \frac{P}{Mc}\frac{\mathrm{d}}{\mathrm{d}t} + \omega_0^2\right)G(t - \tau) = \delta(t - \tau), \qquad G = 0, \quad t < \tau$$

and integrate the oscillator displacement equation (21.9); determine the solution which befits an initially localized velocity without corresponding displacement.

PART II

§30. *Interference*

The particulars of periodic wave forms are manifest in diverse and sensitive ways by the superposition thereof, as touched on briefly in §4, and this suggests effective means of measuring wave parameters. It is thus appropriate to detail some aspects of wave interference and especially those circumstances wherein specific properties of the components are independently combined in arriving at a resultant magnitude or level.

For a pair of waves with the amplitude ratio a and the relative phase φ the sum, at a fixed location, is a time-dependent function

$$y(t) = y_1(t) + y_2(t) = y_1(t)[1 + a\,e^{i\varphi}] \qquad (30.1)$$

that betokens interference and a realizable outcome in a linear theory; its amplitude vanishes, in fact, when averaged over many periods of the oscillatory factor $y_1(t)$, irrespective of the values given to a and φ. The resultant intensity I, on the other hand, or time average of the product of y and its complex conjugate, has the representation

$$I = I_1(1 + a^2 + 2a\cos\varphi) \qquad (30.2)$$

where I_1 is the amount specifically associated with the component wave function $y_1(t)$, and evidently reflects a variable degree of interference through changes of the phase factor $\cos\varphi$. It appears that the extreme values taken by I are $I_1(1 \pm a)^2$, respectively, and that the component intensities are additive, viz.

$$I = I_1 + I_2 = I_1^2(1 + a^2) \qquad (30.3)$$

when a phase average is taken.

For an arbitrary number, N, of constituent waves with a pairwise amplitude ratio and phase difference that both retain an overall constancy, the sum (30.1) is replaced by

$$y(t) = y_1(t)\sum_{n=0}^{N-1} a^n\,e^{in\varphi} = y_1(t)\frac{1 - a^N\,e^{iN\varphi}}{1 - a\,e^{i\varphi}} \qquad (30.4)$$

and the corresponding intensity becomes

$$I_N = I_1 \cdot \frac{1 + a^{2N} - 2a^N\cos N\varphi}{1 + a^2 - 2a\cos\varphi}. \qquad (30.5)$$

The latter expression reduces to (30.2) if $N = 2$, as it must, and when $N \to \infty$,

$$I_\infty = I_1 \cdot \frac{1}{(1 - a)^2 + 4a\sin^2\dfrac{\varphi}{2}} \qquad (30.6)$$

provided that $a < 1$. An approximate version of (30.6),

$$I_\infty = I_1(1 + 2a \cos \varphi),$$

conditional on the hypothesis that a is small compared to unity, indicates a narrow range of intensity variation (explicitly, $\Delta I_\infty / I_1 = 4a$); however, if $a \doteq 1$, the variation of intensity as a function of φ is pronounced, with sharply defined maxima

$$(I_\infty)_{\max} = \frac{I_1}{(1-a)^2}, \qquad \text{at } \varphi = 2m\pi, \qquad m = 0, 1, \ldots$$

since the half-peak values are found at the ends of the small interval

$$\Delta\varphi = 2(1-a)$$

centered about $\varphi = 2m\pi$.

On substituting $a = 1$ in (30.5) we obtain the non-negative form

$$I_N/I_1 = \frac{\sin^2 \frac{1}{2} N\varphi}{\sin^2 \frac{1}{2}\varphi} \tag{30.7}$$

whose profile exhibits principal and secondary maxima as well as zeros. The principal maxima, of intensity

$$(I_N)_{p.\max} = I_1 N^2 \quad \text{at} \quad \varphi = 2m\pi, \qquad m = 0, 1, \ldots$$

have adjacent nulls at $\varphi = 2m\pi \pm 2\pi/N$; between $\varphi = 0$ and $\varphi = 2\pi$ [$m = 0, 1$], in particular, I_N vanishes $(N-1)$ times, corresponding to

$$\varphi = m \cdot \frac{2\pi}{N}, \qquad m = 1, \ldots, N-1$$

and hence $(N-2)$ secondary maxima occur on this interval.

To locate and contrast the relative magnitudes of the various maxima, the condition

$$\frac{1}{N} \tan N\frac{\varphi}{2} - \tan\frac{\varphi}{2} = 0,$$

which expresses a vanishing φ-derivative of (30.7) is pertinent; at a secondary maximum φ must be a root of the latter equation, say $\hat{\varphi}$ $(\neq 2m\pi)$, and taking account of this feature we may readily establish that

$$\frac{(I_N)_{s.\max}}{(I_N)_{p.\max}} = \frac{1}{1 + (N^2 - 1)\sin^2 \frac{\hat{\varphi}}{2}}$$

predicts a small ratio of intensity extrema if $N \gg 1$.

Thus, interference effects are sharply drawn for a large number ($N \gg 1$) of equal amplitude components with a fixed phase difference and also for an unbounded number ($N \to \infty$) whose pairwise amplitude ratio is close to unity. If the relative phase shift φ can be identified in terms of wave propagation over specific geometrical paths, as in the case when the constituent waves arise from multiple reflection, and the separation ΔL of two adjacent principal intensity maxima is accurately measured, the wave length becomes capable of precise determination through the relation $\lambda = 2\Delta L$.

To detail multiple interference effects in a particular context, let us again consider the composite string (§26) formed with a central section ($0 < x < l$) wherein the wave number, k_2, is distinct from that which applies elsewhere ($x < 0, > l$), namely k_1. We envisage that a time-periodic state of vibrations results from incidence on the left ($x < 0$) and focus our attention initially on a point x_1 in the central section; if $y_1(t)$ represents the wave function at this location according to direct transmission through an isolated discontinuity in wave number at $x = 0$, the complete wave function at x_1, taking account of single, double, ... reflection from the endpoints of the section, is given by

$$\begin{aligned} y(t) &= y_1(t)[1 + r_\infty \mathrm{e}^{2\mathrm{i}k_2(l-x_1)} + r_\infty^2 \mathrm{e}^{2\mathrm{i}k_2 l} + r_\infty^3 \mathrm{e}^{2\mathrm{i}k_2 l} + r_\infty^3 \mathrm{e}^{2\mathrm{i}k_2(2l-x_1)} + r_\infty^4 \mathrm{e}^{4\mathrm{i}k_2 l} + \cdots] \\ &= y_1(t)[1 + r_\infty \mathrm{e}^{2\mathrm{i}k_2(l-x_1)}][1 + (r_\infty^2 \mathrm{e}^{2\mathrm{i}k_2 l}) + (r_\infty^2 \mathrm{e}^{2\mathrm{i}k_2 l})^2 + \cdots] \\ &= y_1(t)\frac{1 + r_\infty \mathrm{e}^{2\mathrm{i}k_2(l-x_1)}}{1 - r_\infty^2 \mathrm{e}^{2\mathrm{i}k_2 l}}, \end{aligned} \tag{30.8}$$

where

$$r_\infty = \frac{k_2 - k_1}{k_2 + k_1}.$$

The comparative intensity ratio

$$\frac{I}{I_1} = \frac{1 + r_\infty^2 + 2r_\infty \cos 2k_2(l - x_1)}{1 + r_\infty^4 - 2r_\infty^2 \cos 2k_2 l} \tag{30.9}$$

depends on l, x_1, the local wave length $\lambda_2 = 2\pi/k_2$ and λ_1/λ_2; since x_1 enters only into the numerator of (30.9), we infer that, keeping all other quantities fixed, the limiting values of I/I_1 are proportional to $(1 \pm r_\infty)^2$ according as

$$2\frac{l - x_1}{\lambda_2} = \frac{m}{m + \frac{1}{2}}, \qquad m = 0, 1, \ldots .$$

If $2l > m\lambda_2$ the intensity ratio I/I_1 has more than one maximum on the range $0 \le x_1 \le l$ and its corresponding magnitude rises as r_∞ tends to unity or the difference in value between k_2 and k_1 increases.

The reflected wave intensity in the exterior of the central section, normalized relative to that of the incident wave, is given by (26.12), viz.

$$I_{\text{ref}}/I_{\text{inc}} = 1 - \frac{4n^2}{(1+n^2)^2 \sin^2 nk_1 l + 4n^2 \cos^2 nk_1 l} \tag{30.10}$$

where

$$n = k_2/k_1.$$

For an excitation which does not conform to strict monochromaticity, that is, with incident waves of the same intensity but a spread of wave lengths, the trigonometric factors $\sin nk_1 l$, $\cos nk_1 l$ will take on virtually all values between -1 and $+1$ if $l \gg \lambda_2$; and when we average or integrate (30.10) with respect to the argument of these factors, the result is

$$\begin{aligned} I_{\text{ref}}/I_{\text{inc}} &= 1 - \frac{4n^2}{\pi} \int_0^{\pi} \frac{\mathrm{d}\varphi}{(1+n^2)^2 \sin^2 \varphi + 4n^2 \cos^2 \varphi} \\ &= \frac{(1-n)^2}{1+n^2}. \end{aligned} \tag{30.11}$$

From an expansion of the reflection coefficient (26.8) in powers of $\mathrm{e}^{2ik_2 l}$, namely

$$\begin{aligned} r &= \frac{1-n}{1+n}\left\{1 - \mathrm{e}^{2ik_2 l}\right\}\left\{1 + \left(\frac{1-n}{1+n}\right)^2 \mathrm{e}^{2ik_2 l} + \left(\frac{1-n}{1+n}\right)^4 \mathrm{e}^{4ik_2 l} + \cdots\right\} \\ &= \frac{1-n}{1+n}\left\{1 - \left(1 - \left(\frac{1-n}{1+n}\right)^2\right)\mathrm{e}^{2ik_2 l} - \left(\frac{1-n}{1+n}\right)^2\left(1 - \left(\frac{1-n}{1+n}\right)^2\right)\mathrm{e}^{4ik_2 l}\right. \\ &\quad \left. - \left(\frac{1-n}{1+n}\right)^4\left(1 - \left(\frac{1-n}{1+n}\right)^2\right)\mathrm{e}^{6ik_2 l} - \cdots\right\}, \end{aligned}$$

it may be noted that the intensity contributions due to the multiply reflected wave components herein yield a sum

$$\begin{aligned} &\left(\frac{1-n}{1+n}\right)^2\left[1 + \left(1 - \left(\frac{1-n}{1+n}\right)^2\right)^2\left\{1 + \left(\frac{1-n}{1+n}\right)^4 + \left(\frac{1-n}{1+n}\right)^8 + \cdots\right\}\right] \\ &= \left(\frac{1-n}{1+n}\right)^2\left[1 + \frac{4n^2}{(1+n)^4}\frac{1}{1 - \left(\dfrac{1-n}{1+n}\right)^4}\right] \\ &= \frac{(1-n)^2}{1+n^2} \end{aligned}$$

which is identical with (30.11); once again, therefore, intensities rather than amplitudes are additively combined if variable phase relations necessitate an ensemble average.

§31. *Energy Density and Flux*

It is a familiar aspect of the periodic vibrations of strings that, although each material element executes movements within a confined range, an envelope of such motions can progress along the string in the lengthwise direction. This progressive motion, sustained by the mutual interaction of neighboring elements, is the vehicle for the transport of (mechanical) energy between arbitrarily spaced points on the string; and the transport of energy through extended media, with only local excitations (or displacements, in particular, from an equilibrium state) therein, is generally a hallmark of wave motion.

We may envisage a distribution (and companion density) of energy for any extended pattern of string displacements, and also an energy flux which is linked to the evolution of this pattern. To identify and correlate the respective measures of such a nature, consider an arbitrary section of the string, $x_1 < x < x_2$; after multiplication of the basic dynamical equation (1.1) by $\rho c^2(\partial y/\partial t) = P(\partial y/\partial t)$ and integration over the aforesaid range it follows that

$$\int_{x_1}^{x_2} \rho c^2 \frac{\partial y}{\partial t}\left[\frac{1}{c^2}\frac{\partial^2 y}{\partial t^2} - \frac{\partial^2 y}{\partial x^2}\right] \mathrm{d}x$$

$$= \int_{x_1}^{x_2} \frac{\partial}{\partial t}\left[\frac{1}{2}\rho\left(\frac{\partial y}{\partial t}\right)^2 + \frac{1}{2}P\left(\frac{\partial y}{\partial x}\right)^2\right] \mathrm{d}x - \int_{x_1}^{x_2} \frac{\partial}{\partial x}\left[P\frac{\partial y}{\partial x}\frac{\partial y}{\partial t}\right] \mathrm{d}x = 0,$$

or

$$-\frac{\mathrm{d}E}{\mathrm{d}t} = S\Big]_{x=x_1}^{x=x_2}, \tag{31.1}$$

if

$$E = \int_{x_1}^{x_2} W\,\mathrm{d}x, \qquad W = \frac{1}{2}\rho\left(\frac{\partial y}{\partial t}\right)^2 + \frac{1}{2}P\left(\frac{\partial y}{\partial x}\right)^2, \tag{31.2}$$

$$S = -P\frac{\partial y}{\partial x}\frac{\partial y}{\partial t}. \tag{31.3}$$

The quantity W represents an energy density of the vibrations (per unit length of the string), and includes both kinetic and potential contributions. A kinetic energy component,

$$T = \frac{1}{2}\rho\left(\frac{\partial y}{\partial t}\right)^2 \tag{31.4}$$

is so identified by the squared transverse velocity $(\partial y/\partial t)^2$; to confirm the interpretation and measure of the potential energy density,

$$V = \frac{1}{2}P\left(\frac{\partial y}{\partial x}\right)^2, \tag{31.5}$$

we estimate the expenditure of work that accompanies the displacement of an element of the string from its equilibrium or rectilinear profile, with a consequent extension $ds - dx = [\surd(1+(\partial y/\partial x)^2) - 1]\,dx$. If such a configurational change be effected in the presence of a tensile force P the work involved is

$$P(ds - dx) \doteq \frac{1}{2} P\left(\frac{\partial y}{\partial x}\right)^2 dx$$

approximately, whence the potential energy per unit length acquires the stated form (31.5). It is noteworthy that the potential energy involves a second order magnitude, $(\partial y/\partial x)^2$, hitherto neglected in comparison with unity for the purpose of linearizing the dynamical equation of string motion.

The above interpretation of W makes it clear that E defines the total energy of vibrations for the section of string between $x = x_1$ and $x = x_2$; moreover, in the relation (31.1) we have an accounting for temporal changes in the total energy, namely that these are accompanied by a net difference in the values of the quantity S at the extremities of the section. Such a rate equation for E suggests that S is the magnitude of the energy transfer rate or flux along the string, and it is consistent with this interpretation to remark that the explicit form of S comprises the product of the transverse force, $-P(\partial y/\partial x)$, exerted by an element of the string on its neighbor, with the corresponding velocity of the latter, $(\partial y/\partial t)$. Accordingly, (31.1) describes the overall energy balance for the particular section, equating the time rate of change of energy with the net flux across the extremities thereof; and we may infer the strict conservation of energy $dE/dt = 0$, in the circumstance that the net flux vanishes. The differential form of energy conservation is evidently expressed by

$$\frac{\partial W}{\partial t} + \frac{\partial S}{\partial x} = 0, \tag{31.6}$$

a relation which may otherwise be looked upon as an identity in respect to functions $y(x, t)$ that satisfy the one-dimensional wave equation.

If the motions of the string are impeded by a force proportional to the local velocity, as the equation (5.10) takes into account, we find the substitute for (31.6),

$$\frac{\partial W}{\partial t} + \frac{\partial S}{\partial x} = -2\varepsilon\rho\left(\frac{\partial y}{\partial t}\right)^2$$

and a continual loss of energy is implied by the right-hand member (with fixed sign) herein. The original version of an energy conservation relation, (31.6), retains its validity if a linear restoring force as well as the tension is operative, and the concomitant energy density proves to be

$$W = \frac{1}{2}\rho\left(\frac{\partial y}{\partial t}\right)^2 + \frac{1}{2}P\left(\frac{\partial y}{\partial x}\right)^2 + \frac{1}{2}Ph^2y^2 \tag{31.7}$$

where h is the force constant that appears in the equation of motion (5.1).

When a state of periodic vibrations prevails, the real displacement amplitude can be given the representation

$$y(x, t) = \mathrm{Re}[s(x)\,\mathrm{e}^{-\mathrm{i}\omega t}] = \tfrac{1}{2}[s(x)\,\mathrm{e}^{-\mathrm{i}\omega t} + s^*(x)\,\mathrm{e}^{\mathrm{i}\omega t}], \tag{31.8}$$

the asterisk denoting a complex conjugate of the function to which it is appended. Thus

$$\frac{\partial y}{\partial x} = \mathrm{Re}\left(\frac{\mathrm{d}s}{\mathrm{d}x}\,\mathrm{e}^{-\mathrm{i}\omega t}\right) = \frac{1}{2}\left[\frac{\mathrm{d}s}{\mathrm{d}x}\,\mathrm{e}^{-\mathrm{i}\omega t} + \frac{\mathrm{d}s^*}{\mathrm{d}x}\,\mathrm{e}^{\mathrm{i}\omega t}\right],$$

$$\frac{\partial y}{\partial t} = \omega\,\mathrm{Im}(s\,\mathrm{e}^{-\mathrm{i}\omega t}) = -\frac{\mathrm{i}\omega}{2}[s\,\mathrm{e}^{-\mathrm{i}\omega t} - s^*\,\mathrm{e}^{\mathrm{i}\omega t}].$$

On insertion of the latter expressions into (31.2) and (31.2) we obtain characterizations for the instantaneous energy density and energy flux,

$$W(x, t) = \frac{1}{4}\rho\omega^2|s|^2 + \frac{1}{4}P\left|\frac{\mathrm{d}s}{\mathrm{d}x}\right|^2 - \frac{1}{8}\rho\omega^2\{s^2\,\mathrm{e}^{-2\mathrm{i}\omega t} + s^{*2}\,\mathrm{e}^{2\mathrm{i}\omega t}\}$$

$$+\frac{P}{8}\left\{\left(\frac{\mathrm{d}s}{\mathrm{d}x}\right)^2\mathrm{e}^{-2\mathrm{i}\omega t} + \left(\frac{\mathrm{d}s^*}{\mathrm{d}x}\right)^2\mathrm{e}^{2\mathrm{i}\omega t}\right\},$$

$$S(x, t) = \frac{\mathrm{i}\omega\rho}{4}\left\{-s^*\frac{\mathrm{d}s}{\mathrm{d}x} + s\frac{\mathrm{d}s^*}{\mathrm{d}x}\right\} + \frac{\mathrm{i}\omega P}{4}\left\{s\frac{\mathrm{d}s}{\mathrm{d}x}\,\mathrm{e}^{-2\mathrm{i}\omega t} - s^*\frac{\mathrm{d}s^*}{\mathrm{d}x}\,\mathrm{e}^{2\mathrm{i}\omega t}\right\},$$

which are made up with steady terms and others that oscillate at twice the angular frequency of the displacement itself. The former alone contribute to the time average of the energy density and flux, whose characterizations are, accordingly,

$$\bar{W} = \frac{1}{4}\rho\omega^2|s|^2 + \frac{1}{4}P\left|\frac{\mathrm{d}s}{\mathrm{d}x}\right|^2, \tag{31.9}$$

$$\begin{aligned}\bar{S} &= -\frac{\mathrm{i}\omega P}{4}\left\{s^*\frac{\mathrm{d}s}{\mathrm{d}x} - s\frac{\mathrm{d}s^*}{\mathrm{d}x}\right\}\\ &= \frac{1}{2}\omega P\,\mathrm{Im}\left\{s^*\frac{\mathrm{d}s}{\mathrm{d}x}\right\}.\end{aligned} \tag{31.10}$$

When a tense string has a load of mass M directly attached at the location

$x = 0$ an energy relation of the form

$$\frac{\partial}{\partial t}\left[W + \frac{1}{2}M\left(\frac{\partial y(0, t)}{\partial t}\right)^2 \delta(x)\right] + \frac{\partial S}{\partial x} = 0$$

is implied by the equation of motion

$$P\frac{\partial^2 y}{\partial x^2} - \rho\frac{\partial^2 y}{\partial t^2} = M\frac{\partial^2 y(0, t)}{\partial t^2}\delta(x);$$

and it follows that, for simply periodic motions, the average energy flux varies continuously at the load point, viz.:

$$\bar{S}\Big|_{0-}^{0+} = 0. \tag{31.11}$$

If the representations (16.13), appropriate to an incoming wave train on the range $x < 0$, are employed, we find

$$\bar{S} = \begin{cases} \frac{1}{2}\omega Pk(|A|^2 - |B|^2), & x < 0 \\ \frac{1}{2}\omega Pk|C|^2, & x > 0 \end{cases}$$

and thus (31.11) yields a relationship between the component wave amplitudes,

$$|A|^2 - |B|^2 = |C|^2, \tag{31.12}$$

which is evidently satisfied by the determinations (16.16) for B, C in terms of A, and which attributes the sense of continuity in energy flux at the load to an equality between the net amount arriving at the load from one side (where incident and reflected waves are present) and departing therefrom on the other side (where a transmitted wave obtains).

The average flux manifests a similar behavior at the junction between two sections of a string with unequal densities (where both the velocity and slope remain continuous), and the counterpart of (31.12) takes the form

$$k_1[|A|^2 - |B|^2] = k_2|C|^2 \tag{31.13}$$

when the circumstances of excitation on a composite string are those envisaged in §25.

§32. *Energy Propagation and the Group Velocity*

The ratio of energy flux S and density W is a quantity with the dimensions of a velocity; and the specific velocity defined by the ratio of time averages for S and W, viz.

$$u = \bar{S}/\bar{W} \tag{32.1}$$

may be associated with the rate of energy propagation. For a progressive motion that has a simply periodic wave function

$$y(x, t) = \mathrm{Re}\{A\, \mathrm{e}^{\mathrm{i}(kx-\omega t)}\}, \tag{32.2}$$

whose frequency and wave number satisfy the proportionality

$$\omega = kc \tag{32.3}$$

imposed by the differential equation

$$\frac{\partial^2 y}{\partial x^2} = \frac{1}{c^2}\frac{\partial^2 y}{\partial t^2}, \tag{32.4}$$

we find, on utilizing the forms of $\bar{S}$ and $\bar{W}$ appropriate to a taut string [(31.9), (31.10)],

$$u = \frac{\frac{1}{2}\omega k P|A|^2}{\frac{1}{4}\rho\omega^2|A|^2 + \frac{1}{4}Pk^2|A|^2} = c$$

since $P = \rho c^2$. Thus, the magnitude of u equals that of the phase velocity $v = \omega/k$ for the profile (32.2) and also that with which arbitrary profiles are propagated along the string (when (32.4) applies).

The equality of phase and energy transport velocities no longer prevails if the wave motion belongs to a dispersive category characterized by the functional relation

$$\omega = \omega(k), \qquad \mathrm{d}^2\omega/\mathrm{d}k^2 \not\equiv 0 \tag{32.5}$$

that excludes a direct proportionality between ω and k, or, alternatively, between the period of oscillation and the wave length of a sinusoidal train. A particular representative of this category materializes when the vibrations of a taut string are influenced by a linear restoring force, and the explicit version of (32.5),

$$\omega = c\sqrt{(k^2 + h^2)} \tag{32.6}$$

relates the parameters ω, k of a harmonic wave function (32.2) that satisfies the equation

$$P\frac{\partial^2 y}{\partial x^2} = \rho\frac{\partial^2 y}{\partial t^2} + Ph^2 y. \tag{32.7}$$

The phase velocity deduced from (32.6),

$$v = c\sqrt{(1+(h/k)^2)},$$

is evidently a non-linear function of the wave number k and has a magnitude which exceeds c (unless $\lambda = 2\pi/k = 0$); on the other hand, the energy transport velocity, or ratio between the time average energy flux,

$$\bar{S} = \frac{1}{2}\omega k P |A|^2,$$

and density [cf. (31.7)],

$$\bar{W} = \frac{1}{4}\rho\omega^2|A|^2 + \frac{1}{4}Pk^2|A|^2 + \frac{1}{4}Ph^2|A|^2,$$

turns out to be

$$u = \frac{c}{\sqrt{(1+(h/k)^2)}}$$

with a lesser magnitude than c. It will be observed that the product of phase and energy transport velocities is a constant, namely

$$uv = c^2$$

and that, furthermore,

$$u = \frac{\mathrm{d}\omega}{\mathrm{d}k} \tag{32.8}$$

which form is suggestive of a wider validity.

A wave length dependence of the phase velocity for simply periodic or single frequency wave forms carries with it the implication that any departure from strict monochromaticity will be evidenced through temporal change in the shape of the corresponding aggregate form, which is a gross feature of dispersion. Since an infinite span monochromatic wave profile is literally unattainable, there is compelling reason to construct and analyze wave forms which are in closer accord with the prevalent features of localization and distortion, and thus to elaborate the nature of frequency dispersion; it is a fact, moreover, that virtually all observable wave motions (including the oceanic and atmospheric) exhibit a dispersive character. The prominence of dispersion relations, expressing a functional relationship

$$\mathfrak{F}(\omega, k) = 0 \tag{32.9}$$

between the frequency and wave number of a periodic train, for the theory of wave motions is brought out by noting that the latter need not be specified in terms of partial differential equations; progressive waves of the type (32.2) are compatible, for instance, with the integro-differential equation

$$\frac{\partial y(x,t)}{\partial t} + \int_{-\infty}^{\infty} K(x-\xi)\frac{\partial y(\xi,t)}{\partial \xi}\,\mathrm{d}\xi = 0$$

if

$$\omega = k \int_{-\infty}^{\infty} K(\zeta)\, e^{-ik\zeta}\, d\zeta$$

and this constitutes a dispersion relation in the aforementioned sense.

When the equations that underlie wave phenomena have a linear structure, solutions thereof can be synthesized from those of travelling and periodic type, as in the Fourier integral

$$y(x, t) = \int_{-\infty}^{\infty} f(k)\, e^{i(kx-\omega(k)t)}\, dk \tag{32.10}$$

where $\omega = \omega(k)$ denotes a specific resolution of the dispersion relation (32.9); and a complete solution is achieved after the juxtaposition of similar integrals which involve all the other determinations for $\omega(k)$ that follow from the particular dispersion relation. The reliance on a single (or possibly unique) determination $\omega(k)$ in (32.10) thus signifies that we are able to accomodate but one initial datum, say the profile at $t = 0$,

$$y(x, 0) = \int_{-\infty}^{\infty} f(k)\, e^{ikx}\, dk \tag{32.11}$$

and the weight factor $f(k)$ is then given by the inverse relation to (32.11), namely

$$f(k) = \frac{1}{2\pi} \int_{-\infty}^{\infty} y(x, 0)\, e^{-ikx}\, dx. \tag{32.12}$$

If the initial profile has the property

$$y(-x, 0) = y^*(x, 0)$$

which implies an even symmetry in x for its real part, the function $f(k)$ is real and the variable phase or argument of the exponential factor in (32.10) directly shapes the evolution of the wave form. The latter, as to be expected, moves forward along the x-direction with velocity c and an invariable aspect when $\omega = kc$, since

$$y(x, t) = y(x - ct, 0)$$

in this circumstance.

A comprehensive description of the manner in which the Fourier integral representation (32.10) of a wave function changes with x, t is out of the question if the particulars of $\omega(k)$ and $f(k)$ are only broadly delineated. To secure a degree of generality regarding the properties of said Fourier integral wherein x, t have parametric roles, our attention will be separately focussed on small and large values of t and the hypothesis made, by way of facilitating the acquisition of estimates in the respective circumstances, that variations in one or the other of the functions $f(k)$ and $kx - \omega(k)t$ are decisive. Inasmuch as $f(k)$ and $y(x, 0)$ constitute a transform pair, a narrowly confined distribution of values for the first implies a broadly extended distribution for the second and vice versa. Thus, if $f(k)$ has significant values on the interval $0 \ll k_0 - \frac{1}{2}\Delta k < k < k_0 + \frac{1}{2}\Delta k$, $\Delta k \ll k_0$, the corresponding transform $y(x, 0)$ has a symmetric distribution with appreciable magnitude on the interval $-(1/2\Delta k) < x < (1/2\Delta k)$ of extent $L \sim 1/\Delta k \gg \lambda_0$ ($= 2\pi/k_0$); here $y(x, 0)$ describes a group or packet of simply periodic wave forms whose lengths are clustered round λ_0, with small alteration in the crest amplitude or separation on the scale provided by λ_0 (see Fig. 16).

The influence of dispersion on the early progress of a wave packet can be readily ascertained when the spectrum function $f(k)$ of the initial profile has a narrow range centered at, say, $k = k_0$; thus, if we avail ourselves of a series expansion, relative to $k = k_0$, for the exponential argument in (32.10), i.e.,

$$kx - \omega(k)t = k_0 x - \omega_0 t + \left[x - \left(\frac{\mathrm{d}\omega}{\mathrm{d}k}\right)_0 t\right](k - k_0) - \frac{1}{2}\left(\frac{\mathrm{d}^2\omega}{\mathrm{d}k^2}\right)_0 t(k - k_0)^2 + \cdots \tag{32.13}$$

$$\omega_0 = \omega(k_0), \qquad \left(\frac{\mathrm{d}\omega}{\mathrm{d}k}\right)_0 = \left(\frac{\mathrm{d}\omega}{\mathrm{d}k}\right)_{k=k_0}, \ldots$$

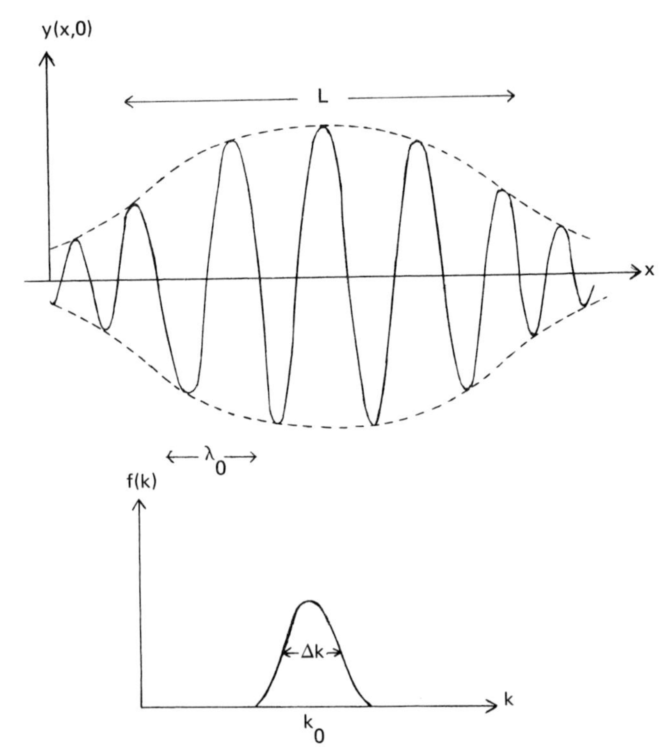

Fig. 16. A wave packet and its spectrum function.

and break off the latter after the term linear in $(k - k_0)$, on the understanding that t remains small enough to justify the omission of further terms, an approximate correlation follows,

$$y(x, t) = \exp[ik_0(u_0 - v_0)t]y(x - u_0t, 0), \tag{32.14}$$

where

$$v_0 = \omega_0/k_0$$

is the phase velocity corresponding to the wave number k_0 and

$$u_0 = \left(\frac{d\omega}{dk}\right)_{k=k_0}. \tag{32.15}$$

Writing $y(x, 0) = F(x)\,e^{ik_0x}$, whence $F(x)$ defines a modulation factor or envelope for the original group profile, the relation (32.14) becomes

$$y(x, t) = \exp(ik_0(x - v_0t))F(x - u_0t) \tag{32.16}$$

and reveals that dispersion is manifest (in this approximation, solely) by a velocity u_0 for the motion of the packet or group of waves which differs from the phase velocity v_0 of the predominant component. It also appears from (32.16) that the local wave length has a magnitude $\lambda = 2\pi/k_0$ relative to any point of observation travelling with the group velocity, though the position of individual crests in the packet is shifted forward or backward in time according as $u_0 < v_0$ or $u_0 > v_0$. On retention of the quadratic term in the expansion (32.13) for the exponential argument a two variable modulation factor is indicated, along with continued enlargement of the wave packet as time passes.

Alternatively, an estimate for the integral (32.10),

$$y(x, t) \doteq f(k_0)\exp[i(k_0x - \omega_0t)]\int_{k_0-1/2(\Delta k)}^{k_0+1/2(\Delta k)} \exp[i(x - u_0t)(k - k_0)]\,dk$$

$$= \frac{\sin\frac{1}{2}\Delta k(x - u_0t)}{\frac{1}{2}(x - u_0t)} f(k_0)\exp[i(k_0x - \omega_0t)],$$

which is consonant with a narrow spread Δk of wave numbers in the group of elementary wave forms, and relies on (32.13) for approximation to their phase when t is small, suggests a variable amplitude factor of the group,

$$\frac{\sin\frac{1}{2}\Delta k(x - u_0t)}{\frac{1}{2}(x - u_0t)},$$

whose maximum value is displaced along the direction of travel at the speed u_0.

It will be appreciated that the expression (32.15) for the group velocity has an antecedent in the previously defined energy transport velocity; to bring out this connection in a more direct manner, we consider the motion of the energy center of gravity of a wave packet, whose instantaneous position is defined by

$$x_E(t) = \frac{1}{E}\int xW(x,t)\,\mathrm{d}x = \frac{\displaystyle\int xW(x,t)\,\mathrm{d}x}{\displaystyle\int W(x,t)\,\mathrm{d}x} \tag{32.17}$$

where $W(x, t)$ designates an energy density, with the specific form

$$W(x,t) = \frac{1}{2}\rho\left\{\left(\frac{\partial y}{\partial t}\right)^2 + c^2\left(\frac{\partial y}{\partial x}\right)^2 + c^2h^2y^2\right\} \tag{32.18}$$

in the case of a taut string acted on by a linear restoring force. Let us suppose that the packet has a considerable breadth from the outset and ascribe this feature to a sharply peaked modulus for the spectrum factor at $k = k_0$; then the integrals in (32.17) can be extended over the entire coordinate range and transformed, without significant loss of accuracy, to others which involve the wave number as a variable of integration and thus evidence the fact of dispersion.

On utilizing the wave function given by the real part of (32.10), together with the 'orthogonality' relation

$$\frac{1}{2\pi}\int_{-\infty}^{\infty} \mathrm{e}^{\mathrm{i}(k-k')x}\,\mathrm{d}x = \delta(k-k'),$$

we obtain an estimate for the contribution from the first term of (32.18) to the total energy of the packet, namely

$$\begin{aligned}
&\frac{1}{2}\rho\int_{-\infty}^{\infty}\left(\frac{\partial y}{\partial t}\right)^2\mathrm{d}x\\
&\quad= \frac{1}{4}\rho\int_{-\infty}^{\infty}\omega(k)\omega(k')[f(k)f^*(k')\exp\{\mathrm{i}[(k-k')x-(\omega(k)-\omega(k'))t]\}\\
&\qquad - \mathrm{Re}(f(k)f(k')\exp\{\mathrm{i}[(k+k')x-(\omega(k)+\omega(k'))t]\})]\,\mathrm{d}x\,\mathrm{d}k\,\mathrm{d}k'\\
&\quad= \frac{\pi\rho}{2}\int_{-\infty}^{\infty}\omega(k)\omega(k')[f(k)f^*(k')\delta(k-k')\exp\{\mathrm{i}(\omega(k)-\omega(k'))t\}\\
&\qquad - \mathrm{Re}(f(k)f(k')\delta(k+k')\exp\{-\mathrm{i}(\omega(k)+\omega(k'))t\})]\,\mathrm{d}k\,\mathrm{d}k'\\
&\quad= \frac{\pi\rho}{2}\int_{-\infty}^{\infty}(\omega(k))^2|f(k)|^2\,\mathrm{d}k,
\end{aligned}$$

since the argument of the second delta function is bounded away from zero in the k, k' neighborhoods where $|f| \neq 0$. The contributions to the total energy of the packet furnished by the remaining terms in (32.18) can be estimated in a similar manner, with the result that

$$\begin{aligned} E = \int W(x,t)\,\mathrm{d}x &= \frac{\pi\rho}{2}\int_{-\infty}^{\infty} [(\omega(k))^2 + c^2(k^2+h^2)]|f(k)|^2\,\mathrm{d}k \\ &= \pi\rho\int_{-\infty}^{\infty} (\omega(k))^2|f(k)|^2\,\mathrm{d}k \end{aligned} \tag{32.19}$$

after account is taken of the dispersion relation (32.6).

A parallel transformation of the other integral in the expression (32.17) for $x_E(t)$ makes use of the representations

$$x\,\mathrm{e}^{\pm ikx} = \mp\mathrm{i}\frac{\partial}{\partial k}\,\mathrm{e}^{\pm ikx}$$

and involves a subsequent integration by parts relative to k; and the consequence is a two part linear time dependence for $x_E(t)$, viz:

$$x_E(t) = x_0 + u_E t, \tag{32.20}$$

where x_0 is the initial value of the coordinate and

$$u_E = \frac{\pi\rho}{E}\int_{-\infty}^{\infty} (\omega(k))^2|f(k)|^2\frac{\mathrm{d}\omega}{\mathrm{d}k}\,\mathrm{d}k. \tag{32.21}$$

It is thus revealed that the energy center of gravity of the wave packet moves with a uniform speed $\mathrm{d}x_E/\mathrm{d}t = u_E$, whose magnitude equals a mean value of the group velocity, $\mathrm{d}\omega/\mathrm{d}k$, averaged over the spectrum of wave numbers comprised therein and weighted by the measure of the energy density in a wave number scale.

§33. *Wave Kinematics and Dispersion*

A distinction brought out in the previous section, during a brief period of movement for a limited wave number aggregate of progressive components, between wave crest or phase velocity and wave number or group velocity, is also in force after a sufficient lapse of time whatever the spectral composition or initial extent of the profile; for the consequence of dispersion is a trend towards overall enlargement of an initial profile and local homogeneity. In this later epoch, wherein a very long and unidirectional train results from the separation of component waves, only small relative changes in the wave number $k(x,t)$, and frequency $\omega(x,t)$, are exhibited on intervals whose size is that of the local wave length and for times measured in terms of the local period. It then becomes possible to envisage the permanence of individual wave crests as a mark of stability in all but the outer fringes of the group, and thus to correlate the change in their total number for a section Δx during the time Δt, namely $(1/2\pi)(\partial k/\partial t)\Delta x \Delta t$, with the net influx

$$\frac{1}{2\pi}[\omega(x, t) - \omega(x+x, t)]\Delta t \doteq -\frac{1}{2\pi}\frac{\partial \omega}{\partial x}\Delta x \Delta t,$$

which implies that

$$\frac{\partial \omega}{\partial x} + \frac{\partial k}{\partial t} = 0 \tag{33.1}$$

or

$$\frac{\partial}{\partial x}(vk) + \frac{\partial k}{\partial t} = 0 \tag{33.2}$$

where

$$v = \frac{\omega}{k} \tag{33.3}$$

is the local phase velocity. If the circumstances underlying wave propagation furnish a dispersion relation, $\omega = \omega(k)$, the conservation statements (33.1), (33.2) are readily converted to the common form

$$\frac{\partial k}{\partial t} + u\frac{\partial k}{\partial x} = 0 \tag{33.4}$$

in which

$$u = \frac{\mathrm{d}\omega}{\mathrm{d}k} = v + k\frac{\mathrm{d}v}{\mathrm{d}k} = v - \lambda\frac{\mathrm{d}v}{\mathrm{d}\lambda} \tag{33.5}$$

represent alternative versions of the hitherto designated group velocity.

The pair of velocities u, v that are linked in analytic fashion by (33.5) find a simple geometrical interrelation through the dispersion curve (see Fig. 17) which associates the phase velocity and wave length; for it appears that the magnitude of the group velocity is specified at a given wave length, by the intercept of the tangent to this curve on the velocity axis. When the phase velocity decreases with increasing wave length, and $u > v$, the dispersion is termed anomalous.

The significance of (33.4) is that k (or λ) remains constant for waves travelling past a given point with the velocity u rather than v; or that waves of a fixed length λ are continually kept in view by an observer who moves along their direction of progress with the requisite value of the group velocity corresponding to this local wave length. For an observer in motion at the phase velocity, v, the time variation of the local wave length is expressible in

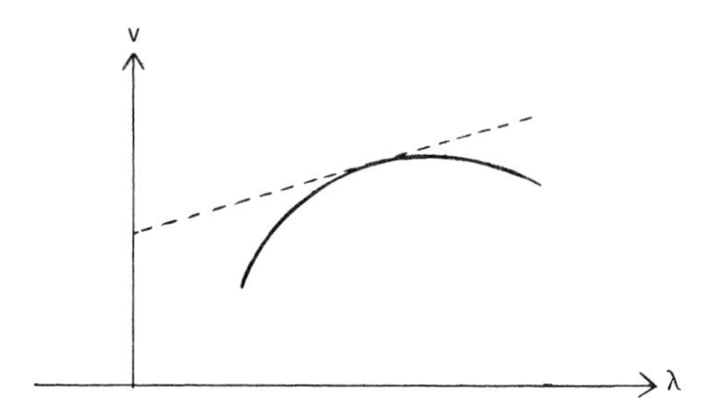

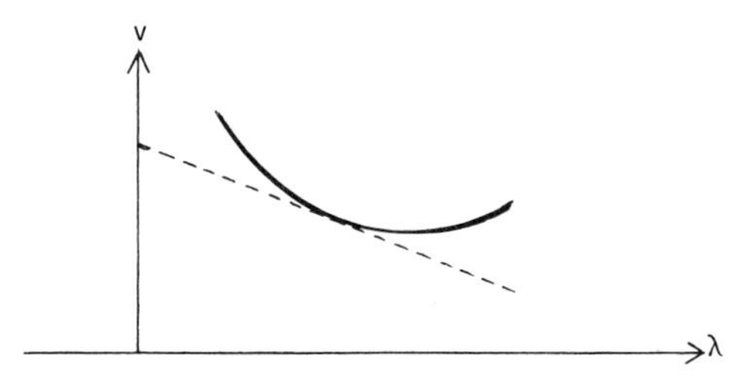

Fig. 17. Dispersion curves.

the alternative manners,

$$\frac{\partial \lambda}{\partial t} + v \frac{\partial \lambda}{\partial x}$$

or

$$\lambda \frac{\partial v}{\partial x} = \lambda \frac{\mathrm{d} v}{\mathrm{d} \lambda} \frac{\partial \lambda}{\partial x},$$

where the latter defines the separation rate for adjacent wave crests; and, from the equivalence thereof, we obtain an equation

$$\frac{\partial \lambda}{\partial t} + \left(v - \lambda \frac{\mathrm{d} v}{\mathrm{d} \lambda} \right) \frac{\partial \lambda}{\partial x} = 0$$

which confirms our prior characterization of, and magnitude for, the group velocity. The frequency, in addition to the wave length, remains invariable for an observer proceeding at the group velocity if wave speed and frequency are uniquely related, since (33.1) can then be recast in the form

$$\frac{\partial \omega}{\partial x} + \frac{1}{v^2} \left[v - \omega \frac{\mathrm{d} v}{\mathrm{d} \omega} \right] \frac{\partial \omega}{\partial t} = 0$$

and the appertaining velocity

$$u = \frac{v^2}{v - \omega \dfrac{dv}{d\omega}} \tag{33.6}$$

proves to be the same as that given by (33.5).

Should the relation between frequency and wave number also incorporate a coordinate dependence, so that $\omega = \omega(k, x)$, the setting for wave propagation is termed inhomogeneous and there obtains, after multiplication of (33.1) by

$$u = \left(\frac{\partial \omega}{\partial k}\right)_{x \text{ constant}} = u(k, x),$$

the result

$$u\frac{\partial \omega}{\partial x} + \left(\frac{\partial \omega}{\partial k}\right)_x \left(\frac{\partial k}{\partial t}\right)_x = \frac{\partial \omega}{\partial t} + u\frac{\partial \omega}{\partial x} = 0 \tag{33.7}$$

which implies that, in this circumstance, waves or packets of given frequency propagate at the variable velocity $u(k, x)$.

A geometrical description of features possessed by solutions of the first order partial differential equations (33.4), (33.7) is simply achieved through attention to their respective families of characteristics, whose equation

$$dx = u\, dt$$

is of the ordinary differential type. For a homogeneous or position independent dispersion relation $\omega = \omega(k)$ the characteristics comprise a family of non-parallel straight lines $x - ut = \text{const.}$ in the x, t plane, each specifying the trajectory of a particular value of k and having a slope which is equal in magnitude to the affiliated group velocity. When the dispersion relation also involves x the characteristic curves or trajectories for ω are curved lines, and their equation has the general form

$$t = \int_0^x \frac{dx}{u} + \text{const.} = \int_0^x \left(\frac{\partial k}{\partial \omega}\right)_x dx + \text{const.}$$

The space-time locus of an individual wave crest that travels with the phase velocity v can be found on integration of the equation

$$\frac{dx}{dt} = \frac{\omega}{k} = v(k(x, t)),$$

whose solutions constitute a one-parameter family of curves which is, in general, distinct from that of the characteristics. An instantaneous crest velocity is thus determined by the value of k on the characteristic curve which is intersected by the crest path; since the crest and wave length or frequency speeds are different, the normals to the family of crest paths do not coincide with the tangent directions at the straight or curved characteristics.

The unidirectional nature of wave number or frequency propagation along the characteristic curves at the pertinent group velocity is linked to the first order differential equation which these quantities satisfy; and contrasts with the feasibility of progressive motions in opposite senses when the basic equations are of higher order.

§34. *Stationary Phase*

Hitherto our analysis of the motion of a wave packet rested on the twin assumptions that only a narrow range of wave numbers enter therein and that the time span is a brief one, with scant opportunity for an alteration of the (broad) initial profile. We next consider a packet of limited original size and develop the analysis suited for describing its aspects after a considerable lapse of time, when dispersion and the attendant wave number resolution are well advanced; and it may be anticipated that features already ascribed to long and slowly varying wave trains will find independent and detailed verification therefrom.

The initial confinement of a packet on a finite portion of the coordinate interval implies a regular or analytic character for the spectral factor $f(k)$ in the Fourier integral representation

$$y(x,t)=\int_{-\infty}^{\infty} f(k)\,\mathrm{e}^{\mathrm{i}(kx-\omega(k)t)}\,\mathrm{d}k=\int_{-\infty}^{\infty} f(k)\,\mathrm{e}^{\mathrm{i}t\psi(k,x,t)}\,\mathrm{d}k \tag{34.1}$$

of the wave function; and it is the phase factor

$$\psi(k,x,t)=k\frac{x}{t}-\omega(k)$$

which plays a decisive role if estimation of the integral is sought when $t\to\infty$ (and the ratio x/t has any assigned value). A sequential process of integration by parts yields, in the second stage,

$$\begin{aligned}\int_{-\infty}^{\infty} f(k)\,\mathrm{e}^{\mathrm{i}t\psi(k,x,t)}\,\mathrm{d}k&=\frac{1}{\mathrm{i}t}\int_{-\infty}^{\infty} f(k)\frac{\partial k}{\partial\psi}\,\mathrm{d}\,\mathrm{e}^{\mathrm{i}t\psi}\\
&=\frac{f(k)}{\mathrm{i}t}\frac{\partial k}{\partial\psi}\,\mathrm{e}^{\mathrm{i}t\psi}\bigg|_{-\infty}^{\infty}-\frac{1}{(\mathrm{i}t)^2}\int_{-\infty}^{\infty}\frac{\partial k}{\partial\psi}\frac{\mathrm{d}}{\mathrm{d}k}\left(f(k)\frac{\partial k}{\partial\psi}\right)\mathrm{d}\,\mathrm{e}^{\mathrm{i}t\psi}\\
&=\left\{\frac{f(k)}{\mathrm{i}t}\frac{\partial k}{\partial\psi}\,\mathrm{e}^{\mathrm{i}t\psi}-\frac{1}{(\mathrm{i}t)^2}\frac{\partial k}{\partial\psi}\frac{\mathrm{d}}{\mathrm{d}k}\left(f(k)\frac{\partial k}{\partial\psi}\right)\mathrm{e}^{\mathrm{i}t\psi}\right\}\bigg|_{-\infty}^{\infty}\\
&\quad+\frac{1}{(\mathrm{i}t)^2}\int_{-\infty}^{\infty}\frac{\mathrm{d}}{\mathrm{d}k}\left(\frac{\partial k}{\partial\psi}\frac{\mathrm{d}}{\mathrm{d}k}\left(f(k)\frac{\partial k}{\partial\psi}\right)\mathrm{e}^{\mathrm{i}t\psi}\right)\mathrm{d}k,\end{aligned}$$

and thus, relying on the behavior of f to ensure a null contribution from the integrated terms, it appears that the magnitude of the original integral falls short of any inverse power of t, in the limit $t \to \infty$, so long as $\partial\psi/\partial k$ maintains a fixed sign. This inference becomes untenable, however, if $\partial\psi/\partial k$ vanishes at one or more values of $k(x, t)$, such zeros being evidently associated with singularities of the integrand for a version of (34.1),

$$y(x, t) = \int f \frac{\partial k}{\partial \psi} e^{it\psi} \, d\psi, \tag{34.2}$$

which reflects the use of ψ, rather than k, as an integration variable (and has appropriate limits of integration). An enhanced magnitude of the integral (34.2) obtains in the latter eventuality, correlated with the local behavior of ψ in the neighborhood of the respective wave numbers that correspond to a stationary value of the phase and, specifically, with the number of higher order derivatives that likewise vanish. The gist of these conclusions, rendered in more qualitative terms, is that interference between waves whose relative phase is a rapidly changing function of the wave number, for large values of t, results in their mutual suppression; whereas those groups for which the phase is nearly constant or stationary at a particular location where x/t is given ultimately determine the level of excitation there.

Accordingly, we may expect to obtain an estimate for the integral (34.1), or wave function at x, t, through contributions from groups whose wave numbers are clustered round a set $k_1(x, t), \ldots, k_n(x, t)$ that render the phase stationary. If we designate the phase of $f(k)$ by $\delta(k)$ that of the whole integrand can be written as $t\Psi(k, x, t)$, where

$$\Psi(k, x, t) = k\frac{x}{t} - \omega(k) + \frac{\delta(k)}{t} \tag{34.3}$$

and the equation indicative of a phase which is stationary with respect to k,

$$\frac{\partial}{\partial k}(t\Psi(k, x, t)) = 0, \tag{34.4}$$

is expressible in the form

$$x = x_0(k) + u(k)t \tag{34.5}$$

with

$$x_0(k) = -\frac{d\delta}{dk}, \qquad u(k) = \frac{d\omega}{dk}. \tag{34.6}$$

A one-parameter family of straight lines in the x, t plane is defined by (34.5) and individual members thereof are specified in terms of their slope $u(k)$ and intercept $x_0(k)$ (see Fig. 18); both k and the slope remain constant on a

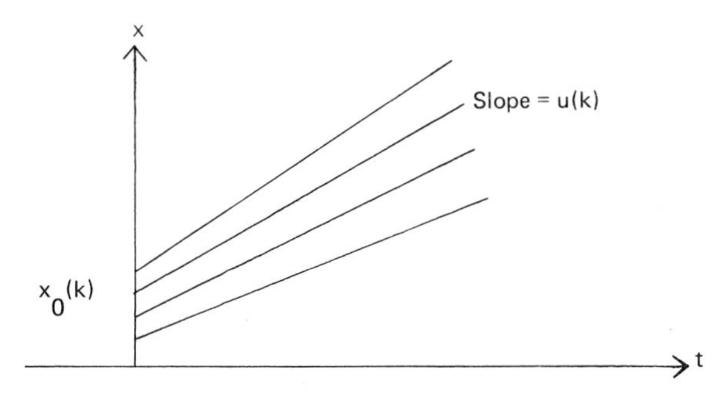

Fig. 18. Ray family.

particular line or ray, whose attributes are thus such as to place it in the family of characteristics referred to earlier.

If the stationary phase condition (34.4) admits but one solution, $k = k_1(x, t)$, a uniquely determined ray is singled out by an assignment of values for x, t. In case the wave profile has a symmetric form (about $x = 0$) initially, so that $f(k)$ is real and $\delta(k)$ vanishes, $k = k_1(x/t)$, and all the rays issue from the origin; a limiting ray with maximum slope, corresponding to the largest magnitude of the group velocity, marks the front beyond which ($x > u_{\max}t$) progressive disturbances do not reach a significant level. Intersecting rays, and multiple determinations of wave numbers compatible with (34.4) are realized when there is a curved front that separates portions of the x, t plane which are covered by, or devoid of, rays [see Fig. 19, after Eckart (1948)].

For the purpose of ascertaining the excitation level achieved by a predominant group a description of the phase appropriate to the small wave number spread therein is indicated. If the phase is stationary at $k = k_1(x, t)$ and $(\partial^2\Psi/\partial k^2)_{k=k_1} > 0$, say, we have

$$\Psi(k, x, t) = \Psi(k_1(x, t), x, t) + \frac{1}{2}\left(\frac{\partial^2\Psi}{\partial k^2}\right)_{k=k_1}(k - k_1)^2 + 0(k - k_1)^3 \qquad (34.7)$$

in accordance with the Taylor series development; following the introduction of a new variable, ζ, defined by

$$\Psi(k, x, t) = \Psi(k_1(x, t), x, t) + \zeta \qquad (34.8)$$

and comparison of (34.7), (34.8) a local version of the transformation,

$$\zeta = \frac{1}{2}\left(\frac{\partial^2\Psi}{\partial k^2}\right)_{k=k_1}(k - k_1)^2\{1 + \alpha(k - k_1) + \cdots\}, \qquad (34.9)$$

is provided, together with the respective inverses,

$$k - k_1 = \pm\left[2\zeta\Big/\left(\frac{\partial^2\Psi}{\partial k^2}\right)_{k=k_1}\right]^{1/2}\{1 + \beta\zeta + \cdots\}, \quad k \gtrless k_1 \qquad (34.10)$$

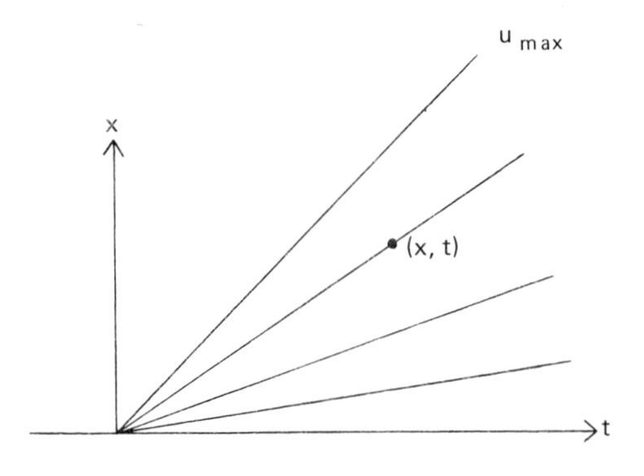

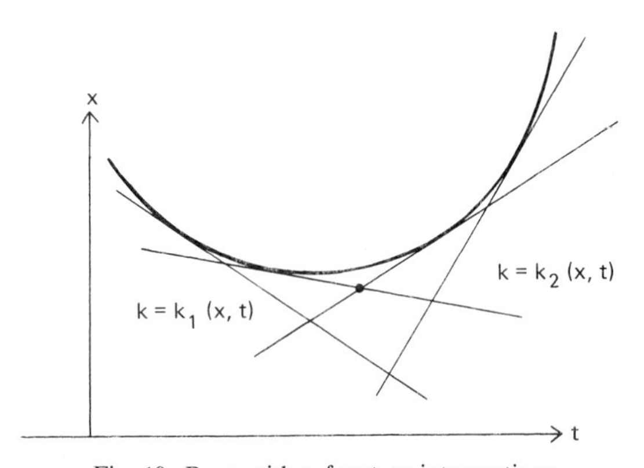

Fig. 19. Rays with a front or intersections.

from which we obtain

$$\frac{\mathrm{d}k}{\mathrm{d}\zeta} = \pm\left[2\left(\frac{\partial^2\Psi}{\partial k^2}\right)_{k=k_1}\right]^{-1/2}\zeta^{-1/2}, \quad \zeta \to 0+, \quad k \gtrless k_1. \tag{34.11}$$

A subgroup defined by that part of the integral (34.1) relating to wave numbers on the range $(k_1, k_1 + \Delta k)$ acquires the form

$$y_+(x, t) = \exp\{\mathrm{i}t\Psi(k_1(x, t), x, t)\}\int_0^{\zeta_+} F(\zeta)\,\mathrm{e}^{\mathrm{i}t\zeta}\,\mathrm{d}\zeta \tag{34.12}$$

in terms of the variable ζ, where the factor $F(\zeta)$ has the representations

$$F(\zeta) = \frac{|f(k)|}{\dfrac{\partial\Psi}{\partial k}} = \frac{|f(k_1)|}{\left(\dfrac{\partial^2\Psi}{\partial k^2}\right)_{k=k_1}}\cdot\frac{1}{k-k_1}\{1+\gamma(k-k_1)+\cdots\}$$

$$= |f(k_1)|\left[2\left(\frac{\partial^2\Psi}{\partial k^2}\right)_{k=k_1}\right]^{-1/2}\zeta^{-1/2}\{1+\delta\zeta+\cdots\} \tag{34.13}$$

and $\zeta_+ = \Psi(k_1 + \Delta k, x, t) - \Psi(k_1, x, t) > 0$. Utilizing the relations

$$\int_0^\eta \zeta^{-1+\mu}\varphi(\zeta) \begin{matrix}\sin\\ \cos\end{matrix} t\zeta \, d\zeta \to \Gamma(\mu)t^{-\mu}\varphi(0+) \begin{matrix}\sin\\ \cos\end{matrix} \frac{\mu\pi}{2}, \quad \begin{matrix}\mu > -1\\ \mu > 0\end{matrix}$$

which hold, in the limit $t \to \infty$, for a function $\varphi(\zeta)$ of bounded variation on the finite interval $(0, \eta)$, and relying on the first term in the development (34.13) of $F(\zeta)$, we find that

$$y_+(x, t) \sim |f(k_1)| \left[\pi/2\left(\frac{\partial^2 t\Psi}{\partial k^2}\right)_{k=k_1}\right]^{1/2} \exp\left\{it\Psi(k_1(x, t), x, t) + i\frac{\pi}{4}\right\},$$

and, inasmuch as the subgroup $y_-(x, t)$ pertaining to the wave number span $(k_1 - \Delta k, k_1)$ furnishes an equal contribution, the resultant estimate for all components that make up a homogeneous phase group, located by means of the condition $(\partial\Psi/\partial k)_{k=k_1} = 0$, becomes

$$y(x, t) \sim |f(k_1)| \left[2\pi\Big/\left[\left|\frac{\partial^2}{\partial k^2}(t\Psi)_{k=k_1}\right|\right]\right]^{1/2} \exp\left\{it\Psi(k_1(x, t), x, t) \pm i\frac{\pi}{4}\right\}$$

$$t \to \infty \qquad (34.14)$$

when the accommodation is made for alternate signs of $(\partial^2\Psi/\partial k^2)_{k=k_1}$.

From the phase function in (34.14),

$$t\Psi(k_1(x, t), x, t) = k_1(x, t)x - \omega(k_1)t + \delta(k_1), \qquad (34.15)$$

whose partial derivatives are

$$\frac{\partial}{\partial x}(t\Psi) = k_1 + [x - u(k_1)t - x_0(k_1)]\frac{\partial k_1}{\partial x} = k_1(x, t) \qquad (34.16)$$

and

$$\frac{\partial}{\partial t}(t\Psi) = -\omega(k_1) + [x - u(k_1)t - x_0(k_1)]\frac{\partial k_1}{\partial t} = -\omega_1(x, t) \qquad (34.17)$$

it is apparent that the profile has the local wave number $k_1(x, t)$ and frequency $\omega_1(x, t)$, both of which are variable and confer the same aspect on the plane velocity $v(x, t) = (\omega_1/k_1)$. A steady magnitude for k_1 is conditional on maintaining a forward rate of progress given by

$$\left(\frac{\partial x}{\partial t}\right)_{k_1} = -\frac{\partial k_1/\partial t}{\partial k_1/\partial x} = u(k_1)$$

since

$$\frac{\partial k_1}{\partial t} = \frac{\partial^2}{\partial x \partial t}(t\Psi) = -\frac{\partial}{\partial x}\omega(k_1) = -u(k_1)\frac{\partial k_1}{\partial x}.$$

In this feature and also the conservation property for wave crests,

$$\frac{\partial \omega_1}{\partial x} + \frac{\partial k_1}{\partial t} = 0, \tag{34.18}$$

which follows directly from (34.16), (34.17), we find an accord with our previous characterizations regarding extended and slowly varying wave trains; that the wave number (and frequency) changes slowly is confirmed, when the parametric assignments requisite to the approximation (34.14) are in force, namely $t \to \infty$ and $x/t \to (\mathrm{d}\omega/\mathrm{d}k)_{k=k_1}$, by the expressions for the logarithmic derivatives of k_1 (and ω_1), viz.

$$\frac{1}{k_1}\frac{\partial k_1}{\partial x} = \frac{1}{k_1\left(\dfrac{\mathrm{d}^2\omega}{\mathrm{d}k^2}\right)_{k=k_1}} \cdot \frac{1}{t}, \qquad \frac{1}{k_1}\frac{\partial k}{\partial t} = -\frac{\left(\dfrac{\mathrm{d}\omega}{\mathrm{d}k}\right)_{k=k_1}}{k_1\left(\dfrac{\mathrm{d}^2\omega}{\mathrm{d}k^2}\right)_{k=k_1}} \cdot \frac{1}{t}, \qquad t \to \infty.$$

The variable amplitude of the wave function (34.14),

$$A(x, t) = |f(k)| \left[2\pi \left| \frac{\partial^2}{\partial k^2}(t\Psi)_{k=k_1} \right| \right]^{1/2} \exp\left\{ \mathrm{i}\frac{\pi}{4} \operatorname{sgn}\left(\frac{\partial^2 \Psi}{\partial k^2}\right)_{k=k_1} \right\}$$

has a squared modulus,

$$|A(x, t)|^2 = \frac{2\pi |f(k_1)|^2}{\left| \dfrac{\partial^2}{\partial k^2}(t\Psi)_{k=k_1} \right|},$$

that points to a diminution of the profile with increasing time, since

$$\left| \frac{\partial^2}{\partial k^2}(t\Psi)_{k=k_1} \right| \sim t \left| \left(\frac{\mathrm{d}^2\omega}{\mathrm{d}k^2}\right)_{k=k_1} \right|, \qquad t \to \infty.$$

It is a simple matter to account for the latter aspect through the overall expansion of the initial profile, reflected by an angular divergence or separation of neighboring rays when t increases; thus, consequent to the uniform growth of the coordinate interval, Δx, between rays that correspond to the wave numbers, k_1, $k_1 + \Delta k$, with the specification

$$\Delta k = \left| \left(\frac{\mathrm{d}\omega}{\mathrm{d}k}\right)_{k=k_1} - \left(\frac{\mathrm{d}\omega}{\mathrm{d}k}\right)_{k=k_1+\Delta k} \right| t \doteq \left| \left(\frac{\mathrm{d}^2\omega}{\mathrm{d}k^2}\right)_{k=k_1} \right| t\Delta k, \qquad t \to \infty$$

there is a concomitant dispersal of energy in the same wave number range. The reduction in magnitude of $|A(x, t)|^2$, in particular, brought about by this measure of scale enlargement for a limited wave number group conforms to the fact that the squared modulus of the wave function is proportional to the local energy density. Alternatively, we may infer from the structure of the amplitude factor $A(x, t)$ that energy conservation, in addition to local subsidence, is assured; for the integral of $|A(x, t)|^2$ on the interval $c_1 t < x < c_2 t$, say,

$$E(t) = 2\pi \int_{c_1 t}^{c_2 t} \frac{|f(k)|^2}{t\left|\dfrac{\partial^2 \Psi}{\partial k^2}\right|^2} \, \mathrm{d}x,$$

retains a steady value if

$$c_1 = \left(\frac{\mathrm{d}\omega}{\mathrm{d}k}\right)_{k=k_1-\Delta k}, \qquad c_2 = \left(\frac{\mathrm{d}\omega}{\mathrm{d}k}\right)_{k=k_1+\Delta k},$$

corresponding to displacement of the endpoints along a pair of rays at the group velocities $u(k_1 - \Delta k)$, $u(k_1 + \Delta k)$, inasmuch as

$$E(t) \sim 2\pi \int_{k_1-\Delta k}^{k_1+\Delta k} |f(k)|^2 \, \mathrm{d}k, \quad t \to \infty.$$

The representation (34.14) provides merely the leading term of an asymptotic development for $t \to \infty$ and fixed x/t,

$$y(x, t) \sim \frac{1}{\sqrt{t}} \exp\{\mathrm{i}t\Psi(k_1(x, t), x, t)\} \sum_{n=0}^{\infty} \frac{\bar{A}_n}{t^n},$$
$$\bar{A}_0 = \left[2\pi \Big/ \left(\frac{\partial^2 \Psi}{\partial k^2}\right)_{k=k_1}\right]^{1/2} \exp\left\{\pm \mathrm{i}\frac{\pi}{4} \operatorname{sgn}\left(\frac{\partial^2 \Psi}{\partial k^2}\right)_{k=k_1}\right\} \tag{34.19}$$

wherein the coefficients $\bar{A}_n$, $n > 0$, pertain to subsequent terms in the expansion (33.13) for $F(\zeta)$ and are formally expressed through derivatives of the latter function at $\zeta = 0$. It is evident that such an expansion for a predominant wave group fails when the second order derivative of the phase function vanishes at $k = k_1$, giving rise to infinite values for the coefficients $\bar{A}_n$.

In the circumstance that the stationary phase condition for a centered initial profile ($\delta(k) = 0$),

$$x = \left(\frac{\mathrm{d}\omega}{\mathrm{d}k}\right)_{k=k_1} t,$$

is supplemented by the additional condition,

$$\left(\frac{\mathrm{d}^2\omega}{\mathrm{d}k^2}\right)_{k=k_1} = 0,$$

we obtain, on employing the phase representation

$$t\Psi(k, x, t) \doteq k_1 x - \omega(k_1)t - \frac{1}{3!} t \left(\frac{\mathrm{d}^3\omega}{\mathrm{d}k^3}\right)_{k=k_1} (k - k_1)^3, \tag{34.20}$$

an estimate for the group wave function

$$y(x,t) \sim f(k_1)\exp\{\mathrm{i}t\Psi(k_1(x,t),x,t)\}\left[6\Big/t\left(\frac{\mathrm{d}^3\omega}{\mathrm{d}k^3}\right)_{k=k_1}\right]^{1/3}\int_{-\infty}^{\infty}\mathrm{e}^{\mathrm{i}\zeta^3}\,\mathrm{d}\zeta$$

$$= f(k_1)\frac{\Gamma\left(\frac{1}{3}\right)}{\sqrt{3}}\left[6\Big/t\left(\frac{\mathrm{d}^3\omega}{\mathrm{d}k^3}\right)_{k=k_1}\right]^{1/3}\exp\left\{\mathrm{i}t\left[k_1\left(\frac{\mathrm{d}\omega}{\mathrm{d}k}\right)_{k=k_1}-\omega(k_1)\right]\right\},\quad t\to\infty \tag{34.21}$$

which decreases less rapidly in time than its counterpart (34.14). To examine the behavior of the wave function in the vicinity of the ray with slope $u(k_1)$, that is, at the locations $x = \Delta + u(k_1)t$, $\Delta \ll x$, an additional term, $(k-k_1)\Delta$, which depends linearly on $(k-k_1)$, is included in the phase representation (34.20); details of this nature are reserved for a later section (§37).

§35. *An Illustrative Example*

It is appropriate, in the first instance, as regards the application and assessment of a stationary phase estimation for integrals, to consider a problem that has an exact solution available for comparison. Accordingly, we shall examine the features of a wave train or approximate solution of the differential equation

$$\frac{\partial^2 y}{\partial x^2} = \frac{1}{c^2}\frac{\partial^2 y}{\partial t^2} + h^2 y \tag{35.1}$$

and, for purposes of contrast, refer to the explicit solution (§5) of the appertaining initial value problem. With a view to concentrating on a particular wave train, whose original specification involves but a single function, we shall assume that the initial distributions

$$y(x,0) = s_0(x), \qquad \frac{\partial}{\partial t}y(x,0) = v_0(x)$$

satisfy the relationship

$$v_0(x) = -c\frac{\mathrm{d}s_0}{\mathrm{d}x}$$

indicative of an excitation with a forward speed c; it follows from (5.15) that

$$y(x,t) = s_0(x-ct) - \frac{h}{2}\int_{x-ct}^{x+ct}\frac{J_1[h\sqrt{(c^2t^2-(x-\xi)^2)}]}{\sqrt{(c^2t^2-(x-\xi)^2)}}[ct+(x-\xi)]s_0(\xi)\,\mathrm{d}\xi$$

and when the localized form

$$s_0(x) = A\delta(x) \tag{35.2}$$

is chosen, the solution

$$y(x,t)=A\delta(x-ct)-\frac{Ah}{2}\frac{ct+x}{\sqrt{(c^2t^2-x^2)}}J_1(h\sqrt{(c^2t^2-x^2)}) \tag{35.3}$$

obtains. The excitation represented by (35.3) comprises a sharp signal or front advancing in the positive direction at the speed c, behind which a wave train extends over the steadily widening range $|x|<ct$, with an amplitude that diminishes as the trailing edge $x=-ct$ is approached. To find the details of this train when $t\to\infty$, independently of the explicit and generally valid wave function representation (35.3), we rely upon the Fourier integral characterization

$$y(x,t)=y_+(x,t)+y_-(x,t)$$

where

$$y_\pm(x,t)=\frac{A}{4\pi}\int_{-\infty}^{\infty}\left[1\pm\frac{kc}{\omega(k)}\right]e^{i(kx\mp\omega(k)t)}\,dk, \tag{35.4}$$

$$\omega(k)=c\sqrt{(k^2+h^2)}.$$

It is readily confirmed that the phase function for $y_+(x,t)$, namely $kx-\omega(k)t$, has a single stationary point

$$k=k_1\left(\frac{x}{t}\right)=\pm h\left\{\left(\frac{ct}{x}\right)^2-1\right\}^{-1/2}=hx\{c^2t^2-x^2\}^{-1/2} \tag{35.5}$$

on the respective intervals $x\gtrless 0$. The ray diagram associated with the stationary phase condition $x-\omega'(k_1)t=0$ is a centered one, encompassing the sector $-\tan^{-1}c\le\tan^{-1}(x/t)\le\tan^{-1}c$, and the ray of maximum slope corresponds to an infinite value for k_1, along with the largest magnitude for the group velocity, namely $u_{max}=c$. A general integral of the equation for wave crest paths,

$$\frac{dx}{dt}=\frac{\omega(k_1)}{k_1}=c^2\frac{t}{x}, \tag{35.6}$$

yields the family of hyperbolas, $c^2t^2-x^2=\text{constant}\ (>0)$, all of which have a common slope at the points of intersection with a particular ray.

Consequent to the determinations

$$k_1x-\omega(k_1)t=-h\sqrt{(c^2t^2-x^2)},$$

$$\omega''(k_1)=\frac{(c^2t^2-x^2)^{3/2}}{hc^2t^3}>0$$

the leading term of the stationary phase estimate for $y_+(x,t)$, deduced from

(35.4), becomes

$$y_+(x,t) \sim \frac{A}{2}\left(\frac{h}{2\pi}\right)^{1/2}(ct+x)(c^2t^2-x^2)^{-3/4}\exp\left\{-\mathrm{i}h\surd(c^2t^2-x^2)-\mathrm{i}\frac{\pi}{4}\right\}, \quad t\to\infty; \tag{35.7}$$

and since the analogous estimate for $y_-(x,t)$ is merely the complex conjugate of the preceding one, we obtain

$$y(x,t) \sim A\left(\frac{h}{2\pi}\right)^{1/2}(ct+x)(c^2t^2-x^2)^{-3/4}\cos\left[h\surd(c^2t^2-x^2)+\frac{\pi}{4}\right], \quad t\to\infty \tag{35.8}$$

where x/t has any given value between $(-c, c)$. The latter estimate is, indeed, consonant with the outcome of substituting for the Bessel function in the exact representation (35.3) an asymptotic form [Hankel, cf. §38]

$$J_1(\xi) \sim [2/\pi\xi]^{1/2}\cos\left(\xi-\frac{3\pi}{4}\right) = -[2/\pi\xi]^{1/2}\cos\left(\xi+\frac{\pi}{4}\right) \tag{35.9}$$

which applies when $\xi \gg 1$.

Differentiation of the variable phase function in the wave train (35.8),

$$\Phi(x,t) = h\surd(c^2t^2-x^2)$$

furnishes the magnitude of the local wave number and frequency, viz.

$$k_1 = -\frac{\partial\Phi}{\partial x}, \qquad \omega(k_1) = \frac{\partial\Phi}{\partial t}$$

and the phase velocity (35.6) is consistent with the expression

$$v^2 = c^2\left(1+\frac{h^2}{k^2}\right) \tag{35.10}$$

that specifies the phase velocity, v, for a time-periodic solution of the differential equation (35.1) which has any magnitude of the wave number. Likewise, the determinations for group velocity and wave number,

$$u(k_1) = \frac{x}{t}, \qquad k_1 = hx(c^2t^2-x^2)^{-1/2},$$

conform to the general interrelation

$$u^2 = c^2\left(1+\frac{h^2}{k^2}\right)^{-1} \tag{35.11}$$

that is associated with the specific dispersion formula (35.10).

If we represent the wave train profile in the form

$$y(x,t) = \mathrm{Re}\{A(x,t)\,\mathrm{e}^{\mathrm{i}\Phi(x,t)}\} \tag{35.12}$$

it follows from (35.8) that

$$|A(x,t)|^2 = \frac{h}{2\pi}\frac{(ct+x)^2}{(c^2t^2-x^2)^{3/2}} = \frac{h}{2\pi t}\frac{(c+u)^{1/2}}{(c-u)^{3/2}}, \qquad u = \frac{x}{t} \tag{35.13}$$

with the anticipated subsidence when $t \to \infty$; inasmuch as

$$\frac{\partial}{\partial t}|A|^2 + \frac{\partial}{\partial x}(u|A|^2) = 0 \tag{35.14}$$

the magnitude of u acquires the significance of a transport speed for the squared modulus of the profile. Relying on (35.8) to calculate the energy density (32.18) and utilizing the fact that variations of A are small compared with those of Φ, we obtain

$$W(x,t) \sim \rho\omega^2|A(x,t)|^2 = \rho(hc)^2\frac{c^2}{c^2-u^2}|A(x,t)|^2, \quad t \to \infty \tag{35.15}$$

apart from terms whose rapid change with time parallels that of the phase function Φ; a link between group velocity and energy transport is revealed by the applicability of the conservation relation to W as well as $|A|^2$.

The stationary phase estimate (35.8) becomes inaccurate when the magnitude of $ct - x$ approaches zero, that is, just behind the front of the advancing train; here the values of k_1 are extremely large and the local wave length is correspondingly short, in keeping with a rise of the group velocity as the wave length diminishes. Since the second and higher order derivatives of $\omega(k)$ vanish in the limit $k \to \infty$ $(x \to ct)$, we cannot avail ourselves of further terms in a Taylor expansion of the phase function $kx - \omega(k)t$ in order to improve on the prior estimate near the front. We may, however, describe the local behavior of the train by means of an expansion for $\omega(k)$ itself that is suitable when k is large, viz.

$$\omega(k) = kc + \frac{1}{2}\frac{h^2c}{k} + \cdots; \tag{35.16}$$

thus, retaining only the first two terms in (35.16) and modifying (35.4) accordingly, it follows that

$$\begin{aligned} y(x,t) \doteq y_+(x,t) &= \frac{A}{\pi}\int_0^\infty \cos\left\{k(ct-x) + \frac{1}{2}\frac{h^2ct}{k}\right\} dk \\ &= -\frac{A}{\sqrt{2}}\left\{\frac{ct}{ct-x}\right\}^{1/2} J_1[h\sqrt{(2ct(ct-x))}], \quad ct - x \ll 1 \end{aligned}$$

which representation evidently obtains after the approximation $ct + x \doteq 2ct$ is employed in the exact formula (35.3).

§36. *Extended Initial Profiles or Ranges of Support*

A wave train that evolves from an initial profile with a finite breadth or effective range bears the imprint of this characteristic length and has some features which are absent in the limit of extreme localization. Let us again refer to the dispersive wave equation (35.1) for an appreciation of the contrasting details and commence with a pair of initial conditions

$$y(x,0)=s_0(x), \qquad \frac{\partial}{\partial t}y(x,0)=0 \tag{36.1}$$

that involve a single datum, $s_0(x)$, whose support is confined to the segment $-l<x<l$. An integral representation of the solution which puts in evidence or isolates the initial datum is suitable for purposes of bringing out its role therein; thus, employing (5.7), (5.8) we write

$$y(x,t)=\int_{-l}^{l} G(x-x',t)s_0(x')\,dx' \tag{36.2}$$

with

$$G(x-x',t)=\frac{1}{2\pi}\int_{-\infty}^{\infty}\cos k(x-x')\cos(\surd(k^2+h^2)ct)\,dk \tag{36.3}$$

$$=-\frac{hct}{2}\frac{J_1[h\surd(c^2t^2-(x-x')^2)]}{\surd(c^2t^2-(x-x')^2)}, \quad ct>|x-x'| \tag{36.4}$$

on the assumption that $ct-|x|>l$, thereby locating the points of observation within the range of influence for the entire segment bearing initial data. If, furthermore, $|x|\gg l$, so that our attention is directed to points far removed from the initial confines of the profile, the inequality $x'\ll x$ applies throughout the domain of the integral (36.2) and the approximations

$$[c^2t^2-(x-x')^2]^{1/2}\doteq[c^2t^2-x^2]^{1/2}\left[1+\frac{2xx'}{c^2t^2-x^2}\right]^{1/2}$$
$$=[c^2t^2-x^2]^{1/2}+xx'[c^2t^2-x^2]^{-1/2}+\cdots \tag{36.5}$$

are indicated, since

$$\frac{2xx'}{c^2t^2-x^2}\leq\frac{l}{ct-x}\cdot\frac{2x}{ct+x}<1.$$

For $t\to\infty$ and a fixed $x/t<c$ the Bessel function in the numerator of (36.4) has an argument large compared with unity and varies more rapidly than the radical in the denominator; thus, the former can be estimated by means of the asymptotic expansion (35.9) together with the two term approximation (36.5), while the first term of (36.5) suffices as regards the latter. The resultant characterization of G (which corresponds to a stationary phase estimate of

the integral (36.3)) is

$$G(x-x',t) \simeq ct\left(\frac{h}{2\pi}\right)^{1/2}(c^2t^2-x^2)^{-3/4}\cos\left[\Phi(x,t)+k\left(\frac{x}{t}\right)x'\right], \quad t\to\infty, \quad x \gg x' \tag{36.6}$$

where

$$\Phi(x,t) = h(c^2t^2-x^2)^{1/2}+\frac{\pi}{4},$$

$$k\left(\frac{x}{t}\right) = h\left(\frac{c^2t^2}{x^2}-1\right)^{-1/2} = -\frac{\partial\Phi}{\partial x}.$$

Accordingly, when we utilize (36.6) and (36.2) in order to describe the train after a considerable lapse of time, there obtains

$$y(x,t) \simeq ct\left(\frac{h}{2\pi}\right)^{1/2}(c^2t^2-x^2)^{-3/4}\left[\cos\{\Phi(x,t)\}\int_{-l}^{l}\cos\left\{k\left(\frac{x}{t}\right)x'\right\}s_0(x')\,\mathrm{d}x' - \sin\{\Phi(x,t)\}\int_{-l}^{l}\sin\left\{k\left(\frac{x}{t}\right)x'\right\}s_0(x')\,\mathrm{d}x'\right] \tag{36.7}$$

save for points near the extremities ($x = \pm ct$). In the case of a symmetric pattern, $s_0(-x) = s_0(x)$, the amplitude factor

$$\mathfrak{A}\left(\frac{x}{t}, l\right) = \int_{-l}^{l}\cos\left\{k\left(\frac{x}{t}\right)x'\right\}s_0(x')\,\mathrm{d}x' \tag{36.8}$$

or cosine transform of $s_0(x)$, conveys the effect of an initial scale for the profile. If the local wave length $\lambda = 2\pi/k$ greatly exceeds the span of the initial profile and $kl \ll 1$, the excitation given by (36.7) duplicates that which is traceable to a concentrated original form, apart from an equivalent or localized 'source strength' of practically fixed amount,

$$\mathfrak{A} \doteq \int_{-l}^{l} s_0(x)\,\mathrm{d}x.$$

In the outer reaches of the train, where small magnitudes for the local wave length are found and $kl \gg 1$, a reduced amplitude or source factor (36.8) obtains, as a consequence of the variable weighting assigned to the given distribution $s_0(x)$. The dependence of the amplitude factor on the ratio x/t, or magnitude of the group velocity which pertains to a steady local wave number, is consistent with the fact that this quantity represents a collective contribution from the whole of the original profile to the later version.

An evaluation of (36.8), based on the particular choice $s_0(x) = A[1-(x/l)^2]$, yields

$$\mathfrak{A}\left(\frac{x}{t}, l\right) = \frac{4Al}{(kl)^2}\left\{\frac{\sin kl}{kl} - \cos kl\right\}$$

$$\doteq \frac{4}{3}Al, \quad kl \to 0$$

$$\simeq -\frac{4Al}{(kl)^2}\cos kl, \quad kl \to \infty;$$

and thus the asymptotic form of the appertaining wave function,

$$y(x, t) \simeq 4Al\left(\frac{h}{2\pi}\right)^{1/2} \frac{ct}{(kl)^2}(c^2t^2 - x^2)^{-3/4}\left[\frac{\sin kl}{kl} - \cos kl\right]\cos \Phi,$$

involves trigonometric factors with the respective arguments Φ, kl. Since $(\partial k/\partial x)/(\partial \Phi/\partial x) = -(1/k)(\partial k/\partial x) \to 0$ as $t \to \infty$, we infer that the size of the source region is manifest through a comparatively broad scale modulation or interference pattern in the extended wave train which develops.

Another initial value problem for a second order differential equation of the form (36.1), wherein the given datum pertains to a time derivative rather than the solution itself, is suggested by the consideration of motions in a heavy fluid of finite depth that rotates about a vertical (say, the z-) axis. More precisely, we have in mind the so-called long wave motions with straight fronts, controlled by gravity and inertia forces alone, for which the vertical acceleration is neglected; and we envisage that the free surface profile, $z = L + \zeta$, reflects small variations ($\zeta \ll L$) about a uniform equilibrium level. When the vertical accelerations are left out and the pressure is assigned the statical value corresponding to the distance from the free surface, the horizontal velocity components and surface elevation constitute the fundamental trio of dependent variables; and a system of three partial differential equations, comprising a pair of dynamical nature along with one that expresses the conservation of fluid matter, is available for their determination.

If u, v designate the horizontal components of fluid velocity relative to x, y axes rotating in their plane with a fixed angular velocity the linearized version of the aforementioned system becomes

$$\frac{\partial u}{\partial t} - \omega v = -g\frac{\partial \zeta}{\partial x}, \qquad \frac{\partial v}{\partial t} + \omega u = -g\frac{\partial \zeta}{\partial y}$$
$$\frac{\partial \zeta}{\partial t} = -L\frac{\partial u}{\partial x} - L\frac{\partial v}{\partial y} \tag{36.9}$$

where ω identifies a Coriolis (rotational) parameter and g is the gravitational constant. Let us suppose that an initial current or streaming motion along the

y-direction, say

$$v_0 = \begin{cases} V, & |x| < l \\ 0, & |x| > l \end{cases} \tag{36.10}$$

provides the stimulus for an unsteady mode of fluid excitation whose coordinate dependence involves x only. After discarding the terms in (36.9) which contain a y-partial derivative and eliminating v, ζ we find that u satisfies the second order equation

$$\frac{\partial^2 u}{\partial t^2} = -\omega^2 u + c^2 \frac{\partial^2 u}{\partial x^2} \tag{36.11}$$

with a velocity $c = \sqrt{(gL)}$ characteristic of long wave motions. Since u and ζ vanish at the initial instant, the appropriate conditions for (36.11) take the form

$$u(x, 0) = 0$$
$$\frac{\partial u(x,0)}{\partial t} = \omega v_0 = \begin{cases} \omega V, & |x| < l \\ 0, & |x| > l \end{cases} \tag{36.12}$$

that is complementary to the specification adopted in (36.1). Consequent to the determination of u, those of v, ζ are secured by means of the representations

$$v = v_0 - \omega \int_0^t u \, dt$$
$$\zeta = -L \int_0^t \frac{\partial u}{\partial x} \, dt, \tag{36.13}$$

whence

$$z = L\left(1 - \int_0^t \frac{\partial u}{\partial x} \, dt\right) \tag{36.14}$$

describes the free surface profile.

An adaptation of the results given in §5 yields

$$u(x, t) = \frac{\omega}{2c} \int_{x-ct}^{x+ct} J_0\left(\frac{\omega}{c}\sqrt{(c^2t^2 - (x - x')^2)}\right) v_0(x') \, dx' \tag{36.15}$$

as a generally valid expression for the solution of the stated initial value problem relative to the differential equation (36.11). After the onset of movement at the location $x > l$, that is, at times $t = t_1(1 + \delta)$, where $ct_1 = x - l$ and $0 < \delta < 2l/(x - l)$, we have

$$u(x, t) = \frac{\omega V}{2c} \int_{l-\delta(x-l)}^{l} J_0\left(\frac{\omega}{c}\sqrt{(c^2t^2 - (x - x')^2)}\right) dx'. \tag{36.16}$$

Utilizing the relation

$$J_0(\sqrt{(\alpha^2-\beta^2)}) = \sum_{n=0}^{\infty} \frac{J_n(\alpha)}{2^n n!\alpha^n}\beta^{2n}$$

the integral form of solution (36.16) may be replaced by a series development

$$u(x,\delta) = \frac{V}{2}\sum_{n=0}^{\infty} \frac{J_n\left(\frac{\omega}{c}(x-l)(1+\delta)\right)}{2^n n!(2n+1)} \left(\frac{\omega}{c}(x-l)\right)^{n+1} \frac{\{(1+\delta)^{2n+1}-1\}}{(1+\delta)^n}$$

which, on the premise that $\delta < 1$, implies an approximately linear time variation for u, inasmuch as

$$u(x,\delta) \doteq \frac{V\delta}{2}\sum_{n=0}^{\infty} \frac{J_n\left(\frac{\omega}{c}(x-l)\right)}{2^n n!} \left(\frac{\omega}{c}(x-l)\right)^{n+1}, \quad \delta \to 0, \quad x-l \gg l.$$

The analogous representations at a location $x < -l$ involve the sum $x+l$, whose sign is opposite to that of the prior combination $x-l$; jointly, these are indicative of upward or downward displacements of the free surface in the opening phases of the motion, according as the point of observation lies to the right or left of the impressed current, which asymmetry is evidently brought about by the fluid rotation.

When $ct > x+l$, the expression

$$u(x,t) = V\int_{-l}^{l} \bar{G}(x-x',t)\,dx'$$

is applicable and the function

$$\bar{G}(x-x',t) = \frac{1}{2\pi c}\int_{-\infty}^{\infty} \cos k(x-x') \frac{\sin\left(k^2+\frac{\omega^2}{c^2}\right)^{1/2}}{\left(k^2+\frac{\omega^2}{c^2}\right)^{1/2}}\,dk$$

$$= \frac{\omega}{2c} J_0\left(\frac{\omega}{c}\sqrt{(c^2t^2-(x-x')^2)}\right)$$

assumes an exclusive role as the propagator of causal effects from the source region. On approximating to the latter in a manner previously described for $G(x-x',t)$, viz.

$$\bar{G}(x-x',t) \simeq \left(\frac{\omega}{2\pi c}\right)^{1/2} (c^2t^2-x^2)^{-1/4} \cos\left\{\frac{\omega}{c}(c^2t^2-x^2)^{1/2} + \frac{\omega x x'}{c(c^2t^2-x^2)^{1/2}} - \frac{\pi}{4}\right\}$$

$$t \to \infty, \quad ct > x \gg l$$

we obtain the estimate

$$u(x,t) \simeq V\left(\frac{2c}{\pi\omega}\right)^{1/2}(c^2t^2-x^2)^{1/4}\sin\left\{\frac{\omega l x}{c(c^2t^2-x^2)^{1/2}}\right\}\cos\left\{\frac{\omega}{c}(c^2t^2-x^2)^{1/2}-\frac{\pi}{4}\right\}$$

$$t\to\infty,\quad \frac{x}{t}<c$$

which contains but a single factor that depends on the breadth of the original stream and assumes a maximal value if the local wave length defined by the cosine factor, namely $\lambda = 2\pi(c/\omega)((ct/x)^2-1)^{1/2}$, is equal to $4l$ [or discrete submultiples thereof].

§37. *An Envelope of Rays or Caustic Curve*

The accuracy of stationary phase estimates for the profile of a travelling disturbance rests on a suitable degree of wave number resolution or separation of the components therein; such a resolution is manifest geometrically through the fanwise spreading or divergence of the rays, and its imprint on the wave function is registered by the amplitude factor [cf. (34.14)]

$$\left[\left|\frac{\partial^2}{\partial k^2}(t\Psi)\right|\right]^{-1/2} = \left[\left|\left(\frac{\partial x}{\partial k}\right)_t\right|\right]^{-1/2}. \tag{37.1}$$

The latter assumes small values when there is an appreciable spacing between rays identified with a given wave number range $\Delta k = k_1 - k_2$, namely

$$\Delta x = \left(\frac{\partial x}{\partial k}\right)_t \Delta k,$$

and this circumstance ultimately prevails since the representation

$$x = x_0(k) + u(k)t \tag{37.2}$$

implies that $|(\partial x/\partial k)_t|$ increases linearly with time along any ray. If the rays are narrowly spaced there is, evidently, less merit in reliance on a particular representative for gauging the level of disturbance and a correspondingly large amplitude factor (37.1) presages the shortcomings of the stationary phase estimate.

A closely knit pattern obtains near the envelope of a family of rays, whose locus is found on elimination of k between the equations (37.2) and

$$\frac{\partial x}{\partial k} = 0;$$

characteristically, this locus specifies a so-called caustic curve or boundary for the region of disturbance in the x, t-plane. Inasmuch as $\partial x/\partial k$ vanishes at the caustic and assumes small values nearby, it follows that (34.14) provides an unsatisfactory local description of the wave function. A superior account of the latter goes together with a change in the prior representation (34.7) of the phase function $t\Psi(k, x, t)$; specifically, there is a need to alter the expression (34.20), which reflects vanishing first and second k-derivatives of Ψ, for better approximation to the wave function on and near the caustic.

As a preliminary to enacting such changes, some qualitative observations which bear on the phase function serve a useful purpose; thus, if

$$t\Psi(k, x, t) = k\frac{x}{t} - \omega(k)$$

and $\mathrm{d}\omega/\mathrm{d}k$ has a maximum, say, at $k = \hat{k}$, that is

$$\Psi''(\hat{k}, x, t) = \frac{\mathrm{d}^2}{\mathrm{d}k^2}\Psi\bigg|_{k=\hat{k}} = 0, \tag{37.3}$$

then, for $x/t > (\mathrm{d}\omega/\mathrm{d}k)|_{k=\hat{k}}$, the first derivative of Ψ vanishes at k_1 and k_2. Schematic variations, relative to k, for $\partial\Psi/\partial k$ and Ψ have the forms indicated in Fig. 20, and it may be noted that with a rise in the magnitude of x/t the separation between the points k_1 and k_2 decreases, implying that Ψ'' is locally small.

When the points k_1 and k_2, or wave number determinations $k = k_1(x, t)$, $k = k_2(x, t)$ consistent with the stationary phase condition $(\partial/\partial k)(t\Psi) = 0$, are not close or comparable in magnitude, it is appropriate to employ separate representations for Ψ, in their respective neighborhoods, which omit the first and contain second degree terms in $(k - k_{1,2})$; and the concomitant estimate for the wave function,

$$\begin{aligned} y(x, t) &= \int_0^\infty f(k)\, \mathrm{e}^{\mathrm{i}t\Psi(k,x,t)}\, \mathrm{d}k \\ &\sim f(k_1)\left(\frac{2\pi}{t|\Psi''(k_1, x, t)|}\right)^{1/2} \exp\{\mathrm{i}t\Psi(k_1, x, t) \pm \mathrm{i}\pi/4\} \\ &\quad + f(k_2)\left(\frac{2\pi}{t|\Psi''(k_2, x, t)|}\right)^{1/2} \exp\{\mathrm{i}t\Psi(k_2, x, t) \pm \mathrm{i}\pi/4\}, \quad t \to \infty \end{aligned} \tag{37.4}$$

[wherein $\pm \to \operatorname{sgn} \Psi''$], belongs to the class (34.14). Proximity of k_1 and k_2, on the other hand, rules out these separate representations and the local behavior of Ψ, whose second k-derivative has a small magnitude at k_1 and k_2, is more aptly rendered by a single third degree polynomial in k, viz.

$$t\Psi(k, x, t) = t\Psi(\hat{k}, x, t) + (k - \hat{k})t\Psi'(\hat{k}, x, t) + \frac{1}{6}(k - \hat{k})^3 t\Psi'''(\hat{k}, x, t) \tag{37.5}$$

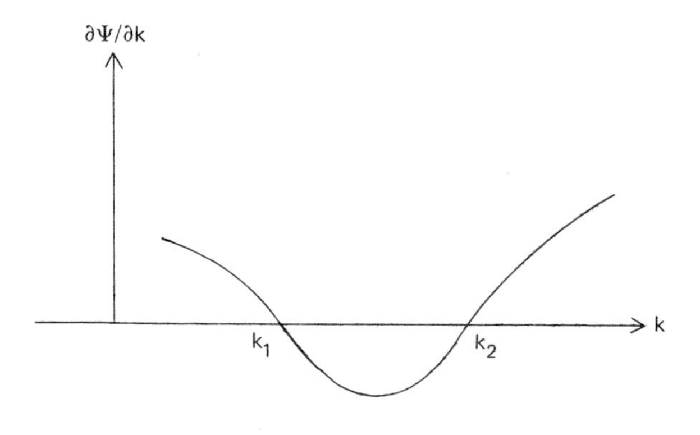

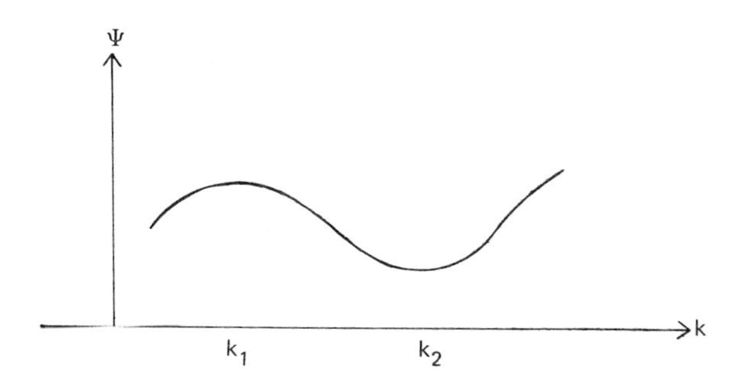

Fig. 20. Phase function and its derivative.

where primes signify derivatives with respect to k and $\hat{k}$ is defined by (37.3).

On the basis of (37.5), there follows an estimate of the wave function,

$$\begin{aligned} y(x, t) &\sim f(\hat{k})\, \mathrm{e}^{\mathrm{i}t\Psi(\hat{k},x,t)} \int_{-\infty}^{\infty} \exp\left\{\mathrm{i}t\kappa\Psi'(\hat{k}, x, t) + \frac{\mathrm{i}}{6} t\kappa^3 \Psi'''(\hat{k}, x, t)\right\} \mathrm{d}\kappa \\ &= 2f(\hat{k})\, \mathrm{e}^{\mathrm{i}t\Psi(\hat{k},x,t)} (2/t\Psi'''(\hat{k}, x, t))^{1/3} \qquad t \to \infty \\ &\quad \times \int_0^{\infty} \cos\left[(2/t\Psi'''(\hat{k}, x, t))^{1/3} t\Psi'(\hat{k}, x, t)\zeta + \frac{1}{3}\zeta^3\right] \mathrm{d}\zeta, \end{aligned}$$

with asymptotic validity for large values of t; this can be displayed in the form

$$\begin{aligned} y(x, t) &\sim 2\surd(\pi) f(\hat{k})\, \mathrm{e}^{\mathrm{i}t\Psi(\hat{k},x,t)} (2/t\Psi'''(\hat{k}, x, t))^{1/3} \\ &\quad \times A\,\mathrm{i}\{(2/t\Psi'''(\hat{k}, x, t))^{1/3} t\Psi'(\hat{k}, x, t)\}, \quad t \to \infty \end{aligned} \tag{37.6}$$

which involves a function

$$A\,\mathrm{i}(z) = \frac{1}{\sqrt{\pi}} \int_0^{\infty} \cos\left(z\zeta + \frac{1}{3}\zeta^3\right) \mathrm{d}\zeta, \tag{37.7}$$

encountered and discussed by Airy (1838) in the study of light intensity variation near a caustic. The Airy function (Fig. 21) constitutes one solution of the homogeneous second order differential equation

$$\left(\frac{d^2}{dz^2} - z\right) A\,i(z) = 0 \tag{37.8}$$

that is regular for all z and, as depicted in the accompanying figure, behaves quite differently on the positive/negative ranges of the (real) independent variable x; the asymptotic formula

$$A\,i(x) \sim \frac{1}{2} x^{-1/4} \exp\left(-\frac{2}{3} x^{3/2}\right)\left\{1 - \frac{5}{48} x^{-3/2} + 0(x^{-3})\right\}, \quad x \gg 1 \tag{37.9}$$

makes plain a rapid and monotonic decrease in value of the Airy function when its argument assumes larger positive magnitudes, and the complementary form

$$\begin{aligned} A\,i(x) \sim x^{-1/4} \sin\left(\frac{2}{3} x^{3/2} + \frac{\pi}{4}\right)\{1 + 0(x^{-3})\} \\ - x^{-1/4} \cos\left(\frac{2}{3} x^{3/2} + \frac{\pi}{4}\right)\left\{\frac{5}{48} x^{-3/2} + 0(x^{-9/2})\right\}, \end{aligned} \quad -x \gg 1 \tag{37.10}$$

reveals a decreasing oscillatory trend when the argument takes on larger negative magnitudes.

An Airy function also figures, it may be noted, in the contribution from the vicinity of an isolated stationary point, say k_1, if the local phase representation includes a third degree term, i.e.,

$$\begin{aligned} t\Psi(k, x, t) = t\Psi(k_1, x, t) + \frac{1}{2}(k - k_1)^2 t\Psi''(k_1, x, t) \\ + \frac{1}{6}(k - k_1)^3 t\Psi'''(k_1, x, t); \end{aligned} \tag{37.11}$$

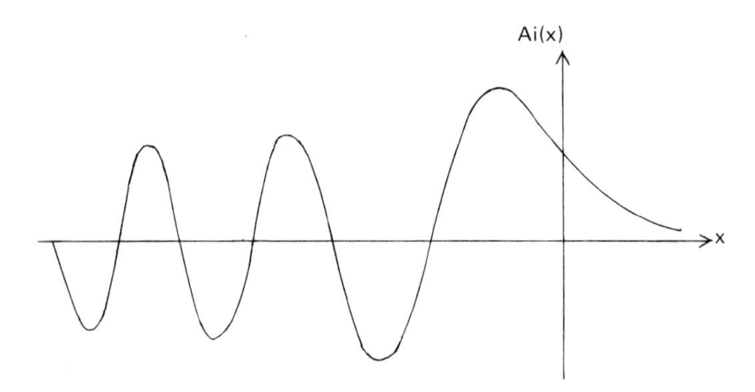

Fig. 21. Profile of the Airy function.

for, consequent to the change of variable $k \to \zeta$ expressed by

$$k - k_1 = (2/t\Psi''')^{1/3}\zeta - \Psi''/\Psi''' \tag{37.12}$$

there obtains

$$t\Psi = t\Psi(k_1, x, t) + \frac{t}{3}\frac{(\Psi'')^3}{(\Psi''')^2} - \left(\frac{t}{2}\right)^{2/3}\frac{(\Psi'')^2}{(\Psi''')^{4/3}}\zeta + \frac{1}{3}\zeta^3 \tag{37.13}$$

with only first and third powers of ζ present. The stationary phase determinations implicit in (37.11) are twofold, viz.

$$k = k_1, \tag{37.14}$$

and

$$k = k_1 - 2\frac{\Psi''(k_1, x, t)}{\Psi'''(k_1, x, t)} \tag{37.15}$$

where the former underlies the very introduction of the development (37.11) and the latter can be presumed to fall within its range of applicability so long as the magnitude of Ψ''/Ψ''' is sufficiently small. Assuming the mutual compatibility of both the determinations (37.14), (37.15) with (37.11), and employing the aforesaid change of variable $k \to \zeta$, we secure the wave function estimate

$$\begin{aligned} y(x,t) &\sim 2f(k_1)\exp\{it\Omega(k_1,x,t)\}(2/t\Psi'''(k_1,x,t))^{1/3} \\ &\quad\times \int_0^\infty \cos\left\{-(t/2)^{2/3}(\Psi''(k_1,x,t))^2(\Psi'''(k_1,x,t))^{-4/3}\zeta + \frac{1}{3}\zeta^3\right\} \mathrm{d}\zeta; \\ &= 2\sqrt{(\pi)}f(k_1)\exp\{it\Omega(k_1,x,t)\}(2/t\Psi'''(k_1,x,t))^{1/3} \\ &\quad\times A\,\mathrm{i}\left(-\left(\frac{t}{2}\right)^{2/3}(\Psi''(k_1,x,t))^2(\Psi'''(k_1,x,t))^{-4/3}\right) \qquad t\to\infty \end{aligned} \tag{37.16}$$

where

$$\Omega(k_1,x,t) = \Psi(k_1,x,t) + \frac{1}{3}(\Psi''(k_1,x,t))^2(\Psi'''(k_1,x,t))^{-3};$$

since the reliability of the preceding formula necessitates a small ratio between the second and third k-derivatives of the phase function, and thus continues in force if $\Psi''(k_1, x, t) = 0$, it appears that (37.16) holds when a quadratic approximation of the phase function near the point k_1 no longer suffices to achieve a satisfactory estimate of y.

Let us next associate the wave numbers $k_1(x, t)$ and $k_2(x, t)$ with a pair of rays which intersect in a point $P(x, t)$ to one side of the caustic or envelope, C, and $\hat{k}(t)$ with a ray tangent to the latter at the point $P_C(x_C, t)$ (see Fig. 22), whose exclusive dependence on t is apparent from the defining relation

$$\left(\frac{\partial x}{\partial k}\right)_t = \frac{\mathrm{d}x_0}{\mathrm{d}k} + \frac{\mathrm{d}u}{\mathrm{d}k}t = 0. \tag{37.17}$$

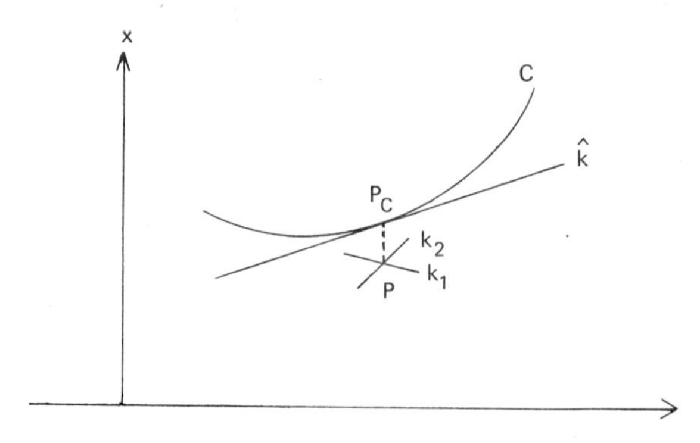

Fig. 22. A caustic or envelope and neighboring rays.

If we express the phase development (37.5) in the form

$$t\Psi(k,x,t)=\hat{\Psi}(x,t)+\alpha(x,t)(k-\hat{k})+\frac{1}{3}\beta^3(t)(k-\hat{k})^3, \tag{37.18}$$

with the identifications

$$\begin{aligned}\hat{\Psi}(x,t)&=\hat{k}(t)x-\omega(\hat{k})t+\delta(\hat{k})\\ \alpha(x,t)&=x-x_0(\hat{k})-u(\hat{k})t\end{aligned} \tag{37.19}$$

and

$$\beta(t)=\left\{-\frac{1}{2}\frac{\mathrm{d}^2}{\mathrm{d}k^2}[x_0(k)+u(k)]_{k=\hat{k}(t)}\right\}^{1/3},$$

the concomitant version of (37.6) becomes

$$y(x,t)\sim\frac{2\sqrt{\pi}}{\beta(t)}f(\hat{k}(t))\exp\{\mathrm{i}\hat{\Psi}(x,t)\}A\,\mathrm{i}\left(\frac{\alpha(x,t)}{\beta(t)}\right),\quad t\to\infty. \tag{37.20}$$

Evidently $\alpha(x,t)=x-x_C$ measures the distance between points $P(x,t)$ and $P_C(x_C,t)$ that share a common coordinate t, and assumes negative/positive values according as the former lies on the side of the caustic where rays are present or not. Thus the argument of the Airy function in (37.20) changes sign with passage across the caustic and distinctive aspects of the appertaining wave function $y(x,t)$ on opposite sides thereof are previewed. A very low excitation level obtains, in conformity with the asymptotic representation (37.9), for large and positive arguments of the Airy function, that is, when $\alpha,\beta>0$; and this corresponds to the 'shadow' side of the caustic. For a symmetric initial profile, the specification given by (37.20) on the latter curve (where $\alpha=0$) agrees with that of the prior approximation (34.12). The form of its differential equation, (37.8), suggests that the transitional role of the Airy function, here exemplified in regard to the caustic wave pattern, has a potentially wider sphere.

A more elaborate procedure is in order if we desire an asymptotic representation for the integral (37.4), as $t \to \infty$, which encompasses a wider range of values for the ratio x/t (and relative disposition of the stationary phase points k_1, k_2) than falls within the scope of either (37.4) or (37.6). This relies on an initial transformation of variable, $\zeta = \zeta(k)$, pursuant to the scheme

$$it\Psi(k, x, t) = \mu\zeta - \tfrac{1}{3}\zeta^3 + \nu \tag{37.21}$$

wherein the disposable parameters μ, ν are assigned by pairing the two distinct first order zeros of the derivative of the cubic in ζ, i.e.,

$$\zeta_{1,2} = \pm\sqrt{\mu},$$

with the corresponding zeros k_1, k_2 of the stationary phase function Ψ; employing the abbreviations

$$\Psi_1 = \Psi(k_1, x, t), \qquad \Psi_2 = \Psi(k_2, x, t) \tag{37.22}$$

and the explicit realization of such pairings,

$$it\Psi_1 = \frac{2}{3}\mu^{3/2} + \nu, \qquad it\Psi_2 = -\frac{2}{3}\mu^{3/2} + \nu.$$

there obtains the individual formulas

$$\frac{4}{3}\mu^{3/2} = \mathrm{i}t(\Psi_1 - \Psi_2), \qquad \nu = \frac{1}{2\,\mathrm{i}t}(\Psi_1 + \Psi_2) \tag{37.23}$$

which fix the parameters. After the exact transformation (37.21) is thus made precise and the condition $\mathrm{d}\zeta/\mathrm{d}k \neq 0$ assured, it follows that

$$\begin{aligned} y(x, t) &= \int_0^\infty f(k)\,\mathrm{e}^{it\Psi(k,x,t)}\,\mathrm{d}k \\ &= \mathrm{e}^{\nu}\int_\Gamma f(k(\zeta))\frac{\mathrm{d}k}{\mathrm{d}\zeta}\,\mathrm{e}^{\mu\zeta - 1/3(\zeta^3)}\,\mathrm{d}\zeta \end{aligned} \tag{37.24}$$

where the straight contour Γ passes through the origin on its way between $\zeta(0)$ and $\infty\,\mathrm{e}^{7\pi\mathrm{i}/6}$.

The principal contributions to the latter integral for large t are fashioned near the stationary points of its exponential argument, or $\pm\sqrt{\mu}$, suggesting an estimate

$$y(x, t) \sim \mathrm{e}^{\nu}\left\{f(k_1)\left(\frac{\mathrm{d}k}{\mathrm{d}\zeta}\right)_{\zeta=\sqrt{\mu}} \times I_1(\mu) + f(k_2)\left(\frac{\mathrm{d}k}{\mathrm{d}\zeta}\right)_{\zeta=-\sqrt{\mu}} \times I_2(\mu)\right\}, \quad t \to \infty \tag{37.25}$$

of additive nature; here

$$I_1(\mu) = \int_{\Gamma_1} \exp\left(\mu\zeta - \frac{1}{3}\zeta^3\right)\mathrm{d}\zeta, \qquad I_2(\mu) = \int_{\Gamma_2} \exp\left(\mu\zeta - \frac{1}{3}\zeta^3\right)\mathrm{d}\zeta$$

characterize individual functions via integrals whose (infinite) paths, $\Gamma_{1,2}$ traverse (separately) one of the stationary points $\zeta_{1,2}$ and whose magnitude depends principally on nearby portions of the respective path. Both I_1 and I_2 are expressible in terms of a common function W after suitable relocation of Γ_1 and Γ_2, viz.

$$I_1(\mu) = \sqrt{(\pi)}\, \mathrm{e}^{-4\pi\, \mathrm{i}/3} W(\mu\, \mathrm{e}^{2\pi\, \mathrm{i}/3}), \qquad I_2(\mu) = \sqrt{(\pi)}\, \mathrm{e}^{-2\pi\, \mathrm{i}/3} W(\mu\, \mathrm{e}^{4\pi\, \mathrm{i}/3})$$

where

$$W(z) = \frac{1}{\sqrt{\pi}} \int_{\Gamma^*} \exp\left(z\zeta - \frac{1}{3}\zeta^3\right) \mathrm{d}\zeta$$

and Γ^* extends from infinity to the origin along the line $\arg \zeta = -2\pi/3$ and thence to infinity along the positive real axis of the ζ-plane; for real argument Im W equals the Airy function integral (37.7). On utilizing the representations $\mu = |\mu|\, \mathrm{e}^{\pi\, \mathrm{i}/3}$ and

$$\left(\frac{\mathrm{d}k}{\mathrm{d}\zeta}\right)_{\zeta=\sqrt{\mu}} = \left(\frac{2|\mu|^{1/2}}{|t\Psi''(k_1)|}\right)^{1/2} \mathrm{e}^{-7\pi\, \mathrm{i}/6}, \qquad \left(\frac{\mathrm{d}k}{\mathrm{d}\zeta}\right)_{\zeta=-\sqrt{\mu}} = \left(\frac{2|\mu|^{1/2}}{|t\Psi''(k_2)|}\right)^{1/2} \mathrm{e}^{-7\pi\, \mathrm{i}/6}$$

in conjunction with properties of W [Fock, 1965], it appears that the estimate (37.25) goes over into the prior versions (37.4), (37.6) when $|\mu|$ assumes large or small values, respectively.

Other representatives of asymptotic developments possessing uniform capability both at and away from (smooth) caustics, have been established by reliance on the differential equation of wave motion; Airy functions and their derivative occur in a particular version which is due to Ludwig (1966).

§38. *Transformation and Estimation of Contour Integrals near Saddle Points*

The principle of stationary phase lends itself directly to seeking estimates for integrals that, as exemplified by (34.1), involve a parameter (t) and an appertaining exponential function of fixed (unit) absolute value; and associates the significant contributions to the aforesaid integrals, when the parameter assumes a large magnitude, with the neighborhood of point(s) on the range of integration at which the derivative of the exponential argument vanishes (other contributions, from endpoints of the range or singular points of a non-parametric integrand factor, may also be important). If the exponential function possesses a complex argument the critical points, fixing the sites of correspondingly localized contributions, are those at which its largest absolute value obtains; and a preliminary transformation of the relevant integral, suggested by function theoretic reasoning, facilitates the attainment of estimates therefor based on such contributions.

Let us refer, by way of illustration, to an integral (encountered in the signalling problem of §6),

$$I(x, t; h) = \int_{\Gamma} \exp\left\{-\mathrm{i}\omega t + \mathrm{i}\left(\frac{\omega^2}{c^2} - h^2\right)^{1/2} x\right\} \frac{\mathrm{d}\omega}{\left(\frac{\omega^2}{c^2} - h^2\right)^{1/2}}, \qquad x, t, h > 0 \tag{38.1}$$

that contains both propagating and attenuating ($|\omega| \gtrless hc$) wave components, relative to the positive x-direction, if the contour Γ follows the real axis in the ω-plane and passes just above each of the two branch points at $\omega = \pm hc$. There is a change of variable, namely $\omega = hc \cosh \vartheta$, which simplifies (38.1) through the removal of square roots; this yields

$$I = c \int_{\Gamma'} \exp\left\{-ihct\left(\cosh \vartheta - \frac{x}{ct} \sinh \vartheta\right)\right\} \mathrm{d}\vartheta \tag{38.2}$$

with a contour Γ' as shown in Fig. 23 (wherein the small arcs about π i and 0 that reflect deviations of the ω-contour above $\pm hc$ are not made explicit).

If $\gamma = x/ct > 1$ and the rearrangement

$$\gamma \sinh \vartheta - \cosh \vartheta = \sqrt{(\gamma^2 - 1)} \sinh(\vartheta - \bar{\varphi}), \qquad \coth \bar{\varphi} = \gamma > 1$$

applies, it follows that

$$I = c \int_{\Gamma'} \exp\{ih\sqrt{(x^2 - c^2t^2)} \sinh(\vartheta - \bar{\varphi})\} \, \mathrm{d}\vartheta, \qquad x > ct;$$

since $\operatorname{Im} \sinh(\vartheta - \bar{\varphi}) > 0$ when $0 < \operatorname{Im} \vartheta < \pi$ and the exponential function herein tends to zero on the segment $\operatorname{Re} \vartheta \to \infty$, $0 < \operatorname{Im} \vartheta < \pi$, an application of Cauchy's integral theorem to the closed contour formed by Γ' and the latter segment reveals that

$$I = 0, \quad x > ct. \tag{38.3}$$

In accordance with the relations

$$\cosh \vartheta - \gamma \sinh \vartheta = \sqrt{(1 - \gamma^2)} \cosh(\vartheta - \varphi), \qquad \tanh \varphi = \gamma < 1,$$

and

$$\tau = \frac{\pi}{2} + \mathrm{i}(\vartheta - \varphi)$$

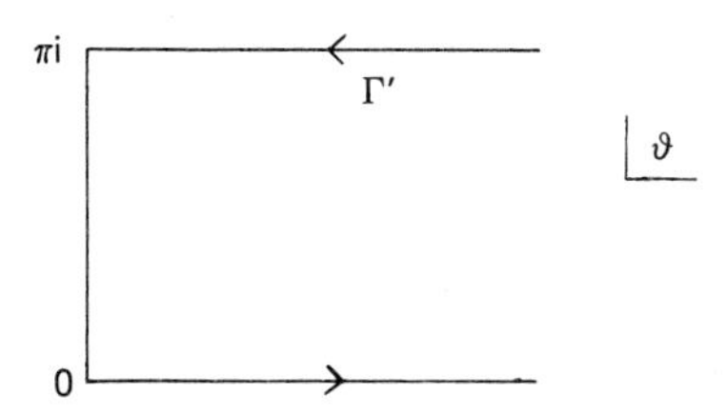

Fig. 23. An integration contour.

another form of (38.2) obtains,

$$I = -\mathrm{i}c \int_{\Gamma''} \exp\left\{ -\mathrm{i}h\sqrt{(c^2t^2 - x^2)} \sin\tau \right\} \mathrm{d}t, \quad ct > x \tag{38.4}$$

wherein the contour Γ'' differs from Γ' principally as regards its orientation (see Fig. 24)

Taking note of the (Sommerfeld) integral representation for the Hankel function of the first and second kinds (see Fig. 25)

$$H_\nu^{(1)}(\alpha) = \frac{1}{\pi} \int_{\Gamma_1} \mathrm{e}^{-\mathrm{i}\alpha \sin\tau + \mathrm{i}\nu\tau} \, \mathrm{d}\tau, \tag{38.5}$$

$$H_\nu^{(2)}(\alpha) = \frac{1}{\pi} \int_{\Gamma_2} \mathrm{e}^{-\mathrm{i}\alpha \sin\tau + \mathrm{i}\nu\tau} \, \mathrm{d}\tau, \tag{38.6}$$

which constitute linearly independent solutions of the differential equation

$$\left(\frac{\mathrm{d}^2}{\mathrm{d}\alpha^2} + \frac{1}{\alpha}\frac{\mathrm{d}}{\mathrm{d}\alpha} + 1 - \frac{\nu^2}{\alpha^2} \right) H_\nu^{(1),(2)}(\alpha) = 0, \tag{38.7}$$

and also of their mean,

$$\begin{aligned} J_\nu(\alpha) &= \tfrac{1}{2}[H_\nu^{(1)}(\alpha) + H_\nu^{(2)}(\alpha)] \\ &= \frac{1}{\pi} \int_{\Gamma_3} \mathrm{e}^{-\mathrm{i}\alpha \sin\tau + \mathrm{i}\nu\tau} \, \mathrm{d}\tau, \end{aligned} \tag{38.8}$$

we are led to the result

$$I = -2\pi \, \mathrm{i}c J_0[h\sqrt{(c^2t^2 - x^2)}], \quad ct > x. \tag{38.9}$$

That the given integral (38.1) admits an exact and explicit reduction is atypical and not deserving of emphasis; on the other hand, considerable importance attaches to the manner – suitable for wave integrals generally – whereby an asymptotic estimation thereof may be found, valuable in this instance when $t \to \infty$ and $x/ct \ll 1$. A particular choice of integration contour underlies the opening stage of an estimation procedure, which thereafter aims

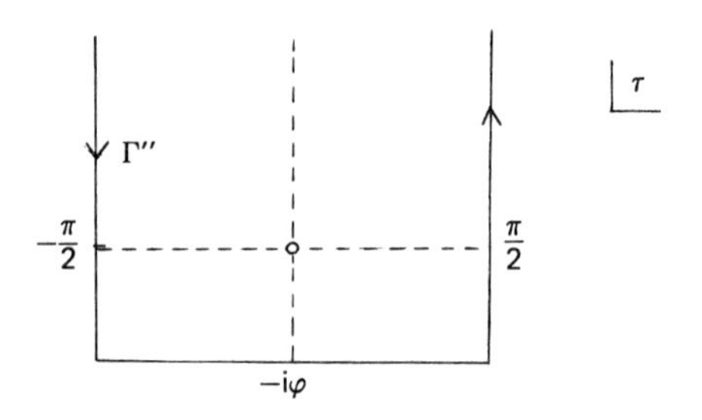

Fig. 24. An integration contour.

at evaluating the integral in terms of selective contributions from predetermined and localized portions of the contour. The salient features are readily exposed in a study of one of the Hankel functions, viz.

$$H_\nu^{(2)}(\alpha) = \frac{1}{\pi} \int_{\Gamma_2} \exp\left\{ -\alpha f\left(\tau, \frac{\nu}{\alpha}\right)\right\} \mathrm{d}\tau, \quad \alpha > 0 \tag{38.10}$$

where

$$f\left(\tau, \frac{\nu}{\alpha}\right) = \mathrm{i}\left(\sin \tau - \frac{\nu}{\alpha}\tau\right), \quad \nu, \alpha \text{ real, say,} \tag{38.11}$$

and the infinite contour Γ_2 in Fig. 25 extends from the lower half of the τ-plane, between the lines $\operatorname{Re} \tau = \mp \pi/2$, to the upper half, between the lines $\operatorname{Re} \tau = \pi/2$ and $3\pi/2$; an appropriate contour is singled out by reference to the conditions

$$\operatorname{Im} f\left(\tau, \frac{\nu}{\alpha}\right) = \operatorname{Im} f\left(\tau^*, \frac{\nu}{\alpha}\right), \tag{38.12}$$

and

$$\frac{\mathrm{d}}{\mathrm{d}\tau} f\left(\tau, \frac{\nu}{\alpha}\right)\bigg|_{\tau=\tau^*} = 0, \quad \text{i.e.,} \quad \cos \tau^* = \frac{\nu}{\alpha}. \tag{38.13}$$

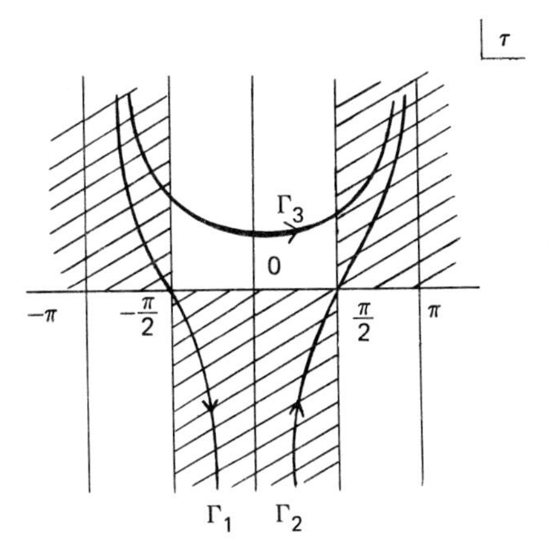

Fig. 25. Integration contours.

These enable us to locate a curve Γ^* and a point τ^* thereupon, such that $\mathrm{Re}\, f$ increases more rapidly along Γ^*, as τ moves away from τ^*, in comparison with any other departure route. If $0 < \nu/\alpha < 1$ the condition (38.13) determines a real τ^* on the interval $(0, \pi/2)$ and the version of (38.12), expressed in terms of the real and imaginary parts ξ, η of τ,

$$\sin \xi \cosh \eta = \sin \tau^* + (\xi - \tau^*) \cos \tau^*, \tag{38.14}$$

characterizes a pair of curves within the strip $0 < \mathrm{Re}\, \tau < \pi$, passing through the point τ^* and asymptotic to the lines $\xi = 0, \pi$ (see Fig. 26). Said curves are mirror images in the real axis as befits the even symmetry of the hyperbolic function in (38.14); and the function $\mathrm{Re}\, f$ manifests a maximum rate of increase starting out (in either direction) from τ^*, along the curve designated by Γ^*, with the opposite or a decreasing behavior relative to displacements along the other curve in Fig. 26. An explanation for these features rests, of course, on the fact that f is a regular (or analytic) function of the complex variable τ, whence its real and imaginary parts define surfaces exhibiting a saddle shape near the point τ^* at which $df/d\tau$ vanishes. It is apparent, furthermore, that an integral along Γ^*, or so-called steepest descent path (insofar as $\exp(-\alpha\, \mathrm{Re}\, f)$ falls off in magnitude with separation from the saddle point) has a narrower range of significant contributions (at and near τ^*), when α becomes larger, than does the integral along Γ_2; and the equality of the respective integrals (taken in the same sense) is assured by the Cauchy integral theorem inasmuch as their common integrand evidences a regular nature throughout the finite τ-plane, with null values in those portions which encompass the termini of both contours.

Consequent to the selection of a steepest descent path there remains the

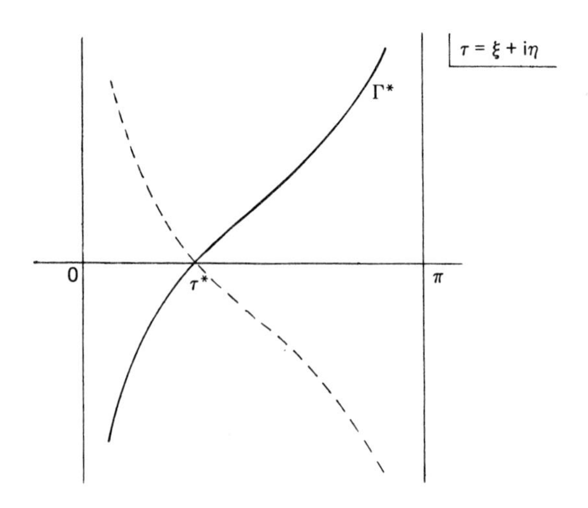

Fig. 26. A steepest descent contour.

procedural task of estimating the integral; this is advanced by employing a transformation of variables,

$$f\left(\tau, \frac{\nu}{\alpha}\right) = f\left(\tau^*, \frac{\nu}{\alpha}\right) + \zeta, \tag{38.15}$$

that possesses a conformal nature everywhere in the strip $0 < \operatorname{Re} \tau < \pi$, excluding the saddle point τ^* (at which $\mathrm{d}\zeta/\mathrm{d}\tau = 0$). Since $\mathrm{d}^2\zeta/\mathrm{d}\tau^2 \neq 0$ in the latter position there is a first order branch point at the origin of the ζ-plane (or image of τ^*); and integration along the two portions of Γ^*, namely, from $-\mathrm{i}\infty$ to τ^* and from τ^* to $\pi + \mathrm{i}\infty$, corresponds to a loop integral round the branch point that proceeds from ∞ to 0 and back. Accordingly,

$$H_\nu^{(2)}(\alpha) = \frac{1}{\pi} \exp\left\{-\alpha f\left(\tau^*, \frac{\nu}{\alpha}\right)\right\}\left[\int_0^\infty \mathrm{e}^{-\alpha\zeta} F_1(\zeta)\,\mathrm{d}\zeta - \int_0^\infty \mathrm{e}^{-\alpha\zeta} F_2(\zeta)\,\mathrm{d}\zeta\right] \tag{38.16}$$

where $F(\zeta) = \mathrm{d}\tau/\mathrm{d}\zeta$ and the subscripts 1, 2 designate values on different branches, or sides of a cut along the positive ζ-axis, with the interconnection

$$F_1(\zeta) = F_2(\zeta\,\mathrm{e}^{-2\pi\mathrm{i}}).$$

Termwise integration in (38.16), after substituting an expansion of $F(\zeta)$, valid in the neighborhood of $\zeta = 0$ and containing fractional powers of the variable, furnishes an asymptotic or semi-convergent development [Copson (1965), Lauwerier (1966)] for the Hankel function, viz.

$$H_\nu^{(2)}(\alpha) \sim \frac{1}{\pi} \exp\{-\mathrm{i}\alpha(\sin\tau^* - \tau^*\cos\tau^*)\}\left[\frac{\mathrm{e}^{\mathrm{i}\pi/4}\Gamma(1/2)}{\left(\dfrac{\alpha}{2}\sin\tau^*\right)^{1/2}} + 0(\alpha^{-3/2})\right], \alpha \to \infty \tag{38.17}$$

which presupposes that the value of $\cos\tau^* = \nu/\alpha$ is bounded away from unity. A similar development for $H_\nu^{(1)}(\alpha)$ – the complex conjugate of $H_\nu^{(2)}(\alpha)$ if α and ν are real – is arrived at on replacing Γ_1 with another steepest descent contour through a saddle point $-\tau^*$ located between $-\pi/2$ and 0; and thus, pursuant to (38.8) or the deformation of Γ_3 into a contour that traverses both of the aforesaid saddle points $\pm\tau^*$, it follows that

$$J_\nu(\alpha) \sim \frac{1}{\pi} \cos\left\{\alpha(\sin\tau^* - \tau^*\cos\tau^*) - \frac{\pi}{4}\right\} \frac{\Gamma^{1/2}}{\left(\dfrac{\alpha}{2}\cos\tau^*\right)^{1/2}} + 0(\alpha^{-3/2}), \quad \begin{array}{c}\alpha \to \infty \\ \dfrac{\nu}{\alpha} \ll 1\end{array}. \tag{38.18}$$

For extensive details bearing on these and other developments that are applicable when $\nu/\alpha \approx 1$ or $\gg 1$, and which jointly reflect a classical stage in the theory of asymptotic representation, reference to the original (Debye,

1909) and subsequent (e.g., Watson, 1922) accounts is indicated. If the order of the Bessel function vanishes and $\cos \tau^* = 0$, or $\tau^* = \pi/2$, there obtains a limiting case of the formula (previously given by Hankel), viz.

$$J_0(\alpha) \sim \left(\frac{2}{\pi\alpha}\right)^{1/2} \cos\left(\alpha - \frac{\pi}{4}\right) + 0(\alpha^{-3/2}), \quad \alpha \to \infty$$

and the concomitant estimate for the integral I in (38.1) becomes

$$I \sim -2\,\mathrm{i}c\sqrt{(2\pi/h)}(c^2t^2 - x^2)^{-1/4} \cos\left\{h(c^2t^2 - x^2)^{1/2} - \frac{\pi}{4}\right\}, \quad \begin{matrix} t \to \infty \\ ct \gg x \end{matrix}. \tag{38.19}$$

It is appropriate to verify that the latter result can be independently established on the basis of the original integral representation (38.1), that is, without a direct intervention of the Bessel function; we need only locate saddle points for the exponential argument in I and add the (principal or leading order) contributions, originating in their respective vicinities, to the integral. On utilizing the characterization of I which emerges after resort to a scale change $\omega = c\tau$, namely

$$I = c\int_\Gamma \exp\{-\mathrm{i}ct\bar{f}(\tau, \gamma)\}\bar{g}(\tau)\,\mathrm{d}\tau$$

wherein

$$\bar{f}(\tau, \gamma) = \tau - \gamma(\tau^2 - h^2)^{1/2}, \qquad \bar{g}(\tau) = (\tau^2 - h^2)^{-1/2}, \qquad \gamma = \frac{x}{ct} > 0,$$

and thence of the affiliated equation

$$\frac{\mathrm{d}}{\mathrm{d}\tau}\bar{f} = 1 - \frac{\gamma t}{(\tau^2 - h^2)^{1/2}} = 0$$

it appears that

$$\tau_\pm = \pm\frac{h}{(1 - \gamma^2)^{1/2}}$$

specify a pair of saddle points on the real axis if $\gamma < 1$ and the choices (that stem from the disposition of the contour Γ)

$$\arg(\tau^2 - h^2)^{1/2} = \begin{matrix} 0, & \tau > h \\ \pi, & \tau < -h \end{matrix}$$

are in force.

A steepest descent path through τ_+ satisfies the condition

$$\bar{f}(\tau, \gamma) - \bar{f}(\tau_+, \gamma) = -\mathrm{i}\sigma^2, \quad \sigma \text{ real}$$

and its approximate version near the saddle point,

$$\frac{1}{2}\left(\frac{\mathrm{d}^2\bar{f}}{\mathrm{d}\tau^2}\right)_{\tau=\tau_+}\cdot(\tau-\tau_+)^2 = -\,\mathrm{i}\sigma^2, \tag{38.20}$$

follows when the Taylor series for $\bar{f}$ is broken off after the term of second degree in $(\tau-\tau_+)$. Since

$$\frac{\mathrm{d}^2\bar{f}}{\mathrm{d}\tau^2} = \frac{\gamma h^2}{(\tau^2-h^2)^{3/2}},$$

and

$$\left(\frac{\mathrm{d}^2\bar{f}}{\mathrm{d}\tau^2}\right)_{\tau=\tau_+} = \pm\frac{(1-\gamma^2)^{3/2}}{h\gamma^2}, \quad 0<\gamma<1$$

the relation (38.20) implies that

$$\tau-\tau_+ \doteq \mathrm{e}^{-\mathrm{i}\pi/4}\frac{\sqrt{(2h)}\gamma}{(1-\gamma^2)^{3/4}}\sigma, \quad \sigma\to 0 \tag{38.21}$$

and we thus form an estimate

$$\begin{aligned} I_+ &= c\,\exp\{-\mathrm{i}ct\bar{f}(\tau_+,\gamma)\}\bar{g}(\tau_+)\,\mathrm{e}^{-\mathrm{i}\pi/4}\frac{\sqrt{(2h)}\gamma}{(1-\gamma^2)^{3/4}}\int_{-\infty}^{\infty}\mathrm{e}^{-ct\sigma^2}\,\mathrm{d}\sigma \\ &= c\sqrt{(2\pi/h)}(c^2t^2-x^2)^{-1/4}\exp\{-\mathrm{i}h(c^2t^2-x^2)^{1/2}-\mathrm{i}\pi/4\}, \quad ct\gg 1 \end{aligned} \tag{38.22}$$

of the local contribution to I. The counterpart of (38.21),

$$\tau-\tau_- \doteq \mathrm{e}^{\mathrm{i}\pi/4}\frac{\sqrt{(2h)}\gamma}{(1-\gamma^2)^{3/4}}\nu, \quad \nu\to 0 \tag{38.23}$$

which holds near the other saddle point, τ_-, is involved in fashioning the comparable related estimate

$$\begin{aligned} I_- &= c\,\exp\{-\mathrm{i}ct\bar{f}(\tau_-,\gamma)\}\bar{g}(\tau_-)\,\mathrm{e}^{\mathrm{i}\pi/4}\frac{\sqrt{(2h)}\gamma}{(1-\gamma^2)^{3/4}}\int_{-\infty}^{\infty}\mathrm{e}^{-ct\nu^2}\,\mathrm{d}\nu \\ &= -\,c\sqrt{(2\pi/h)}(c^2t^2-x^2)^{-1/4}\exp\{\mathrm{i}h(c^2t^2-x^2)^{1/2}+\mathrm{i}\pi/4\}. \end{aligned} \tag{38.24}$$

When (38.22) and (38.24) are added, the outcome

$$\begin{aligned} I_+ + I_- &= -2\,\mathrm{i}c\sqrt{(2\pi/h)}(c^2t^2-x^2)^{-1/4}\sin\{h(c^2t^2-x^2)^{1/2}+\pi/4\} \\ &= -2\,\mathrm{i}c\sqrt{(2\pi/h)}(c^2t^2-x^2)^{-1/4}\cos\{h(c^2t^2-x^2)^{1/2}-\pi/4\}, \end{aligned}$$

manifests full agreement with (38.19).

A general description of the contour that is preferable to Γ for the purpose of estimating I, which links up the sections (cf. (38.21), (38.23)) traversing the saddle points τ_-, τ_+ at angles $\pi/4$, $-\pi/4$ relative to the direction of the positive real axis, becomes necessary in a more systematic calculation.

§39. *Signal Propagation and Dispersion*

The influence of dispersion has thus far been traced in connection with the movement of wave packets or profiles whose spatial characteristics are stipulated at a particular instant of time. A different manifestation of dispersive influences is revealed in the aspects of disturbances which travel away from a particular location where a definite signal or input is prescribed. If the signal strength $F(t) = y(0, t)$ be resolved in terms of its time-periodic components through a Fourier integral

$$F(t) = \int_{-\infty}^{\infty} f(\omega)\, \mathrm{e}^{-\mathrm{i}\omega t}\, \mathrm{d}\omega \tag{39.1}$$

the resulting excitation $y(x, t)$ at any distance $x(>0)$ from the input location is directly expressed by means of the two-parameter integral

$$y(x, t) = \int_{-\infty}^{\infty} f(\omega) \exp\left\{ -\mathrm{i}\omega\left(t - \frac{x}{v(\omega)}\right)\right\} \mathrm{d}\omega \tag{39.2}$$

which superposes these components in accordance with the requisite (and non-negative) value of the phase velocity, $v(\omega)$, for a progressive wave motion at each frequency ω; the formal resemblance between (39.2) and a previous integral (32.10) that expresses the correlation of wave profiles at distinct values of the time is apparent.

For a signal of limited duration and periodic nature, say

$$\begin{aligned} F(t) &= \begin{cases} \sin \omega_1 t, & 0 < t < T = 2N\pi/\omega_1 \\ 0, & t < 0, > T \end{cases} \\ &= \frac{1}{2\pi} \operatorname{Re} \int_{-\infty}^{\infty} [\mathrm{e}^{-\mathrm{i}\omega(t-T)} - \mathrm{e}^{-\mathrm{i}\omega t}] \frac{\mathrm{d}\omega}{\omega - \omega_1} \end{aligned} \tag{39.3}$$

the corresponding transform or frequency distribution function $f(\omega)$ has a maximum modulus at $\omega = \omega_1$, with a proportionately greater value and a narrower spread relative thereto as T increases. If we employ a linear approximation for the exponential argument in (39.2),

$$\begin{gathered} \omega\left(t - \frac{x}{v(\omega)}\right) \doteq \omega_1\left(t - \frac{x}{v_1}\right) + \left(t - \frac{x}{u_1}\right)(\omega - \omega_1) \\ v_1 = v(\omega_1), \qquad \frac{1}{u_1} = \frac{\mathrm{d}}{\mathrm{d}\omega}\left(\frac{\omega}{v(\omega)}\right)_{\omega=\omega_1}, \end{gathered} \tag{39.4}$$

obtained from the leading terms of a Taylor expansion about $\omega = \omega_1$, and a similar approximation for $\omega\{t - T - (x/v(\omega))\}$, the propagated signal

$$y(x, t) = \frac{1}{2\pi} \operatorname{Re} \int_{-\infty}^{\infty} \left[\exp\left\{ -\mathrm{i}\omega\left(t - T - \frac{x}{v(\omega)}\right)\right\} - \exp\left\{ -\mathrm{i}\omega\left(t - \frac{x}{v(\omega)}\right)\right\}\right] \frac{\mathrm{d}\omega}{\omega - \omega_1} \tag{39.5}$$

is found to have, in common with the impressed one, a sharply defined onset (at $t = x/u_1$) and termination (at $t = (x/u_1) + T$); the presence of dispersion is thus registered by a signal velocity

$$u_1 = \frac{v_1^2}{v_1 - \omega_1\left(\dfrac{\mathrm{d}v}{\mathrm{d}\omega}\right)_1} \tag{39.6}$$

whose magnitude equals that of the group velocity at $\omega = \omega_1$, though evidence and details of the expected signal deformation evidently awaits an improvement on the accuracy of the preceding approximation.

We may examine the temporal distortion of the received signal and also bring out related analytical complexities of dispersive wave propagation in a specific model, namely that of a string with elastically (rather than directly) attached loads. A uniform distribution of loads is envisaged, with the number $N\,\mathrm{d}x$ on any section of span $\mathrm{d}x$, and their relative separations are assumed to be negligibly small on the scale of wave lengths for string displacements. If the loads have identical mass M, natural frequency ω_0 and frictional coefficient μ, the equation of motion for any one takes the form

$$\left(\frac{\partial^2}{\partial t^2} + \mu\frac{\partial}{\partial t} + \omega_0^2\right) Y = \omega_0^2 y, \tag{39.7}$$

where $Y(x, t)$ and $y(x, t)$ denote the (collinear) displacements of the load and string, respectively. On the premise that an imprecision in fixing the exact positions of neighboring loads has little significance when relatively long wave lengths are involved, we adopt the dynamical equation

$$\rho\frac{\partial^2 y}{\partial t^2}\,\mathrm{d}x = P\frac{\partial^2 y}{\partial x^2}\,\mathrm{d}x - MN\omega_0^2(y - Y)\,\mathrm{d}x$$

for a short section $\mathrm{d}x$ of the string, and obtain therefrom a local interrelation of displacements

$$\frac{1}{c^2}\frac{\partial^2 y}{\partial t^2} - \frac{\partial^2 y}{\partial x^2} = -\frac{MN\omega_0^2}{\rho c^2}(y - Y). \tag{39.8}$$

Since our interest pertains to the effect of loading on the wave propagation along the string, the elimination of Y from the simultaneous equations

(39.7), (39.8) is indicated; this is a simple matter if y and Y have a synchronous time variation $e^{-i\omega t}$, and leads to the differential equation

$$\frac{1}{c^2}\frac{\partial^2 y}{\partial t^2}-\frac{\partial^2 y}{\partial x^2}=-\frac{MN\omega_0^2}{\rho c^2}\left[1-\frac{\omega_0^2}{\omega_0^2-\omega^2-i\omega\mu}\right]\left[-\frac{1}{\omega_0^2}\frac{\partial^2 y}{\partial t^2}\right]$$

which, following a combination of terms, becomes

$$\left(\frac{\partial^2}{\partial x^2}-\frac{n^2(\omega)}{c^2}\frac{\partial^2}{\partial t^2}\right)y(x,t)=0 \tag{39.9}$$

where

$$n^2(\omega)=1+\frac{MN\omega_0^2}{\rho}\frac{1+i\dfrac{\mu}{\omega}}{\omega_0^2-\omega^2-i\omega\mu}. \tag{39.10}$$

It thus appears that a time-periodic string displacement or solution of (39.9), namely

$$y(x,t)=\exp\left\{-i\omega\left(t-\frac{n(\omega)}{c}x\right)\right\}, \tag{39.11}$$

contains a distinct pair of factors,

$$\exp\left\{-i\omega\left(t-\frac{\operatorname{Re} n(\omega)}{c}x\right)\right\},\quad \exp\left\{-\omega\frac{\operatorname{Im} n(\omega)}{c}x\right\},$$

with a progressive wave character reflected in the former and an absorption implied by the latter. Dispersion is attributable, in the presently conceived circumstances, to the existence of a frequency dependent coefficient in the simplest form of a wave equation, (39.9); the designation of this coefficient, $n(\omega)$, conforms with that customarily given to the optical index of refraction, which serves for comparing the velocities of light in vacuo and a homogeneous material medium. The absorption, manifest by an exponential decrease in the wave function along the direction of propagation, is introduced into the model through a provision for damping of the load movements and is controlled by the frictional parameter μ; if the latter has a very small magnitude the significant values for $\operatorname{Im} n(\omega)$ are realized only when ω is approximately equal to ω_0, and a simplified version of (39.10),

$$n^2(\omega)\doteq 1+\frac{MN}{\rho}\frac{\omega_0^2}{\omega_0^2-\omega^2},\quad |\omega-\omega_0|\gg\mu \tag{39.12}$$

is applicable at all other frequencies. It is, furthermore, implicit in the representation (39.12) that real (and positive) values of $n(\omega)$ do not differ appreciably from unity, as befits a comparatively slight loading of the string.

A phase velocity

$$v=\frac{\omega}{k}=\frac{c}{n(\omega)} \tag{39.13}$$

calculated by means of the expression (39.12) evidently assumes values greater or smaller than c, according as $\omega \gtreqless \omega_0$. The affiliated group velocity, obtained from either of the characterizations

$$u=\frac{\mathrm{d}\omega}{\mathrm{d}k}=\frac{k/\omega}{\dfrac{\mathrm{d}(k^2)}{\mathrm{d}(\omega^2)}}=\frac{nc}{\dfrac{\mathrm{d}}{\mathrm{d}\omega^2}(\omega^2 n^2)}=\frac{nc}{n^2+\omega^2\dfrac{\mathrm{d}}{\mathrm{d}\omega^2}n^2}$$

or

$$u=\frac{v^2}{v-\omega\dfrac{\mathrm{d}v}{\mathrm{d}\omega}}=\frac{c}{n+\omega\dfrac{\mathrm{d}n}{\mathrm{d}\omega}}, \tag{39.14}$$

proves to be

$$u=\frac{nc}{1+\dfrac{MN}{\rho}\left(\dfrac{\omega_0^2}{\omega_0^2-\omega^2}\right)^2} \tag{39.15}$$

and the inequality

$$u \leq c \tag{39.16}$$

is thus assured if

$$\left[1+\frac{MN}{\rho}\left(\frac{\omega_0^2}{\omega_0^2-\omega^2}\right)^2\right]^2 \geq 1+\frac{MN}{\rho}\frac{\omega_0^2}{\omega_0^2-\omega^2}$$

or

$$\left(\frac{\omega_0^2}{\omega_0^2-\omega^2}\right)^2\left[2+\frac{MN}{\rho}\left(\frac{\omega_0^2}{\omega_0^2-\omega^2}\right)^2\right] \geq \frac{\omega_0^2}{\omega_0^2-\omega^2}. \tag{39.17}$$

It appears on inspection that (39.17) holds at frequencies ω in excess of ω_0, since the left- and right-hand members have positive and negative magnitudes, respectively; and, inasmuch as the version of (39.17) which obtains when $\omega<\omega_0$, viz.

$$\frac{\omega_0^2}{\omega_0^2-\omega^2}\left[2+\frac{MN}{\rho}\left(\frac{\omega_0^2}{\omega_0^2-\omega^2}\right)^2\right] \geq 1,$$

is evidently consistent with the appertaining inequality $\omega_0^2/(\omega_0^2-\omega^2) \geq 1$, confirmation of the bound (39.16) for the group velocity follows.

The further property

$$u<v \tag{39.18}$$

is revealed by (39.14) after noting that

$$-\omega\frac{dv}{d\omega}=\frac{\omega c}{n^2}\frac{dn}{d\omega}=\frac{MNc}{\rho}\frac{(\omega\omega_0)^2}{(\omega_0^2-\omega^2)^2}>0.$$

Save for frequencies which have magnitudes in or adjacent to the narrow range

$$(\omega_0,\ \omega_0\sqrt{(\rho_{\text{eff}}/\rho)}),\quad \rho_{\text{eff}}=\rho+MN \tag{39.19}$$

where absorption plays a role, the behavior of phase and group velocities, according to the approximation (39.12) for the index $n(\omega)$, are contrasted in the adjacent plots of c/v and c/u (see Fig. 27); excluding the very high frequencies and correspondingly short wave lengths, when the model formulation hitherto given becomes inaccurate, the group velocity can be seen to assume its largest value at $\omega=0$, and the related expansion

$$\frac{c}{u}=\left(\frac{\rho_{\text{eff}}}{\rho}\right)^{1/2}+\frac{3}{2}\frac{MN}{\rho}\left(\frac{\rho}{\rho_{\text{eff}}}\right)^{1/2}\left(\frac{\omega}{\omega_0}\right)^2+\cdots \tag{39.20}$$

makes clear that

$$\frac{d}{d\omega}\left(\frac{c}{u}\right)=0,\qquad \omega=0. \tag{39.21}$$

To establish the equivalent magnitudes of group and energy transport velocities when absorption is neglected, we utilize a conservation relation for the coupled mechanical system [which is found by combining the simultaneous load and string equations (39.7), (39.8) in the limit $\mu\to 0$]; this takes the anticipated form

$$\frac{\partial W}{\partial t}+\frac{\partial S}{\partial x}=0,$$

with an energy density, W, that includes separate contributions associated

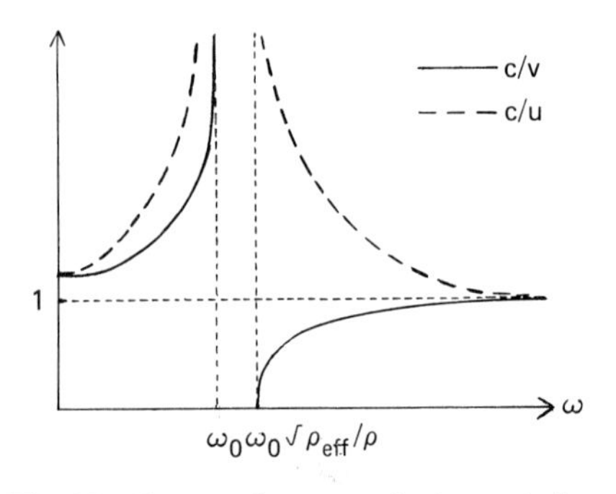

Fig. 27. Phase and group velocity variations.

with string and load displacements,

$$W = \frac{1}{2}\rho\left(\frac{\partial y}{\partial t}\right)^2 + \frac{1}{2}\rho c^2\left(\frac{\partial y}{\partial x}\right)^2 + \frac{1}{2}NM\left\{\left(\frac{\partial Y}{\partial t}\right)^2 + \omega_0^2 Y^2\right\} + \frac{1}{2}NM\omega_0^2(y^2 - 2Yy), \tag{39.22}$$

and energy flux,

$$S = -\rho c^2 \frac{\partial y}{\partial t}\frac{\partial y}{\partial x}, \tag{39.23}$$

which reflects the fact that the string alone acts as a vehicle for energy transport.

The average values, $\bar{S}$ and $\bar{W}$, calculated on the basis of a time-periodic string displacement

$$y(x, t) = \exp\left\{-\mathrm{i}\omega\left(t - \frac{n(\omega)}{c}x\right)\right\} \tag{39.11}$$

and its companion load function

$$Y = \frac{\omega_0^2}{\omega_0^2 - \omega^2}y,$$

are given by

$$\bar{S} = \tfrac{1}{2}\rho c\omega^2 n(\omega)$$

and

$$\begin{aligned}\bar{W} &= \frac{1}{4}\rho\omega^2(1 + n^2(\omega)) + \frac{1}{4}NM(\omega^2 + \omega_0^2)\left(\frac{\omega_0^2}{\omega_0^2 - \omega^2}\right)^2 + \frac{1}{4}NM\omega_0^2\left(1 - \frac{2\omega_0^2}{\omega_0^2 - \omega^2}\right)\\ &= \frac{1}{2}\rho\omega^2 + \frac{1}{2}NM\frac{\omega_0^4\omega^2}{(\omega_0^2 - \omega^2)^2};\end{aligned}$$

accordingly, their ratio

$$u = \frac{\bar{S}}{\bar{W}} = \frac{nc}{1 + \dfrac{NM}{\rho}\left(\dfrac{\omega_0^2}{\omega_0^2 - \omega^2}\right)^2}$$

proves to be in agreement with the group velocity determination (39.15).

Relying on the model of a loaded string for the particulars of dispersion we take up again the matter of signal propagation, whose details are implicit in a representation of the form (39.5) when the time-periodic stimulus at $x = 0$ has a finite duration T; the substance of an earlier and approximate evaluation of the integral (39.5) is that a virtual replica of the input, or main signal, reaches any location $x > 0$ after the lapse of time x/u, and reflects the predominant magnitude of the integrand at the signal frequency ω_1 (which is further enhanced as T increases). If we seek a finer resolution of the transmitted signal it becomes necessary to appraise those contributions to the aforesaid integral which are shaped by the exponential factors therein. It suffices to consider only one term,

$$-\frac{1}{2\pi}\,\mathrm{Re}\int_{-\infty}^{\infty}\exp\left\{-\mathrm{i}\omega\left(t-\frac{x}{v(\omega)}\right)\right\}\frac{\mathrm{d}\omega}{\omega-\omega_1}, \tag{39.24}$$

during the initial epoch of signal reception and, having already accounted for the influence of the factor $(\omega-\omega_1)^{-1}$, there remains the task of identifying other ranges of the integration variable which contribute significantly and of estimating the related amounts. To this end, we invoke the stationary phase condition (when $t \gg 0$),

$$\frac{\partial}{\partial\omega}\left\{\omega\left(t-\frac{x}{v(\omega)}\right)\right\}=t-\frac{x}{u}=0 \tag{39.25}$$

and, recalling the fact that the group velocity itself has a stationary (maximum) value at $\omega=0$,

$$\frac{\mathrm{d}}{\mathrm{d}\omega}\left(\frac{1}{u}\right)=0,\qquad \omega=0, \tag{39.26}$$

our focus can be centered on the (predominant) contribution from the group of low frequency or long wave components; the existence of another stationary point for the group velocity at $\omega=\infty$ is inconsequential, owing to a previously cited inadequacy of the model at extremely short wave lengths.

On utilizing the development (39.20) for c/u that holds in a neighborhood of $\omega=0$, the stationary phase condition (39.25) at $\omega=0$ is found to imply a fixed ratio of x and t,

$$\frac{x}{t}=c\sqrt{(\rho/\rho_{\mathrm{eff}})}, \tag{39.27}$$

equal to the velocity

$$c^*=c\sqrt{(\rho/\rho_{\mathrm{eff}})}$$

that characterizes wave propagation along the string when the loads are uniformly and continuously distributed. If we consider a definite location or observation point x and introduce

$$t'=t+\tau,\quad |\tau|\ll t \tag{39.28}$$

with a view to specifying a short interval of time that includes the instant, t, selected by the condition (39.27), the phase expansion in powers of ω,

$$\omega\left(t'-\frac{x}{v(\omega)}\right)=\omega\tau-\frac{1}{6}x\omega^3\frac{d^3}{d\omega^3}\left(\frac{\omega}{v(\omega)}\right)_{\omega=0}+\cdots$$
$$=\omega\tau-\frac{1}{6}x\omega^3\frac{d^2}{d\omega^2}\left(\frac{1}{u(\omega)}\right)_{\omega=0}+\cdots,$$

allows us to form an estimate of the low frequency contribution to the integral (39.24), viz.

$$\frac{1}{2\pi\omega_1}\operatorname{Re}\int_{-\infty}^{\infty}\exp\left\{-i\omega\tau+\frac{i}{6}x\omega^3\frac{d^2}{d\omega^2}\left(\frac{1}{u(\omega)}\right)_{\omega=0}\right\}d\omega$$
$$=\frac{1}{\surd(\pi)\omega_1}\left[2\Big/x\frac{d^2}{d\omega^2}\left(\frac{1}{u(\omega)}\right)_{\omega=0}\right]^{1/3}A\,i\left(-\tau\left[2\Big/x\frac{d^2}{d\omega^2}\left(\frac{1}{u(\omega)}\right)_{\omega=0}\right]^{1/3}\right) \tag{39.29}$$

where $A\,i(z)$ denotes the Airy integral function (37.7).

Since

$$x\frac{d^2}{d\omega^2}\left(\frac{1}{u(\omega)}\right)_{\omega=0}=3\frac{NMx}{\rho\omega_0^2}\left(\frac{\rho}{\rho_{\text{eff}}}\right)^{1/2}>0$$

the sign of τ determines that of the Airy function argument in (39.29), and we infer from this expression a rising level of signal amplitude when $\tau=0$, or $t=x/c^*=x/u_1(0)$, which precedes the arrival of the main signal at the time $x/u_1(\omega_1)$. The low frequency contribution (39.29) is, on this account, termed a precursor or forerunner of the main signal and its existence indicates a disparity between the input and received signal profiles. An eventual subsidence of the precursor is implied by (39.29) inasmuch as c^*t may be substituted for x in the factors $x^{-1/3}$, according to the stationary phase condition (39.27). The Airy function thus figures – once again – in a transitional representation, namely that of the response, at any distant location x, which encompasses times prior and up to the advent of the main signal.

When the magnitude of the signal frequency, ω_1, is nearly equal to ω_0 and provision for dissipative effects becomes vital if the resulting disturbance is to be held within finite bound, the requisite estimation of the signal integral presents a more exacting task, on account of the intricate frequency dependence of the complex function $n(\omega)$; we omit an account of the collateral details and, save for asserting that the prior agreement between group, and energy transport velocities no longer obtains, cite the investigations of Brillouin (1932) and Baerwald (1930) which deal with the complicated phenomena attributable to the effects of dissipation on progressive wave forms.

§40. *Transient Solutions of a Dispersive Wave Equation and Their Ray Representation*

In the previous section an integral which comprises time-periodic and variable phase velocity progressive wave functions suited to a particular mechanical configuration, and which represents a given transient stimulus at a definite location therein, is directly analyzed for the purpose of characterizing the response elsewhere; there are practical difficulties in the way of calculating this integral to any desired order of accuracy and thus, if we aim at developing the theory of dispersive signals in more general circumstances, it behooves us to consider other analytical procedures. A broadly capable approach, distinguished by the ready geometrical interpretation of its formal structure, expresses the solution with reference to families of rays or curves in space-time, whose individual members are specified through a parametric relation between the x, t variables thereupon. The central role given to rays in this approach alters the objective, from that of estimating an integral form of the solution when x, t have large magnitudes, to that of detailing expansions for the solution of differential equations containing a large parameter.

Let us refer, for an explicit account, to the dispersive wave equation

$$\frac{\partial^2 y}{\partial x^2}-\frac{1}{c^2}\frac{\partial^2 y}{\partial t^2}-h^2 y=\frac{4\pi}{c}\delta(x)a(t)\,\mathrm{e}^{-ih\varphi(t)} \tag{40.1}$$

where the inhomogeneous member describes a source, with variable amplitude $a(t)$ and instantaneous frequency $h\,\mathrm{d}\varphi/\mathrm{d}t$, positioned at the origin. A solution of the linear differential equation (40.1) that vanishes prior to the instant of time $t=0$, at which the homogeneous specifications

$$y(x,0)=\frac{\partial}{\partial t}y(x,0)=0 \tag{40.2}$$

are adopted, is given by the integral

$$y(x,t)=\frac{4\pi}{c}\int_{-\infty}^{\infty}\mathrm{d}x'\int_0^t \mathrm{d}t' G(x-x',t-t')\delta(x')a(t')\,\mathrm{e}^{-ih\varphi(t')} \tag{40.3}$$

wherein the factor $G(x-x',t-t')$ pertains to a source that is localized both in position (x') and time (t'), viz.

$$\left(\frac{\partial^2}{\partial x^2}-\frac{1}{c^2}\frac{\partial^2}{\partial t^2}-h^2\right)G(x-x',t-t')=\delta(x-x')\delta(t-t') \tag{40.4}$$

$$G=0,\quad t<t'; \tag{40.5}$$

the double Fourier integral

$$G(x-x',t-t')=\frac{c^2}{(2\pi)^2}\int_{-\infty}^{\infty}\exp\{ik(x-x')-\mathrm{i}\omega(t-t')\}\frac{\mathrm{d}k\,\mathrm{d}\omega}{\omega^2-(k^2+h^2)c^2} \tag{40.6}$$

provides a solution of (40.4) and complies with the causality condition if the points $\pm c\sqrt{(k^2+h^2)}$ are situated below the contour of integration for ω. Accordingly,

$$G = \frac{c}{4\pi \mathrm{i}} \int_{-\infty}^{\infty} \exp\{\mathrm{i}k(x-x')\}[\exp\{\mathrm{i}c\sqrt{(k^2+h^2)}(t-t')\} - \exp\{-\mathrm{i}c\sqrt{(k^2+h^2)}(t-t')\}] \frac{\mathrm{d}k}{\sqrt{(k^2+h^2)}}$$

and the wave function (40.3) can be displayed in the form

$$y(x,t) = y_+(x,t) + y_-(x,t) \tag{40.7}$$

with

$$y_\pm(x,t) = \mp \mathrm{i} \int_{-\infty}^{\infty} \frac{\mathrm{d}\zeta}{f(\zeta)} \int_0^t a(t')\, \mathrm{e}^{\mathrm{i}h\psi_\pm(\zeta,t')}\, \mathrm{d}t', \tag{40.8}$$

$$\psi_\pm(\zeta,t') = \zeta x \mp cf(\zeta)(t-t') - \varphi(t'), \qquad f(\zeta) = \sqrt{(\zeta^2+1)}. \tag{40.9}$$

A consequence of the generality incorporated in the source description (and expressed via the amplitude and phase functions $a(t)$, $\varphi(t)$, respectively) is that the disturbance arising therefrom has a double integral representation. To cope with the latter, we assume a large magnitude for the parameter h which enters into the exponential function of (40.8) and propose an adaptation of the stationary phase method previously employed in securing estimates of single integrals that contain a rapidly varying factor. The asymptotic evaluation of double integrals in this manner presents a number of features hitherto absent, connected with the variety of critical or stationary points for the exponent. A stationary point of (40.8) is characterized through the simultaneous relations

$$\frac{\partial}{\partial \zeta}\psi_\pm = 0, \qquad \frac{\partial}{\partial t'}\psi_\pm = 0 \tag{40.10}$$

and belongs to the interior variety if the appertaining values of ζ and t' are located within the (infinite) strip domain of integration, viz.: $-\infty < \zeta < \infty$, $0 < t' < t$; after utilizing the explicit forms (40.9) of the phase functions, these relations become

$$x = \pm c\left(\frac{\mathrm{d}f}{\mathrm{d}\zeta}\right)(t-t') = \pm c\frac{\zeta(t-t')}{\sqrt{(\zeta^2+1)}}, \tag{40.11}$$

$$\frac{\mathrm{d}\varphi}{\mathrm{d}t'} = \pm cf(\zeta) = \pm c\sqrt{(\zeta^2+1)}. \tag{40.12}$$

The fact that time-periodic solutions of the homogeneous counterpart to the wave equation (40.1) remain bounded for all values of x so long as $\omega > hc$, and are otherwise endowed with an exponential (or non-propagating) behavior, indicates the relevance of a source frequency condition; specifically, we impose the condition

$$\frac{\mathrm{d}\varphi(t)}{\mathrm{d}t} > c \tag{40.13}$$

to assure that an excitation launched at the source point $x = 0$ in (40.1) may be transmitted, with a proportionate amplitude level, to arbitrarily large distances therefrom. A vanishing t' derivative of ψ_-, expressed by the second of the relations (40.12), is ruled out on the basis of (40.13) and thus the phase function ψ_- has no interior stationary points. Focussing our attention on the range $x > 0$, it appears that (ζ_1, τ) represents an interior stationary point for ψ_+ if $\zeta = \zeta_1(\tau)$ designates the positive solution of (40.12), viz.

$$cf(\zeta_1) = c\sqrt{(\zeta_1^2 + 1)} = \varphi'(\tau), \tag{40.14}$$

and x, t are connected by the relation

$$x = cf'(\zeta_1(\tau))(t - \tau) \tag{40.15}$$

where the primes signify an argument derivative.

To obtain an estimate of the wave function $y_+(x, t)$ or integral (40.8), for which the interior stationary point (ζ_1, τ) bears responsibility, the phase function ψ_+ is approximated (as in the analogous dealing with single integrals) by the first terms of its Taylor expansion, namely

$$\psi_+(\zeta, t') = \psi_+(\zeta_1, \tau) + \tfrac{1}{2}\{(\zeta - \zeta_1)^2(\psi_+)_{\zeta_1\zeta_1} + 2(\zeta - \zeta_1)(t' - \tau)(\psi_+)_{\zeta_1\tau} + (t' - \tau)^2(\psi_+)_{\tau\tau}\} \tag{40.16}$$

where

$$(\psi_+)_{\zeta_1\zeta_1} = \left(\frac{\partial^2}{\partial\zeta^2}\psi_+\right)_{\zeta=\zeta_1} = -\,cf''(\zeta_1)(t - \tau)$$

$$(\psi_+)_{\zeta_1\tau} = \left(\frac{\partial^2}{\partial\zeta\partial t'}\psi_+\right)_{\zeta=\zeta_1, t'=\tau} = cf'(\zeta_1)$$

$$(\psi_+)_{\tau\tau} = \left(\frac{\partial^2}{\partial t'^2}\psi_+\right)_{t'=\tau} = -\,\varphi''(\tau)$$

and

$$\Delta = \begin{vmatrix} (\psi_+)_{\zeta_1\zeta_1} & (\psi_+)_{\zeta_1\tau} \\ (\psi_+)_{\zeta_1\tau} & (\psi_+)_{\tau\tau} \end{vmatrix} = cf''(\zeta_1(\tau))\varphi''(\tau)(t - \tau) - c^2(f'(\zeta_1(\tau)))^2. \tag{40.17}$$

After locating the coordinate origin at the point ζ_1, τ in the domain of the integration variables ζ, t', reducing the second degree polynomial in (40.16) to a diagonal form, assigning the ratio $a(t')/f(\zeta)$ its value at the stationary point and utilizing the integral formula

$$\int_{-\infty}^{\infty} \exp\left\{\frac{\mathrm{i}}{2}\alpha\xi^2\right\} \mathrm{d}\xi = \left(\frac{2\pi}{|\alpha|}\right)^{1/2} \exp\left\{\mathrm{i}\frac{\pi}{4}\operatorname{sgn}\alpha\right\},$$

the leading term of an asymptotic expansion for the integral (40.8) results,

$$y_+(x, t) \sim \frac{2\pi c}{ih} \frac{a(\tau)}{\varphi'(\tau)} \frac{\exp(i\Psi)}{\sqrt{(|\Delta|)}}, \quad h \to \infty \tag{40.18}$$

with

$$\Psi = (t - \tau)[f'(\zeta_1(\tau)) - \varphi'(\tau)] - \varphi(\tau). \tag{40.19}$$

The equations (40.14), (40.15), (40.18) and (40.19) furnish a causal description of the source activity, expressed in terms of the parametric variable τ and with reference to a family of straight lines, or rays, that is specified in (40.15) and shown in Fig. 28. This family spans the region of space-time, or portion of the x-t plane, affected by the source and its individual members may be uniquely correlated with different values of τ. If the source has a periodic time variation, both $\varphi'(t)$ and $\zeta_1(\tau)$ are constant, and the rays form a parallel family; some representatives are indicated in the accompanying figure, along with others which pertain to the range $x < 0$. The ray that corresponds to the parametric value $\tau = 0$ defines a signal front, whose time of arrival at any location $x > 0$ is given by $x/cf'(\zeta_1(0))$. A direct means of calculating the wave function along any ray, as the time varies, is available through the use of the relations (40.18), (40.19), and they also enable us to trace the level of excitation at a fixed point.

For a time-periodic source with a bounded amplitude factor $a(t)$, the wave function estimate (40.18) is likewise bounded, inasmuch as $|\Delta| = c^2(f'(\zeta_1(\tau)))^2 \neq 0$; however, if the phase function of the source is non-linear and $\varphi''(\tau) > 0$, in particular, the foregoing estimate becomes meaningless at the instant of time

$$t = \tau + \frac{c(f'(\zeta_1(\tau))^2}{f''(\zeta_1(\tau))\varphi''(\tau)}$$

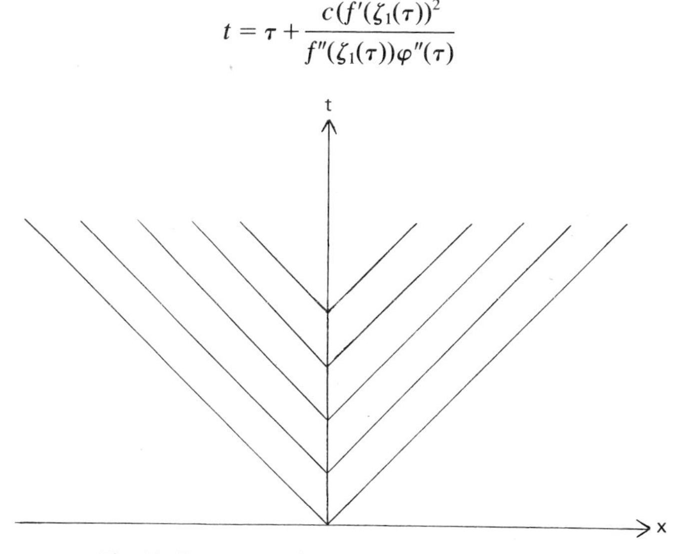

Fig. 28. Rays emanating from a localized source.

since $|\Delta|$ then has a null magnitude. The existence of an envelope (or caustic curve) in the non-parallel ray pattern accounts for this singular behavior and the resultant non-uniform accuracy of the estimate (40.18).

The succeeding (or second) term in the asymptotic expansion for $y_+(x, t)$, and the comparable first term for $y_-(x, t)$, are both associated with contributions from so-called critical boundary points; such a designation refers to points on the boundary of the integration domain where a level curve of the phase function, $\psi_\pm(\zeta, t') = \text{constant}$, has a parallel tangent. Since the relevant boundaries for the case at hand are the straight lines $t' = 0, t$, the appropriate condition is simply that

$$\partial_\zeta \psi_\pm = 0 \tag{40.20}$$

and we infer from (40.11) the absence of critical points (excepting $x = 0$) on the line $t' = t$. To determine the net contribution which the boundary line $t' = 0$ generates we consider the source functions

$$y_\pm^\varepsilon(x, t) = \mp \mathrm{i} \int_{-\infty}^{\infty} \frac{\mathrm{d}\zeta}{f(\zeta)} \int_0^\varepsilon a(t')\, \mathrm{e}^{\mathrm{i}h\psi_\pm(\zeta, t')}\, \mathrm{d}t' \tag{40.21}$$

whose support is located in the adjacent strip $0 \le t' \le \varepsilon$, $-\infty < \zeta < \infty$, or subdomain of the complete integrals (40.8).

An estimate of the double integrals (40.21), attributable to critical points on the line $t' = 0$ which marks the initial stage of signal activity, is achieved through individual handling of the single integrals present therein; specifically, this calls for (as conceived by Lewis, 1964) a stationary phase evaluation of the ζ-integral and a subsequent integration by parts in the t'-integral. The first step yields

$$\int_{-\infty}^{\infty} \frac{\mathrm{d}\zeta}{f(\zeta)}\, \mathrm{e}^{\mathrm{i}h\psi_\pm(\zeta, t')} \simeq \left[\frac{2\pi}{h(t - t')f''(\hat{\zeta}_1(t'))}\right]^{1/2} \frac{\exp\{\mathrm{i}h\psi_\pm(\hat{\zeta}_1(t'), t') \mp \mathrm{i}\pi/4\}}{f(\hat{\zeta}_1(t'))}, \quad h \to \infty \tag{40.22}$$

in accordance with (34.1, 34.14) and the specific relations

$$(\psi_\pm)_{\hat{\zeta}_1\hat{\zeta}_1} = \mp c(t - t')f''(\hat{\zeta}_1(t')),$$

where $\hat{\zeta}_1(t')$ is deduced from the stationary condition (40.20) and has the explicit forms

$$\hat{\zeta}_1(t') = \pm\left[\left(\frac{c(t - t')}{x}\right)^2 - 1\right]^{-1/2}, \quad x > 0 \tag{40.23}$$

relative to $y_\pm^\varepsilon$. On the basis of (40.22), it follows that

$$y_\pm^\varepsilon(x, t) \simeq \mathrm{e}^{\mp 3\mathrm{i}\pi/4}\left(\frac{2\pi}{h}\right)^{1/2} \int_0^\varepsilon [(t - t')f''(\hat{\zeta}_1)]^{-1/2} \frac{a(t')}{f(\zeta_1)}\, \mathrm{e}^{\mathrm{i}h\psi_\pm(\hat{\zeta}_1, t')}\, \mathrm{d}t', \quad h \to \infty \tag{40.24}$$

with

$$\psi_\pm(\hat{\zeta}_1, t') = \hat{\zeta}_1 x \mp cf(\hat{\zeta}_1)(t - t') - \varphi(t')); \tag{40.25}$$

successive integrations by parts, which utilize the relation

$$\exp\{\mathrm{i}h\psi_\pm\} = \left[\mathrm{i}h\frac{\mathrm{d}}{\mathrm{d}t'}\psi_\pm\right]^{-1} \mathrm{d}\exp\{\mathrm{i}h\psi_\pm\},$$

furnish a development for the integral (40.24), containing reciprocal powers of h, whose first term yields an estimate of the excitation due to critical points on the boundary line $t' = 0$,

$$y^0_\pm(x, t) \simeq \mathrm{i}\,\mathrm{e}^{\mp 3\mathrm{i}\pi/4} h^{-3/2} \left(\frac{2\pi}{tf''(\hat{\zeta}_1(0))}\right)^{1/2} \frac{a(0)\,\mathrm{e}^{\mathrm{i}h\psi_\pm(\hat{\zeta}_1(0),\,0)}}{f(\hat{\zeta}_1(0))\left[\dfrac{\mathrm{d}}{\mathrm{d}t'}\psi_\pm(\hat{\zeta}_1, t')|_{t'=0}\right]}, \quad h \to \infty \tag{40.26}$$

after ignoring a contribution from the endpoint $t' = \varepsilon$ [that can be annulled by one from the lower boundary of the contiguous domain of integration, $\varepsilon \le t' \le t$, $-\infty < \zeta < \infty$].

The phase functions $\psi_\pm$ and their t'-derivative which enter into (40.26) acquire a direct dependence on the parametric variable

$$\mu = \hat{\zeta}_1(0), \tag{40.27}$$

viz.:

$$\begin{aligned} \psi_\pm(\hat{\zeta}_1(0), 0) &= \pm c[\mu f'(\mu) - f(\mu)] - \varphi(0), \\ \frac{\mathrm{d}}{\mathrm{d}t'}\psi_\pm(\hat{\zeta}_1(t'), t')|_{t'=0} &= \pm\, cf(\mu) - \varphi'(0) \end{aligned} \tag{40.28}$$

if the stationary conditions (40.11),

$$x = \pm\, cf'(\mu)t, \tag{40.29}$$

are taken into account, and the resultant version of (40.26) becomes

$$y^0_\pm(x, t) \simeq \mathrm{i}\,\mathrm{e}^{\mp 3\mathrm{i}\pi/4} h^{-3/2} \left(\frac{2\pi}{tf''(\mu)}\right)^{1/2} \frac{a(0)\exp\{\pm\,\mathrm{i}hc[\mu f'(\mu) - f(\mu)] - \mathrm{i}h\varphi(0)\}}{f(\mu)[\pm\, cf(\mu) - \varphi'(0)]}, \quad h \to \infty \tag{40.30}$$

A formal likeness is thus shared by the estimates of contributions (y_+ and $y^0_\pm$) arising, respectively, from interior and boundary points of the domain over which the source integral extends; both admit a description in terms of single parameter ray families and provide explicit determinations of the disturbance level along individual rays. The two types of contribution reveal significant differences in detail, however, and notably as regards their respective amplitude factors h^{-1} and $h^{-3/2}$, along with the manner of time variation;

an inevitable disappearance of the boundary point excitation is brought about through the presence in (40.30) of an exclusive time factor $t^{-1/2}$, whereas the expression (40.18) for y_+ exhibits an oscillatory behavior in the case of a time-periodic source and may therefore be said to constitute the main signal.

The ray family $x = cf'(\mu)t > 0$, which has a fanwise pattern relative to the x-t coordinate origin, with its bordering members along the lines $t = 0$ ($\mu = 0$) and $x = ct$ ($\mu = \infty$), includes representatives that traverse any point $0 < x (< ct)$ in advance of those belonging to the family (40.15) [see Fig. 29]; on the sector

$$cf'(\zeta_1(0))t = cf'\left[\left(\left(\frac{\varphi'(0)}{c}\right)^2 - 1\right)^{1/2}\right]t < x < ct$$

wherein rays of the latter family are absent and y_+ vanishes, the excitation represented by y_+^0 is appropriately termed a precursor or forerunner of the main signal. Along the front of the main signal, $x = cf'(\zeta_1(0))t$, which corresponds to the ray parameter $\tau = 0$ in (40.15), there is a coincident ray of the precursor family .with the parametric assignment

$$\mu = \zeta_1(0) = \left(\left(\frac{\varphi'(0)}{c}\right)^2 - 1\right)^{1/2}$$

which implies that

$$cf(\mu) = c\surd((\zeta_1(0))^2 + 1) = \varphi'(0);$$

hence the precursor estimate (40.30), given by y_+^0 in $x > 0$, is deficient at the main signal front and the combination of terms hitherto specified, $y_+ + y_+^0 + y_-^0$, form part of a non-uniform asymptotic expansion of the source function.

To circumvent the latter defect in the characterization of $y(x, t)$, a more refined handling of its integral representation, or an alternative manner of

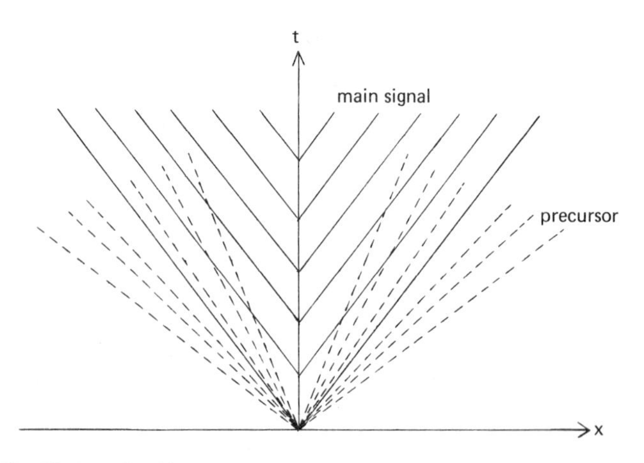

Fig. 29. Ray families descriptive of a main signal and its precursor.

solving the original partial differential equation becomes necessary; both of these objectives are realized by Lewis and others, with the eventual provision of uniformly accurate estimates for asymptotic solutions to a broad class of (constant and variable coefficient) hyperbolic partial differential equations. A judicious initial choice of expansion for the solution (or wave function), whose constituent functions are determined by ordinary differential equations along the rays, lends itself to establishing uniformly accurate estimates.

Problems II(a)

1. Given an initial profile

$$y(x,0) = A\exp\left(-\frac{x^2}{2L^2} + \mathrm{i}k_0 x\right)$$

composed from a product of periodic and modulation factors that depend on two parameters, k_0 and L, the latter characterizing its overall scale. Determine the corresponding spectrum or Fourier transform

$$f(k) = \frac{1}{2\pi}\int_{-\infty}^{\infty} y(x,0)\,\mathrm{e}^{-\mathrm{i}kx}\,\mathrm{d}x$$

and show that if the dispersion relation takes the form

$$\begin{aligned}\omega &= \omega(k)\\ &= \omega(k_0) + \left(\frac{\mathrm{d}\omega}{\mathrm{d}k}\right)_0 (k-k_0) + \frac{1}{2}\left(\frac{\mathrm{d}^2\omega}{\mathrm{d}k^2}\right)_0 (k-k_0)^2,\end{aligned}$$

comprising the first three terms in a Taylor series expansion relative to k_0, the subsequent wave function

$$y(x,t) = \int_{-\infty}^{\infty} f(k)\,\mathrm{e}^{\mathrm{i}(kx-\omega(k)t)}\,\mathrm{d}k, \quad t>0$$

has a squared modulus given by

$$|y(x,t)|^2 = \frac{|A|^2}{\left\{1+\left(\dfrac{t}{L^2}\left(\dfrac{\mathrm{d}^2\omega}{\mathrm{d}k^2}\right)_0\right)^2\right\}^{1/2}} \exp\left\{-\frac{\left(x-\left(\dfrac{\mathrm{d}\omega}{\mathrm{d}k}\right)_0 t\right)^2}{L^2\left[1+\left(\dfrac{t}{L^2}\left(\dfrac{\mathrm{d}^2\omega}{\mathrm{d}k^2}\right)_0\right)^2\right]}\right\}$$

Describe the features of the profile when $t>0$ and, in particular, the roles played by the first and second derivatives $(\mathrm{d}\omega/\mathrm{d}k)_0$, $(\mathrm{d}^2\omega/\mathrm{d}k^2)_0$.

2. The differential equation

$$\rho\frac{\partial^2 y}{\partial t^2} = P\frac{\partial^2 y}{\partial x^2} - a^2\frac{\partial^4 y}{\partial x^4}$$

may be adopted for describing the motion of a taut string whose rigidity is taken into account through a fourth order derivative term. Calculate the group velocity of progressive wave motions along the string from the dispersion relation and confirm that an identical magnitude is yielded by the ratio of average values for the energy flux,

$$S(x,t) = -P\frac{\partial y}{\partial x}\frac{\partial y}{\partial t} + a^2\frac{\partial y}{\partial t}\frac{\partial^3 y}{\partial x^3} - a^2\frac{\partial^2 y}{\partial x \partial t}\frac{\partial^2 y}{\partial x^2},$$

and the energy density

$$W(x,t) = \frac{1}{2}\rho\left(\frac{\partial y}{\partial t}\right)^2 + \frac{1}{2}P\left(\frac{\partial y}{\partial x}\right)^2 + \frac{1}{2}a^2\left(\frac{\partial^2 y}{\partial x^2}\right)^2;$$

what is the relationship between phase and group velocities?

3. Consider an arrangement of two parallel strings with the respective densities ρ_1, ρ_2 and tensions $P, 0$; if the strings are elastically coupled, as implied by a local force on either which is proportional to the difference of their transverse displacements $y_1(x,t)$, $y_2(x,t)$, the simultaneous equations of motion take the form

$$P\frac{\partial^2 y_1}{\partial x^2} - \mu(y_1 - y_2) - \rho_1\frac{\partial^2 y_1}{\partial t^2} = 0$$

and

$$-\mu(y_1 - y_2) + \rho_2\frac{\partial^2 y_2}{\partial t^2} = 0.$$

Obtain the dispersion relation appropriate to harmonic wave profiles travelling along this system and verify the group velocity representation

$$u = \left(\frac{P}{\rho_1}\right)^{1/2}\left[1 + \frac{\rho_2}{\rho_1}\frac{1}{1 - \dfrac{\rho_2\omega^2}{\mu}}\right]^{-1/2}\left[1 + \frac{\rho_2}{\rho_1}\left(\frac{1}{1 - \dfrac{\rho_2\omega^2}{\mu}}\right)^2\right]^{-1},$$

with a real nature save for the frequency range

$$1 + \frac{\rho_2}{\rho_1} > \frac{\rho_2\omega^2}{\mu} > 1.$$

Form the expressions that define the kinetic and potential energies per unit length and the energy flux in terms of the string displacements; and, consequent to determining their average values for a progressive wave motion, characterize an energy propagation velocity in accordance with the group velocity.

4. The voltage in a section normal to the axis of a helical coil, $V(x, t)$, satisfies the equation

$$K\frac{\partial^2 V}{\partial t^2} - C\frac{\partial^4 V}{\partial x^2 \partial t^2} = \frac{1}{L}\frac{\partial^2 V}{\partial x^2}$$

at low frequencies. Find the representations of phase and group velocities in terms of the electrical parameters K, C, and L, utilizing both the dispersion relation and an energy conservation relation for characterizing the latter.

5. Assume that the differential equation

$$\frac{\partial^2 y}{\partial x^2} - \alpha\frac{\partial y}{\partial x} - \frac{1}{c^2}\frac{\partial^2 y}{\partial t^2} = 0$$

describes the local displacement, $y(x, t)$, for vertical motions in a heavy compressible medium whose equilibrium distributions of density and pressure are given by the 'barometric' forms

$$\rho = \rho_0 \, e^{-\alpha x}, \qquad p = p_0 \, e^{-\alpha x}, \quad x > 0$$

with

$$\alpha = g\rho_0/p_0, \qquad c^2 = \gamma p_0/\rho_0,$$

the constants g, γ designating the gravitational acceleration and ratio of specific heats, respectively. Discuss the nature of time-periodic solutions and confirm the expression

$$v = c\left\{1 - \left(\frac{\alpha c}{2\omega}\right)^2\right\}^{-1/2}$$

for the phase velocity when $\omega > \alpha c/2$. Establish the equation of energy conservation

$$\frac{\partial}{\partial t}\left[\frac{1}{2}\rho_0 \, e^{-\alpha x}\left(\frac{\partial y}{\partial t}\right)^2 + \frac{1}{2}\rho_0 c^2 \, e^{-\alpha x}\left(\frac{\partial y}{\partial x}\right)^2\right] + \frac{\partial}{\partial x}\left[-\rho_0 c^2 \, e^{-\alpha x}\frac{\partial y}{\partial x}\frac{\partial y}{\partial t}\right] = 0$$

and show that the velocity obtained from the ratio of time average values for energy flux and density has the magnitude

$$u = c\left\{1 - \left(\frac{\alpha c}{2\omega}\right)^2\right\}^{1/2}$$

which agrees with that of the group velocity.

6. Examine, with regard to the formulas (26.8), (26.9) that specify the reflection and transmission coefficients in a problem of wave scattering on a composite string which is formed by a finite central section (length L and uniform density ρ_2) and surrounding portions of a different though common density, the limit

$$L \to 0$$

wherein

$$\rho_2 L \to M;$$

and verify that the outcome accords with the particulars of scattering attributable to a point load.

7. Suppose that a load of mass M is affixed to an otherwise uniform string of infinite extent and that a time-periodic force, with the magnitude Re $Fe^{-i\omega t}$, is impressed on the point of attachment; determine the resulting string displacement anywhere and verify, through calculation of both the energy flux and the rate of working by the force, that

$$\frac{F^2}{4\rho c} \frac{1}{1+\left(\frac{Mk}{2\rho}\right)^2}$$

expresses the time-average amount of energy which travels outwards along the string.

8. Consider the response of the preceding arrangement, when the string has initially a stationary linear profile and an impulse of magnitude I is communicated to the load. Establish the representation

$$\frac{I}{2\rho c}\left[1-\exp\left(-\frac{2\rho c}{M}t\right)\right]$$

for the ensuing load displacement and compare the amounts of energy possessed by the string and load, respectively, just after the impulse and at a much later time.

9. Let it be assumed that the longitudinal displacements $s(x, t)$ in a bar satisfy the equation

$$\frac{\partial^2 s}{\partial t^2} + \alpha \frac{\partial s}{\partial t} = c^2 \frac{\partial^2 s}{\partial x^2}, \quad \alpha > 0$$

when the surrounding medium offers a resistance which is a multiple $\rho\alpha A$ of the local velocity, where ρ and A specify the density and cross-sectional area of the bar. Derive expressions for the phase velocity and attenuation factor of a wave motion with the frequency ω. If a displacement

$$a \cos \omega_1 t + a \cos \omega_2 t$$

is prescribed at one end of the bar and the component frequencies ω_1, ω_2 satisfy the inequalities

$$\omega_1 - \omega_2 \ll \omega_1 \text{ and } \omega_2, \qquad \omega_1 \text{ and } \omega_2 \gg \alpha$$

deduce that points of locally maximum displacement in the resulting excitation move with the speed

$$c\left\{1 + \frac{1}{8}\left(\frac{\alpha}{\omega}\right)^2 + 0\left(\frac{\alpha}{\omega}\right)^4\right\}, \qquad \omega = \frac{1}{2}(\omega_1 + \omega_2)$$

along the bar.

10. A localized perturbation $\zeta(x, t)$ to the elevation of the surface of water of infinite depth is given by

$$\zeta(x, t) = \operatorname{Re} \int_{-\infty}^{\infty} a(k)\, e^{i(kx - \omega(k)t)}\, dk$$

where $\omega(k) = \surd(g|k|)$. Employ the method of stationary phase to obtain the leading term of an asymptotic series for $\zeta(x, t)$ as $x, t \to \infty$ while x/t is held constant. Determine the time variation of a given wave number amplitude and show how this is consistent with conservation of wave energy.

11. The elevation of the free surface due to a train of capillary waves, in liquid of uniform density ρ and uniform surface tension ρS^2, is

$$\zeta(x, t) = \int_0^{\infty} a(k) \cos\{kx - \omega(k)t\}\, dk,$$

where $\omega(k) = Sk^{3/2}$, $S > 0$. Use the method of stationary phase to obtain an estimate of $\zeta(Ct, t)$ for $t \to \infty$ with C fixed and positive; show that

$$E = \frac{1}{2}\rho S^2 \int_{C_1 t}^{C_2 t} \left(\frac{\partial \zeta}{\partial x}\right)^2 dx \sim \frac{1}{2}\pi\rho S^2 \int_{k_0(C_1)}^{k_0(C_2)} a^2(k) k^2\, dk, \quad t \to \infty$$

if $k_0(C) = (2C/3S)^{1/2}$ and C_1, C_2 are fixed positive magnitudes. Give a physical interpretation of the results, assuming that the total energy between $x = C_1 t$ and $x = C_2 t$ is approximately $2E$.

§41. *Nonlinear Equations for String Motions; Linearization and Other Aspects*

There is an evident utility in juxtaposing and contrasting, whenever possible, both the general features and particular consequences of linear and nonlinear formulations for individual problems; such a comparative process exposes the limitations of a linear basis and helps us to resolve collateral ambiguities (concerning the measure of energy and momentum densities of string vibrations, for example, as discussed in §42). To describe string motions, without an a priori assumption of small amplitude, we may advantageously employ s (length along the string) and t (time) as independent variables. For a string of fixed length which executes planar motions, three fundamental equations are pertinent, namely

$$\left(\frac{\partial x}{\partial s}\right)^2 + \left(\frac{\partial y}{\partial s}\right)^2 = 1, \tag{41.1}$$

$$\rho\frac{\partial^2 x}{\partial t^2} = \frac{\partial}{\partial s}\left(P\frac{\partial x}{\partial s}\right), \tag{41.2}$$

$$\rho\frac{\partial^2 y}{\partial t^2} = \frac{\partial}{\partial s}\left(P\frac{\partial y}{\partial s}\right) \tag{41.3}$$

where the position coordinates $x(s, t)$, $y(s, t)$ and the tension $P(s, t)$ comprise a trio of dependent variables; the first equation, (41.1), registers the fact of inextensibility while the latter pair, involving the (constant) mass density ρ, express component forms of Newton's dynamical law. Points on the string, specified by (x, y), can be situated anywhere in the plane, so far as the nonlinear system (41.1)–(41.3) is concerned; to uncover the relationship between this system and the homogeneous one-dimensional wave equation (1.1) of prior status, however, we shall assume that the string always remains near the x-axis. In such circumstances the transverse coordinate $y(s, t)$ has a small order of magnitude (ε, say) and it is appropriate to write

$$y(s, t) = \varepsilon y_0(s, t). \tag{41.4}$$

Since $\partial y/\partial s$ is likewise of the order ε, the relation (41.1) implies that

$$\frac{\partial x}{\partial s} = 1 + 0(\varepsilon^2)$$

and thence, after integrating, we find

$$x(s, t) = s + \varepsilon^2 x_1(s, t) \tag{41.5}$$

to the lowest non-trivial order. If the representations (41.4), (41.5) and a counterpart for the tension,

$$P(s, t) = P_0(s, t) + \varepsilon^2 P_1(s, t), \tag{41.6}$$

are substituted in (41.1)–(41.3), these become

$$\left(1+\varepsilon^2\frac{\partial x_1}{\partial s}\right)^2+\varepsilon^2\left(\frac{\partial y_0}{\partial s}\right)^2=1 \tag{41.7}$$

$$\varepsilon^2\rho\frac{\partial^2 x_1}{\partial t^2}=\frac{\partial}{\partial s}\left[(P_0+\varepsilon^2P_1)\left(1+\varepsilon^2\frac{\partial x_1}{\partial s}\right)\right], \tag{41.8}$$

$$\varepsilon\rho\frac{\partial^2 y_0}{\partial t^2}=\frac{\partial}{\partial s}\left[(P_0+\varepsilon^2P_1)\varepsilon\frac{\partial y_0}{\partial s}\right]. \tag{41.9}$$

Neglecting all terms with ε factors, the deduction from (41.8),

$$\frac{\partial P_0}{\partial s}=0$$

or

$$P_0(s,t)=P_0(t) \tag{41.10}$$

implies that the lowest order component of the tension has a uniform value, at any instant of time, along the entire length of the string. The magnitude of said component is specified through (boundary) conditions at the extremities of the string, where pulling directly controls the tautness. When we adopt a more restrictive version of (41.10), viz.

$$P_0(s,t)=P_0\ \text{(a constant)} \tag{41.11}$$

the modifications to (41.7)–(41.9) result in a new set of equations

$$\begin{aligned}
2\frac{\partial x_1}{\partial s}+\left(\frac{\partial y_0}{\partial s}\right)^2+\varepsilon^2\left(\frac{\partial x_1}{\partial s}\right)^2&=0\\
\rho\frac{\partial^2 x_1}{\partial t^2}=\frac{\partial}{\partial s}\left[P_0\frac{\partial x_1}{\partial s}+P_1+\varepsilon^2P_1\frac{\partial x_1}{\partial s}\right]&,\\
\rho\frac{\partial^2 y_0}{\partial t^2}=\frac{\partial}{\partial s}\left[(P_0+\varepsilon^2P_1)\frac{\partial y_0}{\partial s}\right]&.
\end{aligned} \tag{41.12}$$

Following up the assumption that ε is small compared with unity and suppressing its appearance in the above equations, there obtains a system

$$2\frac{\partial x_1}{\partial s}+\left(\frac{\partial y_0}{\partial s}\right)^2=0 \tag{41.13}$$

$$\rho\frac{\partial^2 x_1}{\partial t^2}=P_0\frac{\partial^2 x_1}{\partial s^2}+\frac{\partial P_1}{\partial s}, \tag{41.14}$$

$$\rho\frac{\partial^2 y_0}{\partial t^2}=P_0\frac{\partial^2 y_0}{\partial s^2} \tag{41.15}$$

which allows, in principle, the sequential determination of the quantities y_0, x_1, and P_1; thus, if y_0 is ascertained first by means of the linear partial differential equation (41.15) [that possesses the form (1.1)], then x_1, P_1 may be found, in turn, after single integrations of (41.13), (41.14) with respect to s.

Accordingly, linearization of the system (41.7)–(41.9), which describes planar motions of inextensible strings, turns up the homogeneous one-dimensional wave equation if a constant zeroth order tension is predicated. To demonstrate their closer relationship, let us next confirm that the non-linear equations also admit travelling wave solutions characteristic of a linear formulation. If we assume the requisite functional dependences,

$$\begin{aligned} x(s,t) &= s + x(s-ct) \\ y(s,t) &= y(s-ct), \\ P(s,t) &= P(s-ct), \end{aligned} \tag{41.16}$$

where c designates some constant velocity, and substitute in (41.1)–(41.3) it follows that

$$2x' + x'^2 + y'^2 = 0 \tag{41.17}$$

$$\rho c^2 x'' = [P(1+x')]', \tag{41.18}$$

$$\rho c^2 y'' = (Py')' \tag{41.19}$$

with primes signifying argument differentiation. The second and third equations are automatically satisfied when the tension has a fixed value (independent of s and t) such that

$$P = \rho c^2 \tag{41.20}$$

[compare (1.2)], and thus the existence of travelling wave solutions is assured if x and y jointly conform with the first equation, (41.17). A constraining role for (41.17), which aims at guaranteeing the patent reality of x, makes evident that

$$y(s,t) = f(s-ct) \tag{41.21}$$

constitutes an exact solution of the string equations (41.1)–(41.3), so long as $|f'| \leq 1$ and (41.20) holds; furthermore, since these equations are invariant with respect to the time-reversal transformation, $t \to -t$, we may assert that (41.21) and the companion travelling wave function,

$$y(s,t) = g(s+ct), \quad |g'| \leq 1 \tag{41.22}$$

are, individually, solutions of both the nonlinear system (41.1)–(41.3) and its linear affiliate, (41.15).

For a more thoroughgoing comparison of the various differential equations, or formulations, and appreciation of what linearization implies, it is appropriate to enumerate suitable initial and boundary conditions in the respective cases. Inasmuch as the Newtonian pair of equations (41.2), (41.3) contain second order time derivatives, there follows an inference that the string equations, collectively, have a fourth order character and require a like number of initial conditions. A lesser order is, however, suggested by the fact that the position variables x, y satisfy a time-independent constraint (41.1); and this receives support from the prior observation that, after linearization, we need only solve one second order partial differential equation and perform separate integrations with respect to s.

To settle the matter concerning the requisite number of initial conditions, let us display the system (41.1)–(41.3) in a form which evidences just two first order time derivatives. A representation of the latter nature obtains directly after noting that compliance with the equation of constraint (41.1) is assured if

$$\frac{\partial x}{\partial s} = \cos \alpha, \qquad \frac{\partial y}{\partial s} = \sin \alpha \tag{41.23}$$

and that the remaining equations, expressed in terms of the dependent variables $\alpha(s, t)$, $P(s, t)$, are

$$\rho \frac{\partial^2}{\partial t^2} \cos \alpha = \frac{\partial^2}{\partial s^2}(P \cos \alpha),$$

$$\rho \frac{\partial^2}{\partial t^2} \sin \alpha = \frac{\partial^2}{\partial s^2}(P \sin \alpha)$$

or, consequent to explicit differentiation,

$$\rho \frac{\partial^2 \alpha}{\partial t^2} = 2 \frac{\partial P}{\partial s} \frac{\partial \alpha}{\partial s} + P \frac{\partial^2 \alpha}{\partial s^2},$$

$$\rho \left(\frac{\partial \alpha}{\partial t} \right)^2 = -\frac{\partial^2 P}{\partial s^2} + P \left(\frac{\partial \alpha}{\partial s} \right)^2.$$

With a suitable choice of square roots we arrive at the desired relations,

$$\begin{gathered} \sqrt{\rho} \frac{\partial \alpha}{\partial t} = \left[-\frac{\partial^2 P}{\partial s^2} + P \left(\frac{\partial \alpha}{\partial s} \right)^2 \right]^{1/2}, \\ \sqrt{\rho} \frac{\partial}{\partial t} \left[-\frac{\partial^2 P}{\partial s^2} + P \left(\frac{\partial \alpha}{\partial s} \right)^2 \right]^{1/2} = 2 \frac{\partial P}{\partial s} \frac{\partial \alpha}{\partial s} + P \frac{\partial^2 \alpha}{\partial s^2}, \end{gathered} \tag{41.24}$$

which fix the number of initial conditions, namely two, as in the linear approximation.

That the nonlinear description of string motions entails a number of boundary conditions exceeding the pair appropriate for the linear version is brought out by the fact that, to effect a step-by-step s integration of the original equations, we must stipulate four boundary values, viz.

$$x(0, t),\ y(0, t),\ \frac{\partial x(0, t)}{\partial s} \quad \text{and} \quad P(0, t)$$

[while the sign of $\partial y(0, t)/\partial s$ need not be counted as an additional datum]. Specific choices for two boundary conditions thus enter into a reduction scheme that produces the homogeneous linear wave equation, one being implicit in the representation

$$x(s, t) = s + x_0(t) + \varepsilon^2 x_1(s, t)$$

which generalizes a prior integral, (41.5), of (41.1); and it appears, in consequence, that linearization of the string equations (41.1)–(41.3) furnishes the cited wave equation only if we adopt the two boundary conditions

$$x_0(t) = 0 \quad \text{and} \quad P_0(t) = P_0. \tag{41.25}$$

The arrangement depicted schematically in Fig. 30 and designed to realize a time-independent and constant tension, along with null transverse displacement at the extremities of a given length of string, envisages a massless spring attached to one end, whose lateral movement is impeded by a closely fitting ring.

A relaxation of the first condition (41.25) underlies other linearization procedures; when $x_0(t) \neq 0$, the equations (41.7) and (41.9) continue in force, though (41.8) is replaced with

$$\rho\left(\frac{d^2 x_0}{dt^2} + \varepsilon^2 \frac{\partial^2 x_1}{\partial t^2}\right) = \frac{\partial}{\partial s}\left[(P_0 + \varepsilon^2 P_1)\left(1 + \varepsilon^2 \frac{\partial x_1}{\partial s}\right)\right] \tag{41.26}$$

and hence we obtain, to the lowest order,

$$P_0(s, t) = \rho s \frac{d^2 x_0}{dt^2} + P(t) \tag{41.27}$$

instead of (41.10). The trio of linearized equations comprise (41.13) and a pair which supplant (41.14), (41.15), namely

$$\rho \frac{\partial^2 x_1}{\partial t^2} = \frac{\partial}{\partial s}\left[\left\{\rho s \frac{d^2 x_0}{dt^2} + P(t)\right\}\frac{\partial x_1}{\partial s}\right] + \frac{\partial P_1}{\partial s}, \tag{41.28}$$

$$\rho \frac{\partial^2 y_0}{\partial t^2} = \frac{\partial}{\partial s}\left[\left\{\rho s \frac{d^2 x_0}{dt^2} + P(t)\right\}\frac{\partial y_0}{\partial s}\right]. \tag{41.29}$$

Since a string cannot support negative tension, the quantity $P_0(s, t)$ in (41.27) is necessarily positive for all pertinent values of s and t; thus (41.29)

Fig. 30. A string with one end fixed and the other attached to a massless spring of constant tension.

represents a linear hyperbolic partial differential equation with a variable coefficient, whose characteristics are specified by the nonlinear ordinary differential equation,

$$\frac{\mathrm{d}s}{\mathrm{d}t} = \pm \left[s \frac{\mathrm{d}^2 x_0}{\mathrm{d}t^2} + \frac{1}{\rho} P(t) \right]^{1/2}. \tag{41.30}$$

Before concluding this section, we observe that the exact status of travelling wave solutions to the nonlinear string equations suggests an interpretation of string profiles whose admissibility in a linear theory is questionable. Let us refer, in the latter context, to the function

$$y(x, t) = \left\{ \begin{matrix} 0, & x - ct < 0 \\ A, & 0 < x - ct < 1 \\ 0, & 1 < x - ct \end{matrix} \right\} = A[H(x - ct) - H(x - ct - 1)] \tag{41.31}$$

which, on account of a discontinuous nature, has no obvious application to string movements, though it satisfies the one-dimensional wave equation (1.1). In consequence of the fact that (41.31) possesses the form (1.3) and qualifies as an exact solution of the string equations, we are motivated to advance a suitable interpretation therefore, wherein the string has a continuous profile of the shape indicated in Fig. 31 and the concomitant mathematical representations

$x(s, t)$	$y(s, t)$	
s	0	$s - ct \leq 0$
ct	$s - ct$	$0 \leq s - ct \leq A$
$s - A$	A	$A \leq s - ct \leq 1 + A$
$1 + ct$	$1 + 2A - s + ct$	$1 + A \leq s - ct \leq 1 + 2A$
$s - 2A$	0	$1 + 2A \leq s - ct$

along with $P = \rho c^2$. It is easy to confirm that the above expressions for x, y satisfy the equation of constraint and, being of the progressive wave form, constitute an exact solution of the string equations (41.1)–(41.3).

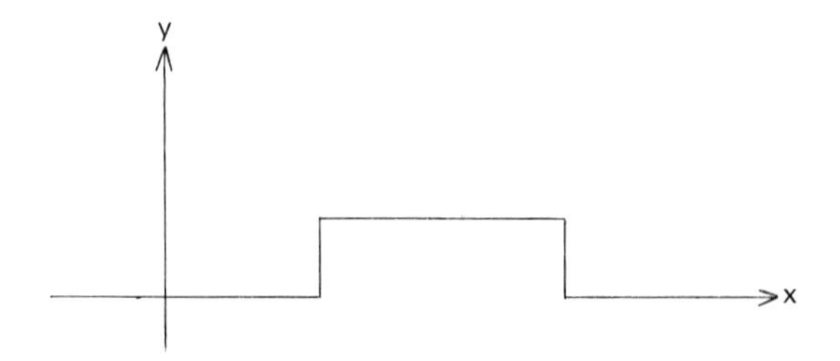

Fig. 31. A particular string profile.

§42. *Conservation Relations and the Pressure Exerted by Waves*

That the differential identities

$$\frac{\partial W}{\partial t}+\frac{\partial S}{\partial x}=0, \tag{42.1}$$

and

$$\frac{\partial G}{\partial t}+\frac{\partial W}{\partial x}=0, \tag{42.2}$$

wherein

$$W(x,t)=\frac{1}{2}\rho\left(\frac{\partial y}{\partial t}\right)^2+\frac{1}{2}\rho c^2\left(\frac{\partial y}{\partial x}\right)^2, \tag{42.3}$$

$$S(x,t)=-\rho c^2\frac{\partial y}{\partial x}\frac{\partial y}{\partial t}=c^2G(x,t), \tag{42.4}$$

are proper mathematical consequences of the homogeneous wave equation

$$\frac{\partial^2 y}{\partial x^2}=\frac{1}{c^2}\frac{\partial^2 y}{\partial t^2} \tag{42.5}$$

is incontestible; however, their physical interpretation, as local balance or conservation relations connected with transverse string displacement, $y(x,t)$, is rendered less certain since (42.5) represents merely a linearized approximation to the dynamical equations of motion, whereas the quantities W, S, and G depend quadratically on y.

The first identity prompts an interpretation in terms of local energy conservation (§31), with W and S as the respective measures of energy density and flux; and the implied constancy in time of the total energy, $E=\int W\,dx$, for a string having fixed endpoints (at which $\partial y/\partial t$ and S vanish) accords with the character of an isolated and dissipationless system. Our prior identification of a potential energy density, viz.

$$V=\frac{1}{2}P\left(\frac{\partial y}{\partial x}\right)^2, \qquad P=\rho c^2, \tag{42.6}$$

comprising the second term in the expression for W and suggested by the energetics of displacing and stretching a taut string, may be called into question when the latter is inextensible; and there are grounds, furthermore, to query the existence of a non-uniform potential energy distribution along the string, manifest by (42.6), with maxima/minima at the positions of greatest/least slope, respectively. For if V represents an elastic energy rooted in the tension $P=\rho c^2$, and both quantities possess a variable nature, we need to reckon on the excitation of longitudinal waves in the string; and, since the latter travel with a speed (controlled by the density and a modulus of elasticity) which is substantially greater than the transverse wave speed c, a uniformization of tension and potential energy is indicated, contrary to the original premise. These considerations disclose, therefore, the inadequacy of a description involving wholly transverse string displacements and oblige us, as emphasized and demonstrated by Broer (1970), to seek a satisfactory interpretation of energy and momentum densities on the basis of a more accurate formulation of the equations governing string movements.

If we contemplate a string of fixed length, which is disposed between flexible supports (or endpoints), the particulars of such a formulation (when the motions are two-dimensional) appear in §41. On multiplying (41.2), (41.3) with $\partial x/\partial t$ and $\partial y/\partial t$, respectively, adding and rearranging terms, we obtain (after utilizing the kinematical condition (41.1))

$$\begin{aligned}\frac{\partial}{\partial t}\left[\frac{1}{2}\rho\left(\frac{\partial x}{\partial t}\right)^2+\frac{1}{2}\rho\left(\frac{\partial y}{\partial t}\right)^2\right]&=\frac{\partial x}{\partial t}\frac{\partial}{\partial s}\left(P\frac{\partial x}{\partial s}\right)+\frac{\partial y}{\partial t}\frac{\partial}{\partial s}\left(P\frac{\partial y}{\partial s}\right)\\&=\frac{\partial}{\partial s}\left[P\left(\frac{\partial x}{\partial t}\frac{\partial x}{\partial s}+\frac{\partial y}{\partial t}\frac{\partial y}{\partial s}\right)\right]-\frac{P}{2}\frac{\partial}{\partial t}\left[\left(\frac{\partial x}{\partial s}\right)^2+\left(\frac{\partial y}{\partial s}\right)^2\right]\\&=\frac{\partial}{\partial s}\left[P\left(\frac{\partial x}{\partial t}\frac{\partial x}{\partial s}+\frac{\partial y}{\partial t}\frac{\partial y}{\partial s}\right)\right]\end{aligned}\tag{42.7}$$

which constitutes an exact energy equation. The version of (42.7), based on the developments (41.4)–(41.6) for (y, x, P) and the retention of the lowest order of magnitude (ε^2) estimates for the separate members, is

$$\frac{\partial}{\partial t}\left[\frac{1}{2}\rho\left(\frac{\partial y_0}{\partial t}\right)^2\right]=\frac{\partial}{\partial s}\left[P_0\frac{\partial x_1}{\partial t}+P_0\frac{\partial y_0}{\partial t}\frac{\partial y_0}{\partial s}\right]\tag{42.8}$$

and, inasmuch as [cf. (41.13)]

$$\frac{\partial^2 x_1}{\partial s\partial t}=-\frac{\partial y_0}{\partial s}\frac{\partial^2 y_0}{\partial s^2}=-\frac{1}{2}\frac{\partial}{\partial s}\left(\frac{\partial y_0}{\partial s}\right)^2,$$

the elimination of x_1 from (42.8) yields

$$\frac{\partial}{\partial t}\left[\frac{1}{2}\rho\left(\frac{\partial y_0}{\partial t}\right)^2+\frac{1}{2}P_0\left(\frac{\partial y_0}{\partial s}\right)^2\right]+\frac{\partial}{\partial s}\left[-P_0\frac{\partial y_0}{\partial s}\frac{\partial y_0}{\partial t}\right]=0$$

or the familiar energy relation (42.1). It thus becomes apparent that the customarily designated potential energy density originates in a component of the energy flux or rate of work associated with displacements along the longitudinal (or x-) direction which is expressible as a time derivative.

Another identity of the conservation type and general status, viz.

$$\frac{\partial}{\partial t}\left[\rho\left(\frac{\partial x}{\partial s}\frac{\partial x}{\partial t}+\frac{\partial y}{\partial s}\frac{\partial y}{\partial t}\right)\right]=\frac{\partial}{\partial s}\left[P+\frac{1}{2}\rho\left(\frac{\partial x}{\partial t}\right)^2+\frac{1}{2}\rho\left(\frac{\partial y}{\partial t}\right)^2\right],\tag{42.9}$$

may be readily established on multiplying (41.2), (41.3) with $\partial x/\partial s, \partial y/\partial s$, respectively, adding and rewriting all terms in the form of either an s- or t-derivative; when orders of magnitude smaller than ε^2 are omitted from estimates of the individual terms, the approximate version

$$\frac{\partial}{\partial t}\left[\rho\left(\frac{\partial x_1}{\partial t}+\frac{\partial y_0}{\partial s}\frac{\partial y_0}{\partial t}\right)\right]=\frac{\partial P_1}{\partial s}+\frac{\partial}{\partial s}\left[\frac{1}{2}\rho\left(\frac{\partial y_0}{\partial t}\right)^2\right] \tag{42.10}$$

emerges and there follows, moreover, after substituting herein the representation (41.14) for $\partial P_1/\partial s$, an equation

$$\frac{\partial}{\partial t}\left[-\rho\frac{\partial y_0}{\partial s}\frac{\partial y_0}{\partial t}\right]+\frac{\partial}{\partial s}\left[\frac{1}{2}\rho\left(\frac{\partial y_0}{\partial t}\right)^2+\frac{1}{2}P_0\left(\frac{\partial y_0}{\partial s}\right)^2\right]=0$$

which has the same makeup as (42.2). That the quantity $G=(1/c^2)S$ in (42.2) relates to a momentum density is suggested by an interpretation of the latter as a mass flux, together with the dimensional equivalence between mass and energy/(velocity)^2. According to (42.10), the longitudinal momentum density (with a lesser order of magnitude than its transverse counterpart, $\rho(\partial y_0/\partial t)$) involves separate contributions from components of the string displacement along and normal to the linear profile, and thus an earlier measure, $G=-\rho(\partial y/\partial x)(\partial y/\partial t)$, is lacking in respect of the former. Finally, it may be appreciated that the relations (42.8), (42.10) provide a self-consistent accounting for changes in the total energy and momentum in terms of the variable tension and velocity at the extremities of the string.

An instance of unsteadiness in the total energy of string motions, from which dynamical aspects of general import follow, is easy to detail for a segment that has a stationary node at one end ($x=0$) and a moving node at the other end ($x=l+vt$); for purposes of realizing the latter constraint, we may imagine that the string passes through a small hole in a mobile screen S, on the far side of which (see Fig. 32) the linear or equilibrium profile obtains. When the energy density $W(x,t)$ is specified exclusively in terms of transverse displacements that conform with the homogeneous linear wave equation, the overall energy

$$E(t)=\int_0^{l+vt}W(x,t)\,\mathrm{d}x=\frac{1}{2}\int_0^{l+vt}\left\{\rho\left(\frac{\partial y}{\partial t}\right)^2+P\left(\frac{\partial y}{\partial x}\right)^2\right\}\mathrm{d}x$$

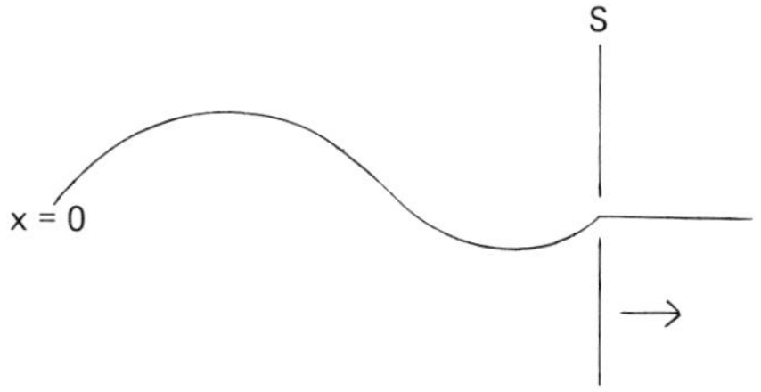

S

x = 0

→

Fig. 32. String with a moving node and variable length.

varies at a rate

$$\begin{aligned}\frac{dE(t)}{dt} &= vW(l+vt,t) + \left[\rho c^2 \frac{\partial y}{\partial x}\frac{\partial y}{\partial t}\right]_{x=l+vt} \\ &= vW(l+vt,t) - c^2 G(l+vt,t).\end{aligned} \tag{42.11}$$

On representing the perturbed string profile by a pair of progressive wave forms

$$y(x,t) = f(x-ct) + g(x+ct), \quad 0 < x < l+vt \tag{42.12}$$

it is found that

$$\frac{dE}{dt} = \rho(v-c)\left(\frac{\partial f}{\partial t}\right)^2_{x=l+vt} + \rho(v+c)\left(\frac{\partial g}{\partial t}\right)^2_{x=l+vt} \tag{42.13}$$

with separate contributions herein from the oppositely directed component waves. If we forego a complete determination of the string displacement, involving a quartet of initial and boundary conditions, and refer to the particular harmonic realizations

$$f(x-ct) = -\frac{A}{k_1 c}\sin k_1(x-ct-l), \qquad g(x+ct) = \frac{B}{k_2 c}\sin k_2(x+ct-l)$$

compliance with the nodal condition at the screen, $y(l+vt,t)=0$, obtains when

$$\begin{gathered} k_1(v-c) = \pm k_2(v+c), \\ k_2 B = \pm k_1 A \quad \text{or} \quad (v+c)B = (v-c)A. \end{gathered}$$

The appertaining vibrations possess an energy density whose average value (relative to time or location),

$$\bar{W} = \frac{\rho}{2}(A^2 + B^2),$$

is a multiple of the time average rate of energy change,

$$\overline{\frac{dE}{dt}} = \frac{\rho}{2}[A^2(v-c) + B^2(v+c)],$$

viz.,

$$\overline{\frac{dE}{dt}} = v\frac{v^2-c^2}{v^2+c^2}\bar{W}. \tag{42.14}$$

It is appropriate to link an energy loss from the string vibrations with the activity of a force on the screen (or transfer point), and if we set

$$-\overline{\frac{dE}{dt}} = \bar{P}v, \tag{42.15}$$

a comparison of (42.14), (42.15) furnishes the magnitude

$$\bar{P} = \frac{c^2 - v^2}{c^2 + v^2} \bar{W} \tag{42.16}$$

for a so-called radiation (or wave) pressure. According to (42.16) there is also a mean pressure exerted on a stationary screen ($v = 0$), whose value equals that of the average density for the adjacent vibrations.

When the screen moves very slowly and the length L of the active part of the string undergoes gradual change, the balance equation

$$-\mathrm{d}W = \bar{P}\,\mathrm{d}L = \frac{W}{L}\mathrm{d}L$$

implies that

$$-\frac{\mathrm{d}W}{W} = \frac{\mathrm{d}L}{L} = \frac{\mathrm{d}\tau}{\tau}$$

or

$$W\tau = \text{const.} \tag{42.17}$$

where τ specifies a period of vibrations (say, $2L/nc$, if there are $n - 1$ nodal points between the extremities of the string); the product of energy density and period in (42.17) is known as an adiabatic invariant.

§43. *Plane Electromagnetic Waves*

A dispersive character is manifest in electromagnetic wave propagation if the medium contains charged particles capable of interacting with the waves. Let it be supposed that the latter have a plane, linearly polarized and time-periodic nature, involving electric and magnetic field components $E(x, t), H(x, t)$ along the y and z directions, respectively, which are transverse to each other and also to the direction of propagation, namely x; then

$$c\frac{\partial E}{\partial x} = -\frac{\partial H}{\partial t} \tag{43.1}$$

expresses the first of a pair of equations which determine the field strengths, where c is the constant electromagnetic wave velocity appropriate to empty space, and a linked pair of progressive wave forms with phase velocity c/n are rendered by

$$E(x, t) = A \cos \omega\left(t - \frac{nx}{c}\right), \qquad H(x, t) = nA \cos \omega\left(t - \frac{nx}{c}\right). \tag{43.2}$$

If $y_i(t)$ designates the displacement of a representative particle with mass m_i and charge q_i, whose equation is that of an undamped though forced oscillator,

$$\frac{d^2 y_i}{dt^2} + \omega_i^2 y_i = \frac{q_i}{m_i} E, \tag{43.3}$$

it follows that

$$y_i = A \frac{\cos \omega \left(t - \dfrac{nx}{c}\right)}{\omega_i^2 - \omega^2}, \qquad \omega \neq \omega_i \tag{43.4}$$

specifies a responsive motion due to the given electric field. Substituting for E, H and y_i in the equation companion to (43.1),

$$-c \frac{\partial H}{\partial x} = \frac{\partial E}{\partial t} + 4\pi \sum_i q_i \frac{dy_i}{dt}, \tag{43.5}$$

yields

$$n^2 = 1 + 4\pi \sum_i \frac{q_i^2}{m_i(\omega_i^2 - \omega^2)} \tag{43.6}$$

and, subsequently, a characterization of the (frequency dependent) phase velocity $v = c/n$. A sequential or two-stage approach to electrodynamic problems entails separate determinations of the fields produced by a given aggregate of charges moving in prescribed fashion and of the motion which a charge executes in a given electromagnetic field. The definitive resolution of any individual problem is achieved when both stages are consistent, that is, when the charges move in such a way that the fields they generate maintain precisely this state of motion. Evidently, the manipulations leading to (43.6) do not constitute a self-consistent analysis and it must therefore be appreciated that the expression for n has relevance only to a tenuous assembly of charges.

Simple calculations reveal that

$$n\omega \frac{dn}{d\omega} = 4\pi \sum_i \frac{q_i^2 \omega^2}{m_i(\omega_i^2 - \omega^2)^2} > 0,$$

$$n + \omega \frac{dn}{d\omega} = \frac{1}{n}\left[1 + 4\pi \sum_i \frac{q_i^2 \omega_i^2}{m_i(\omega_i^2 - \omega^2)^2}\right]$$

whence the group velocity formula (33.6)

$$u = \frac{v^2}{v - \omega \dfrac{dv}{d\omega}} = \frac{c}{n + \omega \dfrac{dn}{d\omega}}$$

provides an inequality

$$u < v \tag{43.7}$$

and the explicit determination

$$u = \frac{cn}{1 + 4\pi \sum_i \frac{q_i^2 \omega_i^2}{m_i(\omega_i^2 - \omega^2)^2}}. \tag{43.8}$$

After multiplying (43.1), (43.3), and (43.5) by H, $\mathrm{d}y_i/\mathrm{d}t$ and E, respectively, and rearranging we obtain the relations

$$cH\frac{\partial E}{\partial x} = -\frac{\partial}{\partial t}\left(\frac{H^2}{2}\right), \qquad -cE\frac{\partial H}{\partial x} = \frac{\partial}{\partial t}\left(\frac{E^2}{2}\right) + 4\pi E \sum_i q_i \frac{\mathrm{d}y_i}{\mathrm{d}t}$$

$$E\sum_i q_i \frac{\mathrm{d}y_i}{\mathrm{d}t} = \frac{\partial}{\partial t}\left[\sum_i \left\{\frac{1}{2}m_i\left(\frac{\mathrm{d}y_i}{\mathrm{d}t}\right)^2 + \frac{1}{2}m_i\omega_i^2 y_i^2\right\}\right]$$

which, consequent to elimination of a sum term and subtraction of the first pair, furnish the differential identity

$$\frac{\partial W}{\partial t} + \frac{\partial S}{\partial x} = 0 \tag{43.9}$$

where

$$S = \frac{c}{4\pi} EH, \tag{43.10}$$

$$W = \frac{E^2 + H^2}{8\pi} + \sum_i \left\{\frac{1}{2}m_i\left(\frac{\mathrm{d}y_i}{\mathrm{d}t}\right)^2 + \frac{1}{2}m_i\omega_i^2 y_i^2\right\}. \tag{43.11}$$

The interpretation of S and W as measures of energy flux and density is prompted by the structure of (43.9); and, having regard for the representations (43.2), we deduce that the amount of energy which traverses unit area normal to the direction of wave propagation in a single period of oscillation is

$$\hat{S} = \frac{c}{4\pi}\int_0^{2\pi/\omega} EH \,\mathrm{d}t = \frac{A^2 cn}{4\omega},$$

with the connection

$$\hat{S} = \frac{2\pi}{\omega}\bar{S}$$

if $\bar{S}$ designates a time average value of the flux. On the basis of (43.11) there follows for the total energy content $\hat{W}$ in a cylindrical domain of unit cross section and elongation equal to the wave length $\lambda = 2\pi c/n\omega$,

$$\begin{aligned}\hat{W} &= \frac{1}{8\pi}\int_0^\lambda \left[E^2 + H^2 + 4\pi \sum_i \left\{m_i\left(\frac{\mathrm{d}y_i}{\mathrm{d}t}\right)^2 + m_i\omega_i^2 y_i^2\right\}\right]\mathrm{d}x \\ &= \frac{A^2 c}{8n\omega}\left[1 + n^2 + 4\pi \sum_i \frac{q_i^2}{m_i}\frac{\omega^2 + \omega_i^2}{(\omega_i^2 - \omega^2)^2}\right] \\ &= \frac{A^2 c}{4n\omega}\left[1 + 4\pi \sum_i \frac{q_i^2\omega_i^2}{m_i(\omega_i^2 - \omega^2)^2}\right],\end{aligned}$$

whose magnitude is in excess of that attributable exclusively to the fields (43.2) by virtue of the dynamical contribution associated with movement of the charges. It is a simple matter to verify, firstly, the connection

$$\hat{W} = \frac{2\pi c}{n\omega}\bar{W}$$

where

$$\bar{W} = \frac{\omega}{2\pi}\int_0^{2\pi/\omega} W\,\mathrm{d}t$$

defines the time average energy density and, secondly, the agreement between the group velocity determination (43.8) and the energy transport velocity defined by

$$u = \frac{\bar{S}}{\bar{W}} = v\frac{\hat{S}}{\hat{W}}. \tag{43.12}$$

If damping plays a role and wave profiles consequently undergo spreading or distortion, there are attendant difficulties relating to the concept of a propagation velocity. The field equations which supplant (43.1), (43.5) in a homogeneous dissipative medium characterized by the complex dielectric constant, $\varepsilon(\omega)$, and magnetic susceptibility, $\mu(\omega)$, viz.

$$c\frac{\partial E}{\partial x} = -\mu(\omega)\frac{\partial H}{\partial t}, \qquad c\frac{\partial H}{\partial x} = \varepsilon(\omega)\frac{\partial E}{\partial t} \tag{43.13}$$

admit the time-periodic plane wave solutions

$$E = \sqrt{(\mu/\varepsilon)}H = A\exp\{\mathrm{i}(\kappa x - \omega t)\} \tag{43.14}$$

where

$$\kappa = \frac{\omega}{c}\sqrt{(\varepsilon\mu)} = \xi + \mathrm{i}\eta. \tag{43.15}$$

A modification of the group velocity formula (33.5), rendered in terms of the derivative of the angular frequency with respect to the real part of the wave number, $\xi = \operatorname{Re} \kappa$, cannot claim theoretical significance when the frequency dependent function $\eta = \operatorname{Im} \kappa$ is different from zero; and it thus becomes appropriate to reexamine the relationship between group velocity and energy in the presence of losses. The circumstance of damping is reflected by a term, R, in the differential equation of energy balance,

$$\frac{\partial W}{\partial t} + \frac{\partial S}{\partial x} = -R,$$

which acquires the form

$$\frac{\partial \bar{W}}{\partial t} + \frac{\partial \bar{S}}{\partial x} = -\bar{R} \tag{43.16}$$

after time-averaging and application of the equality

$$\begin{aligned}\overline{\frac{\partial W}{\partial t}} &= \frac{\omega}{2\pi} \int_t^{t+(2\pi/\omega)} \frac{\partial W}{\partial t} \mathrm{d}t = \frac{\omega}{2\pi} \left[W\left(x, t + \frac{2\pi}{\omega}\right) - W(x, t) \right] \\ &= \frac{\partial}{\partial t} \left\{ \frac{\omega}{2\pi} \int_t^{t+(2\pi/\omega)} W \,\mathrm{d}t \right\} = \frac{\partial \bar{W}}{\partial t}.\end{aligned} \tag{43.17}$$

If $\bar{S}$ is replaced by $u\bar{W}$ in (43.16) we have that

$$\frac{\partial \bar{W}}{\partial t} + u \frac{\partial \bar{W}}{\partial x} = -\bar{R}$$

whence, following multiplication with 1 and x in turn and integration over all x, there obtains (on suppressing an explicit appearance of the infinite limits for the various integrals)

$$\begin{aligned}\frac{\partial}{\partial t} \int \bar{W} \,\mathrm{d}x + \int \bar{R} \,\mathrm{d}x = 0, \\ \frac{\partial}{\partial t} \int x\bar{W} \,\mathrm{d}x - u \int \bar{W} \,\mathrm{d}x + \int x\bar{R} \,\mathrm{d}x = 0\end{aligned} \tag{43.18}$$

providing the disturbance has a finite range, i.e., $\bar{W}, x\bar{W} \to 0$, $|x| \to \infty$. Referring to the coordinates

$$x_E(t) = \frac{\int x\bar{W} \,\mathrm{d}x}{\int \bar{W} \,\mathrm{d}x}, \qquad x_R(t) = \frac{\int x\bar{R} \,\mathrm{d}x}{\int \bar{R} \,\mathrm{d}x} \tag{43.19}$$

which define centers of gravity for energy and losses, respectively, and relying on the equations (43.18), the time derivative of $x_E(t)$ can be exhibited in the alternative forms

$$\frac{\mathrm{d}x_E}{\mathrm{d}t} = u - \frac{\int x\bar{R}\,\mathrm{d}x}{\int \bar{W}\,\mathrm{d}x} + x_E(t)\frac{\int \bar{R}\,\mathrm{d}x}{\int \bar{W}\,\mathrm{d}x} \tag{43.20}$$

or

$$\frac{\mathrm{d}x_E}{\mathrm{d}t} = u + (x_E(t) - x_R(t))\frac{\int \bar{R}\,\mathrm{d}x}{\int \bar{W}\,\mathrm{d}x}. \tag{43.21}$$

When the simpler group velocity representation $u = \mathrm{d}x_E/\mathrm{d}t = \bar{S}/\bar{W}$ retains some measure of aptness, the additional terms in (43.21) are necessarily of small magnitude; and the condition therefor may be stated as

$$\eta L \lesssim 1 \tag{43.22}$$

where L specifies the overall range of the disturbance. A breakdown of (43.22), associated with strong damping on the latter scale, prompts the need for devising other criteria in order to analyse diverse aspects of signal transmission.

§44. *Mechanical/Electrical Analogies and Impedance Concepts for Wave Propagation*

Both the conceptual viewpoint and the concomitant (circuit) formulation that have gained favor in dealing with electrical power transmission, and which constitute a legacy of many years devoted to its theory and practice, can be usefully applied in other instances of wave propagation. Since equivalent circuit representations and electrical analogies are now prominent in the analysis of dynamical systems, it is appropriate to describe their features and utilization in the context of string vibrations.

The conventional engineering theory of a two-conductor (or parallel wire) transmission line schematizes the latter in terms of a distributed parameter circuit which possesses series inductance and shunt capitance (when dissipative effects are left out); for a short section of the line, Δx, there follows the alternative lumped constant representations shown in Fig. 33 where L and C define, respectively, the series inductance and shunt capacitance per unit length. The requisite for a 'short' section is a small ratio between Δx and the wave length λ of waves propagating along the line. Let $V(x, t)$ denote the potential at a point x on the upper line, relative to its value at the corresponding point x on the lower line, and suppose that $I(x, t)$ specifies the current flowing to the right in the upper line at x. The equivalent circuit for

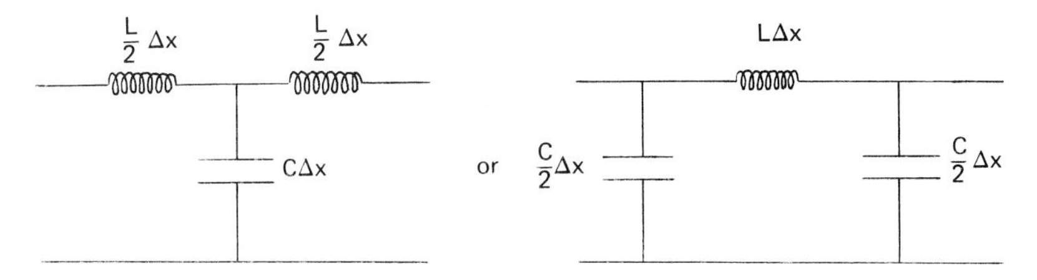

Fig. 33. Lumped constant networks.

the line element Δx implies that

$$\left[V\left(x+\frac{\Delta x}{2}, t\right)-V\left(x-\frac{\Delta x}{2}, t\right)\right]\Delta t=-L[I(x, t+\Delta t)-I(x, t)]\,\Delta x,$$

$$\left[I\left(x+\frac{\Delta x}{2}, t\right)-I\left(x-\frac{\Delta x}{2}, t\right)\right]\Delta t=-C[V(x, t+\Delta t)-V(x, t)]\,\Delta x$$

on neglecting terms of the orders $(\Delta x)^2$, $\Delta x\Delta t$, and $(\Delta t)^2$; hence, in the limits $\Delta x, \Delta t \to 0$, we obtain the transmission line equations

$$\frac{\partial V}{\partial x}=-L\frac{\partial I}{\partial t}, \qquad \frac{\partial I}{\partial x}=-C\frac{\partial V}{\partial t}. \tag{44.1}$$

If the voltage and current oscillate harmonically in time with the angular frequency ω, i.e.,

$$V(x, t)=\text{Re } V(x)\,\mathrm{e}^{-\mathrm{i}\omega t}, \qquad I(x, t)=\text{Re } I(x)\,\mathrm{e}^{-\mathrm{i}\omega t} \tag{44.2}$$

the equations (44.1) give way to a pair

$$\frac{\partial V(x)}{\partial x}=\mathrm{i}\omega L\,I(x), \qquad \frac{\partial I(x)}{\partial x}=\mathrm{i}\omega C\,V(x) \tag{44.3}$$

which possess the solutions

$$V(x)=V\,\mathrm{e}^{\pm\mathrm{i}\kappa x}, \qquad I(x)=\pm\frac{V}{Z}\,\mathrm{e}^{\pm\mathrm{i}\kappa x} \tag{44.4}$$

that correspond to waves propagating in opposite directions along the line. Here the propagation constant and characteristic impedance, κ and Z, respectively, are given by

$$\kappa=\omega\surd(LC)=\frac{\omega}{c}=k, \qquad Z=\left(\frac{L}{C}\right)^{1/2} \tag{44.5}$$

and the particular notation

$$c=\frac{1}{\surd(LC)} \tag{44.6}$$

for the phase velocity of the waves draws attention to the fact that this is the same as the velocity of light. Since the derived quantities Z and κ play a more useful role than L and C, we shall eliminate the latter through the relations

$$\omega L = \kappa Z, \qquad \omega C = \frac{\kappa}{Z} = \kappa Y \tag{44.7}$$

and thereby express the transmission line equations (44.3) in the fundamental form

$$\frac{\partial V(x)}{\partial x} = \mathrm{i}\kappa Z I(x), \qquad \frac{\partial I(x)}{\partial x} = \mathrm{i}\kappa Y V(x) \tag{44.8}$$

where

$$Y = \frac{1}{Z} \tag{44.9}$$

is the characteristic admittance.

A transmission line characterization for taut string movements is contingent on referring to dynamical equations of the system in the canonical or first order coupled differential version (44.1). The transverse velocity,

$$v(x, t) = \frac{\partial y(x, t)}{\partial t}, \tag{44.10}$$

and transverse component of the tension,

$$p(x, t) = -P\frac{\partial y(x, t)}{\partial x} \tag{44.11}$$

constitute an appropriate pair of dependent variables; and their connections

$$\rho\frac{\partial v}{\partial t} = -\frac{\partial p}{\partial x}, \qquad \frac{\partial p}{\partial t} = -P\frac{\partial v}{\partial x}, \tag{44.12}$$

from which the second order linear wave equation (1.1) is a ready consequence, are manifestly in the standard transmission line form. By comparison of (44.1), (44.12) we infer, firstly, the correspondences

$$v \leftrightarrow I, \quad p \leftrightarrow V, \quad \rho \leftrightarrow L, \quad \frac{1}{P} \leftrightarrow C \tag{44.13}$$

and, secondly, the explicit determinations

$$Z = \sqrt{(P\rho)} = \rho c, \qquad \kappa = \frac{\omega}{c} = k, \qquad c = \sqrt{(P/\rho)} \tag{44.14}$$

of the parameters which specify the equivalent line model for the string.

It is noteworthy that the equations (44.8) also apply in more general cases of distributed parameter circuits; thus, for a circuit which possesses distributed series (or longitudinal) impedance Z_l per unit length and distributed shunt (or transverse) admittance Y_t per unit length, we have

$$\frac{dV(x)}{dx} = -Z_l I(x), \qquad \frac{dI(x)}{dx} = -Y_t V(x) \tag{44.15}$$

and, correspondingly, the parameters in (44.8) prove to be

$$i\kappa = \sqrt{((-Y_t)(-Z_l))}, \qquad Z = \sqrt{(Z_l/Y_t)}. \tag{44.16}$$

When each element of a string is acted on by a restoring force, $\rho\omega_0^2 y$, proportional to its displacement from equilibrium ($y = 0$), the appertaining realizations

$$\begin{gathered} Z = -i\omega L + \frac{i}{\omega C'}, \qquad Y_t = -i\omega C, \\ L = \rho, \qquad C = 1/P, \qquad C' = 1/\rho\omega_0^2 \end{gathered} \tag{44.17}$$

signal a distributed series inductance, L, along with series and shunt capacitances, C' and C, respectively. Equivalent networks for a short section Δx of the line have the T or π versions shown in Fig. 34 and the line parameters κ, Z are given by

$$\begin{gathered} \kappa = \frac{1}{c}\sqrt{(\omega^2 - \omega_0^2)} \\ Z = \frac{\kappa P}{\omega} = \rho c\left(1 - \left(\frac{\omega_0}{\omega}\right)^2\right)^{1/2} \qquad c = \sqrt{(P/\rho)} \end{gathered} \tag{44.18}$$

The propagation constant κ evidently differs from $k = \omega/c$ and becomes imaginary at frequencies less than the critical frequency ω_0; in the latter curcumstance wave propagation along the line/string is ruled out and excitations are exponentially attenuated with distance from their source. It appears, furthermore, that a changeover from capacitative (imaginary) to resistive (real) form of the characteristic impedance accompanies the elevation of frequency above the level ω_0. Both of the preceding features are traceable to the resonant character of the series impedance, which gives the circuit the properties of a high pass filter.

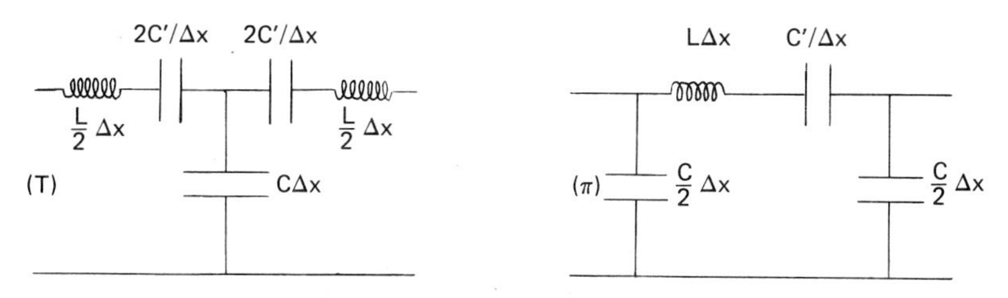

Fig. 34. T and π equivalent networks.

If there is, additionally, a force proportional to the local transverse velocity of the string, $-\varepsilon\rho v$, the series impedance

$$Z_l = R - \mathrm{i}\omega L + \frac{1}{\omega C'}, \qquad R = \varepsilon\rho \tag{44.19}$$

replaces that given in (44.17) and the lumped circuit network representation for a short section Δx takes the form depicted in Fig. 35.
The line constants have a complex nature and it follows, on writing

$$-\mathrm{i}\sqrt{((-Y_t)(-Z_l))} = \kappa + \mathrm{i}\frac{\alpha}{2} \tag{44.20}$$

and thereby retaining the symbol κ for the real part of the propagation constant,, that the voltage, current or string displacement descriptive of waves headed in the positive x-direction contain the factor

$$\exp\left\{\mathrm{i}\left(\kappa + \mathrm{i}\frac{\alpha}{2}\right)x\right\} = \mathrm{e}^{-(\alpha/2)x}\,\mathrm{e}^{\mathrm{i}\kappa x}$$

which involves an attenuation constant, α. If we regard the introduction of the quantity R into the series impedance as equivalent to changing the inductance by a small amount $\mathrm{i}R/\omega$ and employ the relations

$$\delta\kappa = \mathrm{i}\frac{\alpha}{2} = \frac{\partial\kappa}{\partial L}\delta L, \qquad \kappa = \left(\omega^2 LC - \frac{C}{C'}\right)^{1/2}$$

there obtains

$$\alpha = \frac{2R}{\omega}\frac{\partial\kappa}{\partial L} = \frac{\omega}{\kappa}RC = \frac{R}{Z} = \frac{\varepsilon}{c\left(1 - \left(\frac{\omega_0}{\omega}\right)^2\right)^{1/2}} \tag{44.21}$$

and the attenuation constant is thus linked with the ratio of the series resistance per unit length and the (real) characteristic impedance.

The overall effect of any section of line is implicit in a knowledge of the interrelation between voltage and current at its termini; for we may then calculate said quantities at either endpoint, given their values at the other. It is

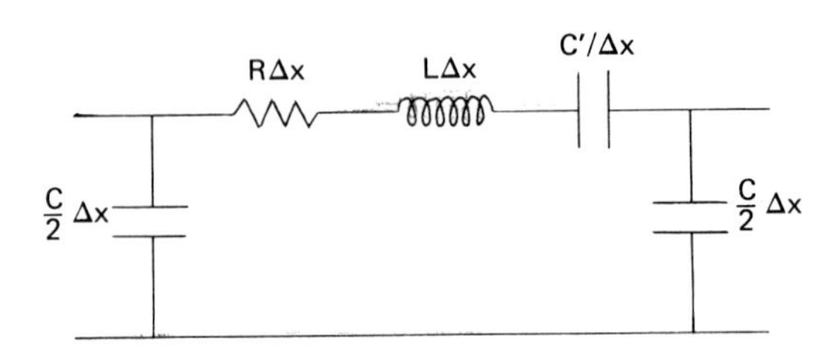

Fig. 35. Network with a resistive element.

a simple matter to obtain such connections in the case of a uniform line on which voltage and current satisfy the common second order differential equation

$$\left(\frac{d^2}{dx^2}+\kappa^2\right)\begin{matrix}V(x)\\I(x)\end{matrix}=0 \tag{44.22}$$

and admit the representations

$$V(x)=V\cos\kappa x+\mathrm{i}Z_0 I\sin\kappa x \tag{44.23}$$

$$I(x)=I\cos\kappa x+\mathrm{i}Y_0 V\sin\kappa x \tag{44.24}$$

in terms of the reference values $V=V(0)$, $I=I(0)$ and the characteristic impedance/admittance, Z_0/Y_0. Adopting the designations $V_1=V(-l/2)$, $I_1=I(-l/2)$, $V_2=V(l/2)$, $I_2=I(l/2)$, and utilizing the consequences of (44.24), viz.

$$I_1+I_2=2I\cos\kappa l/2,\qquad I_1-I_2=-2\,\mathrm{i}Y_0 V\sin\kappa l/2,$$

to eliminate V, I from (44.23) it follows, after the substitutions $x=\mp l/2$ are made therein, that

$$\begin{aligned}V_1&=Z_{11}I_1-Z_{12}I_2\\V_2&=Z_{21}I_1-Z_{22}I_2\end{aligned} \tag{44.25}$$

with

$$Z_{11}=\mathrm{i}\frac{Z_0}{2}(\cot\kappa l/2-\tan\kappa l/2)=Z_{22},$$

$$Z_{12}=\mathrm{i}\frac{Z_0}{2}(\cot\kappa l/2+\tan\kappa l/2)=Z_{21}.$$

The coupling which is set forth in (44.25), between voltage and current pairs at the input-output locations $x=\mp l/2$, is duplicated in the T network arrangement shown in Fig. 36 which thus provides an equivalent circuit representation for the finite length, l, of line. Let us single out the respective values $V=0$ and $I=0$ of midsection voltage and current, associated with an odd symmetry of the corresponding functions $V(x)$ and $I(x)$ about $x=0$; and

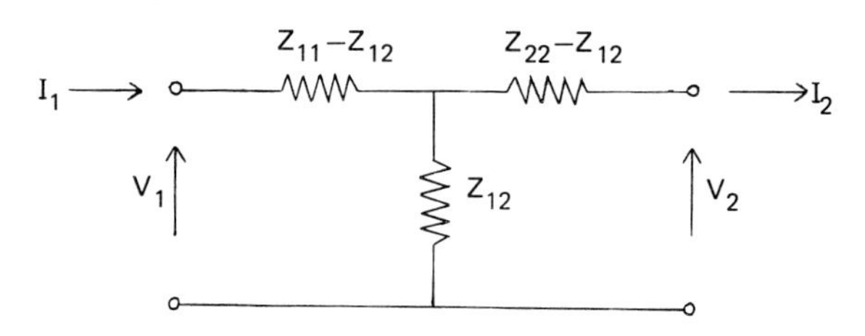

Fig. 36. Voltage and current specifications at the terminals of a network.

refer to the consequent determinations

$$\frac{V(x)}{I(x)} = \mathrm{i}Z_0 \tan kx, \qquad \frac{I(x)}{V(x)} = \mathrm{i}Y_0 \tan \kappa x.$$
$$(V = 0) \qquad\qquad (I = 0)$$

The particular ratios found therefrom,

$$\frac{V_1}{I_1} = -\mathrm{i}Z_0 \tan\frac{\kappa l}{2} = Z_{11} - Z_{12}$$

and

$$\frac{V_1}{I_1} = \mathrm{i}Z_0 \cot\frac{\kappa l}{2} = Z_{11} + Z_{12}$$

define terminal impedances at the input location $x = -l/2$, whose elements are independently specified via the general circuit (Fig. 37), or its alternative realization (Fig. 38) by adopting short and open circuit conditions at the center.

It is appropriate, furthermore, to note that the preceding voltage-current ratios are in harmony with the impedance transformation for a length l of line (see Fig. 39) namely

$$Z_1 = \frac{Z_2 - \mathrm{i}Z_0 \tan \kappa l}{1 - \mathrm{i}Y_0 Z_2 \tan \kappa l}, \tag{44.26}$$

since the input impedance representations supplied by (44.26),

$$Z_1 = -\mathrm{i}Z_0 \tan \kappa l, \qquad V_2 = Z_2 = 0$$

$$\frac{V_1}{I_1} = -\mathrm{i}Z_0 \tan\frac{\kappa l}{2} = Z_{11} - Z_{12}$$

$$\frac{V_1}{I_1} = \mathrm{i}Z_0 \cot\frac{\kappa l}{2} = Z_{11} + Z_{12}$$

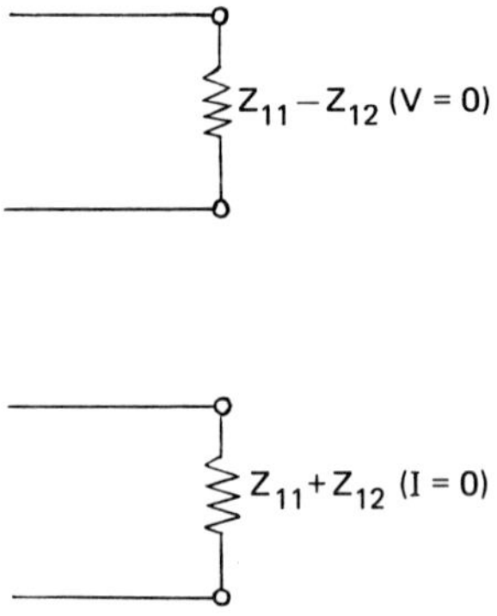

Fig. 37. Terminal impedances.

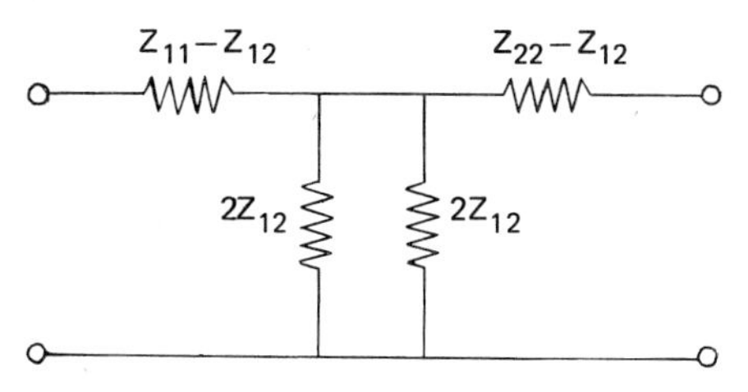

Fig. 38. An equivalent circuit.

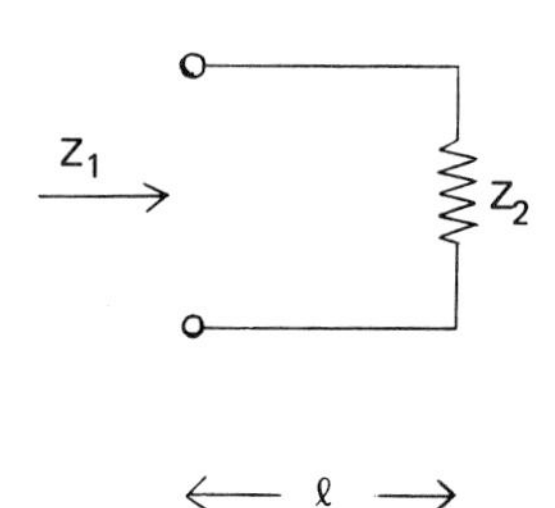

Fig. 39. Input and terminal impedances.

and

$$Z_1 = \mathrm{i}Z_0 \cot \kappa l, \qquad I_2 = Y_2 = 0$$

merely differ from the prior versions insofar as the length scale is concerned ($l/2 \to l$).

The impedance concept serves in a useful capacity for analyzing diverse aspects of excitation in strings and other configurations; thus, to characterize the proper frequencies of a string fixed at both ends ($x = 0, l$), we may arbitrarily select an interior point, $x = \xi$, and equate the output impedance $Z^{(0)}$ from the section $(0, \xi)$ with the input impedance $Z^{(i)}$ of the section (ξ, l), thereby assuring the requisite continuity of displacement and slope everywhere. On the basis of (44.26) it appears that (if $\kappa = k$)

$$Z^{(0)} = -\mathrm{i}Z_0 \cot k\xi, \qquad Z^{(\mathrm{i})} = \mathrm{i}Z_0 \cot k(l - \xi), \qquad Z_0 = \rho c$$

and thence the condition $Z^{(0)} = Z^{(\mathrm{i})}$ assumes the form

$$\frac{\sin kl}{\sin k\xi \sin k(l - \xi)} = 0$$

which leads to the expected frequency relation $kl = n\pi$, $n = 1, 2, \ldots$.

To illustrate the effective use of impedances in determining the proper frequencies of a more complicated system, let us consider the arrangement in Fig. 40 of three coplanar strings that are joined at a common point (0) and maintained in a state of tension by fixing their other extremities (A, B, C). The input impedances at 0, for periodic transverse vibrations of the strings,

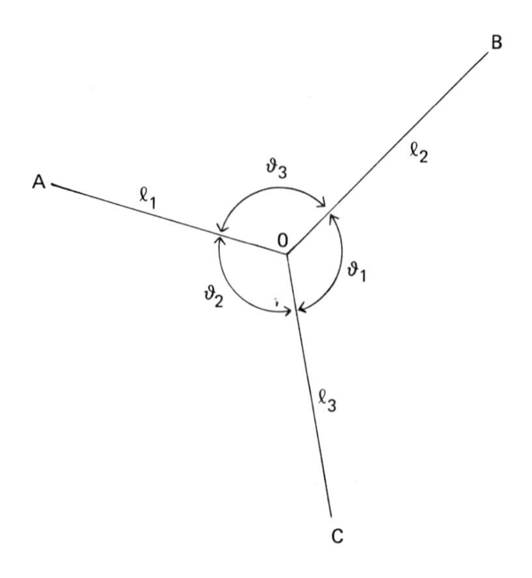

Fig. 40. Connected strings.

assumed to possess the same density and to have unequal lengths l_1, l_2, l_3, are given by

$$Z^{(1)} = \mathrm{i}\sqrt{(P_1\rho)}\cot\left\{\omega l_1\left(\frac{\rho}{P_1}\right)^{1/2}\right\}, \qquad Z^{(2)} = \mathrm{i}\sqrt{(P_2\rho)}\cot\left\{\omega l_2\left(\frac{\rho}{P_2}\right)^{1/2}\right\},$$

$$Z^{(3)} = \mathrm{i}\sqrt{(P_3\rho)}\cot\left\{\omega l_3\left(\frac{\rho}{P_2}\right)^{1/2}\right\}$$

if P_1, P_2, and P_3 denote the respective tensions. On applying the impedance matching condition

$$-Z^{(1)} = Z^{(2)} + Z^{(3)},$$

and the statical balance of tensions (which keeps the junction 0 from undergoing lateral displacement)

$$\frac{P_1}{\sin\vartheta_1} = \frac{P_2}{\sin\vartheta_2} = \frac{P_3}{\sin\vartheta_3} = \mu \text{ (a constant)},$$

the frequency equation becomes

$$\sqrt{(\sin\vartheta_1)}\cot\pi\omega/\omega_1 + \sqrt{(\sin\vartheta_2)}\cot\pi\omega/\omega_2 + \sqrt{(\sin\vartheta_3)}\cot\pi\omega/\omega_3 = 0$$

where ω_1, ω_2, and ω_3 identify the lowest (circular) frequencies that characterize individual vibrations of the strings if the point 0 is also fixed.

A pair of quantities, termed the driving point and transfer impedance/admittance, respectively, are common measures of system response; the former expresses, in a mechanical setting, the ratio between force and velocity at the point of application of the first, while the latter involves the resultant velocity elsewhere. If we have in mind a string which is acted on by the force $F\,e^{-i\omega t}$ at $x = \xi$, and take note of the relation

$$p(\xi + 0, t) - p(\xi - 0, t) = F\,e^{-i\omega t}$$

that links the force with transverse components of the tension in its vicinity, the driving point impedance may be found via the representation

$$\mathscr{Z} = \frac{p(\xi + 0, t) - p(\xi - 0, t)}{v(\xi, t)}. \tag{44.27}$$

When the string has an unlimited extension and the simplest form of wave equation is adopted, the characterizations

$$p(\xi + 0, t)/v(\xi, t) = Z_0, \qquad p(\xi - 0, t)/v(\xi, t) = -Z_0 \tag{44.28}$$

thus enable us to infer that

$$\mathscr{Z} = 2Z_0 = 2\rho c.$$

After multiplying the right-hand side of (44.27) by $v(\xi, t)/v(x, t)$ the corresponding transfer impedance

$$\mathscr{Z}(x, \xi, \omega) = \mathscr{Z}\frac{v(\xi, t)}{v(x, t)} = \frac{F\,e^{-i\omega t}}{v(x, t)} = 2\rho c \exp\left\{-i\frac{\omega}{c}|x - \xi|\right\},$$

obtains. For a string whose ends $(x = 0, l)$ are secured, the driving point impedance at an interior location $x = \xi (0 < \xi < l)$, namely

$$\mathscr{Z} = iZ_0 \frac{\sin\dfrac{\omega}{c}l}{\sin\dfrac{\omega}{c}\xi \sin\dfrac{\omega}{c}(l - \xi)}, \tag{44.29}$$

is based on the characterizations

$$\frac{p(\xi + 0, t)}{v(\xi, t)} = iZ_0 \cot\frac{\omega}{c}(l - \xi), \qquad \frac{p(\xi - 0, t)}{v(\xi, t)} = -iZ_0 \cot\frac{\omega}{c}\xi$$

which replace those of (44.28); the corresponding transfer impedance has alternative forms

$$\mathscr{Z}(x, \xi, \omega) = \begin{cases} iZ_0 \dfrac{\sin\dfrac{\omega}{c}l}{\sin\dfrac{\omega}{c}\xi \sin\dfrac{\omega}{c}(l - x)}, & x > \xi \\[2ex] iZ_0 \dfrac{\sin\dfrac{\omega}{c}l}{\sin\dfrac{\omega}{c}x \sin\dfrac{\omega}{c}(l - \xi)}, & x < \xi \end{cases} \tag{44.30}$$

in accordance with the relative disposition of the points x, ξ.

Extreme values, 0 and ∞, are registered by the driving point impedance at the center of the string ($\xi = l/2$) when $l/\lambda = 1/2$ and 1, respectively, λ being the wave length that corresponds to the frequency ω and wave speed c, i.e., $\lambda = 2\pi c/\omega$; the cited ratios of characteristic lengths l, λ are those pertaining to the fundamental mode and the first overtone (or partial) of string vibrations which evidence, separately, a loop (antinode) and node at the midpoint. Since $v = F/\mathscr{Z} = F\mathscr{Y}$, it may be concluded that the applied force F produces a maximal velocity response in the first case ($\mathscr{Y} = \infty$) and is wholly ineffectual in the second case ($\mathscr{Y} = 0$). The presence of dissipation keeps the magnitude of $\mathscr{Z}$ within finite bounds and thus establishes a lower (non-zero) limit for $\mathscr{Y}$, with the result that some velocity response is always forthcoming. Measures of impedance previously defined, and electrical analogies more generally, lend themselves to good advantage in the analysis of string motions originating with aperiodic forces or the impact of a hammer (cf. Kock, 1937).

§45. *Wave Propagation Along Fluid Boundaries*

A heavy incompressible fluid is capable of propagating waves along its free surface and contiguous ones with different densities can transmit waves along their mutual boundary; excitations of this kind, in which the fluid movements are principally affected by the force of gravity (and secondarily by other forces, such as surface tension), present some distinctive features bearing on dispersion and the influence of superposed flow or streaming motion.

Let us first consider a single fluid with a horizontal or plane free surface (at $y = 0$) in the equilibrium state and a level bottom (at $y = -H$); and give attention to progressive disturbances of the form

$$\zeta = A\, e^{i(kx-\omega t)} \tag{45.1}$$

where ζ denotes the surface displacement (along the y-direction). The latter evidently refer to uni-directional waves whose crests ($\zeta = A$) and troughs ($\zeta = -A$) extend along straight lines parallel to the surface. If the concomitant two-dimensional fluid motion is an irrotational one (that is, initiated from a state of rest) the velocity vector $\mathbf{v}$ has a vanishing curl and its components v_x, v_y along the x, y directions are derivable from a potential function $\Phi(x, y, t)$, i.e.,

$$v_x = \frac{\partial \Phi}{\partial x}, \qquad v_y = \frac{\partial \Phi}{\partial y}; \tag{45.2}$$

by virtue of the equation of continuity or mass conservation within the fluid, $\partial v_x/\partial x + \partial v_y/\partial y = 0$, the velocity potential proves to be a harmonic function, namely

$$\frac{\partial^2 \Phi}{\partial x^2} + \frac{\partial^2 \Phi}{\partial y^2} = 0. \tag{45.3}$$

The vectorial expression of Euler's momentum balance equation for an inviscid fluid

$$\frac{\partial \mathbf{v}}{\partial t} + (\mathbf{v} \cdot \nabla)\mathbf{v} = -\frac{1}{\rho} \nabla p - \nabla(gy), \tag{45.4}$$

where p, ρ represent the fluid pressure and density, and g is the gravitational acceleration, admits a scalar integral

$$\frac{\partial \Phi}{\partial t} + \frac{1}{2}(\nabla \Phi)^2 = -\frac{p}{\rho} - gy, \tag{45.5}$$

known as Bernoulli's equation.

On the free surface, with profile $y = \zeta(x, t)$, the twin requirements of continuity for the pressure and of permanent fluid particle attachment imply that

$$\begin{aligned} &\frac{\partial \Phi}{\partial t} + \frac{1}{2}(\nabla \Phi)^2 = -g\zeta, \\ &\frac{\partial \zeta}{\partial t} - \frac{\partial \Phi}{\partial y} = -\frac{\partial \Phi}{\partial x}\frac{\partial \zeta}{\partial x} \end{aligned} \tag{45.6}$$

if we assign a null value to the atmospheric pressure. Analytical complications arising from the fact that the conditions (45.6) are non-linear and enforceable along an unknown profile make it worthwhile, as a technical matter, to contemplate motions which involve limited displacements of the free surface. A linearized version of the conditions (45.6),

$$\left.\begin{aligned} &g\zeta + \frac{\partial \Phi}{\partial t} = 0, \\ &\frac{\partial \zeta}{\partial t} - \frac{\partial \Phi}{\partial y} = 0, \end{aligned}\right\} \quad \text{at } y = 0 \tag{45.7}$$

is indicated when the wave heights ζ are much smaller than the wave length of a disturbance on the free surface and the latter has a small slope everywhere. Apart from the initial data and asymptotic properties, the velocity potential is governed, at this stage of approximation, by Laplace's equation (45.3) in the fluid, the free surface condition

$$\frac{\partial^2 \Phi}{\partial t^2} + g\frac{\partial \Phi}{\partial y} = 0, \qquad y = 0 \tag{45.8}$$

and the boundary condition

$$\frac{\partial \Phi}{\partial y} = 0, \qquad y = -H \tag{45.9}$$

for a rigid bottom; pursuant to a determination of Φ, those of the surface elevation and pressure at any point in the fluid are achieved through the formulas

$$\zeta = -\frac{1}{g}\left(\frac{\partial \Phi}{\partial t}\right)_{y=0}, \tag{45.10}$$

$$p = -\rho g y - \rho \frac{\partial \Phi}{\partial t}. \tag{45.11}$$

A travelling wave profile of the form (45.1) is described by the complex potential

$$\Phi(x, y, t) = \varphi(x, y)\, \mathrm{e}^{-\mathrm{i}\omega t} \tag{45.12}$$

with a particular harmonic function

$$\varphi(x, y) = \frac{gA}{\mathrm{i}\omega} \frac{\cosh k(H + y)}{\cosh kH} \mathrm{e}^{\mathrm{i}(kx - \omega t)}, \quad 0 \geq y \geq -H \tag{45.13}$$

that has a vanishing y-derivative at $y = -H$. It follows from the mutual compatibility of the potential representation (45.12), (45.13) and the free surface condition (45.8) that gravity waves of small amplitude A propagate on the free surface with the phase velocity

$$v = \frac{\omega}{k} = \left(\frac{g}{k} \tanh kH\right)^{1/2} = \left(\frac{g\lambda}{2\pi} \tanh \frac{2\pi H}{\lambda}\right)^{1/2} \tag{45.14}$$

corresponding to the wave length $\lambda = 2\pi/k$. Evidently, such motions are dispersive in nature unless the fluid depth is very small compared to the wavelength, in which case

$$v \doteq (gH)^{1/2}, \quad \lambda \gg H \tag{45.15}$$

as a small argument approximation of the hyperbolic tangent function brings out.

The group velocity derived from (45.14),

$$u = \frac{\mathrm{d}\omega}{\mathrm{d}k} = v\left[\frac{1}{2} + \frac{kH}{\sinh 2kH}\right] \tag{45.16}$$

has magnitudes consistent with the inequalities

$$\frac{1}{2} < \frac{u}{v} < 1$$

where the lesser and greater of the bounds are realized when $\lambda/H \to 0$ and $\lambda/H \to \infty$. It is significant to observe that the above expression for u results after dividing the time average energy flux across a fixed vertical plane parallel to the wave fronts, viz.

$$\frac{\omega}{2\pi}\int_0^{2\pi/\omega} dt \int_{-L}^{0} dy\ \mathrm{Re}\left(-\rho\frac{\partial\Phi}{\partial t}\right)\cdot \mathrm{Re}\left(\frac{\partial\Phi}{\partial x}\right)$$
$$=\frac{1}{2}g\rho v A^2\left[\frac{1}{2}+\frac{kH}{\sinh 2kH}\right],$$

by the mechanical (potential and kinetic) energy between a pair of vertical planes at unit spacing, namely $1/2(g\rho A^2)$, both measures being associated with a unit length normal to the direction of wave propagation.

When the fluid is infinitely deep, the harmonic function

$$\varphi(x, y)=\frac{gA}{\mathrm{i}\omega}\,\mathrm{e}^{ky}\,\mathrm{e}^{\mathrm{i}(kx-\omega t)},\quad y\leq 0 \tag{45.17}$$

takes the place of (45.13) and the phase velocity is given by

$$v=(g\lambda/2\pi)^{1/2}. \tag{45.18}$$

Since the exponential factor in (45.17) decreases rapidly as the distance from the free surface increases, with a proportionate reduction of the amount $\mathrm{e}^{-2\pi} \doteq 1/500$ for an addition to the depth equal to a wave length, gravity wave movements in a fluid are confined to the vicinity of the surface. Local motions in these surface waves are neither transverse nor longitudinal, and integrals of the equations that determine the coordinates $x(t)$, $y(t)$ of fluid particles,

$$\frac{\mathrm{d}x}{\mathrm{d}t}=\frac{\partial\Phi}{\partial x},\qquad \frac{\mathrm{d}y}{\mathrm{d}t}=\frac{\partial\Phi}{\partial y}, \tag{45.19}$$

indicate circular or elliptical trajectories according as the depth is infinite or finite. The effect of a local disturbance on a quiescent fluid surface, which can be expressed as a superposition of progressive wave forms, is communicated to the remote parts without delay by the long wave components that have correspondingly large phase velocities. It is inevitable that aspects of surface wave motions relating to incompressible fluids differ markedly from those of transverse wave motions along strings or longitudinal wave motions in compressible media since the underlying differential equations (Laplace and wave) belong to distinct types (elliptic and hyperbolic).

Let us next seek to determine, in the framework of a linear approximation, the features of small amplitude gravity waves that are superposed on a uniform streaming (or unperturbed) motion of the fluid. When a velocity potential $\Phi(x, y, t)$ exists, the Laplace equation (45.3) and rigid bottom condition (45.9) remain in force, whereas the free surface condition (45.8) requires modification on account of the assumed flow. If the latter is aligned with the positive x-direction and has the magnitude V everywhere, the pair of surface conditions (45.7) for small disturbances in a fluid at rest are replaced by

$$\left.\begin{aligned}\frac{\partial \Phi}{\partial t}+V\frac{\partial \Phi}{\partial x}&=-g\zeta,\\ \frac{\partial \Phi}{\partial y}&=\frac{\partial \zeta}{\partial t}+V\frac{\partial \zeta}{\partial x}\end{aligned}\right\} \quad y=0; \tag{45.20}$$

it thus follows, after eliminating the surface elevation $\zeta(x, t)$ from the first order system (45.20), that the velocity potential satisfies a free surface condition of the form

$$\left(\frac{\partial}{\partial t}+V\frac{\partial}{\partial x}\right)^2 \Phi + g\frac{\partial \Phi}{\partial y}=0, \qquad y=0. \tag{45.21}$$

On introducing the representation

$$\Phi(x, y, t) = Vx + \operatorname{Re}\{\varphi(x, y)\, e^{-i\omega t}\}, \tag{45.22}$$

whose first term characterizes the unperturbed flow, we obtain from (45.21) a boundary condition

$$-\omega^2\varphi - 2\,i\omega V\frac{\partial \varphi}{\partial x} + V^2\frac{\partial^2\varphi}{\partial x^2} + g\frac{\partial \varphi}{\partial y}=0, \qquad y=0 \tag{45.23}$$

for the plane harmonic function $\varphi(x, y)$.

Consider the particular harmonic functions

$$\left.\begin{aligned}\varphi_+(x, y) &= e^{ikx}\frac{\cosh k(H+y)}{\cosh kH}, \quad k>0,\\ \varphi_-(x, y) &= e^{-ikx}\frac{\cosh k(H+y)}{\cosh kH}, \quad 0\geq y\geq -H\end{aligned}\right\} \tag{45.24}$$

that correspond to surface waves with down/up stream headings, respectively, and the separate dispersion relations

$$\left.\begin{aligned}F_+(k, \omega) &= (\omega - kV)^2 - gk\tanh kH = 0,\\ F_-(k, \omega) &= (\omega + kV)^2 - gk\tanh kH = 0\end{aligned}\right\} \tag{45.25}$$

which are the direct outcome of substituting $\varphi_\pm(x, y)$ in (45.23). If $k \neq 0$ we may rewrite the first of these relations in the form

$$\left(\frac{\omega}{k} - V\right)^2 = \frac{g}{k}\tanh kH = v^2,$$

where v denotes the phase velocity for surface waves in a still fluid, and thence conclude that

$$v^* = \frac{\omega}{k} = V \pm v. \tag{45.26}$$

The composition in (45.26), of a fluid velocity (V) and a relative disturbance velocity (v), suggests that gravity waves are convected and have larger/smaller phase speeds according as they move along with or counter to the prevailing flow.

A more detailed consideration of both the dispersion relations (45.25) is prompted when we regard ω and V as given, with the former characteristic, say, of a source that brings about disturbances of the basic flow. The dispersion relation, $F_+(k, \omega) = 0$, subsumes a pair of others, namely

$$v(\lambda) - \nu\lambda = -V, \tag{45.27}$$

$$v(\lambda) + \nu\lambda = V \tag{45.28}$$

in which $\lambda = 2\pi/k$ and $\nu = \omega/2\pi$ takes the place of k and ω. To examine the matter of solving (45.27), (45.28) for λ, let us follow Becker (1956) and plot in Fig. 41 the dispersion curve

$$v(\lambda) = \left(\frac{g\lambda}{2\pi} \tanh 2\pi \frac{H}{\lambda}\right)^{1/2} \quad \text{vs. } \lambda,$$

along with curves of the functions

$$v(\lambda) - \nu\lambda, \quad v(\lambda) + \nu\lambda.$$

It is apparent from the accompanying figure that the equations (45.27), (45.28) have unique solutions, λ_1, λ_2, respectively, for any assigned values of ν, V. The root λ_1 pertains to waves with a phase velocity $v(\lambda_1)$ relative to the

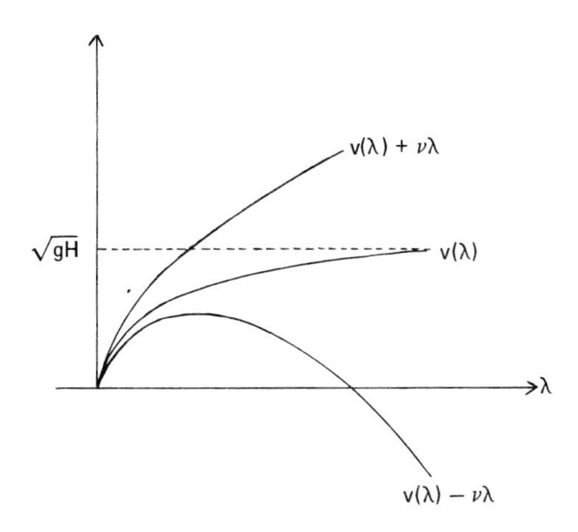

Fig. 41. Dispersion curves.

stream and an absolute velocity $V + v$ in the downstream direction; and the root λ_2 is connected with waves of relative phase velocity $v(\lambda_2) < V$ and absolute downstream velocity $V - v$. A representation of F_+, which exposes the salient features of its dependence on k, is provided by

$$F_+(k) = (k - k_1)(k - k_2)\mathfrak{F}_+(k) \tag{45.29}$$

where $k_1 = 2\pi/\lambda_1$, $k_2 = 2\pi/\lambda_2$ and $\mathfrak{F}_+(k) \neq 0$, $0 < k < \infty$, $\mathfrak{F}_+(k) \to V^2$, $k \to \infty$.

The other dispersion relation, $F_-(k, \omega) = 0$, unites a subsidiary pair,

$$\pm\, v(\lambda) - \nu\lambda = V,$$

in which the second choice of sign for the initial term is ruled out, because their definitions render all terms positive. We are thus led to determine values of λ that satisfy the single equation

$$v(\lambda) - \nu\lambda = V; \tag{45.30}$$

and, as an examination of the figure reveals, the possibilities are threefold, comprising distinct roots ($\lambda_3 < \lambda_4$, say), a double root or no roots. Since the magnitude of the group velocity,

$$u = v - \lambda \frac{dv}{d\lambda},$$

is specified, at any value of λ, through an intercept of the tangent to the dispersion curve, we readily deduce that

$$u < V, \qquad \lambda = \lambda_3, \qquad u > V, \qquad \lambda = \lambda_4$$

in the first case; and there is, accordingly, a transport of energy upstream in waves of length λ_4 and absolute phase velocity $v - V$. A place downstream of the wave maker or source (§46) is indicated for waves of length λ_3 that do not transmit energy upstream. When (45.30) has two distinct roots, $\lambda_3 = 2\pi/k_3$, $\lambda_4 = 2\pi/k_4$, the representation

$$F_-(k) = (k - k_3)(k - k_4)\mathfrak{F}_-(k) \tag{45.31}$$

obtains, with $\mathfrak{F}_-(k) \neq 0$, $0 < k < \infty$ and $\mathfrak{F}_-(k) \to V^2$, $k \to \infty$.

A theory for long gravity waves of small amplitude may be founded on suitable modification of the hydrodynamic equations at the outset; specifically, the vertical acceleration in the fluid is neglected and the horizontal components of velocity are accorded no variation with depth. Pursuant to such assumptions, consider two layers of fluid with the same depth H and unequal densities ρ_1, ρ_2 ($<\rho_1$), which are confined between fixed horizontal planes, the lighter density fluid being uppermost; and let it be required to find the phase velocity of a long straight-crested disturbance which propagates along the mutual interface of the fluids, allowing for parallel flows of the latter at the respective speeds V_1, V_2.

The relevant dynamical equations account for momentum balance along and normal to the flow direction; in the fluid with density ρ_1 the former has a linearized version

$$\frac{\partial v_x^{(1)}}{\partial t} + V_1 \frac{\partial v_x^{(1)}}{\partial x} = -\frac{1}{\rho_1}\frac{\partial p_1}{\partial x} \tag{45.32}$$

and, since $\partial v_y^{(1)}/\partial t$ is presumed to be negligible, the latter possesses an integral

$$p_1 = -\rho_1 g y + \rho_1 F_1(x, t). \tag{45.33}$$

A corresponding pair of relations apply in the fluid with density ρ_2 and the continuity of pressure at the interface between the fluids, whose (small) displacement from the level $y = 0$ is denoted by $\zeta(x, t)$, calls for the interconnection

$$-\rho_1 g\zeta + \rho_1 F_1(x, t) = -\rho_2 g\zeta + \rho_2 F_2(x, t). \tag{45.34}$$

Incompressibility, which requires a vanishing velocity flux across any closed curve in a vertical section of either fluid, signifies that the normal interface velocities and the horizontal velocity gradients are proportional, i.e.,

$$v_y^{(1)} = -H\frac{\partial v_x^{(1)}}{\partial x}, \qquad v_y^{(2)} = H\frac{\partial v_x^{(2)}}{\partial x}, \qquad y = 0; \tag{45.35}$$

the normal velocities can now be eliminated entirely through the substitution in (45.35) of the expressions

$$v_y^{(1)} = \frac{\partial \zeta}{\partial t} + V_1\frac{\partial \zeta}{\partial x}, \qquad v_y^{(2)} = \frac{\partial \zeta}{\partial t} + V_2\frac{\partial \zeta}{\partial x} \tag{45.36}$$

and the resulting equations

$$\left.\begin{aligned} \frac{\partial \zeta}{\partial t} + V_1\frac{\partial \zeta}{\partial x} &= -H\frac{\partial v_x^{(1)}}{\partial x}, \\ \frac{\partial \zeta}{\partial t} + V_2\frac{\partial \zeta}{\partial x} &= H\frac{\partial v_x^{(2)}}{\partial x}, \end{aligned}\right\} \qquad y = 0 \tag{45.37}$$

complete a quintet that determine ζ, $v_x^{(1)}$, $v_x^{(2)}$, F_1, and F_2. Assuming that each of the latter quantities is proportional to $e^{ik(x-vt)}$, with the respective amplitudes $A, \alpha_1, \alpha_2, \beta_1, \beta_2$, there obtains from the momentum equations

$$\alpha_1(v - V_1) = \beta_1, \qquad \alpha_2(v - V_2) = \beta_2 \tag{45.38}$$

while the interface relations yield

$$(\rho_1 - \rho_2)gA = \rho_1\beta_1 - \rho_2\beta_2, \tag{45.39}$$

$$A(v - V_1) = H\alpha_1, \qquad A(v - V_2) = -H\alpha_2. \tag{45.40}$$

The consequence of recasting the right-hand side of (45.39) in terms of A, through successive substitutions from (45.38) and (45.40), is a quadratic equation for the phase velocity, v, of the long wave motions,

$$v^2 - 2v\frac{\rho_1 V_1 + \rho_2 V_2}{\rho_1 + \rho_2} + \frac{\rho_1 V_1^2 + \rho_2 V_2^2}{\rho_1 + \rho_2} - \frac{\rho_1 - \rho_2}{\rho_1 + \rho_2} gH = 0.$$

Hence, the roots of said equation are real or complex, according as

$$\frac{\rho_1 - \rho_2}{\rho_1 + \rho_2} gH \gtrless \frac{\rho_1 \rho_2}{(\rho_1 + \rho_2)^2}(V_1 - V_2)^2,$$

and

$$\hat{v} = \left(\frac{\rho_1 - \rho_2}{\rho_1 + \rho_2} gH\right)^{1/2}$$

evidently characterizes the phase velocity in the absence of streaming motions ($V_1 = V_2 = 0$). A complex value of v implies that the wave function $e^{ik(x-vt)}$ no longer maintains a unit absolute magnitude and can be taken as an indication of instability for the excitation contemplated.

§46. *Source Excited Gravity Waves on a Fluid*

Excitations of the type discussed in the previous section, identifiable with regular (and normalized) functions that satisfy differential equations and boundary conditions, describe the free wave motions of the relevant systems. For linear systems in general, the resultant excitation that is attributable to a source (represented by a singular function of given type and scale) may be expressed wholly, or in part, as a definite linear combination of the free wave modes. In order to develop the representational aspects of such modal forms, let us consider a specific example involving a (line) source situated below the surface of an incompressible fluid. We associate the source itself, in this fluid mechanical context, with a velocity potential of plane harmonic nature that has a logarithmic singularity at the source point; and fix its scale, namely the source strength Q, through an assigned value of the flux or normal velocity along any curve encircling the singular point. Accordingly, the complete velocity potential within the fluid is of the form

$$\Phi(x, y, t) = -\frac{Q}{2\pi} \log r + F(x, y, t), \quad r \to 0 \tag{46.1}$$

at small distances, r, from the source, where F retains a regular harmonic behavior in the limit $r \to 0$. When the fluid streams over a rigid bottom, the additional conditions which Φ is obliged to meet at the boundaries ($y = -H, 0$) are those set out in (45.9), (45.21). If the source, located at $(0, -d)$, has a periodically varying strength, namely

$$Q(t) = \mathrm{Re}(Q_0\, e^{-i\omega t}), \tag{46.2}$$

the representation (45.22) enables us to determine the synchronous excitation of the fluid in terms of a plane harmonic function $\varphi(x, y)$, $-\infty < x < \infty$, $-H < y < 0$, such that [recall (45.23)]

$$-\omega^2\varphi - 2\,i\omega V\frac{\partial \varphi}{\partial x} + V^2\frac{\partial^2\varphi}{\partial x^2} + g\frac{\partial \varphi}{\partial y} = 0, \qquad y = 0, \tag{46.3}$$

$$\frac{\partial \varphi}{\partial y} = 0, \qquad y = -H, \tag{46.4}$$

$$\varphi(x, y) = -\frac{Q_0}{2\pi}\log r + (\text{a regular function of } x, y)$$
$$\text{as } r = (x^2 + (y+d)^2)^{1/2} \to 0. \tag{46.5}$$

It is appropriate from a practical standpoint, because of the numerous stipulations on φ, to adopt a constructive procedure whereby said function is realized through a superposition of others which do not individually comply with all these stipulations and are thus more readily found. We introduce, first of all, the harmonic function

$$\varphi^{(1)}(x, y) = \frac{Q_0}{2\pi}\int_{-\infty}^{\infty} e^{ikx}\begin{Bmatrix}\sinh ky \sinh k(H-d) \\ -\sinh kd \sinh k(H+y)\end{Bmatrix}\frac{dk}{k\cosh kH}, \quad \begin{matrix} 0 > y > -d \\ -H < y < -d\end{matrix} \tag{46.6}$$

which satisfies (46.4), (46.5) and, additionally, the conditions

$$\begin{aligned}\varphi^{(1)}(x, 0) &= 0, \quad -\infty < x < \infty \\ \varphi^{(1)}(x, y) &\to 0, \quad |x| \to \infty, \quad -H < y < 0;\end{aligned} \tag{46.7}$$

this function represents, in fact, an infinite linear array of sources that includes the original one at $(0, -d)$ along with images exterior to the strip domain $-\infty < x < \infty$, $-H < y < 0$ (see Fig. 42).

Next, let us define a regular harmonic function $\varphi^{(2)}(x, y)$, on the same domain as $\varphi^{(1)}(x, y)$ and subject to the conditions

$$-\omega^2\varphi^{(2)} - 2\,i\omega V\frac{\partial \varphi^{(2)}}{\partial x} + V^2\frac{\partial^2\varphi^{(2)}}{\partial x^2} + g\frac{\partial \varphi^{(2)}}{\partial y} = -g\,\frac{\partial \varphi^{(1)}}{\partial y}, \qquad y = 0 \tag{46.8}$$

$$\frac{\partial \varphi^{(2)}}{\partial y} = 0, \qquad y = -H; \tag{46.9}$$

the combination

$$\varphi^{(1)}(x, y) + \varphi^{(2)}(x, y)$$

is thus in accord with the trio of equations (46.3)–(46.5). If we assume that

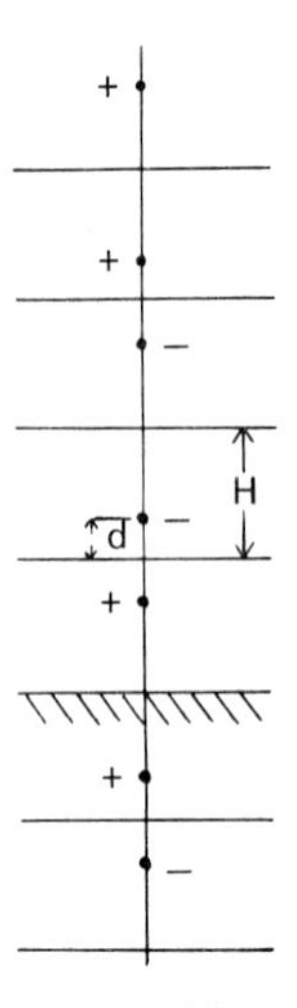

Fig. 42. Source and image system.

$\varphi^{(2)}(x, 0)$ possesses a complex Fourier transform with respect to x and set

$$\varphi^{(2)}(x, y) = \int_{-\infty}^{\infty} \mathrm{e}^{ikx} \frac{\cosh k(H+y)}{\cosh kH} F(k)\, \mathrm{d}k \tag{46.10}$$

the condition (46.9) is fulfilled and (46.8) implies that

$$\begin{aligned} &\int_{-\infty}^{\infty} \mathrm{e}^{ikx}\{-(\omega - kV)^2 + gk \tanh kH\} F(k) \\ &\quad = -g\left(\frac{\partial \varphi^{(1)}}{\partial y}\right)_{y=0} = -\frac{gQ_0}{2\pi} \int_{-\infty}^{\infty} \mathrm{e}^{ikx} \frac{\cosh k(H-d)}{\cosh kH} \mathrm{d}k, \quad -\infty < x < \infty \end{aligned}$$

after utilizing (46.6). Hence

$$F(k) = \frac{gQ_0}{2\pi} \frac{\cosh k(H-d)}{\cosh kH} \frac{1}{(\omega - kV)^2 - gk \tanh kH} \tag{46.11}$$

and it follows from (46.10) that

$$\begin{aligned} \varphi^{(2)}(x, y) &= \frac{gQ_0}{2\pi} \int_{-\infty}^{\infty} \mathrm{e}^{ikx} \frac{\cosh k(H+y) \cosh k(H-d)}{\cosh^2 kH} \frac{\mathrm{d}k}{(\omega - kV)^2 - gk \tanh kH} \\ &= \frac{gQ_0}{2\pi} \int_0^{\infty} \left\{ \frac{\mathrm{e}^{ikx}}{F_+(k)} + \frac{\mathrm{e}^{-ikx}}{F_-(k)} \right\} \frac{\cosh k(H+y) \cosh k(H-d)}{\cosh^2 kH} \mathrm{d}k \end{aligned} \tag{46.12}$$

when the notations (45.25) are in effect. Making use of the representations

(45.29), (45.31) for $F_{\pm}(k)$ we find, by simple rearrangement in (46.12),

$$\varphi^{(2)}(x, y) = \frac{gQ_0}{2\pi}\frac{1}{k_1 - k_2}\int_0^\infty e^{ikx}\left(\frac{1}{k-k_1} - \frac{1}{k-k_2}\right)G_+(k, y)\,dk$$
$$+\frac{gQ_0}{2\pi}\frac{1}{k_3 - k_4}\int_0^\infty e^{-ikx}\left(\frac{1}{k-k_3} - \frac{1}{k-k_4}\right)G_-(k, y)\,dk \tag{46.13}$$

with the integrand factors

$$G_{\pm}(k, y) = \frac{\cosh k(H+y)\cosh k(H-d)}{\mathfrak{F}_{\pm}(k)\cosh^2 kH} \tag{46.14}$$

that are bounded on $0 < k < \infty$ and tend to zero as $k \to \infty$.

The occurrence of poles in (46.13) at the locations k_1, k_2 and k_3, k_4 which identify roots of the dispersion equations (45.25) for free waves headed down or upstream, respectively, necessitate a suitable interpretation of the component integrals; if we adopt the Cauchy principal value of an integral, for instance, then

$$\int_0^\infty \frac{e^{ikx}}{k-k_1}G_+(k, y)\,dk = \operatorname*{Lim}_{\varepsilon\to 0+}\left\{\int_0^{k_1-\varepsilon}\frac{e^{ikx}}{k-k_1}G_+(k, y)\,dk + \int_{k_1+\varepsilon}^\infty \frac{e^{ikx}}{k-k_1}G_+(k, y)\,dk\right\} \tag{46.15}$$

serves as a representative definition. Since the functions which appear in the integrands of (46.13) have analytic continuations from real to complex values of k, we may express each principal value integral as a sum of others, with the same integrand, that are taken along complementary parts of a closed contour. In the case of (46.15) an appropriate contour follows the real axis after leaving the origin, save for a semi-circular indentation (with radius ε) about the pole at $k = k_1$, and returns to the starting point via the particular half plane ($\operatorname{Im} k \gtrless 0$, $x \gtrless 0$) wherein $e^{ikx} \to 0$ as $|k| \to \infty$. When the standard procedures of complex integration are applied, here and in the parallel cases, it turns out that

$$\int_0^\infty \frac{e^{ikx}}{k-k_{1,2}}G_+(k, y)\,dk = \pm\pi i\, e^{ik_{1,2}x}G_+(k_{1,2}, y) + R_+, \qquad x \gtrless 0,$$
$$\int_0^\infty \frac{e^{-ikx}}{k-k_{3,4}}G_-(k, y)\,dk = \mp\pi i\, e^{-ik_{3,4}x}G_-(k_{3,4}, y) + R_-, \qquad x \gtrless 0 \tag{46.16}$$

where residue contributions at the individual poles account for the explicit terms and

$$\operatorname*{Lim}_{|x|\to\infty} R_{\pm} = 0; \tag{46.17}$$

if we restrict our attention to the surface wave motions at great distances from their source, details of the functions $R_\pm$ need not be pursued.

Thus, on the basis of (46.16), every one of the free wave modes is represented, with definite amplitude or scale factor, in the asymptotic expressions for $|x|\to\infty$ of the composite source function $\varphi^{(1)}+\varphi^{(2)}$ that satisfies the ensemble of conditions (46.3)–(46.5). A full complement of modal forms in the source representation is at variance, however, with individual surface wave characteristics previously established (§45) and, specifically, with the fact that waves of length $\lambda_3 = 2\pi/k_3$ are unable to transport energy upstream because their group velocity is smaller than the flow velocity. To secure an appropriate asymptotic behavior in the given source problem, which means that progressive waves with the respective lengths $\lambda_1, \lambda_2, \lambda_3$ are located downstream only and that a single wave length λ_4 is possible upstream, we superimpose on $\varphi^{(1)}+\varphi^{(2)}$ a set of free mode functions and select their coefficients accordingly; this yields for the harmonic functions of the resultant wave complex

$$\varphi(x, y) \simeq \frac{igQ_0}{k_3 - k_4}\, e^{-ik_4x} G_-(k_4, y), \quad x \to -\infty,$$

and

$$\begin{aligned}\varphi(x, y) \simeq{}& \frac{igQ_0}{k_1 - k_2}\left[e^{ik_1x} G_+(k_1, y) - e^{ik_2x} G_+(k_2, y)\right]\\ &- \frac{igQ_0}{k_3 - k_4}\, e^{-ik_3x} G_-(k_3, y), \quad x \to +\infty.\end{aligned} \tag{46.18}$$

If ω and V are such that the dispersion relation (45.30) has no real roots, the contributions associated with k_3 and k_4 disappear from (46.18) and only a pair of downstream surface waves convey indications of the source activity. It becomes necessary to abandon a strictly linear theory, on which the preceding calculation is founded, when $k_3 = k_4$ (and the corresponding group velocity equals the stream velocity V), since unbounded wave amplitudes are forecast in (46.18).

Simple and explicit determinations of the wave parameters and amplitudes obtain in the limiting case of an infinitely deep fluid ($H \to \infty$); these include

$$\begin{aligned} k_{1,2} &= \frac{g + 2\omega V}{2V^2}\left[1 \mp \left(1 - \left(\frac{2\omega V}{g + 2\omega V}\right)^2\right)^{1/2}\right],\\ k_{3,4} &= \frac{g - 2\omega V}{2V^2}\left[1 \pm \left(1 - \left(\frac{2\omega V}{g - 2\omega V}\right)^2\right)^{1/2}\right]\end{aligned} \tag{46.19}$$

whence real and distinct values of k_3, k_4 are contingent on the inequality

$$\omega V < \frac{g}{4}$$

that restricts the magnitudes of both the source frequency and stream velocity. If the stream velocity tends to zero we deduce the estimates

$$k_1 = k_4 \doteq \frac{\omega^2}{g}, \qquad k_2 = k_3 \doteq \frac{g}{V^2}, \qquad k_1 - k_2 = -(k_3 - k_4) \doteq \frac{g}{V^2}, \quad V \to 0$$

and thus, inasmuch as

$$G_+(k, y) = G_-(k, y) = \frac{1}{V^2} \,\mathrm{e}^{-k(d-y)}, \quad H \to \infty \tag{46.20}$$

it follows from (46.18) that

$$\varphi(x, y) \sim \mathrm{i}Q_0 \,\mathrm{e}^{-(\omega^2/g)(d-y)} \,\mathrm{e}^{\mathrm{i}(\omega^2/g)|x|}, \quad |x| \to \infty \tag{46.21}$$

with the expected lateral symmetry relative to the source.

§47. *Wave Propagation in Tubes Having Elastic Walls*

Elasticity of the walls enclosing a fluid or gas is bound to affect the propagation of disturbances therein and its particular consequences have long been matters of interest: to physiologists, for interpreting pulsations in blood flow, to acousticians, in conjunction with measurements of sound speed, and to hydraulic engineers, as regards the transmission of water in turboelectric installations. Boulanger (1913) renders a valuable service in tracing the evolution of theory and experiments, from Euler (1775) and Young (1808) to a contemporary period, and in citing the duplication of effort on the part of numerous contributions with a different focus.

An investigation by Korteweg (1878), aimed at calculating the speed with which a disturbance travels through an initially quiescent column of compressible fluid inside a thin-walled elastic cylinder, constitutes a landmark of the subject; such is the complexity of determining coupled motions for fluids and elastic bodies that the relevant theory has not yet, in fact, attained the ultimate stage of generality and explicitness. Korteweg makes two simplifying assumptions at the outset: first, that the fluid mass originally situated between any pair of planes normal to the axis of the tube thereafter occupies a similar volume, signifying that compression, expansion, or dilation merely alters the spacing between said planes and, second, that for wave lengths large compared to the transverse dimensions (radius, wall thickness) of the tube, motions of its boundary are describable in terms of local dilatation or contraction of independent annular (ring-shaped) sections normal to the axis. When viscosity and heat conduction are left out of account, the parameters on which a velocity of propagation depends include the tube radius R and wall thickness δ, along with the respective densities and elasticity coefficients ρ_f, ρ_w, E_f, E_w of fluid and wall.

A self-consistent dynamical analysis involves three relations, dictating the conservation of fluid mass and of momentum in fluid and wall movements separately. If positions along the tube are specified by the coordinate x and $A(x)$ designates the corresponding area of a normal (plane) section, the equation of continuity or mass conservation is (with the prior assumption) expressible in the form

$$\frac{\partial}{\partial t}(\rho_f A) + \frac{\partial}{\partial x}(\rho_f u A) = 0 \tag{47.1}$$

where u represents the fluid velocity along the x-direction. On correlating small variations of density and fluid pressure, p, through the linear relationship

$$\rho_f = \rho_0\left(1 + \frac{p}{E_f}\right) \tag{47.2}$$

and writing

$$A = \pi R^2 + 2\pi R\eta \tag{47.3}$$

it follows that a first order approximation to (47.1) becomes

$$\frac{\partial u}{\partial x} + \frac{1}{E_f}\frac{\partial p}{\partial t} + \frac{2}{R}\frac{\partial \eta}{\partial t} = 0$$

or

$$\frac{\partial \xi}{\partial x} + \frac{p}{E_f} + \frac{2\eta}{R} = 0 \tag{47.4}$$

in terms of the displacement from equilibrium for any normal section, ξ, whose time derivative equals u.

The stretching or proportionate change η/R in radius of a thin wall filament (see Fig. 43) induces a comparable tension therein; and the resultant of the

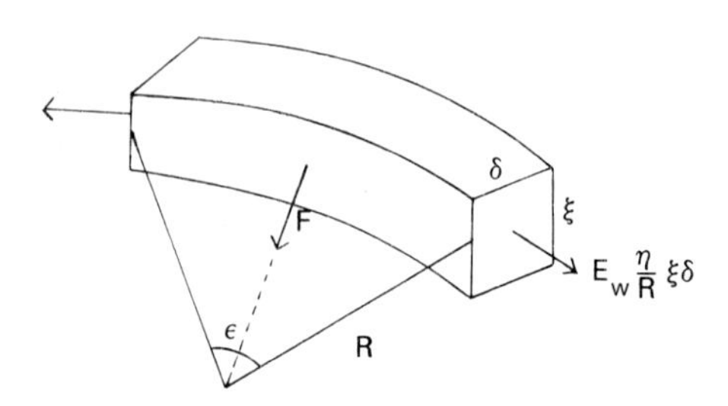

Fig. 43. Curved wall element.

full normal stress, $E_w(\eta/R)\xi\delta$, acting at each end of a filament (with the cross-sectional area $\xi\delta$) that subtends an angle ε at the center of the tube, is a force $F = \varepsilon E_w(\xi\eta\delta/R)$ along the angle bisector or central radius. After division by $\varepsilon R\xi$ the affiliated normal pressure of elastic origin acquires the magnitude

$$p_w = E_w \frac{\eta\delta}{R^2}. \tag{47.5}$$

The first order (linear) approximation to the momentum equation for the fluid is

$$\rho_0 \frac{\partial^2 \xi}{\partial t^2} = -\frac{\partial p}{\partial x}, \tag{47.6}$$

while

$$\rho_w \delta \frac{\partial^2 \eta}{\partial t^2} = p - p_w \tag{47.7}$$

constitutes the analogous dynamical equation for the wall. Utilizing (47.4) and (47.5) to achieve the elimination of p and p_w from (47.6) and (47.7), it follows that

$$\frac{\rho_0}{E_f} \frac{\partial^2 \xi}{\partial t^2} = \frac{\partial^2 \xi}{\partial x^2} + \frac{2}{R} \frac{\partial \eta}{\partial x}, \tag{47.8}$$

$$\frac{\rho_w \delta}{E_f} \frac{\partial^2 \eta}{\partial t^2} + \frac{\partial \xi}{\partial x} + 2\frac{\eta}{R} - \frac{E_w}{E_f} \frac{\eta\delta}{R^2} = 0. \tag{47.9}$$

The direct outcome of applying an x-derivative to the latter equation and replacing $\partial\eta/\partial x$ in accordance with the preceding one is an equation of fourth order which governs the displacement (or axial fluid velocity), viz.

$$\frac{\rho_f \rho_w R^2 \delta}{E_f E_w} \frac{\partial^4 \xi}{\partial t^4} - \frac{\rho_w R^2 \delta}{E_w} \frac{\partial^4 \xi}{\partial x^2 \partial t^2} + \left(\frac{2\rho_f R}{E_w} + \frac{\rho_f \delta}{E_f} \right) \frac{\partial^2 \xi}{\partial t^2} - \delta \frac{\partial^2 \xi}{\partial x^2} = 0. \tag{47.10}$$

If the inertia of the tube wall is negligible ($\rho_w = 0, p = p_w$), a simple wave equation obtains,

$$\frac{\partial^2 \xi}{\partial x^2} = \left(\frac{\rho_f}{E_f} + \frac{2\rho_f R}{E_w \delta} \right) \frac{\partial^2 \xi}{\partial t^2}, \tag{47.11}$$

and its limiting versions

$$\frac{\partial^2 \xi}{\partial x^2} = \frac{\rho_f}{E_f} \frac{\partial^2 \xi}{\partial t^2}, \qquad \frac{\partial^2 \xi}{\partial x^2} = \frac{2\rho_f R}{E_w \delta} \frac{\partial^2 \xi}{\partial t^2} \tag{47.12}$$

befit, in turn, the extremes of perfect wall rigidity ($E_w = \infty$) and fluid incompressibility ($E_f = \infty$). One of the two characteristic wave speeds defined in (47.12), namely $v_f = (E_f/\rho_f)^{1/2}$, expresses Newton's result for small amplitude (or weak) disturbances in a compressible medium, with an elasticity parameter that depends on the local pressure/velocity correlation; while the other, $v_{wf} = (E_w\delta/2\rho_f R)^{1/2}$, contains the essence of Young's later adaptation of same, based on a measure of elasticity whose magnitude is fixed through the pressure/radius correlation at the wall of an incompressible fluid. An evident merit of Korteweg's equation (47.11) is that it furnishes a speed V, given by

$$\frac{1}{V^2} = \frac{1}{v_f^2} + \frac{1}{v_{wf}^2}, \tag{47.13}$$

which interpolates between the aforesaid pair and reveals that elasticity of the wall slows the rate of progress for disturbances in a confined and compressible medium.

On employing the designation $v_w = (E_w/\rho_w)^2$ and referring to an alternative display of the equation (47.10),

$$\frac{R^2}{v_f^2 v_w^2} \frac{\partial^4 \xi}{\partial t^4} - \frac{R^2}{v_w^2} \frac{\partial^4 \xi}{\partial x^2 \partial t^2} + \frac{1}{V^2} \frac{\partial^2 \xi}{\partial t^2} - \frac{\partial^2 \xi}{\partial x^2} = 0, \tag{47.14}$$

wherein the highest order terms contain the mass density of the tube as a coefficient factor, it is found that the wave length $\lambda = 2\pi/k$ and phase velocity v for a propagating wave function,

$$\xi = \operatorname{Re} A\, e^{ik(x-vt)}, \tag{47.15}$$

satisfy the dispersion relation

$$\left(\frac{kR}{v_f v_w}\right)^2 v^4 - \left(\frac{kR}{v_w}\right)^2 v^2 - \frac{1}{V^2} v^2 + 1 = 0; \tag{47.16}$$

the limit property $v(k) \to V$, $k \to 0$, singles out an appropriate determination of v in terms of k.

Dispersion attributable to the wall inertia is also manifest by a different phase velocity representation, arrived at after withdrawing the assumption of exclusively axial fluid motion and admitting, in particular, a radial velocity component $w = \partial\zeta/\partial t$; when the motions possess symmetry round the tube axis, the requisite equation of continuity,

$$\frac{\partial \rho}{\partial t} + \frac{\partial}{\partial x}(\rho u) + \frac{1}{r} \frac{\partial}{\partial r}(\rho r w) = 0,$$

has the linearized form

$$\frac{1}{E_f} \frac{\partial p}{\partial t} + \frac{\partial u}{\partial x} + \frac{w}{r} + \frac{\partial w}{\partial r} = 0$$

or

$$\frac{p}{E_f}+\frac{\partial \xi}{\partial x}+\frac{\zeta}{r}+\frac{\partial \zeta}{\partial r}=0 \tag{47.18}$$

in terms of the axial and radial displacement components $\xi(r, x, t)$, $\zeta(r, x, t)$. The momentum balance (or Euler) equations are, correspondingly and approximately,

$$\rho_f \frac{\partial^2 \xi}{\partial t^2}=-\frac{\partial p}{\partial x}, \qquad \rho_f \frac{\partial^2 \zeta}{\partial t^2}=-\frac{\partial p}{\partial r}; \tag{47.19}$$

and the system of descriptive equations for ξ, ζ, and ρ is completed with a boundary condition

$$p=\frac{E_w \zeta \delta}{R^2}, \quad \text{at } r=R \tag{47.20}$$

which renders in explicit fashion the hypothesis that the tube wall yields radially by an amount proportional to the adjacent pressure.

If we introduce a progressive wave representation

$$\begin{gathered}\xi=\mathrm{Re}\{A(r)\, \mathrm{e}^{\mathrm{i}k(x-vt)}\}, \qquad \zeta=\mathrm{Re}\{\mathrm{i}B(r)\, \mathrm{e}^{\mathrm{i}k(x-vt)}\} \\ p=\mathrm{Re}\{\mathrm{i}C(r)\, \mathrm{e}^{\mathrm{i}k(x-vt)}\}\end{gathered} \tag{47.21}$$

the trio of coupled relations

$$A(r)=-\frac{C(r)}{\rho_f k v^2}, \qquad B(r)=\frac{1}{\rho_f (kv)^2}\frac{\mathrm{d}C}{\mathrm{d}r}$$

and (47.22)

$$\frac{1}{E_f}C(r)+kA(r)+\frac{B(r)}{r}+\frac{\mathrm{d}B}{\mathrm{d}r}=0$$

follow from (47.18), (47.19). It is easy to establish the individual equation

$$\frac{\mathrm{d}^2 C}{\mathrm{d}r^2}+\frac{1}{r}\frac{\mathrm{d}C}{\mathrm{d}r}-k^2\left(1-\frac{v^2}{v_f^2}\right)C=0, \qquad v_f^2=E_f/\rho_f \tag{47.23}$$

and thus to link sequential determinations of A and B with a knowledge of C. Specifically, the solution of (47.23) that exhibits a regular behavior on the range $0 \leq r \leq R$ is given by

$$C(r)=\text{const.}\, I_0\left(\left(1-\frac{v^2}{v_f^2}\right)^{1/2} kr\right) \tag{47.24}$$

where

$$I_0(z)=J_0(\mathrm{i}z)+1+\frac{z^2}{2^2}+\frac{z^4}{2^2 \cdot 4^2}+\cdots \tag{47.25}$$

designates the (zero order) Bessel function of purely imaginary argument. Combining (47.24) and a suitable transcription of the boundary condition (47.20), viz.

$$C(R) = B(R)\frac{E_w\delta}{R^2} = \frac{E_w\delta}{R^2}\cdot\frac{1}{\rho_f(kv)^2}\left(\frac{\mathrm{d}C}{\mathrm{d}r}\right)_{r=R}, \tag{47.26}$$

yields the sought-after dispersion relation

$$I_0\left(\left(1-\frac{v^2}{v_f^2}\right)^{1/2} kR\right) = \frac{E_w\delta}{R^2}\frac{\sqrt{(1-v^2/v_f^2)}}{\rho_f k v^2} I_0'\left(\left(1-\frac{v^2}{v_f^2}\right)^{1/2} kR\right) \tag{47.27}$$

with an argument derivative understood in the last factor.

The phase velocity expression obtained from (47.27) on the basis of the first two terms in the development (47.25),

$$v = V\left\{1-\frac{1}{16}(kR)^2\frac{V^2}{v_w^2}\left(1-\frac{V^2}{v_f^2}\right)-\frac{1}{4}(kR)^2\frac{\delta}{R}\frac{\rho_w}{\rho_f}\frac{V^2}{v_w^2}\right\}, \tag{47.28}$$

embodies corrections to the prior estimate, namely V, which originate in the transverse motion of fluid and wall, respectively, and have small magnitudes for wave lengths $\lambda = 2\pi/k$ considerably greater than both the radius, R, and wall thickness, of the tube. If the aspect ratio δ/R is not small compared with unity, analysis of the wave propagation problem becomes substantially more difficult; this reflects the necessity for a detailed consideration of the tube itself as an elastic continuum and, also, for a description of the coupled motions of tube and neighboring fluid in a self-consistent manner. Korteweg applied the statical or equilibrium equations of elasticity so as to advance an operational definition, befitting a hollow cylindrical tube, of the wall parameter E_w that appears in various expressions above; and Lamb (1898) subsequently obtained a wave velocity formula on the basis of dynamical equations, with the simplifying (albeit unrealistic) assumption that the outer radius of the tube is infinite. It is appropriate to infer from what has been touched on in this section that an accounting for elasticity raises the order of the partial differential equations which govern wave phenomena in extended systems.

Problems II(b)

1. Plane polarized electromagnetic waves travelling along the x-direction at speed c are incident normally on a thin stratum which contains N electrons per unit area, each with mass m and charge e. If $E_y^{\text{inc}}(x, t) = \cos(\omega t - (\omega x/c))$ specifies the electric field in the incident wave train, derive its counterpart

$$E_y^{\text{ref}}(x, t) = \sin\alpha \sin\left(\omega t - \alpha + \frac{\omega}{c}x\right), \qquad \tan\alpha = 2\pi c N \frac{e^2}{m\omega}$$

for the reflected wave train, on the hypothesis that the period of the waves is much smaller than the mean free time of the elastically bound electrons. Deduce the expression

$$\frac{e^2}{2mc}\frac{N}{\omega}(\sin\alpha\cos\alpha - \cos\alpha\sin 2\omega t)$$

for the thrust per unit area on the reflecting layer.

2. Consider a uniform straight tube, of finite length l and cross-sectional area A, and suppose that plane waves of sound are excited therein by periodic motions of a piston at one end ($x = 0$). Let $V\,\mathrm{e}^{-\mathrm{i}\omega t}$ designate the given piston velocity and introduce a resolution for the local (air particle) velocity in terms of waves moving to and fro along the tube, namely

$$v(x)\,\mathrm{e}^{-\mathrm{i}\omega t} = A_1\,\mathrm{e}^{\mathrm{i}(kx-\omega t)} + A_2\,\mathrm{e}^{\mathrm{i}(-kx-\omega t)} + A_3\,\mathrm{e}^{\mathrm{i}(kx-\omega t)} + A_4\,\mathrm{e}^{\mathrm{i}(-kx-\omega t)} + \cdots,$$

with the correspondences

$$A_1 = V, \qquad A_2 = -A_1\delta_l\,\mathrm{e}^{2\mathrm{i}kl}, \qquad A_3 = A_1\delta_0\delta_l\,\mathrm{e}^{2\mathrm{i}kl}, \qquad A_4 = -A_1\delta_0\delta_l^2\,\mathrm{e}^{4\mathrm{i}kl}, \ldots$$

that involve reflection factors δ_0, δ_l at the respective ends of the tube. Sum the series for $v(x)$ and, after achieving similar representations for the pressure $p(x)\,\mathrm{e}^{-\mathrm{i}\omega t}$ with the help of the fundamental equation

$$\rho\frac{\partial}{\partial t}(v(x)\,\mathrm{e}^{-\mathrm{i}\omega t}) = -\frac{\partial}{\partial x}(p(x)\,\mathrm{e}^{-\mathrm{i}\omega t}) \qquad (*)$$

where ρ is the air density, derive the acoustical impedance of the tube at the piston or source,

$$Z = \frac{p(0)}{Av(0)} = \frac{\rho c}{A}\frac{1 + \delta_l\,\mathrm{e}^{2\mathrm{i}kl}}{1 - \delta_l\,\mathrm{e}^{2\mathrm{i}kl}}.$$

Show that the same determination of Z results when the equations

$$\left(\frac{d^2}{dx^2}+k^2\right)v(x)=0, \quad 0<x<l$$

$$v(0)=V, \qquad \frac{1}{v(x)}\frac{dv}{dx}\bigg|_{x=l}=ik\frac{1+\delta_l}{1-\delta_l},$$

and the interrelation (*) between $v(x)$ and $p(x)$, are directly employed.

3. Justify, in the circumstances envisaged above, the characterization

$$p(l)=\frac{cV(1+\delta_l)\,e^{ikl}}{1-\delta_0\delta_l\,e^{2ikl}}$$

for a terminal pressure within the tube, and thence the ratio

$$\frac{p(0)}{p(l)}=\frac{1+\delta_l\,e^{2ikl}}{(1+\delta_l)\,e^{ikl}}$$

between pressures at the respective ends. Let $x=0, l$ mark the open and closed ends of a resonant tube, the reflection factor in the latter site assuming its maximum value, $\delta_l=1$; and represent the total acoustical impedance at the open end, $\mathscr{Z}$, by the sum of two component parts

$$Z=i\frac{\rho c}{A}\cot kl \quad \text{(Cf. problem 2)}$$

and

$$Z'=R-i\omega M \quad \text{(say)},$$

which account, separately, for inertial reactions of the air, inside and outside the tube, to periodic stimulus. Given the proportionality

$$\frac{P}{p(0)}=\frac{\mathscr{Z}}{Z}=\frac{Z+Z'}{Z},$$

wherein P designates a forcing pressure at the open end, show that by suitable choice of the frequency (or tuning)

$$\frac{p(l)}{P}=i\frac{\rho c}{k\bar{V}R}, \quad kl \ll 1$$

in terms of the resonator volume, $\bar{V}=lA$; and hence note the control exercised by the real part of the external impedance, or so-called acoustical radiation resistance, on the comparatively high level of sound pressure within the resonator.

4. Imagine that a straight tube, of length l and sectional area A, is bent into a ring shape, with the consequent joining of its ends. Let $\phi(s)\,\mathrm{e}^{-\mathrm{i}\omega t}$ represent the complex internal velocity potential for time-periodic acoustical excitations which depend solely on the azimuthal (or circumferential) coordinate s; and assume that the equation

$$\left(\frac{\mathrm{d}^2}{\mathrm{d}s^2}+k^2\right)\phi(s)=V\delta(s),\qquad -\frac{l}{2}<s<\frac{l}{2}$$

relates their generation to a prescribed velocity,

$$V=\left.\frac{\mathrm{d}\phi}{\mathrm{d}s}\right|_{s=0-}^{s=0+},$$

at the arbitrarily selected location $s=0$, while the periodic boundary conditions

$$\phi(-l/2)=\phi(l/2),\qquad \frac{\mathrm{d}}{\mathrm{d}s}\phi(-l/2)=\frac{\mathrm{d}}{\mathrm{d}s}\phi(l/2)$$

reflect a closed structure of the tube. Having regard for the relationship $p(s)=\mathrm{i}\omega\rho\phi(s)$, establish the acoustical impedance representation at the source,

$$Z=\frac{p(0)}{AV}=\mathrm{i}\frac{\rho c}{2A}\cot\frac{kl}{2}.$$

what are the travelling wave resolutions of the velocity and pressure?

5. Suppose that a small amplitude disturbance or "earthquake" imparts to the bottom of an incompressible fluid, with mean depth H, the profile $y=H+\varepsilon\cos k(x-ct)$, $\varepsilon\ll H$. Assuming the inviscid fluid executes an irrotational motion and neglecting surface tension, derive the appertaining surface wave amplitude

$$\frac{\varepsilon kc^2}{kc^2\cosh kH-g\sinh kH}$$

and comment on the circumstance wherein the velocity, c, of earthquake propagation equals $\{(g/k)\tanh kH\}^{1/2}$, the speed of free surface waves on the fluid.

6. If water flows with a velocity which is proportional to the distance from its bottom (at the depth H) and V denotes the surface magnitude thereof, prove that the velocity of surface wave propagation in the stream direction, v^*, is specified through the relation

$$(v^* - V)^2 - V(v^* - V)\frac{v^2}{gH} - v^2 = 0,$$

where v is the wave velocity appropriate to still water.

7. When a surface tension, T, is operative the boundary conditions for two-dimensional disturbances involving small displacements $y = \zeta(x, t)$ from a state of rest (wherein $y = 0$), are

$$\frac{\partial \Phi}{\partial y} = \frac{\partial \zeta}{\partial t} \quad \text{and} \quad T\frac{\partial^2 \zeta}{\partial x^2} = \rho\frac{\partial \Phi}{\partial t} + \rho g \zeta, \quad \text{at } y = 0$$

if $\Phi(x, y, t)$ represents the velocity potential. Apply these conditions to obtain a phase velocity characterization of straight-crested surface waves with length λ, namely

$$v^2 = \frac{g\lambda}{2\pi}\left[1 + \frac{4\pi^2 T}{\rho g \lambda^2}\right]\tanh\frac{2\pi H}{\lambda},$$

on the assumption that the fluid has a rigid bottom at the depth H; and describe the manner of its variation with λ in the cases H finite, H infinite.

8. Suppose that two inviscid fluids, having the densities ρ_1 and ρ_2 ($\rho_1 < \rho_2$), occupy the half spaces $y > \zeta(x, t)$, $y < \zeta(x, t)$, respectively, where y is the upward vertical direction. Show that, if Φ_1 and Φ_2 designate the corresponding velocity potentials, the conditions

$$\frac{\partial \Phi_1}{\partial y} = \frac{\partial \zeta}{\partial t} = \frac{\partial \Phi_2}{\partial y} \quad \text{and} \quad T\frac{\partial^2 \zeta}{\partial x^2} = \rho_2\frac{\partial \Phi_2}{\partial t} - \rho_1\frac{\partial \Phi_1}{\partial t} + (\rho_2 - \rho_1)g\zeta \text{ at } y = 0$$

apply to small disturbances $\zeta(x, t)$ of the common boundary, where T is the surface tension between the fluids. Hence show that straight-crested waves can be propagated in the boundary with a speed v and a wave length $2\pi/k$ if

$$v^2 k(\rho_1 + \rho_2) = (\rho_2 - \rho_1)g + k^2 T. \qquad (*)$$

Demonstrate the existence of two possible wave lengths for a given wave speed v in excess of the minimum value

$$v_{\min}^2 = \frac{2}{\rho_1 + \rho_2}[gT(\rho_2 - \rho_1)]^{1/2}.$$

Establish the group velocity representation

$$u = \frac{v}{2} + \frac{kT}{v(\rho_1 + \rho_2)}$$

and examine its behavior for small and large magnitudes of the wave length, respectively. Assuming that a group of waves travel along the common boundary at the speed u, with $u^4 \gg gT/(\rho_2 - \rho_1)$, verify that the wave length of the group corresponds to one of the magnitudes

$$\frac{8\pi u^2}{g}\,\frac{\rho_1+\rho_2}{\rho_2-\rho_1} \quad \text{or} \quad \frac{9\pi T}{2u^2(\rho_1+\rho_2)}.$$

Deduce the extension of (*),

$$k\rho_1(v - V_1)^2 + k\rho_2(v - V_2)^2 = (\rho_2 - \rho_1)g + k^2T,$$

when the fluids are in (parallel) uniform streaming motion with the horizontal velocities V_1, V_2 and the line of each wave crest is perpendicular to the direction of the streams.

9. Two fluids of densities ρ_1 and ρ_2, and of depths h_1 and h_2, are bounded above and below by horizontal rigid planes, the first fluid being uppermost. Prove that the velocity of straight-crested waves at their common surface is given by the equation

$$v^2\left(\rho_1 \coth\frac{2\pi h_1}{\lambda} + \rho_2 \coth\frac{2\pi h_2}{\lambda}\right) = \frac{2\pi T}{\lambda} + \frac{g\lambda}{2\pi}(\rho_2 - \rho_1)$$

where T is the tension and λ the wave length.

10. A liquid layer, of density ρ_1 and thickness h, is situated above an unlimited depth liquid of density ρ_2 ($>\rho_1$). Show that there are two possible values of the phase velocity for progressive waves on the free surface, given any magnitude of the wave length $2\pi/k$, namely

$$v^2 = \frac{g}{k} \quad \text{or} \quad \frac{\rho_2 - \rho_1}{\rho_2 \coth kh + \rho_1}\cdot\frac{g}{k};$$

and find the corresponding ratios of the surface and interface wave amplitude.

Note. For additional problems, see Problems IV on p. 493.

§48. *Scattering Matrices*

The previously examined instances of wave scattering on an infinite string (attributable to mass loading or extended alteration of density) have in common the feature that only one of the distinct wave trains may be labelled as incident or incoming. A significant change of viewpoint and description in these and other one-dimensional configurations is brought about if we allow for a pair of independently assigned and oppositely directed incident waves. Let us imagine an infinite string with uniform outer portions $|x| > L$ and some initially unspecified inhomogeneity in the central section $|x| < L$, which is the basis for scattering; and designate the coordinate factors of the simply periodic incident waves by

$$s^{\text{inc}}(x) = \begin{cases} A_1\,\mathrm{e}^{ikx}, & x < -L \\ A_2\,\mathrm{e}^{-ikx}, & x > L \end{cases} \tag{48.1}$$

while those which correspond to waves that travel outwards from the scattering region are represented by

$$s^{\text{out}}(x) = \begin{cases} B_1\,\mathrm{e}^{-ikx}, & x < -L \\ B_2\,\mathrm{e}^{ikx}, & x > L \end{cases}. \tag{48.2}$$

If the underlying equation for the string displacements is linear, we can superpose contributions to the outgoing waves from each one of the incident pair and write, accordingly,

$$\begin{aligned} B_1 &= S_{11}A_1 + S_{12}A_2 \\ B_2 &= S_{21}A_1 + S_{22}A_2 \end{aligned} \tag{48.3}$$

or

$$B = \begin{pmatrix} B_1 \\ B_2 \end{pmatrix} = S\begin{pmatrix} A_1 \\ A_2 \end{pmatrix} = SA \tag{48.4}$$

in a matrix notation; here the respective amplitudes of the incoming and outgoing waves appear as the elements of two-component column vectors A, B which are coupled by a 2×2 scattering matrix S with the representation

$$S = \begin{pmatrix} S_{11} & S_{12} \\ S_{21} & S_{22} \end{pmatrix}. \tag{48.5}$$

Evidently, the elements of this matrix are independent of the aforesaid amplitudes and the time, their explicit forms being wholly determined by the nature of the scattering region. The diagonal elements S_{11}, S_{22} specify reflection coefficients (on the left and right, respectively), since they relate incoming and outgoing wave amplitudes in common parts of the string, whereas the off-diagonal elements S_{12}, S_{21} correspond to transmission coefficients, for they connect the outgoing and incoming wave amplitudes on opposite sides of the scattering region.

The structure of the matrix embodies both general and specific features, with the former characteristic of an entire class of scatterers (such as the dissipationless variety), while the latter are linked to individual prototypes. We shall establish a first general property of the scattering matrix by exploiting the fact that complex-conjugation of a wave function, followed by time-reversal, merely induces a change in the sense of propagation. Thus,

$$A\,\mathrm{e}^{\mathrm{i}(kx-\omega t)} \quad \text{and} \quad A^*\,\mathrm{e}^{-\mathrm{i}(kx-\omega t)}$$

are equivalent insofar as their real parts specify a single wave function with sense of progression along the positive x-direction; and, following the replacement of t by $-t$ in the second form, the resultant version

$$A^*\,\mathrm{e}^{-\mathrm{i}(kx+\omega t)}$$

manifestly determines another real wave function which contrasts with the previous one as regards the sense of propagation only. It can thus be concluded that a common scattering matrix associates the amplitudes of the respective wave groups

$$(A_1, A_2), \quad (B_1, B_2)$$

and

$$(B_1^*, B_2^*), \quad (A_1^*, A_2^*)$$

where the first pair refer to incoming waves and the second to outgoing waves.

From the concomitant relations

$$B = SA$$

and

$$A^* = SB^*$$

we infer that

$$A^* = SS^*A^*$$

and thence, since the amplitudes A or A^* are arbitrary, that

$$SS^* = S^*S = I \tag{48.6}$$

if I denote the unit matrix.

Another general property of the S matrix is linked with the circumstance of energy conservation and may be established on the basis of the equality between the time-average energy fluxes directed towards and away from the scattering region; the latter equality assumes the form, in terms of the component wave amplitudes,

$$|B_1|^2 + |B_2|^2 = |A_1|^2 + |A_2|^2$$
$$\tilde{B}B^* = \tilde{A}A^* \tag{48.7}$$

where the tilde denotes a transpose (or row) vector. On eliminating B in favor of A the relation (48.7) becomes

$$\widetilde{SA}S^*A^* - \tilde{A}A^* = \tilde{A}(\tilde{S}S^* - I)A^* = 0$$

and thus

$$\tilde{S}S^* = I. \tag{48.8}$$

The conclusion reached through joint reference to (48.6) and (48.8) is that the scattering matrix of a loss-less configuration has a symmetric and unitary character, viz.

$$S = \tilde{S} \quad [S_{12} = S_{21}] \tag{48.9}$$
$$S\tilde{S}^* = I, \tag{48.10}$$

which is expressed by a trio of constituent equations

$$|S_{11}|^2 + |S_{12}|^2 = 1,$$
$$|S_{12}|^2 + |S_{22}|^2 = 1, \tag{48.11}$$
$$S_{11}S_{12}^* + S_{12}S_{22}^* = 0$$

involving elements thereof. If we write

$$S_{11} = r_{11}\,e^{i\varphi_{11}}, \qquad S_{12} = r_{12}\,e^{i\varphi_{12}}, \qquad S_{22} = r_{22}\,e^{i\varphi_{22}} \tag{48.12}$$

and avail ourselves of the equations (48.11), the subsequent relationships

$$r_{11}^2 + r_{12}^2 = 1, \qquad r_{12}^2 + r_{22}^2 = 1, \qquad r_{11}r_{12}\,e^{i(\varphi_{11}-\varphi_{12})} + r_{12}r_{22}\,e^{(\varphi_{12}-\varphi_{22})} = 0 \tag{48.13}$$

allow for a reduction in the number of independent parameters $r_{11}, r_{12}, r_{22}, \varphi_{11}, \varphi_{22}$; the connections (48.13) are satisfied, in particular, when

$$r_{11} = r_{22} = |r|, \qquad r_{12} = \sqrt{(1-|r|^2)}$$
$$\varphi_{11} = \varphi_1, \qquad \varphi_{22} = \varphi_2, \qquad \varphi_{12} = \tfrac{1}{2}(\varphi_1 + \varphi_2 - \pi)$$

and the resulting three parameter representation of the scattering matrix elements,

$$S_{11} = |r|\,e^{i\varphi_1}, \qquad S_{12} = S_{21} = -i\sqrt{(1-|r|^2)}\,e^{i(\varphi_1+\varphi_2)/2}, \qquad S_{22} = |r|\,e^{i\varphi_2}, \tag{48.14}$$

evidently features a single reflection factor $|r|$ along with a pair of phases φ_1, φ_2 which pertain to reflection on the left and right.

To evaluate the latter in the case of a string with a directly attached load (of mass M) at the origin we impose on the displacement factors

$$s_-(x) = A_1\,e^{ikx} + B_1\,e^{-ikx}, \quad x < 0$$
$$s_+(x) = A_2\,e^{-ikx} + B_2\,e^{ikx}, \quad x > 0$$

the conditions

$$s_-(0) = s_+(0)$$
$$\left.\frac{ds_+}{dx}\right|_{x=0+} - \left.\frac{ds_-}{dx}\right|_{x=0-} = -\frac{M\omega^2}{P}\,s_\mp(0)$$

appropriate for the load point. When the forthcoming linear equations in A_1, A_2, B_1 and B_2 are solved for the last named pair an identification of the S matrix elements is immediate, yielding

$$S_{11} = S_{22} = \frac{\mathrm{i}\dfrac{Mk}{2\rho}}{1-\mathrm{i}\dfrac{Mk}{2\rho}}, \qquad S_{12} = S_{21} = \frac{1}{1-\mathrm{i}\dfrac{Mk}{2\rho}}. \tag{48.15}$$

A complete symmetry of the scattering which originates with incidence on either side of the load point is bespoken in the separate equality of both the diagonal and off-diagonal elements of S. Along with the reduction in number of distinct matrix elements from three to two, there is a corresponding decrease in the number of parameters necessary for their representation; thus, only the reflection factor $|r|$ and a single phase φ suffice, and these are jointly expressible in terms of one parameter, viz.

$$|r| = \sin\vartheta, \qquad \varphi = \vartheta + \frac{\pi}{2}$$

where

$$\tan\vartheta = \frac{Mk}{2\rho}, \quad 0 < \vartheta < \pi/2. \tag{48.16}$$

It appears, finally, that the scattering matrix for a point load can be displayed in the form

$$S = \begin{pmatrix} \sin\vartheta\;e^{\mathrm{i}(\vartheta+\pi/2)} & \cos\vartheta\;e^{\mathrm{i}\vartheta} \\ \cos\vartheta\;e^{\mathrm{i}\vartheta} & \sin\vartheta\;e^{\mathrm{i}(\vartheta+\pi/2)} \end{pmatrix}. \tag{48.17}$$

Naturally, the conclusions which follow from the above scheme of analysis encompass those drawn earlier in the circumstance of a lone incoming wave.

It is sometimes convenient to introduce a matrix R that couples the amplitudes of two oppositely directed wave components on the same side of a scattering region with those on the other side, namely

$$\begin{pmatrix} B_2 \\ A_2 \end{pmatrix} = R\begin{pmatrix} A_1 \\ B_1 \end{pmatrix}; \tag{48.18}$$

and a simple calculation confirms both the representation

$$R = \begin{pmatrix} R_{11} & R_{12} \\ R_{21} & R_{22} \end{pmatrix} = \begin{pmatrix} \dfrac{S_{12}^2 - S_{11}S_{22}}{S_{12}} & \dfrac{S_{22}}{S_{12}} \\ -\dfrac{S_{11}}{S_{12}} & \dfrac{1}{S_{12}} \end{pmatrix} \tag{48.19}$$

and the determined property

$$\det R = R_{11}R_{22} - R_{12}R_{21} = 1. \tag{48.20}$$

In the case of a mass-loaded string, for example, the single parameter version of the R matrix which bears correspondence with that of the S matrix, (48.17), proves to be

$$R = \begin{pmatrix} 1 + \mathrm{i} \tan \vartheta & \mathrm{i} \tan \vartheta \\ -\mathrm{i} \tan \vartheta & 1 - \mathrm{i} \tan \vartheta \end{pmatrix}. \tag{48.21}$$

If distinct material or configurational attributes (e.g., density, tension) prevail on different sides of the scattering region, a renormalization of the wave amplitudes is necessary to ensure symmetry of the scattering matrix. By way of illustration, consider a junction of three virtually collinear strings; one of them, with the density ρ_1, stretches from $x = -\infty$ to $x = 0$, while the other two, with the common density ρ_2, extend from $x = 0$ to $x = +\infty$. The requirement of lengthwise equilibrium evidently implies that the tensions in the branched pair each have a magnitude, $P/2$, which equals half of that, P, in the other, and thus the respective wave numbers for small transverse displacements of the individual strings 1, 2, 3 are

$$k_1 = \omega\sqrt{(\rho_1/P)}, \quad x < 0; \qquad k_2 = k_3 = \omega\sqrt{(2\rho_2/P)}, \quad x > 0.$$

When the sets (A_1', A_2', A_3') and (B_1', B_2', B_3') refer to incoming and outgoing wave amplitudes on the various strings, equality of the energy flux directed at and away from their junction is implied by the relationship

$$k_1|A_1'|^2 + \tfrac{1}{2}k_2(|A_2'|^2 + |A_3'|^2) = k_1|B_1'|^2 + \tfrac{1}{2}k_2(|B_2'|^2 + |B_3'|^2) \tag{48.22}$$

and here, in contrast with (48.7), the material parameters occur. If the string displacement factors $s_{1,2,3}(x)$ are fitted together continuously at $x = 0$, both in value and derivative, the resulting equations provide an explicit connection

$$B' = \begin{pmatrix} B_1' \\ B_2' \\ B_3' \end{pmatrix} = S' \begin{pmatrix} A_1' \\ A_2' \\ A_3' \end{pmatrix}$$

that involves an asymmetric scattering matrix $S'(S_{ij}' \neq S_{ji}', i \neq j)$ whose elements depend on k_1, k_2. However, following the introduction of column vectors with the components

$$A = \begin{pmatrix} A_1 \\ A_2 \\ A_3 \end{pmatrix} = \begin{pmatrix} \sqrt{(k_1)}A_1' \\ \sqrt{(k_2/2)}A_2' \\ \sqrt{(k_2/2)}A_3' \end{pmatrix},$$

$$B = \begin{pmatrix} B_1 \\ B_2 \\ B_3 \end{pmatrix} = \begin{pmatrix} \sqrt{(k_1)}B_1' \\ \sqrt{(k_2/2)}B_2' \\ \sqrt{(k_2/2)}B_3' \end{pmatrix},$$

which enables us to express the energy balance in the form

$$(B_1)^2 + (B_2)^2 + (B_3)^2 = (A_1)^2 + (A_2)^2 + (A_3)^2$$

it follows that

$$B = SA$$

where S is a real and symmetric matrix,

$$S = \frac{1}{k_1 + k_2} \begin{pmatrix} k_1 - k_2 & \sqrt{(2k_1k_2)} & \sqrt{(2k_1k_2)} \\ \sqrt{(2k_1k_2)} & -k_1 & k_2 \\ \sqrt{(2k_1k_2)} & k_2 & -k_1 \end{pmatrix}$$

such that $S^2 = I$.

§49. *A Periodically Loaded String*

The matrix analysis of multi-component wave aggregates lends itself to investigating progressive motions along an infinite string which carries a periodic array of equal mass loads. In common with other instances of regular or repetitive spatial configurations, attention can be focussed on a basic cell or geometrical sub-unit, here any linear section of the span l which contains but a single load. Let

$$A_1 \,\mathrm{e}^{ikx} + B_1 \,\mathrm{e}^{-ikx}$$

specify the complete displacement factor for a pair of oppositely directed waves at a point $x = 0+$ alongside the load at $x = 0$; since these waves propagate independently and freely (without scattering) on the interval $0 < x < l$, their amplitudes alongside the adjacent load at $x = l$ are comprised in the vector

$$\begin{pmatrix} A_1 \,\mathrm{e}^{ikl} \\ B_1 \,\mathrm{e}^{-ikl} \end{pmatrix} = T \begin{pmatrix} A_1 \\ B_1 \end{pmatrix}$$

where

$$T = \begin{pmatrix} e^{ikl} & 0 \\ 0 & e^{-ikl} \end{pmatrix} \tag{49.1}$$

is a diagonal translation matrix. Accordingly, the outgoing and incoming wave amplitudes to the right of the load position $x = l$ make up the components B_2, A_2 of a vector such that

$$\begin{pmatrix} B_2 \\ A_2 \end{pmatrix} = R \begin{pmatrix} A_1\, e^{ikl} \\ B_1\, e^{-ikl} \end{pmatrix} = RT \begin{pmatrix} A_1 \\ B_1 \end{pmatrix} \tag{49.2}$$

where R is the transformation matrix (48.21) for an isolated mass scatterer. We encounter in (49.2) a connection, at points separated by the periodicity length l of the configuration, between the sets of wave amplitudes (B_2, A_2) and (A_1, B_1), with the pairings B_2, A_1 and B_1, A_2 insofar as common directions of propagation are concerned. A direct proportionality of the latter amplitudes, stipulated by the relation

$$\begin{pmatrix} B_2 \\ A_2 \end{pmatrix} = \alpha \begin{pmatrix} A_1 \\ B_1 \end{pmatrix} \tag{49.3}$$

with a constant value for α, expresses a phase condition appropriate to either sense of progressive movements along the string. When this kinematic or geometrical condition and the dynamical one embodied in (49.2) are both in force, we obtain the matrix equation

$$\alpha \begin{pmatrix} A_1 \\ B_1 \end{pmatrix} = RT \begin{pmatrix} A_1 \\ B_1 \end{pmatrix} \tag{49.4}$$

which unites a pair of homogeneous linear equations in A_1, B_1, namely

$$[(RT)_{11} - \alpha]A_1 + (RT)_{12}B_1 = 0$$

and (49.5)

$$(RT)_{21}A_1 + [(RT)_{22} - \alpha]B_1 = 0$$

where the subscripts ij ($i, j = 1, 2$) identify elements of the product matrix RT.

A quadratic equation in α,

$$\alpha^2 - [(RT)_{11} + (RT)_{22}]\alpha + (RT)_{11}(RT)_{22} - (RT)_{12}(RT)_{21} = 0, \tag{49.6}$$

is indicative of a vanishing determinant for the coefficients of A_1, B_1 in the system (49.5), and since

$$\det RT = \det R \cdot \det T = 1$$

a simpler version of (49.6) results, viz.

$$\alpha^2 - [(RT)_{11} + (RT)_{22}]\alpha + 1 = 0. \tag{49.7}$$

In compliance with the evident fact that the roots $\alpha_\pm$ of (49.7) have a unit product, we write

$$\alpha_\pm = \mathrm{e}^{\pm \mathrm{i}\kappa l} \tag{49.8}$$

and are thus led to the transcendental relationship

$$\begin{aligned} \cos \kappa l &= \tfrac{1}{2}[(RT)_{11} + (RT)_{22}] \\ &= \cos kl - \frac{Mk}{2\rho} \sin kl \end{aligned} \tag{49.9}$$

on taking account of the explicit determinations

$$(RT)_{11} = (1 + \mathrm{i} \tan \vartheta)\, \mathrm{e}^{\mathrm{i}kl}, \qquad (RT)_{22} = (1 - \mathrm{i} \tan \vartheta)\, \mathrm{e}^{-\mathrm{i}kl}$$

where

$$\tan \vartheta = Mk/2\rho. \tag{49.10}$$

For given values of k $(= \omega/c)$, l and M/ρ the relation (49.9) offers a means of specifying κ, a quantity which may be termed the "effective" wave number, in contradistinction to k, as (49.8) implies that progressive waves accumulate a phase change amounting to κl on passage over any segment of length l. It is fitting to regard (49.9) as a consistency condition which underlies the realizibility of free or self sustaining progressive wave motions on the string. Thus, let us suppose that each load executes simply periodic motions with a common and arbitrary period $2\pi/\omega$; and trace the manner whereby the individual restoring forces, which depend on the tension and relative disposition of neighboring loads, acquire those magnitudes compatible with load oscillations of the preassigned period. Since the tension remains unaltered by mass loading of the string, and adjustments of the restoring forces accompany changes in the profile or wave form, the fact that (49.9) fixes a wave parameter (κ) in terms of the frequency and other assigned configurational or material parameters (l, ρ, M) justifies our looking upon same as an expression of internal consistency for the relevant motions.

The amplitude factors which incorporate the translational property (49.8) and hold anywhere on the string admit the representation

$$s_\pm(x) = \mathrm{e}^{\pm \mathrm{i}\kappa x} \psi(x), \qquad -\infty < x < \infty \tag{49.11}$$

where the function $\psi(x)$ has the period l; and they constitute solutions of the periodic coefficient differential equation

$$\left[\frac{\mathrm{d}^2}{\mathrm{d}x^2} + k^2 + \frac{M\omega^2}{P} \sum_{n=-\infty}^{\infty} \delta(x - nl)\right] s(x) = 0 \tag{49.12}$$

with a discontinuous property for the first derivative, i.e.,

$$\left.\frac{\mathrm{d}s}{\mathrm{d}x}\right|_{nl-0}^{nl+0} = -\frac{M\omega^2}{P} s(nl), \qquad n = 0, \pm 1, \pm 2, \ldots . \tag{49.13}$$

When the right-hand side of (49.9) assumes a magnitude in excess of unity it appears that a real value for κl is ruled out; and the appertaining solutions (49.11) acquire an unbounded behavior in one or the other of the limits $x \to \pm\infty$, which renders them incompatible with the hypothesis of small amplitude vibrations everywhere. As kl varies from 0 to ∞ the contiguous intervals wherein κl is real or not follow in alternating manner; only the former (usually termed pass bounds) delineate the circumstances in which a progressive wave motion results from the superposition of oppositely directed component waves, with a common phase velocity $c = \omega/k$, that are coupled by scattering at the individual load points. Since the cosine function is periodic, specifications for κl by means of (49.9) are modulo 2π, but this ambiguity does not contravene any physical matters; we shall define later, for instance, a velocity of energy propagation equal to $c(\mathrm{d}k/\mathrm{d}\kappa) = \mathrm{d}\omega/\mathrm{d}\kappa$, which conforms with a prior representation of the group velocity and is unaffected by the lack of uniqueness regarding κ. The complex (or purely imaginary) values of κl pertain to the so-called forbidden frequency or stop bands and progressive waves with finitc amplitude over the whole length of the string are then precluded.

To deduce the particulars of the band structure, we recast (49.9) in the form

$$\cos \kappa l = \frac{\cos(kl + \vartheta)}{\cos \vartheta} \tag{49.14}$$

where ϑ is defined by (49.10), and observe that the values of kl which locate the boundaries separating pass and stop intervals are implicit in the relation

$$\cos(kl + \vartheta) = \pm \cos \vartheta \tag{49.15}$$

or

$$\cos kl - \frac{Mk}{2\rho} \sin kl = \pm 1. \tag{49.16}$$

The separate equations comprised in (49.16),

$$1 - \cos kl = 2\sin^2 \frac{kl}{2} = -\frac{Mk}{2\rho} \sin kl, \qquad 1 + \cos kl = 2\cos^2 \frac{kl}{2} = \frac{Mk}{2\rho} \sin kl,$$

yield a trio of conditions

$$\sin kl = 0 \tag{49.17}$$

$$\cot \frac{kl}{2} = \delta \frac{kl}{2}, \qquad \tan \frac{kl}{2} = -\delta \frac{kl}{2}, \qquad \delta = \frac{M}{\rho l} \tag{49.18}$$

that furnish a simple basis for determining the special values of kl. An explicit characterization

$$kl = n\pi, \qquad n = 0, 1, \ldots \tag{49.19}$$

is directly forthcoming from (49.17), while a graphical construction readily provides numerical estimates for the (positive) roots of both the equations (49.18); in specific terms, we have only to plot the curves

$$y = \cot\frac{x}{2} \quad \text{and} \quad y = -\tan\frac{x}{2}$$

(where $x = kl$) and note the respective points of intersection with the straight line

$$y = \frac{1}{2}\delta x.$$

The branches for $\cot(x/2)$ and $\cot(x + \pi/2) = -\tan(x/2)$ are confined to the complementary ranges $2n\pi < x < (2n+1)\pi$, $n = 0, 1, \ldots$ and $(2n+1)\pi < x < 2n\pi$, $n = 1, 2, \ldots$, as shown in Fig. 44 where the abscissae for the corresponding points of intersection bear the labels

$$\alpha_1, \alpha_3, \ldots$$

and

$$\beta_2, \beta_4, \ldots .$$

It is easy to verify, by utilizing (49.15), that an alternative characterization of the implicit relations (49.18) takes the form

$$kl + 2\vartheta = n\pi$$

with even/odd values of the integer n linked to the tan/cot function, respectively. If we assume that

$$kl = n\pi - 2\vartheta - \varepsilon, \quad 0 < \varepsilon \ll 1$$

substitution in (49.9) yields

$$\cos kl \doteq (-1)^n[1 - \varepsilon \tan \vartheta] < 1 \tag{49.21}$$

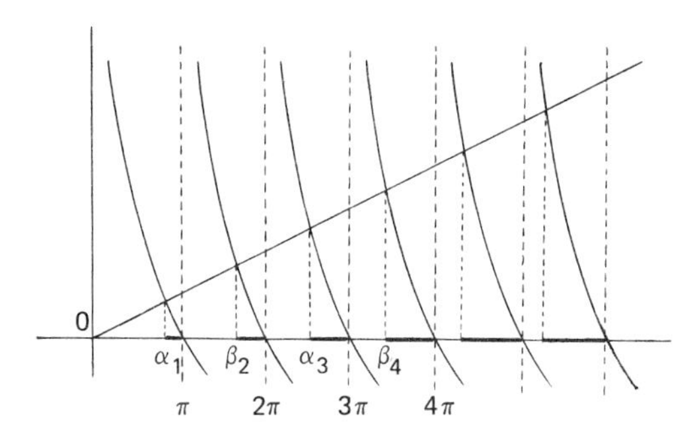

Fig. 44. Graphical resolution of a characteristic equation.

and thus the values

$$kl = \alpha_1, \beta_2, \alpha_3, \dots$$

identify a set of boundaries for the pass bands, with the others given by (49.19) since the inequality (49.21) is also consequent to assuming $kl = n\pi + \varepsilon$. Accordingly, within the nth pass band ($n = 1, 2, \dots$)

$$(n-1)\pi < kl < \begin{cases} \alpha_n, & n = 1, 3, \dots \\ \beta_n, & n = 2, 4, \dots \end{cases} \tag{49.22}$$

or

$$(n-1)\pi < kl < n\pi - 2\vartheta_n$$

and it appears, from both the steepening slope of the straight line $y = 1/2(\delta x)$ in the figure and the expression for the band breadth

$$\Delta_n = \pi - 2\vartheta_n = \pi - 2\tan^{-1}\left(\frac{M}{\rho l} \cdot kl\right), \tag{49.23}$$

that the latter diminishes as the mass of the loads (or kl) increases.

Relying upon series expansions of the trigonometric functions in the right-hand side of (49.9) we find (at the low frequency end of the first pass band)

$$\cos \kappa l \doteq 1 - \frac{1}{2}(kl)^2\left(1 + \frac{M}{\rho l}\right), \qquad kl \to 0$$

and thus

$$\kappa l \doteq kl\left(1 + \frac{M}{\rho l}\right)^{1/2} = kl\sqrt{(1+\delta)}, \quad kl \ll 1 \tag{49.24}$$

renders an approximate form of the effective wave number when the corresponding wave length ($2\pi/\kappa$) greatly exceeds the separation between loads and their sole effect is to increase the average density of the string by the factor $1 + \delta$. If we retain terms with the succeeding (and smaller) order of magnitude in these expansions, it follows that

$$(\kappa/k)^2 = 1 + \delta + \tfrac{1}{12}\delta^2(kl)^2, \quad kl \ll 1 \tag{49.25}$$

which makes evident the dispersive character of the wave motions on the periodically loaded string.

§50. *A Periodic Coefficient Differential Equation*

An alternative investigation of simply periodic wave motions along the periodically loaded string, based on the differential equation (49.12), permits both a useful contrast with and extension of the results furnished by the prior matrix analysis. In order to represent a solution of (49.12) or, for that matter, of similar equations containing other forms of the periodic coefficient, we have recourse to a Fourier expansion of the latter function. Let us replace, in the first instance, the distribution or generalized function

$$F(x) = \sum_{n=-\infty}^{\infty} \delta(x - nl) \tag{50.1}$$

with a Gaussian version composed of individually continuous functions,

$$F_\varepsilon(x) = \frac{1}{\varepsilon} \sum_{n=-\infty}^{\infty} \exp\{-\pi(x-nl)^2/\varepsilon^2\}, \tag{50.2}$$

such that

$$\operatorname*{Lim}_{\varepsilon \to 0} F_\varepsilon(x) = F(x).$$

The periodicity property

$$F_\varepsilon(x + l) = F_\varepsilon(x)$$

facilitates an individual determination of the coefficients C_m in the Fourier expansion

$$F_\varepsilon(x) = \sum_{m=-\infty}^{\infty} C_m \, \mathrm{e}^{2\pi\, imx/l}, \tag{50.3}$$

inasmuch as

$$\begin{aligned} C_m &= \frac{1}{\varepsilon l} \int_0^l \sum_{n=-\infty}^{\infty} \exp\{-\pi(x-nl)^2/\varepsilon^2\}\, \mathrm{e}^{-2\pi\, im\pi x/l}\, \mathrm{d}x \\ &= \frac{1}{\varepsilon l} \int_{-\infty}^{\infty} \mathrm{e}^{-\pi x^2/\varepsilon^2}\, \mathrm{e}^{-2\pi\, imx/l}\, \mathrm{d}x \end{aligned}$$

and therefore

$$C_m = \frac{1}{l}\, \mathrm{e}^{-\pi m^2 \varepsilon^2/l^2}, \qquad m = 0, \pm 1, \ldots . \tag{50.4}$$

When (50.2) is introduced into the differential equation

$$\left(\frac{\mathrm{d}^2}{\mathrm{d}x^2} + k^2 + \frac{M\omega^2}{P} F_\varepsilon(x)\right) s(x) = 0, \tag{50.5}$$

along with a modified type of Fourier expansion

$$s(x) = e^{i\kappa x} \sum_{n=-\infty}^{\infty} A_n \, e^{2in\pi x/l} \tag{50.6}$$

that endows $s(x)$ with the translation property

$$s(x+l) = e^{i\kappa l} s(x), \tag{50.7}$$

and terms possessing a common exponential function argument are thereafter identified, it follows that

$$A_n\left[k^2 - \left(\kappa + \frac{2n\pi}{l}\right)^2\right] + \frac{M\omega^2}{P} \sum_{m=-\infty}^{\infty} A_{n-m} C_m = 0, \qquad n = 0, \pm 1, \dots . \tag{50.8}$$

The requirement for other than a trivial solution of this homogeneous system of linear equations in the quantities A_n, expressed by a null value of their coefficient determinant, with elements

$$D_{pq} = [(\kappa l + 2p\pi)^2 - (kl)^2]\delta_{pq} - \frac{M\omega^2 l^2}{P} C_{|p-q|}$$

$$\delta_{pq} = \begin{matrix} 1, & p = q \\ 0, & p \neq q \end{matrix}, \qquad p, q = 0, \pm 1, \dots , \tag{50.9}$$

affords a means of specifying κl (mod 2π). In the limit $\varepsilon \to 0$, when (50.4) reveals that the Fourier coefficients C_m tend to a common value, $1/l$, independent of m, the characteristic equation for κ can be given a simpler, non-determinantal structure; the key to such a conversion rests on a sum property obtained from (50.6), namely

$$\sum_{n=-\infty}^{\infty} A_n = s(0),$$

which enables us to represent the quantities A_n, interconnected by (50.8), in the individual fashion

$$A_n = \frac{Mk^2 l/\rho}{(\kappa l + 2n\pi)^2 - (kl)^2} s(0), \qquad n = 0, \pm 1, \dots . \tag{50.10}$$

Substitution from (50.10) into (50.6) yields a relationship between the amplitude functions at a fixed ($x = 0$) and arbitrary ($x \neq 0$) location on the periodically loaded string, viz.

$$s(x) = \frac{M}{\rho l}(kl)^2 s(0) \sum_{n=-\infty}^{\infty} \frac{\exp\left\{i\left(\kappa + \frac{2n\pi}{l}\right)x\right\}}{(\kappa l + 2n\pi)^2 - (kl)^2}, \qquad -\infty < x < \infty \tag{50.11}$$

which, as a mark of self-consistency at the fixed point, implies that

$$1 = \frac{M}{\rho l}(kl)^2 \sum_{n=-\infty}^{\infty} \frac{1}{(\kappa l + 2n\pi)^2 - (kl)^2}. \tag{50.12}$$

To establish the equivalence of this characteristic equation for κl and the trigonometric one (49.9), an evaluation of the sum becomes necessary. Consider, therefore, the function

$$f(z) = \frac{\pi \cot \pi z}{(z-A)^2 - B^2},$$

which has simple poles at $z = n$ (integral or zero) with residue $[(n-A)^2 - B^2]^{-1}$, and also simple poles at $z = A \pm B$ with the residues

$$\frac{\pi}{2B} \cot \pi(A+B), \quad -\frac{\pi}{2B} \cot \pi(A-B),$$

respectively. On a circle of radius $|z| = n - 1/2$ the absolute value of $\cot \pi z$ remains bounded and $|f(z)|$ is arbitrarily small for sufficiently large n, whence

$$\oint f(z)\, \mathrm{d}z < \varepsilon;$$

accordingly,

$$\sum_{n=-\infty}^{\infty} \frac{1}{(n-A)^2 - B^2} + \frac{\pi}{2B}[\cot \pi(A+B) - \cot \pi(A-B)] = 0$$

or

$$\sum_{n=-\infty}^{\infty} \frac{1}{(n-A)^2 - B^2} = \frac{\pi}{B} \frac{\sin 2\pi B}{\cos 2\pi B - \cos 2\pi A}. \tag{50.13}$$

A particular version of the latter sum formula, namely

$$\sum_{n=-\infty}^{\infty} \frac{1}{(\kappa l + 2n\pi)^2 - (kl)^2} = \frac{1}{2kl} \frac{\sin kl}{\cos kl - \cos \kappa l}, \tag{50.14}$$

provides immediate confirmation of the anticipated equivalence between (49.9) and (50.12).

The expansion (50.12), which constitutes an implicit characterization for the effective wave number, lends itself to uncovering diverse aspects of the relationships between κl and kl; thus, if $kl \ll 1$ and only the $n = 0$ term be retained therein, we find

$$\kappa^2 = k^2\left(1 + \frac{M}{\rho l}\right), \quad kl \ll 1$$

in agreement with (49.24). Further, when the amount of loading is very slight ($M/\rho l \ll 1$), a pair of terms in the expansion (50.12) take precedence over all others; these are the ones with comparable and small magnitudes of the denominator, corresponding to the indicial values $n = 0, -p$. Relying exclusively on said pair, there obtains a connection between kl and κl which is simply a quadratic form in $(kl)^2$, though a non-degenerate quartic form relative to κl. The twin determinations $(kl)^2_\pm$ signal a discontinuity in the functional dependence of κl on kl and their difference

$$\Delta(kl)^2 = (kl)^2_+ - (kl)^2_- \doteq 2\frac{M}{\rho l}(kl)^2$$

points to a small range for kl,

$$\Delta(kl) = kl \cdot \frac{M}{\rho l} = \frac{Mk}{\rho},$$

adjoining the location $kl = p\pi$, wherein real values of κl are excluded. This feature is in accord with our earlier description of the band structure, inasmuch as (49.23) predicts a stop band width

$$\Delta(kl) = 2\vartheta \doteq \frac{Mk}{\rho}, \quad \vartheta \ll 1.$$

§51. *Green's Functions and the Periodically Loaded String*

A self-consistency which underlies the existence of an unforced mode of vibration on a regularly loaded string is brought out clearly through the constructive role of a Green's function corresponding to a single discontinuity in slope or load point. According to (17.16),

$$\frac{M\omega^2}{P} s(x')G(x - x') = \frac{\mathrm{i}}{2k} \cdot \frac{M\omega^2}{P} s(x')\, \mathrm{e}^{\mathrm{i}k|x-x'|}$$

represents the string displacement factor affiliated with a lone discontinuity in slope, of the magnitude $(M\omega^2/P)s(x')$, appropriate to a mass load at x'; thus, in consequence of the linearity of the motions contemplated, it follows that

$$s(x) = \frac{M\omega^2}{P} \sum_{n=-\infty}^{\infty} s(nl)G(x - nl) \tag{51.1}$$

species a displacement profile for a periodic array of equal loads at $x = nl$, $n = 0, \pm 1, \ldots$. Let us envisage a correlated pattern of load displacement factors given by

$$s(nl) = A\, \mathrm{e}^{\mathrm{i}\kappa nl}, \quad n = 0, \pm 1, \ldots \tag{51.2}$$

with the implied relationship

$$s(ml) = \exp\{i\kappa(m-n)l\}s(nl)$$

that involves an as yet unspecified parameter κ. The necessity for reproducing, via (51.1), any postulated set of load displacements is expressed by the equations

$$s(ml) = \frac{M\omega^2}{P} \sum_{n=-\infty}^{\infty} s(nl)G(ml-nl), \quad m = 0, \pm 1, \ldots \tag{51.3}$$

and with the choice (51.2), in particular, these take the form

$$e^{i\kappa ml} = \frac{M\omega^2}{P} \sum_{n=-\infty}^{\infty} e^{i\kappa nl} G(ml-nl), \quad m = 0, \pm 1, \ldots. \tag{51.4}$$

To substantiate the evident inference that the latter must determine κ a transformation of the related series (51.1), namely

$$\mathfrak{G}(x) = \sum_{n=-\infty}^{\infty} e^{i\kappa nl} G(x-nl), \tag{51.5}$$

is indicated; however, for the purpose of dealing with a properly convergent expansion, we consider at the outset

$$\mathfrak{G}_\varepsilon(x) = \sum_{n=-\infty}^{\infty} e^{i\kappa nl} G_\varepsilon(x-nl) \tag{51.6}$$

where G_ε satisfies the differential equation

$$\left(\frac{d^2}{dx^2} + k^2\right) G_\varepsilon(x-nl) = -\frac{1}{\varepsilon} \exp\{-\pi(x-nl)^2/\varepsilon^2\} \tag{51.7}$$

whose inhomogeneous member is a (temporary) substitute for $\delta(x-nl)$. Let

$$\sum_{n=-\infty}^{\infty} G_\varepsilon(x-nl) = \sum_{m=-\infty}^{\infty} B_m\, e^{2im\pi x/l} \tag{51.8}$$

and observe that the outcome of applying the differential operator $d^2/dx^2 + k^2$ thereto yields, after (50.2), (50.3), and (50.4) are utilized

$$-\frac{1}{\varepsilon} \sum_{n=-\infty}^{\infty} \exp\{-\pi(x-nl)^2/\varepsilon^2\} = -\frac{1}{l} \sum_{m=-\infty}^{\infty} \exp\{-\pi m^2\varepsilon^2/l^2\}\, e^{2im\pi x/l}$$

$$= -\sum_{m=-\infty}^{\infty} B_m \left[\left(\frac{2m\pi}{l}\right)^2 - k^2\right] e^{2im\pi x/l},$$

whence

$$\begin{aligned} B_m &= \frac{1}{l}\int_0^l \sum_{n=-\infty}^{\infty} G_\varepsilon(x-nl)\,\mathrm{e}^{-2\,\mathrm{i}m\pi x/l}\,\mathrm{d}x \\ &= \frac{1}{l}\int_{-\infty}^{\infty} G_\varepsilon(x)\,\mathrm{e}^{-2\,\mathrm{i}m\pi x/l}\,\mathrm{d}x = \frac{1}{l}\,\frac{\exp\{-(2m\pi x/l)^2\varepsilon^2/4\pi\}}{\left(\dfrac{2m\pi}{l}\right)^2 - k^2}. \end{aligned} \tag{51.9}$$

The latter relation makes it apparent that if we write

$$\begin{aligned} \mathfrak{G}_\varepsilon(x) &= \mathrm{e}^{\mathrm{i}\kappa x}\sum_{n=-\infty}^{\infty} \mathrm{e}^{\mathrm{i}\kappa(nl-x)}G_\varepsilon(x-nl) \\ &= \mathrm{e}^{\mathrm{i}\kappa x}\sum_{m=-\infty}^{\infty} C_m\,\mathrm{e}^{2\,\mathrm{i}m\pi x/l}, \end{aligned} \tag{51.10}$$

the desired Fourier expansion coefficients

$$\begin{aligned} C_m &= \frac{1}{l}\int_0^l \mathfrak{G}_\varepsilon(x)\,\mathrm{e}^{-\mathrm{i}(\kappa+2m\pi/l)x}\,\mathrm{d}x \\ &= \frac{1}{l}\int_0^l \sum_{n=-\infty}^{\infty} \mathrm{e}^{\mathrm{i}\kappa nl}G_\varepsilon(x-nl)\,\mathrm{e}^{-\mathrm{i}(\kappa+2m\pi/l)x}\,\mathrm{d}x \\ &= \frac{1}{l}\int_{-\infty}^{\infty} G_\varepsilon(x)\,\mathrm{e}^{-\mathrm{i}(\kappa+2m\pi/l)x}\,\mathrm{d}x \end{aligned}$$

possess the values

$$C_m = \frac{\exp\left\{-\left(\kappa+\dfrac{2m\pi}{l}\right)^2\varepsilon^2\Big/4\pi\right\}}{l\left[\left(\kappa+\dfrac{2m\pi}{l}\right)^2 + k^2\right]}, \quad m = 0, \pm 1, \ldots. \tag{51.11}$$

After a passage to the limit $\varepsilon \to 0$ the convergent series representation

$$\mathfrak{G}(x) = l\sum_{n=-\infty}^{\infty} \frac{\exp\{\mathrm{i}(\kappa+2n\pi/l)x\}}{(\kappa l+2n\pi)^2-(kl)^2} \tag{51.12}$$

obtains, and we deduce that the compatibility relations (51.4) are equivalent to the characteristic equation (50.12) for κ.

The expression (51.1), which constitutes a particular integral of the differential equation (49.12), can also be viewed as direct proportionality between the displacement factor at an arbitrary point and a Green's function with source coordinate at the origin, that is

$$s(x) = \frac{M\omega^2}{P}\mathfrak{G}(x, 0; \kappa, k)s(0) \tag{51.13}$$

where

$$\begin{aligned}\mathfrak{G}(x-x';\kappa,k)&=\frac{\mathrm{i}}{2k}\sum_{n=-\infty}^{\infty}\mathrm{e}^{\mathrm{i}\kappa nl}\exp\{\mathrm{i}k|x-x'-nl|\}\\&=l\sum_{n=-\infty}^{\infty}\frac{\exp\{\mathrm{i}(\kappa+2n\pi/l)(x-x')\}}{(\kappa l+2n\pi)^2-(kl)^2};\end{aligned}\tag{51.14}$$

the absence herein of an incident wave or forcing term, such as appears in the analogous representation (17.16) pertinent to scattering by a single load, reflects our current preoccupation with a free or self-sustaining mode of oscillation for the regularly loaded string. It is noteworthy that the second of the characterizations (51.14) involves everywhere regular and differentiable functions (of x or x'), whereas each term of the first has a discontinuous derivative (at $x=x'+nl$ or $x'=x-nl$); a singular behavior of the term with index n in the latter development, which the condition

$$(\kappa\pm k)l=2n\pi,\quad n \text{ integral}$$

suggests, is excluded by the restriction imposed on the joint magnitudes of κ, k in any specific problem.

Among the ready consequences of (51.14) are the translation properties

$$\begin{aligned}\mathfrak{G}(x+l,x';\kappa,k)&=\mathrm{e}^{\mathrm{i}\kappa l}\mathfrak{G}(x,x';\kappa,k)\\\mathfrak{G}(x,x'+l;\kappa,k)&=\mathrm{e}^{-\mathrm{i}\kappa l}\mathfrak{G}(x,x';\kappa,k)\end{aligned}\tag{51.15}$$

which lend support to the designation of $\mathfrak{G}$ as a Green's function with the modulation length l, and the rule for coordinate exchange,

$$\mathfrak{G}(x',x;-\kappa,k)=\mathfrak{G}(x,x';\kappa,k),\tag{51.16}$$

that contrasts with the corresponding symmetry of G.

To secure differential relations descriptive of the modulated Green's function we employ the operator $(\mathrm{d}^2/\mathrm{d}x^2)+k^2$ in (51.14) and thereby arrive at the results

$$\begin{aligned}&\left(\frac{\mathrm{d}^2}{\mathrm{d}x^2}+k^2\right)\mathfrak{G}(x,x';\kappa,k)\\&\qquad=-\sum_{n=-\infty}^{\infty}\delta(x-x'-nl)\,\mathrm{e}^{\mathrm{i}\kappa nl}=-\frac{1}{l}\sum_{n=-\infty}^{\infty}\exp\left\{\mathrm{i}\left(\kappa+\frac{2n\pi}{l}\right)(x-x')\right\}\end{aligned}\tag{51.17}$$

which show that the sources of $\mathfrak{G}$ form a regularly phased and spaced sequence. Fixing attention on the interval $(n-\frac{1}{2})l<x,x'<(n+\frac{1}{2})l$, where the same differential equation characterizes $\mathfrak{G}$ and G, viz.

$$\left(\frac{\mathrm{d}^2}{\mathrm{d}x^2}+k^2\right)\mathfrak{G}(x,x';-\kappa,k)=-\delta(x-x'),$$

and utilizing the equation

$$\left(\frac{d^2}{dx^2}+k^2\right)s(x)=-\frac{M\omega^2}{P}s(nl)\delta(x-nl)$$

which takes account of a load at the midpoint we find, after multiplication with the respective functions $s(x)$, $\mathfrak{G}(x, x'; -\kappa, k)$, subtraction and integration over said interval, that

$$s(x')-\frac{M\omega^2}{P}s(nl)\mathfrak{G}(nl, x'; -\kappa, k)$$

$$=\left[s(x)\frac{d}{dx}\mathfrak{G}(x, x'; -\kappa, k)-\mathfrak{G}(x, x'; -\kappa, k)\frac{ds}{dx}\right]\Bigg|_{x=\left(n-\frac{1}{2}\right)l}^{x=\left(n+\frac{1}{2}\right)l}$$

If $s(x)$ is altered by the factor $e^{i\kappa l}$ following a translation of the magnitude l, as for $\mathfrak{G}$, then $s(d\mathfrak{G}/dx)$ and $\mathfrak{G}(ds/dx)$ have identical values at the endpoints of the integration range, and

$$s(x)=\frac{M\omega^2}{P}\mathfrak{G}(nl, x; -\kappa, k)s(nl) \qquad \left(n-\frac{1}{2}\right)l<x<\left(n+\frac{1}{2}\right)l$$

$$=\frac{M\omega^2}{P}\mathfrak{G}(x, nl; \kappa, k)s(nl);$$

furthermore, on employing the translation properties of the factors on the right-hand side, there obtains for any x,

$$s(x)=\frac{M\omega^2}{P}\mathfrak{G}(x, 0; \kappa, k)s(0),$$

which is just the representation (51.13). Thus, the usefulness of the modulated Green's function for purposes of generating like patterns of excitation in periodic arrangements is demonstrated.

The Fourier expansion (51.14) of $\mathfrak{G}(x, x'; \kappa, k)$ is explicitly summable, i.e.,

$$l\sum_{n=-\infty}^{\infty}\frac{\exp\{i(\kappa+2n\pi/l)(x-x')\}}{(\kappa l+2n\pi)^2-(kl)^2}=\frac{i}{2k}\left[\frac{e^{ik(x-x')}}{1-e^{i(k-\kappa)l}}-\frac{e^{-ik(x-x')}}{1-e^{i(k+\kappa)l}}\right],$$
$$0\leq x-x'\leq l \tag{51.18}$$

and the versions suited to other ranges of the coordinates x, x' are obtained by continuation in accordance with (51.15). It may be noted that the above result emerges after rewriting the first of the representations (51.14) in the form

$$\frac{\mathrm{i}}{2k}\,\mathrm{e}^{\mathrm{i}k(x-x')}\sum_{n=-\infty}^{0}\mathrm{e}^{\mathrm{i}(\kappa-k)nl}+\frac{\mathrm{i}}{2k}\,\mathrm{e}^{-\mathrm{i}k(x-x')}\sum_{n=1}^{\infty}\mathrm{e}^{\mathrm{i}(\kappa+k)nl}$$

and utilizing the geometric sum formula

$$\sum_{n=0}^{\infty}\mathrm{e}^{\mathrm{i}n\alpha}=\frac{1}{1-\mathrm{e}^{\mathrm{i}\alpha}},\qquad \alpha\neq 2m\pi. \tag{51.19}$$

Application of the same formula yields, when $[x/l]$ denotes the integral part of x/l,

$$\sum_{n=-\infty}^{\infty}\exp\{\mathrm{i}k|x-nl|+\mathrm{i}\kappa nl\}=\mathrm{e}^{\mathrm{i}kx}\sum_{n=-\infty}^{[x/l]}\mathrm{e}^{-\mathrm{i}(k-\kappa)nl}+\mathrm{e}^{-\mathrm{i}kx}\sum_{n=[x/l]+1}^{\infty}\mathrm{e}^{\mathrm{i}(k+\kappa)nl}$$

$$=\mathrm{i}\,\mathrm{e}^{\mathrm{i}\kappa l[x/l]}\,\frac{\mathrm{e}^{\mathrm{i}\kappa l}\sin k\left\{x-\left[\frac{x}{l}\right]l\right\}-\sin k\left\{x-\left[\frac{x}{l}\right]l-l\right\}}{\cos\kappa l-\cos kl}$$

and hence (51.13) becomes

$$s(x)=-\frac{Mk}{2\rho}\,s(0)\,\mathrm{e}^{\mathrm{i}\kappa x}\Psi(x) \tag{51.20}$$

where

$$\Psi(x)=\mathrm{e}^{-\mathrm{i}\kappa\{x-[x/l]l\}}\,\frac{\mathrm{e}^{\mathrm{i}\kappa l}\sin k\left\{x-\left[\frac{x}{l}\right]l\right\}-\sin k\left\{x-\left[\frac{x}{l}\right]l-l\right\}}{\cos\kappa l-\cos kl}$$

has the period l; if we set $x=0$ in (51.20) the trigonometric form (49.9) of the equation which specifies κ is forthcoming.

§52. *Selective Reflection*

A continuing significance for the basic relation (49.9) which determines conditions favorable to (κ real), or discriminating against (κ imaginary), the progress of a wave along a periodically loaded infinite string, appears in other circumstances; specifically, its wider role is revealed in connection with wave reflection at a one-sided periodic array of loads on an unbounded string. If loads are affixed at the origin and at the locations $x=l, 2l, \ldots, nl, \ldots$ the dynamical equation for small amplitude transverse motions has the form

$$\left[\rho+M\sum_{m=0}^{\infty}\delta(x-ml)\right]\frac{\partial^2 y}{\partial t^2}=P\frac{\partial^2 y}{\partial x^2} \tag{52.1}$$

and when a time-periodic state prevails, wherein

$$y(x,t)=\operatorname{Re}[s(x)\,e^{-i\omega t}] \tag{52.2}$$

the coordinate factor $s(x)$ appropriate to an incoming or primary wave from the left ($x<0$) is rendered by

$$\begin{aligned} s(x) &= A\,e^{ikx}+\frac{M\omega^2}{P}\sum_{m=0}^{\infty} s(ml)G(x,ml) \\ &= A\,e^{ikx}+\frac{iMk}{2\rho}\sum_{m=0}^{\infty} s(ml)\,e^{ik|x-ml|}, \qquad -\infty<x<\infty. \end{aligned} \tag{52.3}$$

Giving attention, initially, to the latter representation on the interval $nl<x<(n+1)l$ of the loaded range, where

$$x=(n+\xi)l, \quad 0<\xi<1 \tag{52.4}$$

affords a parametric coordinate representation, and introducing the correlation

$$s(ml)=e^{i\kappa ml}s(0), \quad m\ge 0 \tag{52.5}$$

it follows that

$$\begin{aligned} s(x) &= A\,e^{ikx}+\frac{iMk}{2\rho}s(0)\left[e^{ikx}\sum_{m=0}^{n} e^{i(\kappa-k)ml}+e^{-ikx}\sum_{m=n+1}^{\infty} e^{i(\kappa+k)ml}\right] \\ &= A\,e^{ikx}+\frac{iMk}{2\rho}s(0)\left[e^{ikx}\left(\frac{1-e^{i(\kappa-k)(n+1)l}}{1-e^{i(\kappa-k)l}}\right)+e^{-ikx}\left(\frac{e^{i(\kappa+k)(n+1)l}}{1-e^{i(\kappa+k)l}}\right)\right] \\ &= \left[A+i\frac{Mk}{2\rho}\frac{s(0)}{1-e^{i(\kappa-k)l}}\right]e^{ikx}+i\frac{Mk}{2\rho}s(0)\,e^{i\kappa x}\left[\frac{e^{i(\kappa+k)(1-\xi)l}}{1-e^{i(\kappa+k)l}}-\frac{e^{i(\kappa-k)(1-\xi)l}}{1-e^{i(\kappa-k)l}}\right]. \end{aligned} \tag{52.6}$$

A different behavior is evidenced by the components in (52.6), featuring one or the other of the factors $e^{ikx}, e^{i\kappa x}$, where respective values at locations spaced a distance l apart stand in the distinct ratios $e^{ikl}, e^{i\kappa l}$; the former makes no reference to the parametric variable ξ and thus retains a uniform validity in $x>0$, while the latter may be continued beyond the original range if n or x is correspondingly altered and the span for ξ is preserved. Since the character of the first component is only suited to the unloaded part of the string ($x<0$), we must arange for its removal from (52.6), thereby obtaining a primary wave extinction relation

$$A+i\frac{Mk}{2\rho}\frac{s(0)}{1-e^{i(\kappa-k)l}}=0. \tag{52.7}$$

In accordance with the residual expression for $s(x)$ on the loaded half of the string,

$$s(x)=\frac{\mathrm{i}Mk}{2\rho}s(0)\,\mathrm{e}^{\mathrm{i}\kappa x}\left\{\frac{\mathrm{e}^{\mathrm{i}(\kappa+k)(1-\xi)l}}{1-\mathrm{e}^{\mathrm{i}(\kappa+k)l}}-\frac{\mathrm{e}^{\mathrm{i}(\kappa-k)(1-\xi)l}}{1-\mathrm{e}^{\mathrm{i}(\kappa-k)l}}\right\},\quad x>0 \tag{52.8}$$

it appears that continuity of the displacement factor at the origin $(x, \xi=0)$ is assured if

$$1=\frac{\mathrm{i}Mk}{2\rho}\left\{\frac{\mathrm{e}^{\mathrm{i}(\kappa+k)l}}{1-\mathrm{e}^{\mathrm{i}(\kappa+k)l}}-\frac{\mathrm{e}^{\mathrm{i}(\kappa-k)l}}{1-\mathrm{e}^{\mathrm{i}(\kappa-k)l}}\right\}=-\frac{Mk}{2\rho}\frac{\sin kl}{\cos\kappa l-\cos kl},$$

which constitutes the familiar relation (49.9) for determining the parameter κ.

Utilizing once more the generally applicable representation (52.3) and focussing attention on the unloaded half of the string $(x<0)$, we deduce therefrom, with use of the correlation (52.5), that

$$s(x)=A\,\mathrm{e}^{\mathrm{i}kx}+\frac{\mathrm{i}Mk}{2\rho}s(0)\frac{\mathrm{e}^{-\mathrm{i}kx}}{1-\mathrm{e}^{\mathrm{i}(\kappa+k)l}},\quad x\le 0 \tag{52.9}$$

whence

$$s(0)=\frac{A}{1-\dfrac{\mathrm{i}Mk/2\rho}{1-\mathrm{e}^{\mathrm{i}(\kappa+k)l}}}. \tag{52.10}$$

The knowledge of $s(0)$ is a prerequisite to explicit comparison of the incoming and outgoing wave factors in $x<0$ and their amplitude ratio at $x=0$, or the reflection coefficient, which follows from (52.9) has the form

$$r=\frac{\mathrm{i}Mk}{2\rho}\frac{s(0)}{A}\frac{1}{1-\mathrm{e}^{\mathrm{i}(\kappa+k)l}}. \tag{52.11}$$

When the expression for $s(0)/A$ furnished by the extinction relation (52.7) is substituted in (52.11) we find

$$r=-\frac{1-\mathrm{e}^{\mathrm{i}(\kappa-k)l}}{1-\mathrm{e}^{\mathrm{i}(\kappa+k)l}}=\frac{\sin(k-\kappa)l/2}{\sin(k+\kappa)l/2}\,\mathrm{e}^{-\mathrm{i}kl} \tag{52.12}$$

and the same result obtains after employing the determination (52.10) for $s(0)/A$ along with the equation

$$\cos\kappa l=\cos kl-\frac{Mk}{2\rho}\sin kl \tag{52.13}$$

to rewrite the consequent version of (52.11), viz.

$$r=\frac{\mathrm{i}Mk/2\rho}{1-\mathrm{i}\dfrac{Mk}{2\rho}-\mathrm{e}^{\mathrm{i}(k+\kappa)l}}.$$

It is apparent that the reflection coefficient (52.12) remains unaffected by the lack of uniqueness inherent in the totality of equivalent specifications $\kappa l + 2n\pi$, n integral. If the wave length $\lambda = 2\pi/k$ for the incoming wave is much larger than the load spacing, and thus $kl \ll 1$, the approximation (49.24), namely

$$\kappa l \doteq kl\left(1+\frac{M}{\rho l}\right)^{1/2}$$

becomes available and there follows, with the concomitant simplification of (52.12), an estimate for the reflection coefficient

$$r \doteq \frac{k-\kappa}{k+\kappa} = \frac{\sqrt{\rho} - \left(\rho + \dfrac{M}{l}\right)^{1/2}}{\sqrt{\rho} + \left(\rho + \dfrac{M}{l}\right)^{1/2}}$$

that manifests the loading in terms of a uniform increment, M/l, to the string density. More generally, if the right-hand side of (52.13) takes a magnitude in excess of unity, so that κ is complex or purely imaginary, the reflection coefficient (52.12) assumes unit absolute value; in other words, when the values of κl pertain to any of the stop-bands of an unbounded periodic configuration there is total reflection of a wave incident on the latter from the exterior. The reflection is only partial if the conditions dictate that κl is real, as befits any of the pass bands.

If we are, at the outset, prepared to adopt the correlation (52.5) between load displacements, along with the fundamental relation (52.13) which fixes the effective wave number κ, a straightforward calculation of the reflection coefficient can be based on the representations

$$s(x) = A\,\mathrm{e}^{ikx} + B\,\mathrm{e}^{-ikx}, \quad x \le 0 \tag{52.14}$$

and

$$s(ml) = s(0)\,\mathrm{e}^{i\kappa ml}, \quad m \ge 0 \tag{52.15}$$

in complementary portions of the string. The first of a pair of relations on which a determination of the unknowns B and $s(0)$ rests,

$$A + B = s(0), \tag{52.16}$$

expresses a continuous behavior for the string displacement at the point $x = 0$ which is common to the representations (52.14), (52.15). A second relation originates in the equation of motion for the load at $x = 0$, i.e.,

$$-M\omega^2 s(0) = P\frac{\mathrm{d}s}{\mathrm{d}x}\bigg|_{x=0-}^{x=0+}, \tag{52.17}$$

and, if the evident characterization

$$s(x) = s(0)\cos kx + \frac{s(l) - s(0)\cos kl}{\sin kl}\sin kx, \quad 0 \leqslant x \leqslant l$$

be employed to specify $\mathrm{d}s/\mathrm{d}x$ in the limit $x \to 0+$, there obtains from (52.17)

$$\begin{aligned} A - B &= \frac{s(0)}{\mathrm{i}\sin kl}\left[\frac{M\omega^2}{kP}\sin kl + \mathrm{e}^{\mathrm{i}\kappa l} - \cos kl\right] \\ &= \frac{\cos kl - \mathrm{e}^{-\mathrm{i}\kappa l}}{\mathrm{i}\sin kl}s(0) \end{aligned} \tag{52.18}$$

after making use of (52.14), (52.15), and (52.13). Solution of the linear equations (52.16) and (52.18) for B and $s(0)$ yields

$$r = \frac{B}{A} = -\frac{\mathrm{e}^{-\mathrm{i}kl} - \mathrm{e}^{-\mathrm{i}\kappa l}}{\mathrm{e}^{\mathrm{i}kl} - \mathrm{e}^{-\mathrm{i}\kappa l}} = -\frac{1 - \mathrm{e}^{\mathrm{i}(\kappa - k)l}}{1 - \mathrm{e}^{\mathrm{i}(\kappa + k)l}}$$

in agreement with (52.12).

§53. *Average Energy Flux Along the Periodically Loaded String*

Given an infinite string which carries regularly spaced and equal loads whose displacements bear phase relations appropriate to a progressive form of excitation with the wave number κ, the local representation obtained from (51.13), (51.14),

$$s(x) = \mathrm{i}\frac{Mk}{2\rho}s(0)[\mathrm{e}^{\mathrm{i}k|x|} + A\,\mathrm{e}^{\mathrm{i}kx} + B\,\mathrm{e}^{-\mathrm{i}kx}], \quad -\frac{l}{2} \leq x \leq \frac{l}{2} \tag{53.1}$$

has one component that accounts for the slope discontinuity at the interior load point ($x = 0$), while the others represent smooth wave factors that convey the presence of exterior loads at $x = \pm l, \pm 2l, \ldots$. The latter components possess complex coefficients

$$A = \frac{\mathrm{e}^{\mathrm{i}(k-\kappa)l}}{1 - \mathrm{e}^{\mathrm{i}(k-\kappa)l}}, \quad B = \frac{\mathrm{e}^{\mathrm{i}(k+\kappa)l}}{1 - \mathrm{e}^{\mathrm{i}(k+\kappa)l}} \tag{53.2}$$

such that

$$\operatorname{Re} A = \operatorname{Re} B = -\frac{1}{2}, \tag{53.3}$$

$$|A|^2 - |B|^2 = \frac{\sin kl \sin \kappa l}{(\cos \kappa l - \cos kl)^2} = \left(\frac{2}{Mk}\right)^2 \frac{\sin \kappa l}{\sin kl} \tag{53.4}$$

by virtue of (49.9).

To specify a time-average energy flux in the positive x-direction we fix our attention on the point $x = -l/2$ and utilize the expression

$$\bar{S} = \frac{\omega P}{2} \operatorname{Im}\left[s^* \frac{\mathrm{d}s}{\mathrm{d}x} \right]$$

in conjunction with (53.1); this yields, having regard for (53.3),

$$\begin{aligned}
\bar{S} &= \frac{\omega P k}{2}\left(\frac{Mk}{2\rho}\right)^2 |s(0)|^2 \operatorname{Im}\{-\mathrm{i}[\mathrm{e}^{-\mathrm{i}kl/2} + A^*\,\mathrm{e}^{\mathrm{i}kl/2} + B^*\,\mathrm{e}^{-\mathrm{i}kl/2}] \\
&\qquad\qquad \times [\mathrm{e}^{\mathrm{i}kl/2} - A\,\mathrm{e}^{-\mathrm{i}kl/2} + B\,\mathrm{e}^{\mathrm{i}kl/2}]\} \\
&= \frac{\omega P k}{2}\left(\frac{Mk}{2\rho}\right)^2 |s(0)|^2 \operatorname{Im}\{-\mathrm{i}[1 + 2\operatorname{Re} A + |B|^2 - |A|^2]\} \\
&= \frac{\omega P k}{2}\left(\frac{Mk}{2\rho}\right)^2 |s(0)|^2 \{|A|^2 - |B|^2\},
\end{aligned}$$

where the final version evidently comprises a difference of fluxes pertaining to the oppositely directed partial wave terms in (53.1), and thus

$$\bar{S} = \frac{\omega P k}{2} |s(0)|^2 \frac{\sin \kappa l}{\sin kl} \tag{53.5}$$

after reliance on (53.4). The observations that (53.5) makes no reference to the point ($x = -l/2$) chosen for evaluating $\bar{S}$, and that its reality is coupled with a like aspect of κl, are noteworthy; furthermore, the indicated energy balance relation in the problem (§52) of reflection from a half-loaded string, viz.

$$\frac{\omega P k}{2}(1 - |r|^2)|A|^2 = \frac{\omega P k}{2}|s(0)|^2 \frac{\sin \kappa l}{\sin kl},$$

whose validity is upheld by the determinations (49.9), (52.10), and (52.12), lends support to the particulars of (53.5) when a progressive wave can be sustained in the periodic configuration.

For a characterization of the time-average energy content on the periodicity interval $(-l/2, l/2)$ we require the magnitudes of

$$\bar{T} = \frac{1}{4}\rho\omega^2 \int_{-l/2}^{l/2} |s(x)|^2\,\mathrm{d}x + \frac{1}{4}M\omega^2 |s(0)|^2, \tag{53.6}$$

and

$$\bar{V} = \frac{1}{4}P \int_{-l/2}^{l/2} \left| \frac{\mathrm{d}s}{\mathrm{d}x} \right|^2 \mathrm{d}x, \tag{53.7}$$

where the first, or kinetic, part involves a supplemental contribution due to the mass load at the origin. The substitution herein of the representation (53.1) leads to the expressions

$$\bar{T} = \frac{\rho\omega^2}{4}\left(\frac{Mk}{2\rho}\right)^2 |s(0)|^2\left[l(|A|^2+|B|^2)+\mathrm{Re}\left\{\frac{1}{\mathrm{i}k}(1-\mathrm{e}^{-\mathrm{i}kl})(A+B+A^*B+AB^*)\right\}\right] + \frac{1}{4}M\omega^2|s(0)|^2,$$

and

$$\bar{V} = \frac{Pk^2}{4}\left(\frac{Mk}{2\rho}\right)^2 |s(0)|^2\left[l(|A|^2+|B|^2)-\mathrm{Re}\left\{\frac{1}{\mathrm{i}k}(1-\mathrm{e}^{-\mathrm{i}kl})(A+B+A^*B+AB^*)\right\}\right],$$

whose equivalence is established with the help of the result

$$\mathrm{Re}\left\{\frac{1}{\mathrm{i}k}(1-\mathrm{e}^{-\mathrm{i}kl})(A+B+A^*B+AB^*)\right\} = \frac{\sin kl}{k}\frac{1}{\cos \kappa l - \cos kl} = -\frac{2\rho}{Mk^2}$$

inasmuch as

$$A+B = -\left(1+\mathrm{i}\frac{2\rho}{Mk}\right).$$

Thus

$$\begin{aligned}\bar{T} = \bar{V} &= \frac{\rho\omega^2}{4}\left(\frac{Mk}{2\rho}\right)^2 |s(0)|^2[l(|A|^2+|B|^2)] + \frac{1}{8}M\omega^2|s(0)|^2 \\ &= \frac{1}{4}\rho\omega^2|s(0)|^2 l\left(1+\frac{Mk}{2\rho}\cot kl\right)+\frac{1}{8}M\omega^2|s(0)|^2, \end{aligned} \tag{53.8}$$

and the total energy of this or any similar portion of the string becomes

$$\bar{E} = l\frac{\rho\omega^2}{2}|s(0)|^2\left(1+\frac{M}{2\rho l}\right)+l\frac{M\omega^2 k}{4}|s(0)|^2\cot kl. \tag{53.9}$$

If we form the ratio between energy flux $\bar{S}$ and average energy content $\bar{E}/l$ on a periodicity interval the velocity which obtains,

$$u = \frac{\bar{S}}{\bar{E}/l} = c\frac{\sin \kappa l}{\left(1+\dfrac{M}{2\rho l}\right)\sin kl + \dfrac{Mk}{2\rho}\cos kl} = c\frac{\mathrm{d}k}{\mathrm{d}\kappa} = \frac{\mathrm{d}\omega}{\mathrm{d}\kappa}, \tag{53.10}$$

has a differential representation appropriate to the energy transport or group velocity and a value distinct from the common phase and energy velocities on the unloaded string. Applying the relationship (49.9) between κ and k it follows that

$$u = c\frac{\sin kl \sin \kappa l}{1+\dfrac{M}{2\rho l}\sin^2 kl - \cos kl \cos kl}$$

whence

$$\left|\frac{u}{c}\right| \leq \left|\frac{\sin kl \sin \kappa l}{1 - \cos kl \cos \kappa l}\right| \leq 1$$

if both kl and κl assume real values belonging to any pass band.

§54. *Forced and Free Motions of a Regularly Loaded String*

Our analysis has hitherto dwelt upon the nature of time-periodic displacement profiles which travel along an infinite string with regularly spaced loads, leaving aside (save for the instance of reflection on a string that features a one-sided array of loads) the particulars of excitation. In this section we consider the string motions attributable to a pair of different stimuli; specifically, these involve the application of a force, which varies harmonically in time, to one of the loads and, separately, the stipulation of an initial displacement pattern.

If a fluctuating force acts on the load at $x = 0$, and the overall equation of string motion assumes the form

$$\left[\rho + M \sum_{n=-\infty}^{\infty} \delta(x - nl)\right] \frac{\partial^2 y}{\partial t^2} - P \frac{\partial^2 y}{\partial x^2} = \mathrm{Re}\left\{F \delta(x)\, \mathrm{e}^{-\mathrm{i}\omega t}\right\}, \tag{54.1}$$

we investigate the synchronous displacements, expressed by

$$y(x, t) = \mathrm{Re}\{s(x)\, \mathrm{e}^{-\mathrm{i}\omega t}\}, \tag{54.2}$$

in terms of the ordinary differential equation which follows from (54.1), viz.

$$\frac{\mathrm{d}^2 s}{\mathrm{d}x^2} + k^2 s = -\frac{F}{P} \delta(x) - \frac{M\omega^2}{P} \sum_{n=-\infty}^{\infty} \delta(x - nl) s(nl) \tag{54.3}$$

where

$$k = \frac{\omega}{c} = \omega \surd(\rho/P).$$

The solution of (54.3), incorporating an outward wave aspect relative to each of the source (or load) points $x = 0, \pm l, \pm 2l, \ldots$ is

$$s(x) = \frac{\mathrm{i}F}{2kP} \mathrm{e}^{\mathrm{i}k|x|} + \frac{\mathrm{i}Mk}{2\rho} \sum_{n=-\infty}^{\infty} s(nl)\, \mathrm{e}^{\mathrm{i}k|x - nl|} \tag{54.4}$$

and, consistent with the fact that $s(x)$ has an even symmetry, we employ the individual characterizations

$$s(nl) = \begin{matrix} e^{i\kappa nl} s(0), & n \geq 0 \\ e^{-i\kappa nl} s(0), & n < 0 \end{matrix} \tag{54.5}$$

for the load displacements. In consequence of their joint usage, (54.4) and (54.5) yield

$$s(x) = \frac{iF}{2kP} e^{ikx} + \frac{iMk}{2\rho} s(0) \left\{ \sum_{n=0}^{\infty} e^{i\kappa nl} e^{ik|x-nl|} + \sum_{n=1}^{\infty} e^{i\kappa nl} e^{ik(x+nl)} \right\} \tag{54.6}$$

at any positive value of x; for purposes of evaluating the first sum herein, we set

$$x = (n + \xi)l, \quad 0 < \xi < 1$$

and thereafter deduce that (recalling (52.3), (52.4))

$$s(x) = \frac{iF}{2kP} e^{ikx} + \frac{iMk}{2\rho} s(0) \left[e^{ikx} \frac{1 - e^{i(\kappa-k)(n+1)l}}{1 - e^{i(\kappa-k)l}} + e^{-ikx} \frac{e^{i(\kappa+k)(n+1)l}}{1 - e^{i(\kappa+k)l}} \right]$$
$$+ \frac{iMk}{2\rho} s(0) \left[\frac{1}{1 - e^{i(\kappa+k)l}} - 1 \right] e^{ikx}, \quad x > 0$$

where the three constituent parts are in direct correspondence with those of (54.6). There follows, by rearrangement,

$$s(x) = \left[\frac{iF}{2kP} + \frac{iMk}{2\rho} s(0) \left\{ \frac{1}{1 - e^{i(\kappa+k)l}} + \frac{1}{1 - e^{i(\kappa-k)l}} - 1 \right\} \right] e^{ikx}$$
$$+ \frac{iMk}{2\rho} s(0)\, e^{i\kappa x} \left\{ \frac{e^{i(\kappa+k)(1-\xi)l}}{1 - e^{i(\kappa+k)l}} - \frac{e^{i(\kappa-k)(1-\xi)l}}{1 - e^{i(\kappa-k)l}} \right\}, \quad x > 0$$

and the necessity for eliminating the term proportional to e^{ikx}, whose structure is inappropriate on a loaded string, yields the connection

$$\frac{iF}{2kP} + \frac{iMk}{2\rho} s(0) \left\{ \frac{1}{1 - e^{i(\kappa+k)l}} + \frac{1}{1 - e^{i(\kappa-k)l}} - 1 \right\} = 0 \tag{54.7}$$

between F and $s(0)$, together with the relationship

$$s(x) = \frac{iMk}{2\rho} s(0)\, e^{i\kappa x} \left\{ \frac{e^{i(\kappa+k)(1-\xi)l}}{1 - e^{i(\kappa+k)l}} - \frac{e^{i(\kappa-k)(1-\xi)l}}{1 - e^{i(\kappa-k)l}} \right\}.$$

The latter reproduces an earlier representation [cf. (52.8)] of the properly constituted wave factor for the periodic arrangement and its continuity at $x = 0$ furnishes the equation (52.17) which determines κ. If the equation (52.17) be applied in the simplification of (54.7) we find that

$$s(0) = \frac{iF}{2kP} \frac{\sin kl}{\sin \kappa l} \tag{54.8}$$

and hence, using the expression (53.5) and taking account of the symmetry about the load subjected to the force F, the time-average energy which travels away from this load has the magnitude

$$2\bar{S} = \omega P k |s(0)|^2 \frac{\sin \kappa l}{\sin kl} = \frac{F^2}{4\rho c} \frac{\sin kl}{\sin \kappa l}. \tag{54.9}$$

The energy balance relation,

$$\frac{\partial}{\partial t}\left[\frac{1}{2}\left(\rho + M \sum_{n=-\infty}^{\infty} \delta(x - nl)\right)\left(\frac{\partial y}{\partial t}\right)^2 + \frac{1}{2}P\left(\frac{\partial y}{\partial x}\right)^2\right] + \frac{\partial}{\partial x}\left[-P\frac{\partial y}{\partial x}\frac{\partial y}{\partial t}\right] = F\delta(x)\frac{\partial y(0, t)}{\partial t} e^{-i\omega t},$$

suggests an alternative recipe for determination of the outgoing energy flux, namely, in terms of the rate at which the force does work

$$F\frac{\partial y(0, t)}{\partial t} e^{-i\omega t},$$

whose time average equals

$$\frac{\omega F}{2} \operatorname{Im} s(0) = \frac{F^2}{4\rho c} \frac{\sin kl}{\sin \kappa l}.$$

A more intricate analysis becomes necessary when the load displacements are not simply correlated and the string executes non-periodic motions, such as those consequent to assigning initial values for the displacement and velocity. Let us suppose that the load at $x = 0$ has the initial displacement A, with the concomitant string profile

$$y(x, 0) = \begin{cases} A\left(1 - \frac{|x|}{l}\right), & |x| \le l \\ 0, & |x| > l \end{cases} \tag{54.10}$$

and envisage a null velocity everywhere along its length at the same time, viz.:

$$\frac{\partial y(x, 0)}{\partial t} = 0, \quad -\infty < x < \infty. \tag{54.11}$$

Conversion of the dynamical equation of motion for the string,

$$\left[\rho + M \sum_{n=-\infty}^{\infty} \delta(x - nl)\right]\frac{\partial^2 y}{\partial t^2} = P\frac{\partial^2 y}{\partial x^2},$$

by a Laplace transform with respect to the time variable yields, on accomodating both the specific initial conditions (54.10) and (54.11), an inhomogeneous ordinary differential equation

$$\frac{d^2\bar{y}}{dx^2} - \frac{p^2}{c^2}\bar{y} = -\frac{Mp}{\rho c^2}A\delta(x) - \frac{p}{c^2}y(x, 0) + \frac{Mp^2}{\rho c^2}\sum_{n=-\infty}^{\infty} \bar{y}(nl, p)\delta(x - nl) \tag{54.12}$$

in which the general transform function of string displacement,

$$\bar{y}(x, p) = \int_0^\infty e^{-pt} y(x, t) \, dt \tag{54.13}$$

and the particular transforms, $\bar{y}(nl, p)$, pertaining to the load displacements, appear jointly.

Taking note of the relation

$$\left(\frac{d^2}{dx^2} - \frac{p^2}{c^2}\right)\left\{\frac{c}{2p} e^{-(p/c)|x-x'|}\right\} = -\delta(x - x')$$

we express a bounded solution of (54.12) in the fashion

$$\bar{y}(x, p) = Y(x, p) - \frac{Mp}{2\rho c} \sum_{n=-\infty}^{\infty} \bar{y}(nl, p) \, e^{-(p/c)|x-nl|} \tag{54.14}$$

where

$$Y(x, p) = \frac{MA}{2\rho c} e^{-(p/c)|x|} + \frac{1}{2c} \int_{-l}^{l} y(x', 0) \, e^{-(p/c)|x-x'|} \, dx' \tag{54.15}$$

constitutes a known function of x and p. After assigning the specific set of values $x = ml$, $m = 0, \pm 1, \ldots$ in (54.14) the transforms of all the load displacements, $\bar{y}(nl, p)$, are found to satisfy the linear system of equations

$$\bar{y}(ml, p) = Y(ml, p) - \sum_{n=-\infty}^{\infty} K(|m - n|l, p)\bar{y}(nl, p)$$

with

$$K(|n|l, p) = \frac{Mp}{2\rho c} e^{-(p/c)|n|l}. \tag{54.16}$$

The resolution of (54.16) and characterization of the affiliated string profile may be left aside for study elsewhere [cf. Problems II(c)].

§55. *Wave Motions in a Linear Chain*

A variant of the loaded string configuration previously studied, in which the inertia of the string itself is left out of account, has independent claims to historical significance; for the model of a massless string, with a finite or infinite number of equal and regularly spaced loads, whose planar motions are capable of explicit and detailed description, has been employed to obtain results for a homogeneous string with continuously distributed mass (through a suitable limiting process) and also to initiate the theory of wave propagation along periodic systems. If particles, each with mass M, are attached to a light string of unlimited extent at equal intervals l, the representative equation for their small transverse displacements $y_n(t)$, $n = 0, \pm 1, \ldots$, viz.:

$$\frac{d^2 y_n}{dt^2} = \alpha(y_{n+1} - 2y_n + y_{n-1}), \qquad \alpha = P/Ml, \tag{55.1}$$

is of a difference-differential type, and the intervening sections of the string are without curvature. The same functional relation applies to the longitudinal motions in a linear chain of identical masses which are elastically or spring coupled to their nearest neighbors; when the (restoring) force between any pair of masses is a constant multiple, h, of their relative separation the appropriate value of the parameter α in (55.1) equals h/M. In the latter circumstances our attention is focussed on the displacements of the masses themselves (and not at the intervening points) and a progressive, time-periodic, excitation features a common phase difference between the simultaneous displacements of any two neighboring masses. Thus, if

$$y_n(t) = nl + \xi_n(t) \tag{55.2}$$

specifies the instantaneous position of a mass with the rest coordinate nl, a progressive 'wave' is characterized by the sequence of displacements

$$\xi_n(t) = s(n)\, e^{-i\omega t} = A\, e^{i(\kappa nl - \omega t)}, \qquad n = 0, \pm 1, \ldots \tag{55.3}$$

with the direct correlation

$$s(n+1) = e^{i\kappa l} s(n). \tag{55.4}$$

Since the totality of values expressed by $\kappa l + 2p\pi$, p integral, are equivalent insofar as (55.3) is concerned, the compatibility requirement for a displacement pattern of such type, which finds expression in a relationship between ω and κ, must necessarily manifest a periodic dependence on κl.

A difference equation for the amplitude factor $s(n)$ obtains after the substitution in (55.1) of the representations (55.2), (55.3), namely

$$\omega^2 s(n) + \alpha[s(n+1) - 2s(n) + s(n-1)] = 0 \tag{55.5}$$

and this is consistent with (55.4) if

$$\omega^2 = 4\alpha \sin^2(\kappa l/2)$$

or

$$\omega = \omega_0 \left| \sin \frac{\kappa l}{2} \right|, \qquad \omega_0 = 2\sqrt{(\alpha)}. \tag{55.6}$$

It may be observed that the dispersion relation (55.6) follows directly from the one which takes account of string inertia, i.e.,

$$\cos \kappa l = \cos \frac{\omega l}{c} - \frac{1}{2\alpha} \left[\frac{\omega c}{l} \sin \frac{\omega l}{c} \right], \tag{55.7}$$

and which is based on the counterpart of the difference equation (55.5),

$$\omega^2 s(n) + \alpha \frac{\omega l/c}{\sin \omega l/c}\left[s(n+1) - 2\cos\frac{\omega l}{c}\, s(n) + s(n-1)\right] = 0, \qquad (55.8)$$

if we let ρ and therefore $(1/c)\sqrt{(\rho/P)} \to 0$. The salient features of (55.6) include an evident periodicity relative to κl and, on the supposition that the latter has a real magnitude, the existence of an upper limit, ω_0, for the frequency spectrum. If $\omega > \omega_0$ a complex nature of κl is implied by (55.6), viz.

$$\kappa l = \mathrm{i}\mu + \pi \qquad (55.9)$$

where

$$\mu = \mathrm{Im}\, \kappa l = 2 \cosh^{-1}\left(\frac{\omega}{\omega_0}\right),$$

and the corresponding displacements (55.3),

$$\xi_n(t) = A\, \mathrm{e}^{-\mu n}\, \mathrm{e}^{-\mathrm{i}\omega t + \mathrm{i}n\pi},$$

either fall away to nothing or else increase without limit at the distant load sites ($|n| \to \infty$). Accordingly, propagation along the linear chain is restricted to a single, low-frequency, pass band $0 < \omega < \omega_0$, in contrast with the recurring bands that typify wave motion along a string whose uniform density is altered by regularly spaced point loads.

Choosing a particular range $-\pi < \kappa l < \pi$ of the periodic function in (55.6), the dispersion relation acquires a graphical form through the plot of ω vs $1/\lambda = \kappa/2\pi$, as shown in Fig. 45; another mark of dispersive behavior for the load movements appears in the wave length dependence of the phase velocity

$$v = \frac{\omega}{\kappa} = v_\infty \frac{\sin \pi l/\lambda}{\pi l/\lambda}, \qquad (55.10)$$

whose long wave limit has the magnitude

$$v_\infty = \lim_{\lambda \to \infty} v = \omega_0 l/2.$$

To examine an instance of forced vibrations in the linear chain and appreciate the simplicity of its analysis, as compared with that (§54) for a dense string, let us adopt the coupled dynamical equations

$$\frac{\mathrm{d}^2 \xi_n}{\mathrm{d}t^2} - \frac{1}{4}\omega_0^2(\xi_{n+1} - 2\xi_n + \xi_{n-1}) = \frac{F}{M}\delta_{n0}\, \mathrm{e}^{-\mathrm{i}\omega t} \qquad (55.11)$$

which involve the direct action of a harmonically varying force $F\,\mathrm{e}^{-\mathrm{i}\omega t}$ on a single mass ($n = 0$). The synchronous mass displacements

$$\xi_n(t) = s(n)\, \mathrm{e}^{-\mathrm{i}\omega t}, \qquad (55.12)$$

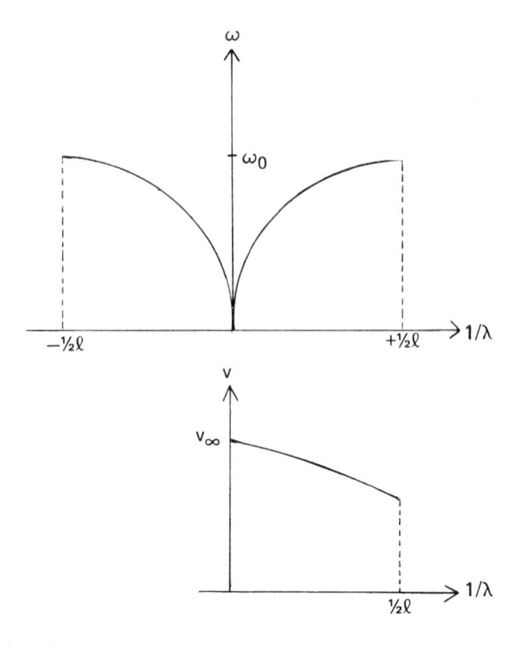

Fig. 45. Dispersion curves for a linear chain of masses.

with coefficients $s(n)$ that obey the correlations

$$s(n) = \begin{array}{ll} e^{i\kappa nl} s(0), & n \geq 1 \\ e^{-i\kappa nl} s(0), & n \leq -1 \end{array} \tag{55.13}$$

indicative of their stimulus at the center of the chain, are suited to the case at hand; and it is apparent that substitution of the forms (55.12), (55.13) into any of the homogeneous representatives ($n \neq 0$) of the equations (55.11) merely reproduce the basic dispersion relation (55.6). On employing the latter, together with the inhomogeneous representative ($n = 0$) of the equations (55.11), we arrive at the determinations

$$\begin{aligned} s(0) &= \frac{2\,iF}{M\omega_0^2 \sin \kappa l} = \frac{iF}{M\omega \sqrt{(\omega_0^2 - \omega^2)}}, \qquad 0 < \omega < \omega_0, \\ \xi_n(t) &= \frac{iF}{M\omega \sqrt{(\omega_0^2 - \omega^2)}}\, e^{i(|n|\kappa l - \omega t)}, \qquad n = 0, \pm 1, \dots \end{aligned} \tag{55.14}$$

for the resultant motion of elements in the chain.

The time average amount of energy outwardly directed from the mass subject to the action of the impressed force is given by

$$\bar{S} = \overline{\left(F\, e^{-i\omega t} \frac{d\xi_0}{dt} \right)} = \frac{\omega F}{2} s(0) \tag{55.15}$$

$$= \frac{F^2}{2M\sqrt{(\omega_0^2 - \omega^2)}}, \qquad 0 < \omega < \omega_0 \tag{55.16}$$

or alternately, by

$$\bar{S} = 2u\frac{\bar{E}}{l},$$

where

$$u = \frac{\mathrm{d}\omega}{\mathrm{d}\kappa} = \frac{\omega_0 l}{2}\cos\frac{\kappa l}{2} = \frac{l}{2}\sqrt{(\omega_0^2 - \omega^2)} \tag{55.17}$$

defines the group velocity and

$$\bar{E} = \tfrac{1}{2}M\omega^2|s(0)|^2$$

represents the average energy content for each subsection of the chain containing a single mass.

If $\omega > \omega_0$, on the other hand,

$$\sin\kappa l = -\mathrm{i}\sinh\mu = -2\,\mathrm{i}\frac{\omega\sqrt{(\omega^2 - \omega_0^2)}}{\omega_0^2}$$

and the expressions which replace (55.14) are

$$\begin{aligned} s(0) &= \frac{-F}{M\omega\sqrt{(\omega^2 - \omega_0^2)}}, \qquad \omega > 0, \\ \xi_n(t) &= \frac{-F}{M\omega\sqrt{(\omega^2 - \omega_0^2)}}\,\mathrm{e}^{-\mu|n| - \mathrm{i}\omega t + \mathrm{i}|n|\pi}; \end{aligned} \tag{55.18}$$

from the fact that Im $s(0)$ vanishes we deduce, via (55.15), a null value of the outward energy flux $\bar{S}$, in conformity with the localized nature of the displacement pattern (55.18). There is a finite average value of the whole excitation energy of the chain in these circumstances, namely

$$\begin{aligned} 2\sum_{n=-\infty}^{\infty}\frac{M}{2}\overline{\left(\frac{\mathrm{d}\xi_n}{\mathrm{d}t}\right)^2} &= \frac{1}{2}\frac{F^2}{M^2(\omega^2 - \omega_0^2)}\sum_{n=-\infty}^{\infty}\mathrm{e}^{-2\mu|n|} \\ &= \frac{F^2(2\omega^2 - \omega_0^2)}{4M\omega(\omega^2 - \omega_0^2)^{3/2}}, \end{aligned}$$

since the kinetic and potential contributions are equal.

The analysis of transient or initial value problems for an infinite chain is also served by a preliminary conversion of the equations of motion into a recurrence that typifies cylinder functions. Schrödinger (1914), who first suggested this approach, adopts the dimensionless independent variable $\tau = \omega_0 t$ and the dependent variables

$$z_{2n}(\tau) = \frac{\mathrm{d}y_n}{\mathrm{d}t} = \omega_0\frac{\mathrm{d}y_n}{\mathrm{d}\tau} \tag{55.19}$$

$$z_{2n+1}(\tau) = \tfrac{1}{2}\omega_0(y_n - y_{n+1}), \qquad \alpha = \tfrac{1}{4}\omega_0^2 \tag{55.20}$$

with the interrelation

$$\frac{dz_{2n+1}}{d\tau} = \frac{1}{2}(z_{2n} - z_{2n+2});$$

and observes that, since the equations of motion (55.1) take the form

$$\frac{dz_{2n}}{d\tau} = \frac{1}{2}(z_{2n-1} - z_{2n+1})$$

in terms of the new variables, the recurrence

$$\frac{dz_n}{d\tau} = \frac{1}{2}(z_{n-1} - z_{n+1}) \tag{55.21}$$

applies for all integers n. The latter possesses a bounded solution

$$z_n = \text{const. } J_{n-m}(\tau), \quad m \text{ integral}$$

where $J_{n-m} = (-1)^{n-m} J_{m-n}$ denotes a Bessel function, and its general solution, answering to the set of initial values $z_m(0)$,

$$z_n(\tau) = \sum_{m=-\infty}^{\infty} z_m(0) J_{n-m}(\tau), \tag{55.22}$$

is predicated on the fact that $J_p(0) = \delta_{p0} \begin{array}{l} = 1, \quad p = 0 \\ = 0, \quad p \neq 0 \end{array}.$

By way of utilizing this formulation, let us consider an initial state of rest in the chain, with a given displacement for one mass; from the specifications

$$y_0 = A, \quad y_n = 0, \quad n \neq 0; \quad \frac{dy_n}{dt} = 0, \quad \text{all } n, \quad t = 0$$

we find, according to (55.19), (55.20),

$$z_{\pm 1}(0) = \pm \frac{\omega_0}{2} A, \quad z_{2n+1}(0) = 0, \quad n \neq 0, -1; \qquad z_{2n}(0) = 0$$

and thence (55.22) yields, if $n = 2p$,

$$z_{2p}(\tau) = \omega_0 \frac{dy_p}{d\tau} = \omega_0 A \frac{d}{d\tau} J_{2p}(\tau)$$

or

$$y_p(t) = A J_{2p}(\omega_0 t).$$

§56. *Green's Functions for the Linear Chain and Applications*

Let us define the function $G(n, m; t)$, whose arguments include the discrete (or integer) pair n, m and a continuous variable t, to be a solution of the equations

$$\frac{d^2}{dt^2}G(n, m; t) - \frac{\omega_0^2}{4}[G(n+1, m; t) - 2G(n, m; t) + G(n-1, m; t)] = \delta_{nm}\delta(t),$$

$$-\infty < n, m < \infty, \quad t > 0 \tag{56.1}$$

which reflect a stimulus (force) on the site m of a linear chain at the initial instant of time $t = 0$. Introducing the Fourier integral representation

$$G(n, m; t) = \frac{1}{2\pi}\int_{-\pi}^{\pi} \hat{G}(\vartheta; t)\, \mathrm{e}^{\mathrm{i}(n-m)\vartheta}\, \mathrm{d}\vartheta \tag{56.2}$$

and noting the affiliated version of (56.1),

$$\frac{1}{2\pi}\int_{-\pi}^{\pi}\left[\frac{\mathrm{d}^2\hat{G}}{\mathrm{d}t^2} + \omega_0^2 \sin^2\frac{\vartheta}{2}\hat{G}\right]\mathrm{e}^{\mathrm{i}(n-m)\vartheta}\, \mathrm{d}\vartheta = \delta_{nm}\delta(t),$$

we may conclude that $\hat{G}$ obeys the differential equation

$$\frac{\mathrm{d}^2\hat{G}}{\mathrm{d}t^2} + \omega_0^2 \sin^2\frac{\vartheta}{2}\hat{G} = \delta(t) \tag{56.3}$$

and is, subject to the proviso that $\hat{G} = 0$, $t < 0$, uniquely given by

$$\hat{G}(\vartheta; t) = H(t)\frac{\sin\left(\omega_0 t \sin\dfrac{\vartheta}{2}\right)}{\omega_0 \sin\dfrac{\vartheta}{2}} \tag{56.4}$$

where H designates the Heaviside function (§20). The original, or causal, Green's function of the linear chain thus acquires the characterizations

$$\begin{aligned} G(n, m; t) &= \frac{H(t)}{2\pi\omega_0}\int_{-\pi}^{\pi}\frac{\sin\left(\omega_0 t \sin\dfrac{\vartheta}{2}\right)}{\sin\dfrac{\vartheta}{2}}\mathrm{e}^{\mathrm{i}(n-m)\vartheta}\, \mathrm{d}\vartheta \\ &= \frac{2H(t)}{\pi\omega_0}\int_0^{\pi/2}\frac{\sin(\omega_0 t \sin\psi)}{\sin\psi}\cos 2(n-m)\psi\, \mathrm{d}\psi \\ &= \frac{H(t)}{\omega_0}\int_0^{\omega_0 t} J_{2(n-m)}(\zeta)\, \mathrm{d}\zeta, \end{aligned} \tag{56.5}$$

where the latter follows readily from the fact that the function

$$I(x) = \int_0^{\pi/2} \frac{\sin(x \sin\psi)}{\sin\psi} \cos 2p\psi \, \mathrm{d}\psi, \quad I(0) = 0$$

has a first derivative

$$\frac{\mathrm{d}I}{\mathrm{d}x} = \frac{\pi}{2} J_{2p}(x)$$

expressible in terms of the Bessel function with even order $2p$. It is a consequence, moreover, of the formula

$$\int_0^{\infty} J_\nu(x)\,\mathrm{d}x = 1, \quad \nu > -1$$

and the property

$$J_{2(n-m)}(x) = J_{-2(n-m)}(x),$$

that

$$G(n, m; t) \to \frac{1}{\omega_0}, \quad t \to \infty. \tag{56.6}$$

Utilizing the Green's function we readily construct a particular solution of the equations of motion (55.11) for a chain in which one mass (with the location index $m = 0$) is subject to a harmonically varying force $F\,\mathrm{e}^{-\mathrm{i}\omega t}$, viz.

$$\xi_n(t) = \frac{F}{M} \int_{-\infty}^{t} G(n, 0; t - t')\,\mathrm{e}^{-\mathrm{i}\omega t'}\,\mathrm{d}t'; \tag{56.7}$$

the mass displacements (56.7) have null values in the limit $t \to -\infty$, signifying a state of rest in the chain, if $\operatorname{Im}\omega = -\varepsilon$. Substitution from (56.5) into (56.7) and integration by parts yields

$$\begin{aligned}
\xi_n(t) &= \frac{\mathrm{i}F}{M\omega} \int_{-\infty}^{t} J_{2n}(\omega_0(t - t'))\,\mathrm{e}^{-\mathrm{i}\omega t'}\,\mathrm{d}t' \\
&= \frac{\mathrm{i}F}{M\omega}\,\mathrm{e}^{-\mathrm{i}\omega t} \int_0^{\infty} J_{2n}(\omega_0\tau)\,\mathrm{e}^{\mathrm{i}\omega\tau}\,\mathrm{d}\tau
\end{aligned} \tag{56.8}$$

whence, referring to the Bessel function integrals

$$\int_0^{\infty} J_{2n}(\omega_0\tau)\cos\omega\tau\,\mathrm{d}\tau = \begin{cases} (\omega_0^2 - \omega^2)^{-1/2} \cos\left(2n \sin^{-1}\dfrac{\omega}{\omega_0}\right), & \omega < \omega_0, \\ \\ & \omega > \omega_0, \end{cases}$$

$$\int_0^\infty J_{2n}(\omega_0\tau)\sin\omega\tau\,\mathrm{d}t = \begin{cases} (\omega_0^2-\omega^2)^{-1/2}\sin\left(2n\sin^{-1}\dfrac{\omega}{\omega_0}\right), & \omega<\omega_0 \\ (-1)^{|n|}(\omega^2-\omega_0^2)^{-1/2}\left(\dfrac{\omega}{\omega_0}-\left(\dfrac{\omega^2}{\omega_0^2}-1\right)^{1/2}\right)^{2|n|}, & \omega>\omega_0 \end{cases}$$

we arrive at specifications of the mass displacements $\xi_n(t)$ which are fully equivalent with the prior forms (55.14), (55.18).

Let us next indicate a different usage for a source or Green's function of the linear chain, having reference to simply-periodic motions therein; in particular, we shall assume that conditions are favorable to propagation along an infinite chain with equal masses M, namely

$$-\pi<\varphi\overset{\text{def}}{=}\kappa l<\pi, \tag{56.9}$$

and consider the displacement pattern, in the presence of a substitutional 'impurity' with mass M' at the lattice site ml, which includes a primary (incident) wave component

$$\xi_n^{inc}(t)=\mathrm{e}^{in\varphi-i\omega t}=s^{inc}(n)\,\mathrm{e}^{-i\omega t} \tag{56.10}$$

directed from left to right if $\varphi>0$. The requisite dynamical equations are

$$\begin{aligned} M\frac{\mathrm{d}^2\xi_n}{\mathrm{d}t^2}&=h[\xi_{n+1}-2\xi_n+\xi_{n-1}], \qquad n\neq m, \\ M'\frac{\mathrm{d}^2\xi_m}{\mathrm{d}t^2}&=h[\xi_{m+1}-2\xi_m+\xi_{m-1}], \end{aligned} \tag{56.11}$$

with the coefficient inequality

$$4\frac{h}{M}=\omega_0^2>\omega^2$$

dictated by (56.9).

If

$$S(n)=\mathrm{e}^{in\varphi}+s(n) \tag{56.12}$$

designates the composite (incident and scattered) amplitude factor of the load displacements, it is a simple matter to verify that the equations of motion (56.11) are collectively expressed in the form

$$M\omega^2 s(n)+h[s(n+1)-2s(n)+s(n-1)]=(M-M')\omega^2[\mathrm{e}^{in\varphi}+s(n)]\delta_{nm} \tag{56.13}$$

which makes clear the connection between a mass irregularity in the chain and the appearance of a secondary (or scattered) wave. A solution of (56.13) is evidently proportional to that of the companion difference equation

$$M\omega^2 G(n, m) + h[G(n+1), m) - 2G(n, m) + G(n-1, m)] = \delta_{nm}, \tag{56.14}$$

viz.

$$s(n) = 4h(1-\delta)\left(\frac{\omega}{\omega_0}\right)^2 G(n, m)[e^{im\varphi} + s(m)] \tag{56.15}$$

where

$$\delta = M'/M. \tag{56.16}$$

Consequent to the special choice $n = m$ in (56.15) a characterization of $s(m)$ follows, namely

$$s(m) = \frac{\gamma}{1-\gamma} e^{im\varphi}$$

and thence

$$s(n) = \frac{\gamma}{1-\gamma} \frac{G(n, m)}{G(m, m)} e^{im\varphi}, \quad n = 0, \pm 1, \ldots \tag{56.17}$$

with

$$\gamma = 4h(1-\delta)\left(\frac{\omega}{\omega_0}\right)^2 G(m, m). \tag{56.18}$$

We introduce, for purposes of determining $G(n, m)$, the integral representation

$$G(n, m) = \frac{1}{2\pi}\int_{-\pi}^{\pi} \hat{G}(\vartheta)\, e^{i(n-m)\vartheta}\, d\vartheta \tag{56.19}$$

and observe that the difference equation (56.14) is satisfied if

$$\hat{G}(\vartheta) = \frac{1}{M\omega^2 - 2h(1-\cos\vartheta)} = \frac{1}{2h}\frac{1}{\cos\vartheta - \cos\varphi} \tag{56.20}$$

where $\sin\varphi/2 = \omega/\omega_0$. Since the Green's function transform $\hat{G}(\vartheta)$ manifests a singular behavior at $\vartheta = \pm\varphi, 0 < \varphi < \pi$, it becomes necessary to supply an interpretation for the integral (56.19) or a version of same,

$$G(n, m) = \frac{1}{2\pi i h}\oint \frac{z^{n-m}\, dz}{(z - e^{i\varphi})(z - e^{-i\varphi})}, \tag{56.21}$$

whose contour is the circumference of the unit circle $|z| = 1$. A deformation or displacement of the contour away from the singular points of the integrand, $z = e^{\pm i\varphi}$, is required; by analogy with prior considerations (§17) bearing upon the Green's function for a string of uniform density, we effect the twin displacements in such manner that only the pole at $z = e^{i\varphi}$ lies inside the closed contour, so as to secure an outgoing or source behavior of the Green's function. There obtains, accordingly, a representation

$$G(n, m) = \frac{e^{i|n-m|\varphi}}{2\,ih \sin \varphi} = \frac{e^{i|n-m|\kappa l}}{2\,ih \sin \kappa l} \tag{56.22}$$

which leads, in conjunction with the formulas (56.12), (56.17), to the resultant mass amplitude factors on either side of the impurity,

$$S(n) = \begin{cases} e^{in\kappa l} + \dfrac{\gamma}{1-\gamma}\, e^{2im\kappa l}\, e^{-in\kappa l}, & n \le m \\[2ex] \dfrac{1}{1-\gamma}\, e^{in\kappa l}, & n \ge m \end{cases} \tag{56.23}$$

wherein

$$\gamma = \frac{2}{i}(1-\delta)\left(\frac{\omega}{\omega_0}\right)^2 \operatorname{cosec} \kappa l = \frac{1}{i}(1-\delta) \tan \frac{\kappa l}{2}. \tag{56.24}$$

From (56.23) we deduce the reflection and transmission coefficients

$$\begin{aligned} r_- &= \frac{s(m-1)}{s^{\text{inc}}(m-1)} = \frac{\gamma}{1-\gamma}\, e^{2i\kappa l}, \\ t_- &= \frac{S(m)}{s^{\text{inc}}(m-1)} = \frac{e^{i\kappa l}}{1-\gamma} \end{aligned} \tag{56.25}$$

appropriate to an incident wave on the left, and a similar analysis yields the pair of coefficients

$$\begin{aligned} r_+ &= \frac{s(m)}{s^{\text{inc}}(m)} = \frac{\gamma}{1-\gamma} \\ t_+ &= \frac{S(m-1)}{s^{\text{inc}}(m)} = \frac{e^{i\kappa l}}{1-\gamma} \end{aligned} \tag{56.26}$$

that correspond to a wave incident upon the impurity from the right. Hence, if A_1, A_2 designate the incoming wave amplitude factors at the locations $(m-1)l$, ml, respectively, and B_1, B_2 identify their outgoing wave counterparts at the same locations,

$$\begin{pmatrix} B_1 \\ B_2 \end{pmatrix} = \begin{pmatrix} r_- & t_+ \\ t_- & r_+ \end{pmatrix} \begin{pmatrix} A_1 \\ A_2 \end{pmatrix} \tag{56.27}$$

where the elements of the scattering matrix for the impurity are specified in (56.25), (56.26). The scattering matrix elements have unbounded values if

$\gamma = 1$ or, as follows from (56.24),

$$\tan\frac{\kappa l}{2} = \frac{\mathrm{i}}{1-\delta},$$

and this condition may be linked to the existence of a localized mode of impurity vibration that does not require an incident wave stimulus. Writing $\kappa l = \pi + \mathrm{i}\mu$ we find that

$$\coth\frac{\mu}{2} = \frac{1}{1-\delta} \quad \text{or} \quad \mu = \log\frac{2-\delta}{\delta}, \quad \delta < 1 \tag{56.28}$$

expresses the value of μ in terms of the mass ratio, $\delta = M'/M$; moreover, the characteristic frequency of the localized impurity vibration, ω^*, is given by

$$\omega^{*2} = \omega_0^2 \frac{1+\cosh\mu}{2}$$

and thus has a magnitude in excess of the cut-off frequency ω_0 for wave propagation along the regular chain.

§57. *Causality and Dispersion Relations*

Both the dispersion of waves, revealed in a frequency dependent phase velocity, and their absorption, implied by a complex form of the wave number, represent intertwined aspects of excitations in general settings. If the analytical description for such excitations rests on a linear basis and a causal aspect (or exclusive dependence on their particulars prior to the instant of observation) prevails, the aforementioned indicators have a deep-seated relationship; in essence, a knowledge of either the phase velocity or absorption at all frequencies is sufficient for determining the overall variation of the other. The explicit formulae that detail relationships of the latter type, whose structure is rooted in connection between the real and imaginary parts of functions of a complex variable, are known as dispersion relations. An account of dispersion relations, which is without reference to specific instances of excitation and thus underscores their general character, precedes our eventual use of same for the dynamical configuration involving a periodic array of equal masses.

The twin aspects of linearity (superposition) and causality are incorporated into the relation

$$F(t) = \int_{-\infty}^{\infty} K(t-t')G(t')\,\mathrm{d}t' \tag{57.1}$$

between a pair of functions $F(t)$ (the output) and $G(t)$ (the input), if the time delay function K vanishes for negative values of its argument, i.e.,

$$K(\tau) = 0, \quad \tau < 0; \tag{57.2}$$

moreover, owing to the fact that K is assigned an argument of the difference type in the convolution integral (57.1), input and response possess a temporal correlation,

$$G(t+T) \rightarrow F(t+T),$$

which expresses the shift of the latter consequent to an arbitrary translation of the support range for the whole input.

Allied with the foregoing evolutionary representation of a linear system is the relation

$$F(\omega) = K(\omega) G(\omega) \tag{57.3}$$

between Fourier transforms of the individual functions which occur in (57.1), e.g.,

$$\begin{aligned} K(\omega) &\overset{\text{def}}{=} \int_{-\infty}^{\infty} e^{i\omega\tau} K(\tau)\, d\tau \\ &= \int_{0}^{\infty} e^{i\omega\tau} K(\tau)\, d\tau \end{aligned} \tag{57.4}$$

where the latter version obtains as a result of the causal requirement (57.2). Special importance attaches to the transform $K(\omega)$, which conveys an intrinsic property of the system, and it may be concluded that, since a half-infinite range figures in the integral (57.4), the latter, or $K(\omega)$, defines a regular function of ω in the (upper) half-plane ($\text{Im}\, \omega > 0$). If we assume that $K(\omega)$ is a square integrable function on the real axis,

$$\int_{-\infty}^{\infty} |K(\omega)|^2\, d\omega < \infty, \quad (\omega \text{ real}) \tag{57.5}$$

the same attribute applies to any parallel contour in the upper half-plane, viz.,

$$\int_{-\infty}^{\infty} |K(\omega + i\delta)|^2\, d\omega < \infty, \quad (\delta > 0) \tag{57.6}$$

and there follows an asymptotic behavior of the transform,

$$\lim_{|\omega| \to \infty} K(\omega + i\delta) = 0, \quad \delta > 0. \tag{57.7}$$

Through an application of Cauchy's residue theorem [along with (57.7)] the function $K(\omega)$ acquires, in terms of its values on the real axis, a representation

$$K(\omega) = \frac{1}{2\pi\,\mathrm{i}} \int_{-\infty}^{\infty} \frac{K(\omega')}{\omega' - \omega}\,\mathrm{d}\omega', \quad \operatorname{Im}\omega > 0 \tag{57.8}$$

at any point in the upper half-plane; and for real ω the corresponding version is

$$K(\omega) = \frac{1}{\pi\,\mathrm{i}} \int_{-\infty}^{\infty} \frac{K(\omega')}{\omega' - \omega}\,\mathrm{d}\omega', \quad \operatorname{Im}\omega = 0 \tag{57.9}$$

with a principal value interpretation of the integral. According to (57.9), the real and imaginary parts of $K(\omega)$ satisfy the interrelations

$$\begin{aligned} \operatorname{Re} K(\omega) &= \frac{1}{\pi} \int_{-\infty}^{\infty} \frac{\operatorname{Im} K(\omega')}{\omega' - \omega}\,\mathrm{d}\omega', \\ \operatorname{Im} K(\omega) &= -\frac{1}{\pi} \int_{-\infty}^{\infty} \frac{\operatorname{Re} K(\omega')}{\omega' - \omega}\,\mathrm{d}\omega' \end{aligned} \tag{57.10}$$

which are variously identified as the Plemelj formulas (in the theory of functions of a complex variable), or the Kramers–Kronig dispersion relations (in the realm of physical theories). The designation of a causal transform is conferred on each representative from the class of functions $K(\omega)$ which fulfill the conditions (57.6), (57.10) and are such that

$$\begin{aligned} K(t) &= \frac{1}{2\pi} \int_{-\infty}^{\infty} \mathrm{e}^{-\mathrm{i}\omega t} K(\omega)\,\mathrm{d}\omega \\ &= 0, \quad t < 0; \end{aligned} \tag{57.11}$$

a theorem established by Titchmarsh rigorously demonstrates the equivalence of these conditions, which is to say that any one implies the others, and thus the postulate of causality, in particular, is sufficient for assuring the existence of the dispersion relations (57.10). Modified forms of the latter are indicated when the characteristic function, $K(t)$, of the linear system has a distributional sense, or the transform $K(\omega)$ ceases to be square integrable.

The pair of frequency dependent functions $v(\omega)$, $\alpha(\omega)$ which determine the coordinate behavior of the simply periodic excitation

$$y(x, t) = \mathrm{e}^{-\alpha(\omega)x} \exp\left\{-\mathrm{i}\omega\left(t - \frac{x}{v(\omega)}\right)\right\}, \tag{57.12}$$

and constitute measures of the phase velocity and absorption or attenuation coefficient, respectively, enter into the complex combination

$$K(\omega) = -\frac{\partial y/\partial x}{\partial y/\partial t} = \frac{1}{v(\omega)} + \mathrm{i}\frac{\alpha(\omega)}{\omega}. \tag{57.13}$$

Thus, if y represents a displacement function, $K(\omega)$ expresses a proportionality between the local velocity, $\partial y/\partial t$, and a dynamical magnitude, such as a component of tension or the pressure, that varies linearly with the local slope or strain, $\partial y/\partial x$. The combination (57.13) is also relevant to longitudinal motions along a chain of identical and equally spaced point masses which are elastically coupled in nearest neighbor fashion; there is no absorption in this system at frequencies less than a characteristic magnitude, ω_0, controlled by the strength of the coupling and the common value of the mass. For a progressive wave in said arrangement, with the pattern of mass displacements

$$\begin{aligned}\xi_n &= \exp\{\mathrm{i}(\kappa n l - \omega t)\}\\ &= \exp\left\{-\mathrm{i}\omega\left(t - \frac{nl}{v(\omega)}\right)\right\}\end{aligned}$$

at the locations $x = nl$, $n = 0, \pm 1, \ldots$, the relationship between frequency and effective wave number $\kappa = 2\pi/\lambda$ is [cf. (55.6)]

$$\omega^2 = \omega_0^2 \sin^2\left(\frac{\kappa l}{2}\right) \tag{57.14}$$

and the concomitant phase velocity has the representations

$$v = \frac{\omega}{\kappa} = v_\infty \frac{\sin(\kappa l/2)}{\kappa l/2} = v_\infty \frac{\sin(\pi l/\lambda)}{\pi l/\lambda} \tag{57.15}$$

where $v_\infty = \omega_0 l/2$ designates the long wave ($\lambda \to \infty$) magnitude therefor. The shortest wave length

$$\lambda_0 = 2l,$$

that is compatible with the periodic expression (57.14) for ω on the fundamental range $-\pi \leq \kappa l \leq \pi$ corresponds to the critical frequency ω_0; when $|\omega| > \omega_0$, a complex form of κl obtains, viz.

$$\kappa l = \pi + \mathrm{i}\alpha l$$

and this yields the following determinations of phase velocity and attenuation factor,

$$v = \frac{\omega l}{\pi}, \tag{57.16}$$

$$\alpha = \frac{2}{l} \cosh^{-1}\left|\frac{\omega}{\omega_0}\right|. \tag{57.17}$$

Let us rely upon the explicit characterizations

$$\frac{\alpha(\omega)}{\omega}=\begin{cases}0, & |\omega|<\omega_0\\[2ex] \dfrac{2}{\omega l}\cosh^{-1}\left|\dfrac{\omega}{\omega_0}\right|, & |\omega|>\omega_0\end{cases}\tag{57.18}$$

for Im $K(\omega)$ in (57.13) and utilize one of the dispersion relations (57.10) to calculate Re $K(\omega)$, or the reciprocal of the phase velocity. After a few transformations we find

$$\begin{aligned}\operatorname{Re} K(\omega)&=\frac{2}{\pi l}\int_{-\infty}^{\infty}\cosh^{-1}\left|\frac{\omega'}{\omega_0}\right|\frac{\mathrm{d}\omega'}{\omega'(\omega'-\omega)},\quad |\omega'|>\omega_0\\ &=\frac{2}{\pi l}\int_{\omega_0}^{\infty}\cosh^{-1}\left(\frac{\omega'}{\omega_0}\right)\left[\frac{1}{\omega'}\left(\frac{1}{\omega'-\omega}+\frac{1}{\omega'+\omega}\right)\right]\mathrm{d}\omega'\\ &=\frac{2}{\pi\omega l}\int_{\omega_0}^{\infty}\cosh^{-1}\left(\frac{\omega'}{\omega_0}\right)\left[\frac{1}{\omega'+\omega}-\frac{1}{\omega'-\omega}\right]\mathrm{d}\omega'\end{aligned}$$

and thus an integration by parts provides the formula

$$\begin{aligned}\operatorname{Re} K(\omega)&=\frac{2}{\pi\omega_0\omega l}\int_{\omega_0}^{\infty}\left\{\left(\frac{\omega'}{\omega_0}\right)^2-1\right\}^{-1/2}\log\left|\frac{\omega'+\omega}{\omega'-\omega}\right|\mathrm{d}\omega'\\ &=\frac{2}{\pi\omega l}\int_0^{\pi/2}\log\left|\frac{\delta+\sin\vartheta}{\delta-\sin\vartheta}\right|\frac{\mathrm{d}\vartheta}{\sin\vartheta}\end{aligned}\tag{57.19}$$

if $\omega'/\omega_0=1/\sin\vartheta$ and

$$\delta=\frac{\omega_0}{\omega}.\tag{57.20}$$

When $\delta>1$ the change of variable $\tau=\tan\vartheta$ brings (57.19) to the form

$$\operatorname{Re} K(\omega)=\frac{2}{\pi\omega l}I(\delta)\tag{57.21}$$

where

$$I(\delta)=\int_0^{\infty}\log\left(\frac{\delta\sqrt{(1+\tau^2)}+\tau}{\delta\sqrt{(1+\tau^2)}-\tau}\right)\frac{\mathrm{d}\tau}{\tau\sqrt{(1+\tau^2)}}.\tag{57.22}$$

and a particular value

$$\begin{aligned}I(1)&=\int_0^{\infty}\log(\sqrt{(1+\tau^2)}+\tau)^2\frac{\mathrm{d}\tau}{\tau\sqrt{(1+\tau^2)}}\\ &=2\int_0^{\infty}\log(\mathrm{e}^{\zeta})\frac{\mathrm{d}\zeta}{\sinh\zeta}=2\int_0^{\infty}\frac{\zeta\,\mathrm{d}\zeta}{\sinh\zeta}=\pi^2/2\end{aligned}\tag{57.23}$$

is noted. Differentiation of (57.22) gives

$$\frac{dI}{d\delta} = -\frac{2}{\delta^2}\int_0^\infty \frac{d\tau}{1+\left(1-\dfrac{1}{\delta^2}\right)\tau^2} = -\frac{\pi}{\delta\sqrt{(\delta^2-1)}}$$

whence

$$I(\delta) = \frac{\pi^2}{2} - \pi\int_1^\delta \frac{d\eta}{\eta\sqrt{(\eta^2-1)}} = \pi \sin^{-1}\left(\frac{1}{\delta}\right),$$

and

$$\mathrm{Re}\, K(\omega) = \frac{2}{\omega l}\sin^{-1}\left(\frac{\omega}{\omega_0}\right), \quad \omega < \omega_0.$$

On eliminating ω in favor of κ, by recourse to (57.14), it follows that

$$\mathrm{Re}\, K(\omega) = \frac{2}{\omega_0 l}\frac{\kappa l/2}{\sin\dfrac{\kappa l}{2}} = \frac{1}{v},$$

as is consistent with the phase velocity determination, (57.15), below the cutoff frequency ω_0.

If $\omega > \omega_0$ and $\delta < 1$, we have

$$\mathrm{Re}\, K(\omega) = \frac{2}{\pi\omega l} J(\varphi)$$

where

$$\delta = \cos\varphi,$$

and

$$J(\varphi) = \int_0^{\pi/2} \log\left|\frac{\cos\varphi + \cos\vartheta}{\cos\varphi - \cos\vartheta}\right| \frac{d\vartheta}{\cos\vartheta};$$

use of the expansion

$$\log\left|\frac{\cos\varphi + \cos\vartheta}{\cos\varphi - \cos\vartheta}\right| = 4\sum_{m=0}^{\infty} \frac{\cos(2m+1)\vartheta\cos(2m+1)\varphi}{2m+1}$$

yields, after termwise integration,

$$J(\varphi) = 2\pi\sum_{m=0}^{\infty}(-1)^m\frac{\cos(2m+1)\varphi}{2m+1} = \frac{\pi^2}{2}, \quad -\pi/2 < \varphi < \pi/2$$

whence

$$\mathrm{Re}\, K(\omega) = \frac{\pi}{\omega l} = \frac{1}{v}, \quad \omega > \omega_0$$

in keeping with (57.16).

Problems II(c)

1. Consider a string to which a load M is transversely and elastically coupled, with their respective (parallel) displacements $y(x, t)$ and $z(t)$ from positions of rest governed by the system of equations

$$\left(\frac{\mathrm{d}^2}{\mathrm{d}t^2}+\gamma\frac{\mathrm{d}}{\mathrm{d}t}+\omega_0^2\right)z(t)=\omega_0^2 y(0,t)$$

and

$$\rho\frac{\partial^2 y}{\partial t^2}-P\frac{\partial^2 y}{\partial x^2}=-M\omega_0^2[y(0,t)-z(t)]\delta(x),$$

if $x=0$ marks the site of attachment. Determine the elements of a scattering matrix for this configuration and indicate how those appropriate to a load which is directly affixed to the string can be recovered. Examine the particulars of energy flux on the string and affiliated measures of energy for the load if there is only a single incoming wave train.

2. Show that when a string of line density ρ and tension $P=\rho c^2$ is loaded at equal intervals l by particles of mass M, the scale factor in time-periodic displacements $s_n\, \mathrm{e}^{-i\omega t}$ of the nth particle satisfies the recurrence relation

$$-M\omega^2 s_n=\frac{kP}{\sin kl}(s_{n+1}-2\cos kl\cdot s_n+s_{n-1}),\quad k=\omega/c \tag{*}$$

or

$$s_{n+1}-2\cos\kappa l\cdot s_n+s_{n-1}=0$$

where

$$\cos\kappa l=\cos kl-\frac{Mk}{2\rho}\sin kl. \tag{**}$$

Progressive motions along the string, with an effective wave number κ are feasible if $|\cos\kappa l|<1$, as is the case for sufficiently small values of kl [cf. §49]. Let the particles be acted on by linear restoring forces, in addition to the tensile ones, by adding a term $-M\omega_0^2 s_n$ on the right-hand side of (*); find the (dispersion) relation which supplants (**) and verify that progressive disturbances are excluded at the longest wave lengths or lowest frequencies. Discuss the analogous features when the particles have elastic connections to the string.

3. Given a set $1, 2, \ldots, N$ of equidistant parallel planes each of which reflects and transmits the same fraction rA, tA of the amplitude A for a normally incident wave. A particular ensemble of waves for this configuration comprises one that is incident from the exterior on the first plane and oppositely directed pairs between neighboring planes, which are the outcome of multiple reflections; let the forward and backward going waves of the ensemble have the amplitudes $A_1, A_2, \ldots, A_N$ and $B_1, B_2, \ldots, B_N$ at the respective planes, as shown in the accompanying figure.

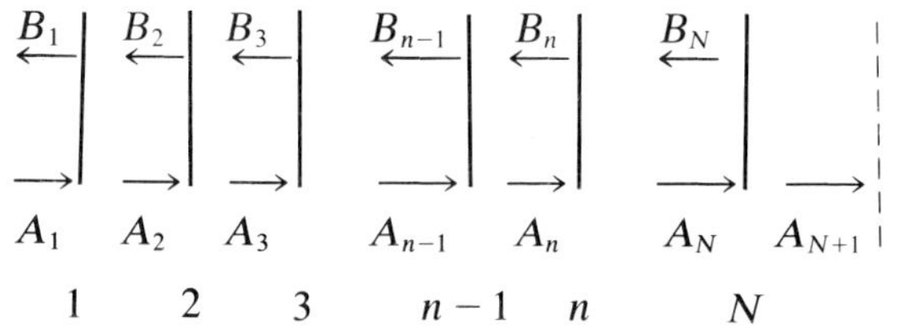

B_1 B_2 B_3 B_{n-1} B_n B_N

A_1 A_2 A_3 A_{n-1} A_n A_N A_{N+1}

1 2 3 $n-1$ n N

Use of the phase factors $\mathrm{e}^{\mp ikl}$ to shift forward and backward wave amplitudes from any plane to the preceding one at a distance l brings out the fact that waves of amplitude A_{n-1} and $B_n\,\mathrm{e}^{ikl}$ are incident on the plane $n-1$, while those of amplitude B_{n-1} and $A_n\,\mathrm{e}^{-ikl}$ are headed away; hence

$$B_{n-1} = rA_{n-1} + tB_n\,\mathrm{e}^{ikl}$$

and

$$\mathrm{e}^{ikl}A_n = rB_n\,\mathrm{e}^{ikl} + tA_{n-1}.$$

Form the corresponding connections at the plane n and then derive the common recurrence satisfied by A_n, B_n. Introducing the transmitted wave amplitude A_{N+1} at a distance l beyond the plane N, and assigning a null value to B_{N+1} for exclusion of locally incoming waves, establish on the basis of these recurrences that

$$A_{N-m} = A_{N+1}\left(\frac{P_{m+1}(\gamma)}{t\,\mathrm{e}^{ikl}} - P_m(\gamma)\right), \qquad \gamma = t\,\mathrm{e}^{ikl} + \frac{1}{t\,\mathrm{e}^{ikl}} - \frac{r^2}{t}\mathrm{e}^{ikl}$$

$$B_{N-m} = A_{N+1}\frac{rP_{m+1}(\gamma)}{t\,\mathrm{e}^{ikl}}$$

where the (Mauguin) polynomials $P_m(\gamma)$ are specified by

$$P_{m+1} + P_{m-1} = \gamma P_m, \qquad P_0 = 0, \qquad P_1 = 1$$

and possess the explicit representation

$$P_m = \frac{\zeta^m - \zeta^{-m}}{\zeta - \zeta^{-1}}, \qquad \gamma = \zeta + \frac{1}{\zeta}.$$

Show that the reflection and transmission coefficients of the aggregate of planes admit the characterizations

$$\frac{B_1}{A_1} = r\frac{P_N(\gamma)}{P_N(\gamma) - t\,\mathrm{e}^{ikl}P_{N-1}(\gamma)}, \qquad \frac{A_{N+1}}{A_1} = \frac{t\,\mathrm{e}^{ikl}}{P_N(\gamma) - t\,\mathrm{e}^{ikl}P_{N-1}(\gamma)}.$$

If the scattering of individual planes is non-dissipative and rendered by the coefficients

$$t = \cos\alpha\,\mathrm{e}^{i\beta}, \qquad r = \mathrm{i}\sin\alpha\,\mathrm{e}^{i\beta}, \qquad |r|^2 + |t|^2 = 1, \qquad \alpha, \beta \text{ real}$$

confirm that

$$\gamma = 2\frac{\cos(kl+\beta)}{\cos\alpha} \tag{*}$$

and

$$\left|\frac{B_1}{A_1}\right|^2 = \frac{P_N^2\cdot\sin^2\alpha}{P_N^2\cdot\sin^2\alpha + \cos^2\alpha}, \qquad \left|\frac{A_{N+1}}{A_1}\right|^2 = \frac{\cos^2\alpha}{P_N^2\cdot\sin^2\alpha + \cos^2\alpha},$$

whence

$$|B_1|^2 + |A_{N+1}|^2 = |A_1|^2$$

and the overall balance of energy obtains. Contrast the behavior of both the polynomials P_m and the solutions in the cases $|\gamma| > 2, < 2$. Indicate how (*) may be transformed, and thus linked, to the (dispersion) relation (49.14) by suitable identification of the parameters α, β.

4. Consider an unbounded string with a finite number N of equal and regularly spaced loads M at the locations $x = 0, l, \ldots, (N-1)l$; adopting the characterizations

$$s(x) = \begin{matrix} A\,\mathrm{e}^{ikx} + B\,\mathrm{e}^{-ikx}, & x \le 0 \\ E\,\mathrm{e}^{ikx}, & x \ge (N-1)l \end{matrix}$$

$$s(ml) = C\,\mathrm{e}^{i\kappa ml} + D\,\mathrm{e}^{-i\kappa ml}, \quad m = 0, 1, \ldots, N-1$$

which place in evidence a wave of amplitude A incident on the loaded range, and an expression which relates κ and k, show that

$$r_N = \frac{B}{A} = \frac{\mathrm{i}(\cos kl - \cos\kappa l)\sin N\kappa l\cdot\mathrm{e}^{-ikl}}{\sin kl\sin\kappa l\cos N\kappa l - \mathrm{i}(1-\cos kl\cos\kappa l)\sin N\kappa l}$$

and

$$t_N = \frac{E}{A} = \frac{\sin kl\sin\kappa l\cdot\mathrm{e}^{iNkl}}{\sin kl\sin\kappa l\cos N\kappa l - \mathrm{i}(1-\cos kl\cos\kappa l)\sin N\kappa l}$$

represent the secondary wave amplitudes (or reflection and transmission coefficients) outside scattering configuration. Confirm the energy balance relationship

$$|r_N|^2 + |t_N|^2 = 1$$

and establish that, when $N \gg 1$, the mean values of $|r_N|^2, |t_N|^2$ relative to narrow ranges of the incident wave length, $\lambda = 2\pi/k$, satisfy the equations

$$\overline{|t_N|^2} = 1 - \overline{|r_N|^2} = \left|\frac{\sin kl \sin \kappa l}{1 - \cos kl \cos \kappa l}\right| = \frac{1 - |r_\infty|^2}{1 + |r_\infty|^2}$$

where $|r_\infty|$ is the modulus of the reflection coefficient (cf. §52) for an infinite (one-sided) array of loads.

Derive, and comment upon, the results implied by the preceding if $N \to \infty$, $l, M \to 0$ and the limits $Nl \to L$, $M/l \to \hat{\rho}$ prevail.

5. Examine and discuss the particulars of reflected, internal and/or transmitted wave displacement factors, for both the finite and semi-infinite aggregates of regularly spaced loads on a string, when the inequality (cf. §49)

$$\left|\cos kl - \frac{Mk}{2l} \sin kl\right| > 1$$

obtains, and the effective wave number κ in the periodic range is complex-valued. Calculate and balance the time average energy fluxes attributable to the primary (incident) and secondary (reflected/transmitted) waves.

6. A massless cord of infinite length is stretched with tension P; on it at the points $x = l, 2l, \ldots$ are fixed particles of equal mass m: on it at the points $x = -l', -2l', \ldots$ are fixed other particles of equal mass m': the former particles, when displaced, are attracted to their positions of equilibrium by forces each equal to $m\omega_1^2$ multiplied by the displacement, the latter by forces each equal to $m'\omega_2^2$ multiplied by their displacement: at the origin is fixed a particle of mass M attracted towards the origin with a force $M\omega_3^2$ multiplied by the displacement. Prove that, c denoting $(P/m\omega)^{1/2}$, transverse waves of permanent type and period $2\pi/\omega$ will be propagated along the positive half of the cord with velocity $\omega l/\theta$ where $\sin(1/2)\theta = (\omega^2 - \omega_1^2)^{1/2}/2c$.

An infinite train of waves advances towards the origin along the positive half of the cord; show how to determine the amplitude and phase of the part of it reflected back along the same half of the cord. Show that there will be reflection even when the velocities of propagation are the same in the two halves of the cord; and prove that when in addition $l = l'$, and the natural period of M is equal to that of the waves, the amplitude of the reflected wave is the fraction $\sin(1/2)\theta$ of that of the incident wave while its phase is accelerated by the fraction $1/4 + \theta/4\pi$ of a period.

7. Consider a massless cord on which are attached an infinite series of point loads spaced at unit distance apart. The loads repeat their relations in successive tetrads: in each tetrad their masses are m_1, m_2, m_3, m_4 and transverse forces $\mu_1 s_1$, $\mu_2 s_2$, $\mu_3 s_3$, $\mu_4 s_4$ act on them, s_1, s_2, s_3, s_4 being their small transverse displacements. Prove that, if $\alpha_r - 2 = (\mu_r - m_r\omega^2)/P$, where P designates the tension of the cord, the velocity of propagation along the cord of simple wave trains of frequency $\omega/2\pi$ is of the form $4\omega/\theta$, with

$$2\cos\theta = 2 - \alpha_1\alpha_2 - \alpha_3\alpha_4 - \alpha_1\alpha_4 - \alpha_2\alpha_3 + \alpha_1\alpha_2\alpha_3\alpha_4.$$

From a study of the graph of $\cos\theta$ in terms of ω deduce that such a linear configuration may exhibit, even when there are no dissipative mechanisms, two broad "absorption bands" corresponding to ranges of frequency within which waves cannot penetrate into it.

8. A string of length $(n+1)l$ whose inertia may be neglected carries n equal loads of mass m at intervals l, the action of gravity being negligible. The tension of the string is mlc^2. Find the velocity of propagation of waves of length λ; and show that if $(l/\lambda)^4$ be neglected the wave velocity differs from that in a uniform string of the same length and total mass by the fraction $\pi^2 l^2/6\lambda^2$ of the latter velocity. Investigate the nature of forced oscillations when one end of the string is fixed and the other is given a small transverse motion varying as $\sin\omega t$, distinguishing between the cases according as ω is greater or less than $2c$.

9. Imagine that the identical members of a linear array of point masses M, regularly spaced along the x-axis, experience in the course of small longitudinal (or collinear) movements, 1) the action of nearest neighbors, with a restoring force whose coefficient equals h, and 2) elastic actions, centering on pairs of fixed points, $\ldots(P_{n-1}, Q_{n-1}), (P_n, Q_n), (P_{n+1}, Q_{n+1}), \ldots$, with the tension P [see figure]. Apply the equation of motion for the nth mass, say, to

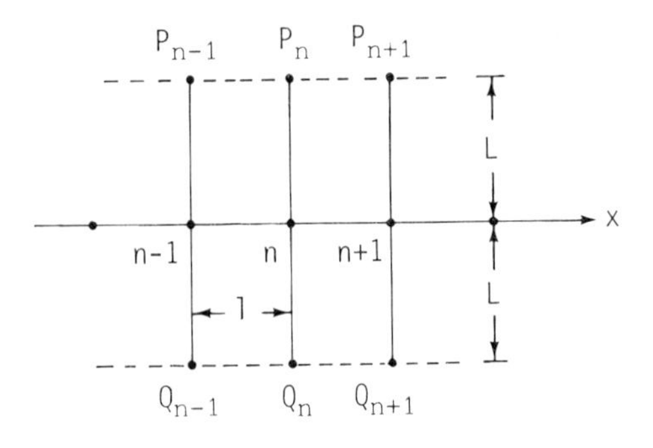

Fig. P1.

demonstrate that a progressive disturbance is feasible in this arrangement, provided the frequency is restricted to a single (pass) band,

$$2\left(\frac{P}{2ML}\right)^{1/2} < \omega < 2\left(\frac{1}{2M}\left(\frac{P}{L}+2h\right)\right)^{1/2}.$$

Adapt the dispersion analysis given in §57 for the limiting case $P = 0$ to cover the present circumstance wherein $P > 0$.

10. Envisage a group of N parallel-sided homogeneous layers, with the alternating densities ρ_1 and ρ_2 in the units $1, 3, \ldots$ and $2, 4, \ldots$, respectively; let the coordinates $(x_0, x_1), (x_1, x_2), \ldots, (x_{N-1}, x_N)$ mark the faces of the individual layers and suppose that their breadths

$$l_1 = x_1 - x_0 = x_3 - x_2 = \cdots$$
$$l_2 = x_2 - x_1 = x_4 - x_3 = \cdots$$

have recurring fixed magnitudes. If we contemplate acoustical excitations of frequency ω, which depend on the single coordinate x normal to the layers, and express the time reduced wave equation for the velocity potential ψ in the forms

$$\frac{d^2}{dx^2}\psi_{2n+1} + k_1^2\psi_{2n+1} = 0, \quad x_{2n} < x < x_{2n+1}$$
$$\frac{d^2}{dx^2}\psi_{2n+2} + k_2^2\psi_{2n+2} = 0, \quad x_{2n+1} < x < x_{2n+2} \qquad (*)$$

where $c_1 = \omega/k_1$ and $c_2 = \omega/k_2$ are the characteristic sound speeds in the layers, then suitable interface or boundary conditions ensuring continuity of pressure and velocity, are

$$\rho_1\psi_{2n+1} = \rho_2\psi_{2n+2}$$

$$\text{at } x = x_{2n+1},\ n = 0, 1, \ldots, \quad \text{integral part of } ((N-3)/2)$$

$$\frac{d}{dx}\psi_{2n+1} = \frac{d}{dx}\psi_{2n+2}.$$

On choosing as integrals of (*)

$$\psi_{2n+1} = A_{2n+1}\,e^{ik_1(x-x_{2n})} + B_{2n+1}\,e^{-ik_1(x-x_{2n})}$$

and

$$\psi_{2n+2} = A_{2n+2}\,e^{ik_2(x-x_{2n+1})} + B_{2n+2}\,e^{-ik_2(x-x_{2n+1})}$$

show that the appertaining coefficients (with even or odd indices) satisfy a common recurrence, namely

$$A_{2n+5} - 2\cos\kappa(l_1+l_2)A_{2n+3} + A_{2n+1} = 0, \qquad (**)$$

where

$$\cos \kappa(l_1+l_2) = \cos k_1 l_1 \cos k_2 l_2 - \frac{1}{2}\left(\frac{k_1\rho_2}{k_2\rho_1}+\frac{k_2\rho_1}{k_1\rho_2}\right) \sin k_1 l_1 \sin k_2 l_2. \qquad (***)$$

Assume, in accordance with (**), the coefficient representations

$$\begin{aligned} A_{2n+1} &= \alpha\, e^{in\kappa(l_1+l_2)} + \beta\, e^{-in\kappa(l_1+l_2)} \\ &= \alpha\, e^{i\kappa(x_{2n}-x_0)} + \beta\, e^{-i\kappa(x_{2n}-x_0)} \\ B_{2n+1} &= \alpha'\, e^{i\kappa(x_{2n}-x_0)} + \beta'\, e^{-i\kappa(x_{2n}-x_0)} \\ A_{2n+2} &= \alpha''\, e^{i\kappa(x_{2n+1}-x_0)} + \beta''\, e^{-i\kappa(x_{2n+1}-x_0)} \\ B_{2n+2} &= \alpha'''\, e^{i\kappa(x_{2n+1}-x_0)} + \beta'''\, e^{-i\kappa(x_{2n+1}-x_0)} \end{aligned}$$

and obtain a system of six independent linear equations which link the eight factors $\alpha, \alpha', \alpha'', \alpha'''$ and $\beta, \beta', \beta'', \beta'''$. Determine therefrom the ratios $\alpha'/\alpha, \alpha''/\alpha, \alpha'''/\alpha, \beta'/\beta, \beta''/\beta, \beta'''/\beta$ and establish the definitive forms of solution to the equations (*), namely

$$\begin{aligned} \psi_{2n+1} = {} & \alpha\left[e^{ik_1(x-x_{2n})} + \frac{\alpha'}{\alpha}\, e^{-ik_1(x-x_{2n})}\right] e^{i\kappa(x_{2n}-x_0)} \\ & + \beta\left[e^{ik_1(x-x_{2n})} + \frac{\beta'}{\beta}\, e^{-ik_1(x-x_{2n})}\right] e^{-i\kappa(x_{2n}-x_0)} \qquad x_{2n} < x < x_{2n+1} \end{aligned}$$

$$\begin{aligned} \psi_{2n+2} = {} & \alpha''\left[e^{ik_2(x-x_{2n+1})} + \frac{\alpha'''}{\alpha}\, e^{-ik_2(x-x_{2n+1})}\right] e^{i\kappa(x_{2n+1}-x_0)} \\ & + \beta\left[\frac{\beta''}{\beta}\, e^{ik_2(x-x_{2n+1})} + \frac{\beta'''}{\beta}\, e^{-ik_2(x-x_{2n+1})}\right] e^{-i\kappa(x_{2n+1}-x_0)}, \qquad x_{2n+1} < x < x_{2n+1} \end{aligned}$$

with the ultimate specifications of α, β achieved through the assignment of boundary conditions at the terminal planes x_0, x_N.

Express the corresponding forms of solution which are appropriate to a space-filling layered medium ($N=\infty$) and comment on the role of the parameter κ. Study the equation (***) for the latter, or its alternative version,

$$\tan^2 \frac{\kappa(l_1+l_2)}{2} = \frac{\tan^2 \dfrac{k_1 l_1}{2} + \tan^2 \dfrac{k_2 l_2}{2} + \left(\dfrac{k_1\rho_2}{k_2\rho_1}+\dfrac{k_2\rho_1}{k_1\rho_2}\right) \tan \dfrac{k_1 l_1}{2} \tan \dfrac{k_2 l_2}{2}}{1+\tan^2 \dfrac{k_1 l_1}{2} \cdot \tan^2 \dfrac{k_2 l_2}{2} - \left(\dfrac{k_1\rho_2}{k_2\rho_1}+\dfrac{k_2\rho_1}{k_1\rho_2}\right) \tan \dfrac{k_1 l_1}{2} \tan \dfrac{k_2 l_2}{2}},$$

with a view to delineating the circumstances that favor, or preclude, the long range transmission of disturbances in the medium.

Note. For additional problems, see Problems IV on p. 493.

PART III

§58. *A String with Continuously Variable Density*

In the examples treated thus far, the departures from uniformity of the string density have either been localized (as at a load point) or marked by discontinuous change (as at a junction between distinct, albeit uniform, sections of the string). An abrupt change of density gives rise to partial reflection of progressive wave motions on the string, though complete transmission is found in special circumstances when more than one such point of discontinuity is present, as for a composite string having a finite central portion with a different value of the density (§26). These general features are altered if the density is continuously variable and then the controlling aspect proves to be the rapidity of its variation, measured in terms of the local wave length whose magnitude, for time-periodic motions, is proportional to the ratio of local wave speed against frequency. It will appear that a 'slowly varying' material parameter (in this and other contexts) presages small reflection over a continuous range of frequencies, so long as the preceding appelation remains valid.

We may evolve a continuously variable density via an intermediate stage in which the string is partitioned into a multitude of finite sub-sections, each having its own uniform density. Let the sections be ordered in integer fashion, with the terminal coordinates of the nth section equal to x_{n-1} and x_n; if $k_n = \omega / c_n$ denotes the wave number in this section, of breadth $d_n = x_n - x_{n-1}$, a general state of time-periodic excitation therein has the displacement factor

$$s(x) = A_n\, \mathrm{e}^{\mathrm{i}k_n(x-x_n)} + B_n\, \mathrm{e}^{-\mathrm{i}k_n(x-x_n)}, \quad x_{n-1} < x < x_n \tag{58.1}$$

which includes oppositely directed wave components. The boundary conditions at the extremities of the section call for a smooth fit of both $s(x)$ and its derivative,

$$s'(x) = \mathrm{i}k_n[A_n\, \mathrm{e}^{\mathrm{i}k_n(x-x_n)} - B_n\, \mathrm{e}^{-\mathrm{i}k_n(x-x_n)}], \tag{58.2}$$

with the corresponding quantities in each of the adjacent sections. These requirements are satisfied at $x = x_{n-1}$, in particular, if the values of s and s' at x_{n-1} and x_n, bearing the designations s_{n-1}, s_n, s'_{n-1}, s'_n, obey the recurrences

$$\begin{aligned} s_{n-1} &= s_n \cos k_n d_n - s'_n \frac{\sin k_n d_n}{k_n}, \\ s'_{n-1} &= s_n k_n \sin k_n d_n + s'_n \cos k_n d_n. \end{aligned} \tag{58.3}$$

A compact statement of the latter in matrix fashion, viz.

$$\begin{pmatrix} s_{n-1} \\ s'_{n-1} \end{pmatrix} = M^{(n)} \begin{pmatrix} s_n \\ s'_n \end{pmatrix} \tag{58.4}$$

where

$$M^{(n)} = \begin{pmatrix} \cos k_n \, \mathrm{d}_n & -\dfrac{1}{k_n} \sin k_n \, \mathrm{d}_n \\ k_n \sin k_n \, \mathrm{d}_n & \cos k_n \, \mathrm{d}_n \end{pmatrix}, \quad \det M^{(n)} = 1 \tag{58.5}$$

affords a ready means of expressing the relationship between values of s, s' at any pair of points which locate section boundaries; thus,

$$\begin{pmatrix} s_0 \\ s_0' \end{pmatrix} = M^{(1)} M^{(2)} \ldots M^{(n)} \begin{pmatrix} s_n \\ s_n' \end{pmatrix}, \tag{58.6}$$

if the subscripts $0, n$ refer to $x = x_0$ and $x = x_n > 0$, and a matrix product is involved. By means of (58.6) we may connect, furthermore, the ratios $s_0'/\mathrm{i}k_0 s_0$ and $s_n'/\mathrm{i}k_n s_n$ or the reflection coefficients $r_0 = B_0/A_0$ and $r_n = B_n/A_n$, since

$$\frac{s_n'}{\mathrm{i}k_n s_n} = \frac{1 - r_n}{1 + r_n}. \tag{58.7}$$

Suppose that a uniform section $0 < x < l$, with freely assignable span l and wave number k, is interposed between a pair of others wherein the wave numbers are $k_1(x < 0)$, $k_2(x > l)$; the absence of reflection for a progressive disturbance which advances on the central section from the left is implicit in the conditions $r_0 = r_l = 0$, or

$$\frac{s_0'}{\mathrm{i}k_1 s_0} = \frac{s_l'}{\mathrm{i}k_2 s_l} = 1$$

and, subsequent to the elimination of these displacement factors from coupled equations of the type (58.3), there obtains a single complex relationship

$$k^2 \sin kl + \mathrm{i}kk_2 \cos kl = k_1 k_2 \sin kl + \mathrm{i}kk_1 \cos kl.$$

Accordingly, it is possible to suppress the partial wave reflection characteristic of a direct linkage between distinct, though uniform, strings if the parameters k, l of an intervening section are such that

$$k = \surd(k_1 k_2), \tag{58.8}$$
$$\cos kl = 0, \qquad kl = \pi/2, 3\pi/2, \ldots.$$

We progress towards a continuously variable density (and wave number) by augmenting the number of sections indefinitely and diminishing their breadth to infinitesimal scale, in accordance with the hypothesis that $\mathrm{d}_n = \mathrm{d}x$ (all n). The set of algebraic connections (58.4) at integer values of n is then supplanted by a pair of differential relations

$$\begin{pmatrix} s(x) \\ s'(x) \end{pmatrix} - \frac{\mathrm{d}}{\mathrm{d}x} \begin{pmatrix} s(x) \\ s'(x) \end{pmatrix} \mathrm{d}x = \begin{pmatrix} 1 & -\mathrm{d}x \\ k^2(x)\,\mathrm{d}x & 1 \end{pmatrix} \begin{pmatrix} s(x) \\ s'(x) \end{pmatrix} \tag{58.9}$$

wherein the displacement factor appears with a continuous argument rather than a discrete index. One of the two separate equations contained in (58.9) has the status of an identity, while the other takes the form

$$\frac{d^2 s}{dx^2} + k^2(x)s = 0, \qquad k^2(x) = \frac{\omega^2 \rho(x)}{P} \tag{58.10}$$

that may be readily inferred from direct dynamical considerations. Thus, the analysis of small amplitude, time-periodic, motions in a non-uniform string centers on a linear and variable coefficient ordinary differential equation of the second order.

If we define a reflection factor $r(x)$ by means of the relationship

$$\frac{s'(x)}{s(x)} = ik(x)\frac{1 - r(x)}{1 + r(x)} \tag{58.11}$$

which is patterned after (58.7), and note that

$$w(x) = \frac{s'(x)}{s(x)} = \frac{d}{dx} \log s(x) \tag{58.12}$$

satisfies the equation

$$\frac{dw}{dx} + w^2(x) + k^2(x) = 0 \tag{58.13}$$

in consequence of (58.10), there follows

$$\frac{dr}{dx} = -2\,ik(x)r + \frac{1}{2}\left\{\frac{d}{dx} \log k(x)\right\}(1 - r^2) \tag{58.14}$$

on eliminating w from (58.12)–(58.14).

The non-linear differential equation (58.14) for $r(x)$ is a Ricatti type and, being of the first order, its integral connects the values of the reflection factor at any pair of points, say x_1 and x_2. If k remains constant over the interval between these points only the first term on the right-hand side of (58.14) differs from zero, and the appropriate solution

$$r(x_2) = r(x_1)\exp\{-2\,ik(x_2 - x_1)\}$$

confirms that a phase difference alone distinguishes the respective values of r. The other terms are significant when k is variable and, by virtue of their real coefficients, make a contribution to the amplitude ratio of the reflection factors at different locations. We may presume that the amount of amplitude change for $r(x)$ depends on the values of the function

$$\nu(x) = \frac{d}{dx} \log k(x) = \frac{k'(x)}{k(x)} \tag{58.15}$$

which characterizes the logarithmic derivative of the wave number; since k is inversely proportional to the local wave length a small magnitude for $r(x)$ is consonant with limited change in the wave number on a scale set by the wave length.

If the reflection factor is small and its square, along with the non-linear term in (58.14) be neglected, an integral of the latter equation which vanishes at $x = x_2$ has the form

$$r(x_1) = -\frac{1}{2}\int_{x_1}^{x_2}\left(\frac{d}{dx'}\log k(x')\right)\exp\left(2\,i\int_{x_1}^{x'} k(x'')\,dx''\right)dx' \tag{58.16}$$

at $x = x_1$. The approximate expression (58.16) suggests an absence of reflection in continuously variable settings when the magnitude of k (or the frequency) tends to infinity, with an opposite trend of the wave length, since rapid oscillations of the exponential function bring about an effective decrease in value of the integral.

Since the local behavior of the wave number enters into the differential equation (58.14) which determines the appertaining reflection factor, we are free to alter the wave number elsewhere for purposes of rederiving said equation. Consider, therefore, a sectioned string whose inhomogeneity or variable wave number pertains to the interval $\xi < x < L$ only, and outside of which the uniform assignments $k = k_1(x < \xi)$, $k_2(x > L)$ apply. Suppose that a unit amplitude wave train is incident upon the inhomogeneous section from the left, and that the concomitant representations of the displacement factor on either side thereof take the forms

$$s(x) = \begin{cases} e^{ik_1(x-\xi)} + r(\xi)\,e^{-ik_1(x-\xi)}, & x \leq \xi \\ t(\xi)\,e^{ik_2(x-L)}, & x \geq L \end{cases}$$

wherein an eventual variability of ξ is previewed by its argument status for both the coefficients r, t. On the interval $\xi < x < L$ we employ a formal representation

$$s(x) = C_1\psi_1(x) + C_2\psi_2(x)$$

in which ψ_1, ψ_2 designate a pair of linearly independent solutions of the differential equation (58.10) and satisfy, moreover, the particular conditions

$$\psi_1(L) = \psi_2'(L) = 1$$
$$\psi_1'(L) = \psi_2(L) = 0.$$

The continuity requirements for $s(x), s'(x)$ at $x = L$ and $x = \xi$ underlie individual specifications of the quartet $r(\xi)$, $t(\xi)$, C_1, C_2, and those of the first pair follow from the simultaneous algebraic equations

$$\begin{aligned} 1 + r(\xi) &= t(\xi)[\psi_1(\xi) + ik_2\psi_2(\xi)] \\ ik_1(1 - r(\xi)) &= t(\xi)[\psi_1'(\xi) + ik_2\psi_2'(\xi)]. \end{aligned} \tag{58.17}$$

If we postulate that $k_1 = k(\xi)$, so as to exclude a local discontinuity in the wave number, and thereafter regard ξ as a variable entity, the ratio of equations (58.17) yields

$$\frac{\mathrm{d}}{\mathrm{d}\xi}\log[\psi_1(\xi)+\mathrm{i}k_2\psi_2(\xi)]=\mathrm{i}k(\xi)\frac{1-r(\xi)}{1+r(\xi)}; \tag{58.18}$$

inasmuch as the logarithmic derivative defines a function which obeys the equation (58.13), we may conclude from the simlarity of (58.11), (58.18) that the reflection coefficient satisfies the equation (58.14) when k is differentiable. Furthermore, after the relation that emerges from (58.17) on eliminating ψ_1, ψ_2, viz.

$$r'(\xi)=\mathrm{i}k(\xi)(1-r(\xi))+\frac{t'(\xi)}{t(\xi)}(1+r(\xi))$$

is combined with (58.14), there obtains

$$t'(x)=\left[-\mathrm{i}k(x)+\frac{1}{2}(1-r(x))\frac{\mathrm{d}}{\mathrm{d}x}\log k(x)\right]t(x) \tag{58.19}$$

as the differential equation for the transmission factor which is companion to (58.14).

A partial integration of the non-linear equation (58.14) satisfied by $r(x)$ becomes possible through the substitution

$$r(x)=|r(x)|\,\mathrm{e}^{\mathrm{i}\varphi(x)} \tag{58.20}$$

therein and the consequent separation of real and imaginary parts; such a procedure yields a pair of coupled equations for the absolute magnitude and phase of the reflection coefficient, namely

$$\begin{aligned}\frac{\mathrm{d}}{\mathrm{d}x}|r(x)|&=\frac{1}{2}\nu(x)[1-|r(x)|^2]\cos\varphi(x),\\ \frac{\mathrm{d}}{\mathrm{d}x}\varphi(x)&=2k(x)-\frac{1}{2}\nu(x)\left[\frac{1}{|r(x)|}+|r(x)|\right]\cos\varphi(x),\end{aligned} \tag{58.21}$$

of which the first is separable and thus leads to the correlation

$$\frac{1-|r(x_2)|}{1+|r(x_2)|}=\frac{1-|r(x_1)|}{1+|r(x_1)|}\exp\left\{-\int_{x_1}^{x_2}\nu(x)\cos\varphi(x)\,\mathrm{d}x\right\} \tag{58.22}$$

between magnitudes of the reflection coefficient at an arbitrary pair of locations. If we let $x_2\to\infty$ and set $|r(x_2)|=0$, then

$$|r(x_1)|^2=\tanh^2\left(\frac{1}{2}\int_{x_1}^{\infty}\nu(x)\cos\varphi(x)\,\mathrm{d}x\right) \tag{58.23}$$

which indicates that the 'orthogonality' condition

$$\int_{x_1}^{\infty}\nu(x)\cos\varphi(x)\,\mathrm{d}x=0 \tag{58.24}$$

is sufficient for a vanishing value of $|r(x_1)|$.

§59. *An Inhomogeneous Segment*

Consider a prototype of settings with limited inhomogeneity wherein the wave number manifests a variable form $k(x)$ on the range $(-L/2, L/2)$ and assumes a uniform (common) value elsewhere. If the differential equation (58.10) for the pertinent wave factor $s(x)$ in time-periodic states is rewritten as

$$\left(\frac{d^2}{dx^2}+k^2\right)s(x) = -\kappa(x)s(x), \quad -\infty<x<\infty \tag{59.1}$$

the function introduced thereby,

$$\kappa(x) = k^2(x) - k^2 \tag{59.2}$$

has a range identical with that of the inhomogeneity, i.e., $\kappa(x) \neq 0, -L/2<x< L/2$. An integral of (59.1) which incorporates the primary or incident wave excitation factor e^{ikx} on the left $(x<-L/2)$ is given by

$$s(x) = e^{ikx} + \int_{-L/2}^{L/2} G(x-x')\kappa(x')s(x')\, dx', \quad -\infty<x<\infty \tag{59.3}$$

and the explicit Green's function factor of the integrand,

$$G(x-x') = \frac{i}{2k}\, e^{ik|x-x'|},$$

helps to shape the appropriate representations outside the inhomogeneous sector,

$$s(x) = \begin{matrix} e^{ikx} + r\, e^{-ikx}, & x<-L/2 \\ e^{ikx} + (t-1)\, e^{ikx}, & x>L/2 \end{matrix} \tag{59.4}$$

and also to provide the formal characterizations

$$\begin{aligned} r &= \frac{i}{2k}\int_{-L/2}^{L/2} \kappa(x)s(x)\, e^{ikx}\, dx \\ t-1 &= \frac{i}{2k}\int_{-L/2}^{L/2} \kappa(x)s(x)\, e^{-ikx}\, dx \end{aligned} \tag{59.5}$$

of the exterior reflection and transmission coefficients. The explicit determinations of r, t are achieved, in principle, after the solution of (59.3), that is, an integral equation for $s(x)$ in the domain of inhomogeneity.

A relationship between the absolute magnitudes of r and t, whose significance is bound up with energy conservation and thus encompasses an entire class of functions $k(x)$, is a ready consequence of the fact that both $s(x)$ and its complex conjugate $s^*(x)$ satisfy the same differential equation (59.1) when $k(x)$ is real. The constancy of the Wronskian for the latter functions thus leads to the condition

$$\left[s^*(x)\frac{ds}{dx} - s(x)\frac{ds^*}{dx}\right]_{x=-X}^{x=+X} = 0, \quad X > L/2$$

and, by utilizing the representation (59.3) along with its complex conjugate, we deduce therefrom that

$$2\,ik|t|^2 = 2\,ik(1-|r|^2)$$

or

$$|r|^2 + |t|^2 = 1, \tag{59.6}$$

which assures the equality of the time-average energy fluxes arriving at and departing from the inhomogeneous segment.

Another general inference is forthcoming on consideration of the Wronskian for functions $s_\pm(x)$ which pertain to oppositely directed primary waves $[e^{\pm ikx}]$ and thus involve two pairs of coefficients r_+, t_+ and r_-, t_-; neither reflection coefficient makes a contribution to the value of the Wronskians at $x = \pm X$ and equality of the latter reveals that

$$t_+ = t_- \tag{59.7}$$

which signals a complete reversibility as regards transmission through the inhomogeneous segment. The identity of r_+ and r_- is assured when $k(x)$ has a symmetric profile relative to the midpoint of its range, for the representation (59.3) furnishes a means of verifying that $s(x)$ and $s(-x)$ are solutions which befit oppositely directed excitations with the same magnitude and phase at $x = 0$.

If we recast (59.3) in the form

$$s(x) = e^{ikx}\left[1 + \frac{i}{2k}\int_{-L/2}^{L/2} \kappa(x)s(x)\,e^{-ikx}\,dx\right]$$
$$+\frac{i}{2k}\left[e^{-ikx}\int_x^{L/2} \kappa(x)s(x)\,e^{ikx}\,dx - e^{ikx}\int_x^{L/2} \kappa(x)s(x)\,e^{-ikx}\,dx\right]$$

and call upon the transformation formulas

$$\int_x^{L/2} \kappa(x)s(x)\,e^{\pm ikx}\,dx = -\int_x^{L/2} e^{\pm ikx}\frac{d^2s}{dx^2}\,dx - k^2\int_x^{L/2} e^{\pm ikx}s(x)\,dx$$
$$= \left[e^{\pm ikx}\left\{-\frac{ds}{dx} \pm ikx\right\}\right]_x^{L/2} \tag{59.8}$$

which apply to solutions of (59.1), it follows that

$$0 = e^{ikx}\left[1 + \frac{i}{2k}\int_{-L/2}^{L/2} \kappa(x)s(x)\,e^{-ikx}\,dx + \frac{i}{2k}\left(s'\left(\frac{L}{2}\right) + iks\left(\frac{L}{2}\right)\right)e^{-ikL/2}\right]$$
$$-\frac{i}{2k}\left(s'\left(\frac{L}{2}\right) - iks\left(\frac{L}{2}\right)\right)e^{-ikx+ikL/2},$$

whence

$$s'\left(\frac{L}{2}\right) = \mathrm{i}ks\left(\frac{L}{2}\right), \tag{59.9}$$

$$1 + \frac{\mathrm{i}}{2k}\int_{-L/2}^{L/2} \kappa(x)s(x)\,\mathrm{e}^{-\mathrm{i}kx}\,\mathrm{d}x = -\frac{\mathrm{i}}{2k}\left(s'\left(\frac{L}{2}\right) + \mathrm{i}ks\left(\frac{L}{2}\right)\right)\mathrm{e}^{-\mathrm{i}kL/2}$$
$$= s\left(\frac{L}{2}\right)\mathrm{e}^{-\mathrm{i}kL/2}. \tag{59.10}$$

We encounter in (59.9) a homogeneous condition for the wave factor at the right-hand end ($x = L/2$) of the inhomogeneous segment that typifies a single outgoing wave, which is of course none other than the transmitted wave. Extinction of the primary wave is secured by the condition (59.10) that, according to (59.4), (59.5), involves the formal characterization of the transmission coefficient. Pursuant to a transformation of the integral in (59.10) by the scheme (59.8), and use of (59.9), the primary wave extinction relation takes the form

$$s'\left(-\frac{L}{2}\right) + \mathrm{i}ks\left(-\frac{L}{2}\right) = 2\,\mathrm{i}k\,\mathrm{e}^{-\mathrm{i}kL/2}, \tag{59.11}$$

that is, an inhomogeneous boundary condition for the wave factor at the left-hand end ($x = -L/2$) of the variable segment. With the help of the extinction relation (59.10), in which the level of the function $s(x)$ is fixed by that of the primary wave, we may exhibit both the reflection and transmission coefficients in a scale-independent fashion, namely

$$r = \frac{\dfrac{\mathrm{i}}{2k}\displaystyle\int_{-L/2}^{L/2} \kappa(x)s(x)\,\mathrm{e}^{\mathrm{i}kx}\,\mathrm{d}x}{\mathrm{e}^{-\mathrm{i}kL/2}s\left(\dfrac{L}{2}\right) - \dfrac{\mathrm{i}}{2k}\displaystyle\int_{-L/2}^{L/2} \kappa(x)s(x)\,\mathrm{e}^{-\mathrm{i}kx}\,\mathrm{d}x} \tag{59.12}$$

$$t = \frac{\mathrm{e}^{-\mathrm{i}kL/2}s\left(\dfrac{L}{2}\right)}{\mathrm{e}^{-\mathrm{i}kL/2}s\left(\dfrac{L}{2}\right) - \dfrac{\mathrm{i}}{2k}\displaystyle\int_{-L/2}^{L/2} \kappa(x)s(x)\,\mathrm{e}^{-\mathrm{i}kx}\,\mathrm{d}x}. \tag{59.13}$$

Let the real functions $\psi_1(x), \psi_2(x)$ denote a pair of linearly independent solutions to the equation (59.1) on the interval $(-L/2, L/2)$, with the specifications

$$\psi_1\left(\frac{L}{2}\right) = 1, \qquad \psi_2\left(\frac{L}{2}\right) = 0$$
$$\psi_1'\left(\frac{L}{2}\right) = 0, \qquad \psi_2'\left(\frac{L}{2}\right) = 1 \tag{59.14}$$

and the concomitant Wronskian

$$\psi_1(x)\psi_2'(x) - \psi_1'(x)\psi_2(x) = 1, \quad -L/2 \le x \le L/2. \tag{59.15}$$

The combination $\psi_1(x) + \mathrm{i}k\psi_2(x)$ is evidently in accord with the boundary condition (59.9) and may be adopted as a representation of $s(x)$ for purposes of evaluating the expressions (59.12), (59.13); this yields, after recourse again to (59.8),

$$r = -\frac{\psi_1'\left(-\frac{L}{2}\right) + k^2\psi_2\left(-\frac{L}{2}\right) - \mathrm{i}k\left[\psi_1\left(-\frac{L}{2}\right) - \psi_2'\left(-\frac{L}{2}\right)\right]}{\psi_1'\left(-\frac{L}{2}\right) - k^2\psi_2\left(-\frac{L}{2}\right) + \mathrm{i}k\left[\psi_1\left(-\frac{L}{2}\right) + \psi_2'\left(-\frac{L}{2}\right)\right]}\,\mathrm{e}^{-\mathrm{i}kL},$$

$$t = \frac{2\,\mathrm{i}k\,\mathrm{e}^{-\mathrm{i}kL}}{\psi_1'\left(-\frac{L}{2}\right) - k^2\psi_2\left(-\frac{L}{2}\right) + \mathrm{i}k\left[\psi_1\left(-\frac{L}{2}\right) + \psi_2'\left(-\frac{L}{2}\right)\right]}. \tag{59.16}$$

It is a simple matter to verify that these expressions for the reflection and transmission coefficients satisfy the conservation relation (59.6), given the joint property (59.15) of the functions ψ_1, ψ_2.

§60. *An Inhomogeneous Layer*

If the wave number $k(x)$ varies on a finite range, $-L/2 < x < L/2$, and assumes uniform though distinct values on the adjacent ranges, say $k_1(x < -L/2)$ and $k_2(x > L/2)$, we may refer to an inhomogeneous (transition) layer. The analysis of reflection from and transmission through such a layer can be undertaken along lines described in the preceding section, with a slightly greater complexity of detail. Let us postulate an incident train on the left, characterized by the wave factor $\mathrm{e}^{\mathrm{i}k_1x}$, and focus attention alternately on the intervals $I_-:(-\infty, L/2)$, $I_+:(-L/2, \infty)$ whose common or overlapping portion is the site of the layer itself.

After multiplying the equations

$$\left(\frac{\mathrm{d}^2}{\mathrm{d}x^2} + k^2(x)\right)s_-(x) = 0, \qquad \left(\frac{\mathrm{d}^2}{\mathrm{d}x^2} + k_1^2\right)G_1(x-x') = -\delta(x-x') \tag{60.1}$$

with the respective functions

$$G_1(x-x') = \frac{\mathrm{i}}{2k_1}\,\mathrm{e}^{\mathrm{i}k_1|x-x'|}, \quad s_-(x),$$

subtracting and integrating over I_-, we arrive at the representation

$$
\begin{aligned}
s_-(x) = \mathrm{e}^{\mathrm{i}k_1x} + \frac{\mathrm{i}}{2k_1}\,\mathrm{e}^{\mathrm{i}k_1((L/2)-x)}s'_-\left(\frac{L}{2}\right) + \frac{1}{2}\,\mathrm{e}^{\mathrm{i}k_1((L/2)-x)}s_-\left(\frac{L}{2}\right) \\
+ \int_{-L/2}^{L/2} \kappa_1(x')s_-(x')G_1(x-x')\,\mathrm{d}x', \quad -\infty < x < \frac{L}{2}
\end{aligned}
\tag{60.2}
$$

where

$$
\begin{aligned}
\kappa_1(x) &= k^2(x) - k_1^2 \\
&\neq 0, \quad (-L/2, L/2);
\end{aligned}
\tag{60.3}
$$

when $x < -L/2$, in particular, the last three terms in (60.2) collectively make up an outgoing wave factor and thus permit a formal identification of the reflection coefficient,

$$
r = \frac{\mathrm{i}}{2k_1}\left\{\left(s'_-\left(\frac{L}{2}\right) - \mathrm{i}k_1 s_-\left(\frac{L}{2}\right)\right)\mathrm{e}^{\mathrm{i}k_1L/2} + \int_{-L/2}^{L/2} \kappa_1(x)s_-(x)\,\mathrm{e}^{\mathrm{i}k_1x}\,\mathrm{d}x\right\}. \tag{60.4}
$$

At any point within the layer, $-L/2 < x < L/2$, a sequential conversion of the integral in (60.2), aided by relations of the type (59.8), viz.

$$
\begin{aligned}
&\int_{-L/2}^{L/2} \kappa_1(x')s_-(x')G_1(x-x')\,\mathrm{d}x' \\
&= \frac{\mathrm{i}}{2k_1}\,\mathrm{e}^{\mathrm{i}k_1x}\int_{-L/2}^{L/2} \kappa_1(x)s_-(x)\,\mathrm{e}^{-\mathrm{i}k_1x}\,\mathrm{d}x \\
&+ \frac{\mathrm{i}}{2k_1}\left[\mathrm{e}^{-\mathrm{i}k_1x}\int_x^{L/2} \kappa_1(x)s_-(x)\,\mathrm{e}^{\mathrm{i}k_1x}\,\mathrm{d}x - \mathrm{e}^{\mathrm{i}k_1x}\int_x^{L/2} \kappa_1(x)s_-(x)\,\mathrm{e}^{-\mathrm{i}k_1x}\,\mathrm{d}x\right] \\
&= \frac{\mathrm{i}}{2k_1}\,\mathrm{e}^{\mathrm{i}k_1x}\int_{-L/2}^{L/2} \kappa_1(x)s_-(x)\,\mathrm{e}^{-\mathrm{i}k_1x}\,\mathrm{d}x \\
&+ s_-(x) - \frac{\mathrm{i}}{2k_1}\left[s'_-\left(\frac{L}{2}\right) - \mathrm{i}k_1 s_-\left(\frac{L}{2}\right)\right]\mathrm{e}^{\mathrm{i}k_1((L/2)-x)} + \frac{\mathrm{i}}{2k_1}\left[s'_-\left(\frac{L}{2}\right) + \mathrm{i}k_1 s_-\left(\frac{L}{2}\right)\right]\mathrm{e}^{-\mathrm{i}k_1((L/2)-x)},
\end{aligned}
$$

enables us to bring (60.2) into the form

$$
\begin{aligned}
0 = \mathrm{e}^{\mathrm{i}k_1x}\Bigg[1 + \frac{\mathrm{i}}{2k_1}\int_{-L/2}^{L/2} \kappa_1(x)s_-(x)\,\mathrm{e}^{-\mathrm{i}k_1x}\,\mathrm{d}x \\
+ \frac{\mathrm{i}}{2k_1}\left(s'_-\left(\frac{L}{2}\right) + \mathrm{i}k_1 s_-\left(\frac{L}{2}\right)\right)\mathrm{e}^{-\mathrm{i}k_1L/2}\Bigg], \quad -\frac{L}{2} < x < \frac{L}{2}.
\end{aligned}
$$

Accordingly, the primary wave extinction condition becomes

$$
\frac{\mathrm{i}}{2k_1}\int_{-L/2}^{L/2} \kappa_1(x)s_-(x)\,\mathrm{e}^{-\mathrm{i}k_1x}\,\mathrm{d}x + \frac{\mathrm{i}}{2k_1}\left(s'_-\left(\frac{L}{2}\right) + \mathrm{i}k_1 s_-\left(\frac{L}{2}\right)\right)\mathrm{e}^{-\mathrm{i}k_1L/2} = -1 \tag{60.5}
$$

and, following a reduction of the integral herein, there obtains

$$s'_-\left(-\frac{L}{2}\right)+\mathrm{i}k_1 s_-\left(-\frac{L}{2}\right)=2\,\mathrm{i}k_1\,\mathrm{e}^{-\mathrm{i}k_1 L/2}, \tag{60.6}$$

which constitutes an appropriate revision of (59.11).

To establish analogous results on the range I_+ we employ the differential equations

$$\left(\frac{\mathrm{d}^2}{\mathrm{d}x^2}+k^2(x)\right)s_+(x)=0, \qquad \left(\frac{\mathrm{d}^2}{\mathrm{d}x^2}+k_2^2\right)G_2(x-x')=-\delta(x-x')$$

and deduce therefrom, retaining the prior hypothesis about the direction of incidence, a representation

$$\begin{aligned} s_+(x)=&-\frac{\mathrm{i}}{2k_2}\,\mathrm{e}^{\mathrm{i}k_2(x+(L/2))}s'_+\left(-\frac{L}{2}\right)+\frac{1}{2}\,\mathrm{e}^{\mathrm{i}k_2(x+(L/2))}s_+\left(-\frac{L}{2}\right)\\ &+\int_{-L/2}^{L/2}\kappa_2(x')s_+(x')G_2(x-x')\,\mathrm{d}x', \qquad -\frac{L}{2}<x<\infty \end{aligned} \tag{60.7}$$

with the pair of functions

$$\begin{aligned} \kappa_2(x)&=k^2(x)-k_2^2, \\ G_2(x-x')&=\frac{1}{2k_2}\,\mathrm{e}^{\mathrm{i}k_2|x-x'|}; \end{aligned} \tag{60.8}$$

outside of the layer, where $x>L/2$, a transmission coefficient

$$t=\frac{\mathrm{i}}{2k_2}\left\{\int_{-L/2}^{L/2}\kappa_2(x)s_+(x)\,\mathrm{e}^{-\mathrm{i}k_2 x}\,\mathrm{d}x-\left(s'_+\left(-\frac{L}{2}\right)+\mathrm{i}k_2 s_+\left(-\frac{L}{2}\right)\right)\mathrm{e}^{\mathrm{i}k_2 L/2}\right\} \tag{60.9}$$

is provided by (60.7) and within the layer the sole consequence of transforming the integral which occurs in (60.7), as accomplished above, is a relation

$$s'_+\left(\frac{L}{2}\right)=\mathrm{i}k_2 s_+\left(\frac{L}{2}\right) \tag{60.10}$$

that furnishes a homogeneous boundary condition and expresses the counterpart of (59.9).

We may henceforth dispense with the subscripts $\mp$ on the closed interval $(-L/2, L/2)$ which is common to I_-, I_+ and observe that either of the representations (60.2), (60.7) constitutes an integral equation for the wave factor $s(x)$ in the layer. The outcome of jointly utilizing the boundary and extinction conditions (60.10), (60.6), along with the representation (60.4), is a scale independent version of the reflection coefficient, namely

$$r=\frac{\mathrm{i}(k_1-k_2)s\left(\dfrac{L}{2}\right)\mathrm{e}^{\mathrm{i}k_1 L/2}-\displaystyle\int_{-L/2}^{L/2}\kappa_1(x)s(x)\,\mathrm{e}^{\mathrm{i}k_1 x}\,\mathrm{d}x}{\mathrm{i}(k_1+k_2)s\left(\dfrac{L}{2}\right)\mathrm{e}^{-\mathrm{i}k_1 L/2}+\displaystyle\int_{-L/2}^{L/2}\kappa_1(x)s(x)\,\mathrm{e}^{-\mathrm{i}k_1 x}\,\mathrm{d}x}, \tag{60.11}$$

or

$$r = \mathrm{e}^{-\mathrm{i}k_1 L} \frac{\mathrm{i}k_1 - \left[s'\left(-\frac{L}{2}\right) \Big/ s\left(-\frac{L}{2}\right)\right]}{\mathrm{i}k_1 + \left[s'\left(-\frac{L}{2}\right) \Big/ s\left(-\frac{L}{2}\right)\right]} \tag{60.12}$$

after a reduction of the integral in (60.11) on the basis of the differential equation $s'' + k^2(x)s = 0$ and the aforementioned boundary condition. According to (60.12) the reflection coefficient can be calculated in terms of the logarithmic derivatives for the complete and incident wave factors at the entrance face of the layer, if the conditions within and at the exit face are satisfied. A similar pattern of manipulations leads to the companion formula for the transmission coefficient,

$$t = -2\,\mathrm{i}k_1 \frac{\left[s\left(\frac{L}{2}\right) \Big/ s\left(-\frac{L}{2}\right)\right]}{\mathrm{i}k_1 + \left[s'\left(-\frac{L}{2}\right) \Big/ s\left(-\frac{L}{2}\right)\right]} \mathrm{e}^{-\mathrm{i}(k_1+k_2)L/2}. \tag{60.13}$$

By way of application, let us select the particular form of variable wave number

$$k(x) = \frac{\alpha}{\beta + x}, \qquad -\frac{L}{2} < x < \frac{L}{2} \tag{60.14}$$

and choose the parameters α, β so that $k(-L/2) = k_1$, $k(L/2) = k_2$, viz.

$$\alpha = \frac{k_1 k_2}{k_1 - k_2} L, \qquad \beta = \frac{k_1 + k_2}{k_1 - k_2} \cdot \frac{L}{2}. \tag{60.15}$$

A general solution of the differential equation

$$\left(\frac{\mathrm{d}^2}{\mathrm{d}x^2} + \frac{\alpha^2}{(\beta + x)^2}\right) s(x) = 0$$

is available in explicit and elementary fashion

$$s(x) = (\beta + x)^{1/2} \{C_1 \mathrm{e}^{\gamma \log(\beta + x)} + C_2 \mathrm{e}^{-\gamma \log(\beta + x)}\}, \tag{60.16}$$

where

$$\gamma = \left(\frac{1}{4} - \alpha^2\right)^{1/2}$$

and C_1, C_2 are constants. Consequent to fixing the coefficient ratio,

$$\frac{C_2}{C_1} = -\frac{\frac{1}{2} + \gamma - \mathrm{i}k_2\left(\beta + \frac{L}{2}\right)}{\frac{1}{2} - \gamma - \mathrm{i}k_2\left(\beta + \frac{L}{2}\right)} \mathrm{e}^{2\gamma \log(\beta + (L/2))},$$

through the boundary condition $s'(L/2) = ik_2 s(L/2)$ and substituting in (60.12), we find

$$r = -\frac{i}{2} e^{-ik_1 L} \frac{\sinh\left(\gamma \log \frac{k_1}{k_2}\right)}{\alpha \sinh\left(\gamma \log \frac{k_1}{k_2}\right) + i\gamma \cosh\left(\gamma \log \frac{k_1}{k_2}\right)} \tag{60.17}$$

and thence

$$|r|^2 = \frac{\sinh^2\left(\left(\frac{1}{4} - \alpha^2\right)^{1/2} \log \frac{k_1}{k_2}\right)}{\sinh^2\left(\left(\frac{1}{4} - \alpha^2\right)^{1/2} \log \frac{k_1}{k_2}\right) + 1 - 4\alpha^2}, \quad \alpha < \frac{1}{2}. \tag{60.18}$$

The value of $|r|^2$ decreases from a maximum,

$$|r|^2_{\max} = \left(\frac{k_1 - k_2}{k_1 + k_2}\right)^2,$$

which is attained in the limit of vanishingly small layer breadth ($L, \alpha \to 0$) and the appropriate continuation of (60.18) when α exceeds 1/2, or $L > (1/2)(k_1 - k_2/k_1 k_2)$,

$$|r|^2 = \frac{\sin^2\left(\left(\alpha^2 - \frac{1}{4}\right)^{1/2} \log \frac{k_1}{k_2}\right)}{\sin^2\left(\left(\alpha^2 - \frac{1}{4}\right)^{1/2} \log \frac{k_1}{k_2}\right) + 4\alpha^2 - 1}, \quad \alpha > \frac{1}{2} \tag{60.19}$$

evidences an oscillatory behavior, superposed on a diminishing magnitude, as L increases. Applying the general representation (60.13) for the transmission coefficient, we find in the case at hand

$$t = -i\left(\frac{k_1}{k_2}\right)^{1/2} \gamma \frac{e^{-i(k_1+k_2)L/2}}{\alpha \sinh\left(\gamma \log \frac{k_1}{k_2}\right) + i\gamma \cosh\left(\gamma \log \frac{k_1}{k_2}\right)}, \tag{60.20}$$

$$|t|^2 = 4\frac{k_1}{k_2} \frac{\frac{1}{4} - \alpha^2}{\sinh^2\left(\left(\frac{1}{4} - \alpha^2\right)^{1/2} \log \frac{k_1}{k_2}\right) + 1 - 4\alpha^2}, \quad \alpha < \frac{1}{2} \tag{60.21}$$

with the expected relation

$$k_1(1 - |r|^2) = k_2|t|^2.$$

§61. *Variable Wave Numbers with a Periodic Nature*

The matrix formulation previously given (§58) in connection with wave motions on a string which is composed of homogeneous sections also lends itself to the analysis for more general types of density stratification. Suppose that $\psi_1(x), \psi_2(x)$ represent a particular pair of solutions to the variable coefficient differential equation (58.10), whose independence is evidenced by the 'initial' conditions

$$\begin{aligned} \psi_1(0) = 1, \qquad \psi_2(0) = 0 \\ \psi_1'(0) = 0, \qquad \psi_2'(0) = 1 \end{aligned} \tag{61.1}$$

and the consequent form of their invariant relationship (or Wronskian)

$$\psi_1(x)\psi_2'(x) - \psi_1'(x)\psi_2(x) = 1. \tag{61.2}$$

These attributes of ψ_1, ψ_2 enable us to specify an integral of (58.10),

$$s(x) = s(0)\psi_1(x) + s'(0)\psi_2(x) \tag{61.3}$$

with preassigned value, $s(0)$, and derivative, $s'(0)$; and thereafter to express the latter in the reciprocal fashion

$$\begin{aligned} s(0) &= s(x)\psi_2'(x) - s'(x)\psi_2(x) \\ s'(0) &= -s(x)\psi_1'(x) + s'(x)\psi_1(x). \end{aligned} \tag{61.4}$$

Hence, an overall description of the wave factor is furnished by the square matrix

$$M(x) = \begin{pmatrix} \psi_2'(x) & -\psi_2(x) \\ -\psi_1'(x) & \psi_1(x) \end{pmatrix} \tag{61.5}$$

which links amplitude and slope at variable and fixed locations, respectively, and has the property

$$\det M(x) = 1. \tag{61.6}$$

If the analogous matrices $M^{(1)}(x), M^{(2)}(x), \dots, M^{(n)}(x), \dots$ pertain to complementary intervals $x_1 - 0, x_2 - x_1, \dots, x_n - x_{n-1}, \dots$, where the wave numbers are independently assigned, the transformation between s, s' at $x = 0$ and $x > x_n$ with the form

$$\begin{pmatrix} s_0 \\ s_0' \end{pmatrix} = M^{(1)}(x_1)M^{(2)}(x_2) \dots M^{(n)}(x_n)M^{(n+1)}(x)\begin{pmatrix} s(x) \\ s'(x) \end{pmatrix} \tag{61.7}$$

guarantees the continuity of their individual values at all intervening points.

In the circumstance that the function $k(x)$ has a strictly repetitive nature, the transfer matrix for an integer number, N, of periods with span l is simply the Nth power of the matrix for a single period, viz.

$$M^{(1)}(l)M^{(2)}(2l) \dots M^{(N)}(Nl) = (M^{(1)}(l))^N,$$

since the functions $\psi_1(x), \psi_2(x)$ are obtained from those on the interval $nl \leq x \leq (n+1)l$ are obtained from those on the interval $0 \leq x \leq l$ by the translation $x \to x - nl$. When the elements of $M^{(1)}(l)$ are denoted by m_{ij}, $i, j = 1, 2$, their counterparts for the Nth power of the matrix prove to be

$$(M^{(1)}(l))^N = \begin{pmatrix} m_{11}S_{N-1}(\xi) - S_{N-2}(\xi) & m_{12}S_{N-1}(\xi) \\ m_{21}S_{N-1}(\xi) & m_{22}S_{N-1}(\xi) - S_{N-2}(\xi) \end{pmatrix} \tag{61.8}$$

where

$$\xi = m_{11} + m_{22} = 2\cos\vartheta \tag{61.9}$$

and

$$S_N(\xi) = \frac{\sin(N+1)\vartheta}{\sin\vartheta}, \quad n \geq 0 \tag{61.10}$$

represents a (Chebyshev) polynomial of degree $N-1$ in ξ; if $|\xi| > 2$ the characterizations

$$\xi = 2\cosh\vartheta, \tag{61.11}$$

and

$$S_N(\xi) = \frac{\sinh(N+1)\vartheta}{\sinh\vartheta} \tag{61.12}$$

take the place of (61.9), (61.10).

The result (61.8) is directly forthcoming from an expression for the nth power of a 2×2 matrix M, with elements M_{ij}, in terms of M itself and the unit matrix I of the same dimension, i.e.,

$$(M)^n = \frac{\lambda_1^n - \lambda_2^n}{\lambda_1 - \lambda_2} M - \lambda_1\lambda_2 \frac{\lambda_1^{n-1} - \lambda_2^{n-1}}{\lambda_1 - \lambda_2} I \tag{61.13}$$

where λ_1, λ_2 are proper values of M and satisfy the quadratic equation

$$\lambda^2 - (M_{11} + M_{22})\lambda + M_{11}M_{22} - M_{12}M_{21} = 0. \tag{61.14}$$

Inasmuch as the transfer matrix $M^{(1)}$ has a unit determinant the consequent relation for its proper values

$$\lambda_1\lambda_2 = m_{11}m_{22} - m_{12}m_{21} = 1$$

allows us to write

$$\lambda_1 = e^{i\vartheta}, \qquad \lambda_2 = e^{-i\vartheta} \tag{61.15}$$

and thus obtain the representations

$$\xi = m_{11} + m_{22} = 2\cos\vartheta$$

$$\frac{\lambda_1^n - \lambda_2^n}{\lambda_1 - \lambda_2} = \frac{\sin n\vartheta}{\sin\vartheta} \overset{\text{def}}{=} S_{n-1}(\xi).$$

An immediate deduction from (61.8)–(61.10) is that the N-section transfer matrix acquires a diagonal character, viz.

$$(M^{(1)}(l))^N = \begin{pmatrix} (-1)^m & 0 \\ 0 & (-1)^m \end{pmatrix} \tag{61.16}$$

when

$$m_{11} + m_{22} = 2\cos\frac{m\pi}{N}, \quad m = 1, 2, \ldots, N-1 \tag{61.17}$$

and evidently the appertaining excitations at $x = 0$ and $x = Nl$ are either identical or distinguished by a phase shift of the amount π. Since a pair of linearly independent solutions to the equation (61.10) with a periodic coefficient function take the forms

$$e^{i\kappa x}\varphi(x), \qquad e^{-i\kappa x}\varphi(-x),$$

where $\varphi(x)$ has the period l, we may construct therefrom combinations which satisfy the conditions prescribed for $\psi_1(x), \psi_2(x)$ and afterwards verify that

$$\begin{aligned} \xi = m_{11} + m_{22} &= \psi_1(l) + \psi_2'(l) \\ &= 2\cos\kappa l; \end{aligned}$$

accordingly, the stipulations (61.17) are expressed by

$$\kappa l = \frac{m\pi}{N}, \quad m = 1, 2, \ldots, N-1$$

if reference is made to an effective wave number for disturbances in the periodic setting.

Let the N-period section be interposed between homogeneous strings which possess the distinct wave numbers $k_1(x < 0)$ and $k_2(x > Nl)$; and assume that

$$s(x) = \begin{matrix} e^{ik_1x} + r_N\, e^{-ik_1x}, & x \le 0 \\ t_N\, e^{ik_2(x-Nl)}, & x \ge Nl \end{matrix} \tag{61.18}$$

where r_N, t_N denote reflection and transmission coefficients of the inhomogeneous section for an incoming wave train on the left. Then

$$\begin{pmatrix} s(0) \\ s'(0) \end{pmatrix} = \begin{pmatrix} 1 + r_N \\ ik_1(1 - r_N) \end{pmatrix}, \qquad \begin{pmatrix} s(Nl) \\ s'(Nl) \end{pmatrix} = \begin{pmatrix} t_N \\ ik_2 t_N \end{pmatrix}$$

and it follows with the help of the representation (61.8) for the matrix which relates these column vectors that

$$r_N = \frac{[k_1(im_{11} - m_{12}k_2) - (m_{21} + ik_2m_{22})]S_{N-1} - i(k_1 - k_2)S_{N-2}}{[k_1(im_{11} - m_{12}k_2) + (m_{21} + ik_2m_{22})]S_{N-1} - i(k_1 + k_2)S_{N-2}}, \tag{61.19}$$

$$t_N = \frac{2\,ik_1}{[k_1(im_{11} - m_{12}k_2) + (m_{21} + ik_2m_{22})]S_{N-1} - i(k_1 + k_2)S_{N-2}}. \tag{61.20}$$

A more compact version of the preceding equation is achieved on introducing the analogous reflection and transmission coefficients r_1, t_1 for a single period, whose characterizations are implicit in (61.19), (61.20) with the choices $N = 1$, $S_{N-1} = 1$ and $S_{N-2} = 0$; thus

$$r_N = \frac{r_1 - t_1 \dfrac{k_1 - k_2}{2k_1} \dfrac{S_{N-2}}{S_{N-1}}}{1 - t_1 \dfrac{k_1 + k_2}{2k_1} \dfrac{S_{N-2}}{S_{N-1}}}, \tag{61.21}$$

and

$$t_N = \frac{1}{S_{N-1}} \frac{t_1}{1 - t_1 \dfrac{k_1 + k_2}{2k_1} \dfrac{S_{N-2}}{S_{N-1}}} \tag{61.22}$$

whence

$$\frac{r_N}{t_N} = S_{N-1} \frac{r_1}{t_1} - \frac{k_1 - k_2}{2k_1} S_{N-2}. \tag{61.23}$$

If the wave number is real throughout, with equal values $k_1 = k_2$ in the uniform outer sections, there obtains from (61.23) and the energy conservation relation,

$$|r_N|^2 + |t_N|^2 = 1,$$

a result

$$|r_N|^2 = \frac{|r_1|^2 |S_{N-1}|^2}{1 + |r_1|^2 (|S_{N-1}|^2 - 1)} \tag{61.24}$$

that specifies the reflected wave intensity in terms of the corresponding amount ($|r_1|^2$) for the smallest subunit of the configuration and a structure factor (S_{N-1}); the latter features are, in fact, typical for the exterior scattering of all periodic arrangements with a finite extension.

§62. *Reflection from a Periodically Composed Semi-infinite Range*

An integrated approach to the analysis of wave motions in settings which combine uniform and periodic sectors has a practical utility along with theoretical significance. In support of this assertion, we shall examine wave incidence from a uniform range ($x < 0$) on a periodically constituted adjacent range ($x > 0$) and derive, inter alia, an efficient means of calculating the reflection coefficient. Let us adopt, for time-harmonic states of excitation in the composite domain, the wave factor differential equation

$$\left(\frac{d^2}{dx^2} + k^2(1 + \eta(x)) \right) s(x) = 0, \quad -\infty < x < \infty \tag{62.1}$$

and postulate a continuous overall behavior of s and s' $(= \mathrm{d}s/\mathrm{d}x)$; the intrinsic properties of the half-ranges $x \lessgtr 0$ are summarized in the relations

$$\begin{aligned} \eta(x) &= 0, \qquad x<0, \\ \eta(x+l) &= \eta(x), \quad x>0 \end{aligned} \tag{62.2}$$

which do not necessarily presuppose a continuous variation for $\eta(x)$ at $x = 0$. If the excitation is attributed to a primary or incident wave train in the uniform sector, with the normalized complex wave factor

$$s^{inc} = \mathrm{e}^{ikx},$$

the concomitant solutions of (62.1) are such that

$$s(x) = \mathrm{e}^{ikx} + r\,\mathrm{e}^{-ikx}, \quad x \le 0, \tag{62.3}$$

$$s(x+l) = \mathrm{e}^{i\kappa l} s(x), \quad x \ge 0. \tag{62.4}$$

Here r designates the reflection coefficient and the translation property (62.4) conforms to that for integrals of a periodic coefficient differential equation (§48); the effective wave number κ requires specification in terms of its freely assignable counterpart, k, on the uniform range and the particulars of $\eta(x)$.

The representation

$$s(x) = \mathrm{e}^{ikx} + \frac{ik}{2}\int_0^\infty \mathrm{e}^{ik|x-x'|}\eta(x')s(x')\,\mathrm{d}x', \quad -\infty < x < \infty \tag{62.5}$$

which incorporates the given primary wave factor and assures both the continuity and differential properties of $s(x)$, provides a formula for the reflection coefficient,

$$r = \frac{ik}{2}\int_0^\infty \mathrm{e}^{ikx}\eta(x)s(x)\,\mathrm{d}x, \tag{62.6}$$

that involves a half-range Fourier transform of the product $\eta(x)s(x)$. If we apply the stipulations (62.2), (62.4) to recast the integral in (62.6) over a single period of the function $\eta(x)$, it follows that

$$\int_0^\infty \mathrm{e}^{ikx}\eta(x)s(x)\,\mathrm{d}x = \frac{1}{1-\mathrm{e}^{\mathrm{i}(\kappa+k)l}}\int_0^l \mathrm{e}^{ikx}\eta(x)s(x)\,\mathrm{d}x$$

and thence

$$r = \frac{ik}{2}\frac{\varphi(k)}{\mathrm{i}-\mathrm{e}^{\mathrm{i}(\kappa+k)l}} \tag{62.7}$$

where

$$\varphi(\zeta) = \int_0^l \mathrm{e}^{\mathrm{i}\zeta x}\eta(x)s(x)\,\mathrm{d}x. \tag{62.8}$$

On eliminating the function $\eta(x)$ from (62.8) with the help of (62.1), there obtains

$$\varphi(\zeta) = -\frac{1}{k^2}\int_0^l \mathrm{e}^{\mathrm{i}\zeta x}\left(\frac{\mathrm{d}^2 s}{\mathrm{d}x^2} + k^2 s\right)\mathrm{d}x$$

and, following successive integration by parts,

$$\begin{aligned}\varphi(\zeta) &= -\frac{1}{k^2}[\mathrm{e}^{\mathrm{i}\zeta l}s'(l) - s'(0)] + \frac{\mathrm{i}\zeta}{k^2}[\mathrm{e}^{\mathrm{i}\zeta l}s(l) - s(0)] + \frac{\zeta^2 - k^2}{k^2}\int_0^l \mathrm{e}^{\mathrm{i}\zeta x}s(x)\,\mathrm{d}x \\ &= \frac{1}{k^2}[\mathrm{e}^{\mathrm{i}(\zeta+\kappa)l} - 1][\mathrm{i}\zeta s(0) - s'(0)] + \frac{\zeta^2 - k^2}{k^2}\int_0^l \mathrm{e}^{\mathrm{i}\zeta x}s(x)\,\mathrm{d}x \end{aligned} \tag{62.9}$$

when use is made of the connections, implicit in (62.4), between the values of s and s' at the locations $x = 0, l$. Moreover, since $s(x)$ and $s'(x)$ are continuous at $x = 0$, the representation (62.3) indicates that

$$s(0) = 1 + r, \qquad s'(0) = \mathrm{i}k(1 - r)$$

and, accordingly, (62.9) takes the form

$$\varphi(\zeta) = \frac{\mathrm{i}}{k^2}[\mathrm{e}^{\mathrm{i}(\zeta+\kappa)l} - 1][\zeta - k + r(\zeta + k)] + \frac{\zeta^2 - k^2}{k^2}\int_0^l \mathrm{e}^{\mathrm{i}\zeta x}s(x)\,\mathrm{d}x. \tag{62.10}$$

Pursuant to the choices $\zeta = \pm k$ in (62.10), we establish a pair of relations,

$$\varphi(k) = \frac{2\,\mathrm{i}r}{k}[\mathrm{e}^{\mathrm{i}(k+\kappa)l} - 1], \tag{62.11}$$

$$\varphi(-k) = \frac{2\,\mathrm{i}}{k}[1 - \mathrm{e}^{\mathrm{i}(\kappa-k)l}] \tag{62.12}$$

where the former agrees with (62.7) and the latter has to do with primary wave extinction on the periodic sector. From their ratio we arrive at a representation

$$r = -\frac{1 - \mathrm{e}^{\mathrm{i}(\kappa-k)l}}{1 - \mathrm{e}^{\mathrm{i}(\kappa+k)l}}\cdot\frac{\varphi(k)}{\varphi(-k)} = \mathrm{e}^{-\mathrm{i}kl}\frac{\sin(k-\kappa)l/2}{\sin(k+\kappa)l/2}\cdot\frac{\varphi(k)}{\varphi(-k)} \tag{62.13}$$

that is manifestly independent of a scale factor for $s(x)$.

A resolution of the basic differential equation (62.1) becomes necessary in order to advance the joint determinations of $\varphi(\zeta)$ and r; let us introduce, for such purposes, an expansion of the periodic coefficient therein,

$$\begin{aligned}\eta(x) &= \sum_{m=-\infty}^{\infty} B_m\,\mathrm{e}^{-2\pi\,\mathrm{i}mx/l} = B_0 + 2\sum_{m=1}^{\infty}\mathrm{Re}\{B_m\,\mathrm{e}^{-2\pi\,\mathrm{i}mx/l}\} \\ &= B_0 + 2\sum_{m=1}^{\infty}\left[a_m\cos\frac{2m\pi x}{l} + b_m\sin\frac{2m\pi x}{l}\right]\end{aligned} \tag{62.14}$$

with

$$a_m = \operatorname{Re} B_m, \qquad b_m = \operatorname{Im} B_m,$$

$$B_m = B^*_{-m} = \frac{1}{l}\int_0^l \eta(x)\, \mathrm{e}^{2\pi\, \mathrm{i}mx/l}\, \mathrm{d}x \tag{62.15}$$

provided that $\eta(x)$ is real. Employing (62.14) and the development

$$s(x) = \mathrm{e}^{\mathrm{i}\kappa x} \sum_{n=-\infty}^{\infty} C_n\, \mathrm{e}^{-2\,\mathrm{i}n\pi x/l}, \quad x \geq 0 \tag{62.16}$$

whose coefficients are given by

$$C_n = \frac{1}{l}\int_0^l s(x)\, \mathrm{e}^{-\mathrm{i}\kappa x + 2\,\mathrm{i}n\pi x/l}\, \mathrm{d}x, \tag{62.17}$$

the outcome of substitution in (62.1) and identification of like exponential terms is a linear system for the C_n,

$$\left[k^2 - \left(\kappa - \frac{2n\pi}{l}\right)^2\right] C_n + k^2 \sum_{m=-\infty}^{\infty} C_{n-m} B_m = 0, \quad n = 0, \pm 1, \ldots. \tag{62.18}$$

If we utilize another set of coefficients, A_n, which are defined by

$$A_n = \frac{l}{k^2}\left[\left(\kappa - \frac{2n\pi}{l}\right)^2 - k^2\right] C_n \tag{62.19}$$

the system (62.18) acquires the form

$$A_n = (kl)^2 \sum_{m=-\infty}^{\infty} \frac{A_m B_{n-m}}{(\kappa l - 2m\pi)^2 - (kl)^2}, \quad n = 0, \pm 1, \ldots \tag{62.20}$$

wherein

$$B_{n-m} = \frac{1}{l}\int_0^l \eta(x)\, \mathrm{e}^{2\pi\, \mathrm{i}(n-m)x/l}\, \mathrm{d}x. \tag{62.21}$$

It may be shown that the A_n have a direct association with particular values of the transform $\varphi(\zeta)$, viz.

$$A_n = \varphi\left(\frac{2n\pi}{l} - \kappa\right) \tag{62.22}$$

and are thus, taking account of (62.9), proportional to the transform of $s(x)$ itself, namely

$$A_n = \frac{1}{k^2}\left[\left(\frac{2n\pi}{l}\right)^2 - k^2\right] \int_0^l \mathrm{e}^{\mathrm{i}((2n\pi/l)-\kappa)x} s(x)\, \mathrm{d}x. \tag{62.23}$$

The homogeneity of the system of linear equations (62.20) is a feature shared with the differential equation (62.1), and the affiliated determinantal condition

$$\left\| \delta_{nm} - (kl)^2 \frac{B_{n-m}}{(\kappa l - 2m\pi)^2 - (kl)^2} \right\| = 0 \tag{62.24}$$

serves for the specification of $\kappa l (\operatorname{mod} 2\pi)$. If $kl \ll 1$ and only the $m = n = 0$ entry is retained (i.e., $A_0 \neq 0$, $A_n = 0$, $n \neq 0$), it follows that

$$\kappa^2 \doteq k^2 \left(1 + \frac{1}{l} \int_0^l \eta(x)\, \mathrm{d}x \right) = k^2 (1 + B_0), \tag{62.25}$$

whence the average value of $\eta(x)$ on the periodicity interval accounts for the variance between κ and k at this stage of approximation. When $|\eta(x)| \ll 1$ and $kl \doteq p\pi$, p integral, the comparable and small magnitudes for the denominator in (62.20), $(\kappa l - 2m\pi)^2 - (kl)^2$, at the values $m = 0, p$ suggests the predominant status of the related coefficients A_0, A_p; moreover, after neglecting all others, there obtains from (62.24) a connection between kl and κl which is quadratic in $(kl)^2$ and (a non-degenerate) quartic in κl. The explicitly achieved determinations $(kl)^2_\pm$ are indicative of a discontinuous relationship between kl and κl, and their difference

$$\Delta (kl)^2 = (kl)^2_+ - (kl)^2_- \doteq 2(kl)^2 |B_p|$$

specifies a narrow interval,

$$\Delta (kl) \doteq kl |B_p| \doteq p\pi \left| \frac{1}{l} \int_0^l \eta(x)\, \mathrm{e}^{-2\pi \mathrm{i} px/l}\, \mathrm{d}x \right|, \tag{62.26}$$

adjoining the location $kl = p\pi$, wherein real values of κl are excluded and hence the excitation in the periodic half-range $x > 0$ diminishes exponentially with increasing values of x.

A formal means of determining the reflection coefficient r becomes available once there exists detailed information about the wave factor in the periodic range $x > 0$; in particular, if we assign a scale factor t to the representation [cf. (62.16), (62.19)]

$$s(x) = \sum_{n=-\infty}^{\infty} \frac{A_n}{(\kappa l - 2n\pi)^2 - (kl)^2}\, \mathrm{e}^{\mathrm{i}\kappa x - 2\mathrm{i} n\pi x/l}, \quad x \geq 0 \tag{62.27}$$

and utilize its counterpart (62.3) on the uniform range $x < 0$, the twin requirements of continuity for $s(x)$, $s'(x)$ at $x = 0$ imply that

$$1 + r = t \sum_{n=-\infty}^{\infty} \frac{A_n}{(\kappa l - 2n\pi)^2 - (kl)^2}$$

along with

$$k(1-r)=t\sum_{n=-\infty}^{\infty}\frac{\left(\kappa-\dfrac{2n\pi}{l}\right)A_n}{(\kappa l-2n\pi)^2-(kl)^2}.$$

After t is eliminated from the latter relations we find

$$r=-\frac{\displaystyle\sum_{n=-\infty}^{\infty}\frac{A_n}{\kappa l+kl-2n\pi}}{\displaystyle\sum_{n=-\infty}^{\infty}\frac{A_n}{\kappa l-kl-2n\pi}},\tag{62.28}$$

which expression, taken in conjunction with the system of equations (62.20) for the A_n, affords a basis for the exact calculation of the reflection coefficient.

On the other hand, the relations (62.8), (62.13) and (62.27) provide an alternative characterization

$$\begin{aligned}r&=\mathrm{e}^{-\mathrm{i}kl}\frac{\sin(k-\kappa)l/2}{\sin(k+\kappa)l/2}\,\frac{\displaystyle\int_0^l \mathrm{e}^{\mathrm{i}kx}\eta(x)s(x)\,\mathrm{d}x}{\displaystyle\int_0^l \mathrm{e}^{-\mathrm{i}kx}\eta(x)s(x)\,\mathrm{d}x}\\&=\mathrm{e}^{-\mathrm{i}kl}\frac{\sin(k-\kappa)l/2}{\sin(k+\kappa)l/2}\,\frac{\displaystyle\sum_{n=-\infty}^{\infty}\frac{A_n}{(\kappa l-2n\pi)^2-(kl)^2}\int_0^l\eta(x)\,\mathrm{e}^{\mathrm{i}(\kappa+k-(2n\pi/l))x}\,\mathrm{d}x}{\displaystyle\sum_{n=-\infty}^{\infty}\frac{A_n}{(\kappa l-2n\pi)^2-(kl)^2}\int_0^l\eta(x)\,\mathrm{e}^{\mathrm{i}(\kappa-k-(2n\pi/l))x}\,\mathrm{d}x}\end{aligned}\tag{62.29}$$

which involves the same set of coefficients A_n. It is evident that (62.29) enjoys a marked advantage over (62.28) insofar as the features making for rapidity of convergence of the constituent expansions. The structural aptness of (62.29) is underscored, especially, when $\eta(x)$ has the localized form

$$\eta(x)=\alpha\sum_{p=0}^{\infty}\delta\left(x-(2p+1)\frac{l}{2}\right)$$

and the concomitant expansions differ only by the constant phase factors $\mathrm{e}^{\pm\mathrm{i}kl/2}$, whence

$$r=\frac{\sin(k-\kappa)l/2}{\sin(k+\kappa)l/2},$$

with the value of $\kappa l (\mathrm{mod}\, 2\pi)$ implicit in (62.24); to arrive at this result from the representation (62.28) entails use of the determinations

$$B_{n-m}=\frac{\alpha}{l}(-1)^{n-m},\qquad A_n=\text{const.}\,(-1)^n$$

and explicit evaluation of the pertinent sum,

$$S = \sum_{n=-\infty}^{\infty} \frac{(-1)^n}{a + bn} = -\frac{a}{n^2} \sum_{n=-\infty}^{\infty} \frac{(-1)^n}{n^2 - \left(\frac{a}{b}\right)^2}$$

$$= \frac{\pi}{b} \operatorname{cosec}\left(\pi \frac{a}{b}\right).$$

The interrelation between the separate versions of the reflection coefficient set out in (62.28), (62.29) may be readily exposed, and it is appropriate to observe at the outset that both acquire a substantive character in terms of the coefficient ratios A_n/A_0 which follow from the homogeneous system (62.20). In fact, the pursuit of such information from (62.20) forms an aspect that is wholly independent of any intended usage in the particular fashion demanded by the first of the expressions for r, whereas the second already incorporates specific knowledge of the linear system. To make the latter point explicit and also display the link between (62.28), (62.29) we merely substitute for the coefficient A_n in (62.28) the expansion which appears on the right-hand side of (62.20); then, the other characterization for r is directly consequent to an interchange of summation orders and use of the relation

$$\sum_{n=-\infty}^{\infty} \frac{e^{2\pi ian}}{b - 2n\pi} = \frac{1}{b} - 2b \sum_{n=1}^{\infty} \frac{\cos 2\pi an}{(2n\pi)^2 - b^2} - 2i \sum_{n=1}^{\infty} \frac{2n\pi \sin 2\pi an}{(2n\pi)^2 - b^2}$$

$$= i \frac{e^{iab}}{e^{ib} - 1}, \qquad 0 < a < 1.$$

Accordingly, the distinction between the pair of formulas that specify the reflection coefficient stems from the role played by the fundamental system of linear equations in shaping their formal appearance.

§63. *Variable Wave Number Profiles with a Discontinuous Derivative or a Null Point*

We examine, in the present and following section, some time-periodic excitations for inhomogeneous configurations, based on specific wave number profiles that permit an explicit analysis and thence inferences of wider scope. A wave number variation over part of the coordinate interval is featured in the complementary assignments

$$k^2(x) = \begin{cases} k_1^2, & x < 0 \\ k_2^2 + (k_1^2 - k_2^2)\, e^{-2x/L}, & x > 0 \end{cases} \tag{63.1}$$

which have a common value at the point of contact, $x=0$, and imply a discontinuous first derivative for $k(x)$ there, viz.

$$\Delta k'(0)=\left.\frac{\mathrm{d}k}{\mathrm{d}x}\right|_{x=0+}-\left.\frac{\mathrm{d}k}{\mathrm{d}x}\right|_{x=0-}=\frac{k_2^2-k_1^2}{k_1L}, \tag{63.2}$$

given any finite value of the parameter or scale factor L; when $L>0$ it appears from (63.1) that $k^2(x)$ undergoes a continuous change between the extreme values k_1^2 and $k_2^2=\operatorname{Lim}_{x\to\infty}k^2(x)$, verging on the discontinuous as $L\to 0$.

After the replacement of x by $\xi=\mathrm{e}^{-x/L}$, the wave factor differential equation

$$\left[\frac{\mathrm{d}^2}{\mathrm{d}x^2}+k_2^2\left(1+\frac{k_1^2-k_2^2}{k_2^2}\,\mathrm{e}^{-2x/L}\right)\right]s(x)=0,\quad 0<x<\infty \tag{63.3}$$

becomes

$$\left[\frac{\mathrm{d}^2}{\mathrm{d}\xi^2}+\frac{1}{\xi}\frac{\mathrm{d}}{\mathrm{d}\xi}+(k_2L)^2+\frac{k_1^2-k_2^2}{\xi^2}\right]s(\xi)=0,\quad 0<\xi<1; \tag{63.4}$$

and the latter is brought, following a change of scale for the independent variable, to the form

$$\left(\frac{\mathrm{d}^2}{\mathrm{d}z^2}+\frac{1}{z}\frac{\mathrm{d}}{\mathrm{d}z}+1-\frac{\nu^2}{z^2}\right)Z_\nu(z)=0 \tag{63.5}$$

that characterizes cylinder functions $Z_\nu(z)$. The regular solutions of (63.4), in particular, are expressible in terms of the Bessel functions $J_\nu(z)$, with an ascending power series expansion

$$J_\nu(z)=\sum_{n=0}^{\infty}(-1)^n\frac{(z/\nu)^{\nu+2n}}{n!\,\Gamma(\nu+n+1)},\qquad \Gamma(z+1)=z\Gamma(z), \tag{63.6}$$

namely,

$$s(\xi)=J_{\pm\mathrm{i}k_2L}(\surd(k_1^2-k_2^2)L\xi)$$

and their counterparts for (63.3) become

$$s(x)=J_{\pm\mathrm{i}k_2L}(\surd(k_1^2-k_2^2)L\,\mathrm{e}^{-x/L}). \tag{63.7}$$

When $x\to\infty$ and the Bessel function arguments in (63.7) have uniformly small magnitudes we obtain, through the application of (63.6), the respective asymptotic behaviors

$$s(x)\sim\mathrm{e}^{\mp\mathrm{i}k_2x}(\surd(k_1^2-k_2^2)L/2)^{\pm\mathrm{i}k_2L}\cdot\frac{1}{\Gamma(1\pm\mathrm{i}k_2L)},\quad x\to\infty$$

which display the phase factors $\mathrm{e}^{\mp\mathrm{i}k_2x}$ appropriate to oppositely directed waves.

Thus, for an incoming or primary wave of amplitude A in the homogeneous section $x<0$, the requisite solutions of $s''+k^2(x)s=0$ are given by

$$s(x)=\begin{cases} A\,e^{ik_1x}+B\,e^{-ik_1x}, & x<0 \\ CJ_{-ik_2L}(\sqrt{(k_1^2-k_2^2)}L\,e^{-x/L}), & x>0 \end{cases} \tag{63.8}$$

with secondary wave coefficients B, C that may be related to A on imposing the requirements of continuity for s and ds/dx at $x=0$. The subsequent expression for the ratio $r=B/A$, or reflection coefficient relative to $x=0$, is

$$r=\frac{J_{-ik_2L}(\alpha)-\dfrac{i\alpha}{k_1L}J'_{-ik_2L}(\alpha)}{J_{-ik_2L}(\alpha)+\dfrac{i\alpha}{k_1L}J'_{-ik_2L}(\alpha)} \tag{63.9}$$

where

$$\alpha=\sqrt{(k_1^2-k_2^2)}L$$

and the prime signifies differentiation with respect to the argument of the Bessel function. A comparison between the amplitudes of similarly directed waves at $x=\infty, 0$ is involved in defining the transmission coefficient

$$t=\lim_{x\to+\infty}(e^{-ik_2x}s(x))/s^{inc}(0)=\frac{C}{A}\frac{\alpha^{-ik_2L}}{\Gamma(1-ik_2L)}$$

and after the ratio C/A is determined we find

$$t=\frac{2\alpha^{-ik_2L}[\Gamma(1-ik_2L)]^{-1}}{J_{-ik_2L}(\alpha)+\dfrac{i\alpha}{k_1L}J'_{-ik_2L}(\alpha)}. \tag{63.10}$$

On availing ourselves of the functional relations

$$\begin{aligned} J_\nu(z)J'_{-\nu}(z)-J'_\nu(z)J_{-\nu}(z)&=-\frac{2\sin\nu\pi}{\pi z}, \\ \Gamma(1+ix)\Gamma(1-ix)&=\frac{\pi x}{\sinh\pi x}, \quad x \text{ real} \end{aligned} \tag{63.11}$$

it is readily verified that

$$k_1(1-|r|^2)=k_2|t|^2,$$

which confirms the equality of the net energy fluxes arriving at the origin and (ultimately) departing therefrom on the other side.

Since the argument of the Bessel functions entering into (63.9), (63.10) decreases proportionally with L, we use the power series expansion (63.6) for these functions and thus recover the familiar limiting expressions

$$r \doteq \frac{k_1 - k_2}{k_1 + k_2}, \qquad t \doteq \frac{2k_1}{k_1 + k_2}, \qquad L \to 0$$

that pertain to a step function profile of the wave number. If L greatly exceeds the wave lengths $\lambda_1 = 2\pi/k_1$, $\lambda_2 = 2\pi/k_2$ and both the argument and order of the Bessel functions assume magnitudes large compared with unity, we have recourse to the (Debye, 1909) asymptotic formula

$$J_{-i\nu}(x) = \frac{1}{\sqrt{(2\pi)}} \frac{\exp\left(i\frac{\pi}{4} + \nu\frac{\pi}{2}\right)}{(\nu^2 + x^2)^{1/4}} \exp\left\{-i\sqrt{(\nu^2 + x^2)} + i\nu \log\left(\left(1 + \frac{\nu^2}{x^2}\right)^{1/2} + \frac{\nu}{x}\right)\right\} \times \left(1 + 0\left(\frac{1}{x}\right)\right) \tag{63.12}$$

from which it follows that the leading terms in the real and imaginary parts of the logarithmic derivative of $J_{-i\nu}(x)$ are given by

$$\frac{J'_{-i\nu}(x)}{J_{-i\nu}(x)} = \frac{x}{2} \frac{1}{\nu^2 + x^2} - \frac{i}{x}\sqrt{(\nu^2 + x^2)}.$$

Accordingly,

$$\frac{J'_{-ik_2L}(\alpha)}{J_{-ik_2L}(\alpha)} = -\frac{1}{2}\frac{\sqrt{(k_1^2 - k_2^2)}}{k_1^2 L} - i\frac{k_1}{\sqrt{(k_1^2 - k_2^2)}}$$

and an estimate for the reflection coefficient based upon (63.9) proves to be

$$r = \frac{\dfrac{i}{2}\dfrac{k_1^2 - k_2^2}{k_1^3 L}}{2 - \dfrac{i}{2}\dfrac{k_1^2 - k_2^2}{k_1^3 L}} \doteq \frac{i}{4}\frac{k_1^2 - k_2^2}{k_1^3 L} = -\frac{i}{4}\frac{\Delta k'(0)}{k_1^2}, \qquad k_1 L \gg 1. \tag{63.13}$$

The estimate (63.13), applicable when the local wave length is short compared to the length scale for wave number changes, accords with prior intimations of small reflection if the wave parameters are slowly varying; its latter version, wherein the magnitude of an isolated discontinuity in first derivative of the wave number number enters, has a more general significance and also contrasts markedly, as will soon become apparent, with the analogous result for wave number profiles whose derivatives are continuous in all orders.

A distinctive aspect of reflection is exposed in connection with the overall wave number assignment

$$k^2(x) = k^2(1 - e^{2x/L}), \qquad -\infty < x < \infty$$

such that

$$k(0) = 0, \tag{63.14}$$

and

$$k^2 \gtrless 0, \qquad x \gtrless 0;$$

we may, for conceptual purposes, associate the occurrence of a null or transition point for the wave number with electromagnetic excitations of ionized media, rather than with wholly mechanical wave supporting arrangements. It appears from (63.14) that proper wave motions are excluded on the range $x > 0$, where $k(x)$ has imaginary values, and this fact suggests total reflection, in the complementary range $x < 0$, of waves incident upon the transition point $x = 0$. To confirm such anticipated behavior, let us first bring the differential equation

$$\left(\frac{\mathrm{d}^2}{\mathrm{d}x^2} + k^2[1 - \mathrm{e}^{2x/L}]\right) s(x) = 0 \tag{63.15}$$

to the standard cylinder function form

$$\left(\frac{\mathrm{d}^2}{\mathrm{d}\xi^2} + \frac{1}{\xi}\frac{\mathrm{d}}{\mathrm{d}\xi} + 1 - \left(\frac{\mathrm{i}kL}{\xi}\right)^2\right) s(\xi) = 0 \tag{63.16}$$

through a change of independent variable,

$$\xi = \mathrm{i}kL\,\mathrm{e}^{x/L}, \qquad \arg \mathrm{i} = \pi/2. \tag{63.17}$$

The particular solutions of (63.5) designated as Hankel functions of the first and second kinds, viz.

$$\begin{aligned} H_\nu^{(1)}(z) &= \frac{\mathrm{i}}{\sin \nu\pi}[\mathrm{e}^{-\nu\pi\mathrm{i}} J_\nu(z) - J_{-\nu}(z)], \\ H_\nu^{(2)}(z) &= -\frac{\mathrm{i}}{\sin \nu\pi}[\mathrm{e}^{\nu\pi\mathrm{i}} J_\nu(z) - J_{-\nu}(z)] \end{aligned} \tag{63.18}$$

furnish, in linear combination, the general solution of (63.16), namely

$$s(\xi) = C_1 H_{\mathrm{i}kL}^{(1)}(\xi) + C_2 H_{\mathrm{i}kL}^{(2)}(\xi). \tag{63.19}$$

A bounded nature for $s(\xi)$ in the limit $\xi \to \mathrm{i}\infty$, $\arg \xi = \pi/2$, corresponding to $x \to +\infty$, is conditional on the choice $C_2 = 0$, since the asymptotic representations

$$\begin{aligned} H_\nu^{(1)}(z) &\sim (2/\pi z)^{1/2} \exp\left[\mathrm{i}\left(z - \frac{2\nu+1}{4}\pi\right)\right], \quad -\pi < \arg z < 2\pi \\ &\qquad |z| \gg 1, \quad |z/\nu| > 1 \\ H_\nu^{(2)}(z) &\sim (2/\pi z)^{1/2} \exp\left[-\mathrm{i}\left(z - \frac{2\nu+1}{4}\pi\right)\right], \quad -2\pi < \arg z < \pi \end{aligned} \tag{63.20}$$

imply that

$$H_\nu^{(1)}(\xi)\to 0, \qquad H_\nu^{(2)}(\xi)\to\infty, \qquad |\xi|\to\infty, \qquad \arg\xi=\pi/2.$$

Thus, dropping an unessential scale factor and reverting to the original coordinate variable x, we obtain the solution of (63.15),

$$\begin{aligned} s(x) &= H_{\mathrm{i}kL}^{(1)}(\mathrm{i}kL\,\mathrm{e}^{x/L}) \\ &= \frac{1}{\sinh kL}[\mathrm{e}^{\pi kL}J_{\mathrm{i}kL}(\mathrm{i}kL\,\mathrm{e}^{x/L}) - J_{-\mathrm{i}kL}(\mathrm{i}kL\,\mathrm{e}^{x/L})] \end{aligned} \tag{63.21}$$

which has a regular character for all x. The latter form enables us to achieve a progressive wave decomposition when $x\to-\infty$, on the basis of the power series development (63.6), with the result that

$$s(x)\simeq\frac{1}{\sinh kL}\left[\frac{(kL/2)^{\mathrm{i}kL}}{\Gamma(1+\mathrm{i}kL)}\mathrm{e}^{\mathrm{i}kx}-\frac{(kL/2)^{-\mathrm{i}kL}}{\Gamma(1-\mathrm{i}kL)}\mathrm{e}^{-\mathrm{i}kx}\right], \quad x\to-\infty. \tag{63.22}$$

From the amplitude ratio of the respective terms in (63.22) we arrive at the reflection coefficient

$$r=-(kL/2)^{-2\mathrm{i}kL}\frac{\Gamma(1+\mathrm{i}kL)}{\Gamma(1-\mathrm{i}kL)}=(kL/2)^{-2\mathrm{i}kL}\frac{\Gamma(\mathrm{i}kL)}{\Gamma(-\mathrm{i}kL)}, \tag{63.23}$$

whose absolute value equals unity for all kL.

§64. *A Multi-parameter Family of Smooth Wave Number Profiles*

The instances wherein exact analytical solutions of the variable coefficient time-reduced wave equation (58.10) may be attained are all too meager and, apart from their intrinsic significance, such solutions evidently afford a valuable basis for testing the accuracy of broadly aimed approximation procedures. In the special and soluble cases of inhomogeneity hitherto described, a continuous variation for the wave number prevails, though its first derivative has only a piecewise continuous behavior; by way of contrast, as regards both the style and outcome of the analysis, we devote attention next to an important representative from the class of functions $k(x)$ which possess continuous derivatives of all orders. Let us, following de Hoop (1965), consider the related second order differential equations

$$[a_2+b_2\exp x]\frac{\mathrm{d}^2\bar{s}}{\mathrm{d}x^2}+[a_1+b_1\exp x]\frac{\mathrm{d}\bar{s}}{\mathrm{d}x}+[a_0+b_0\exp x]\bar{s}=0, \tag{64.1}$$

$$\frac{\mathrm{d}^2 s}{\mathrm{d}x^2}+k^2(x)s=0, \tag{64.2}$$

with the link between their respective dependent variables,

$$\bar{s}(x) = \left[1 + \frac{b_2}{a_2} \exp x\right]^{\frac{1}{2}((a_1/a_2)-(b_1/b_2))} \left[\frac{b_2}{a_2} \exp x\right]^{-\frac{1}{2}(a_1/a_2)} s(x), \tag{64.3}$$

and the characterization

$$k^2(x) = \frac{a_2^2 k_1^2 + a_2 b_2 (k_1^2 + k_2^2 + K) \exp x + b_2^2 k_2^2 \exp 2x}{(a_2 + b_2 \exp x)^2} \tag{64.4}$$

wherein

$$k_1^2 = \left(\frac{a_0}{a_2} - \frac{a_1^2}{4a_2^2}\right), \qquad k_2^2 = \left(\frac{b_0}{b_2} - \frac{b_1^2}{4b_2^2}\right)$$

$$K = \frac{1}{2}\left(\frac{a_1}{a_2} - \frac{b_1}{b_2}\right)\left[1 + \frac{1}{2}\left(\frac{a_1}{a_2} - \frac{b_1}{b_2}\right)\right], \tag{64.5}$$

An integral of (64.1) can be more readily secured than is the case for (64.2), as a consequence of the fact that the coefficients in the former are linear functions of e^x whereas the latter contains a rational coefficient function.

It is stipulated that the various arbitrary constants a_i, b_i have positive magnitudes and that $k(x)$, k_1, k_2 assume either positive real or imaginary values; moreover, the limit relations

$$k^2(x) \sim \begin{cases} k_2^2 & x \to +\infty \\ k_1^2 & x \to -\infty \end{cases}, \tag{64.6}$$

suggest the existence of solutions to (64.2) and (64.1) with the asymptotic behaviors

$$s(x) \sim \begin{cases} A_1 e^{ik_1 x} + B_1 e^{-ik_1 x}, & x \to -\infty \\ A_2 e^{-ik_2 x} + B_2 e^{ik_2 x}, & x \to +\infty, \end{cases} \tag{64.7}$$

and

$$\bar{s}(x) \sim \begin{cases} (b_2/a_2)^{-\frac{1}{2}(a_1/a_2)}[A_1 e^{\alpha x} + B_1 e^{\alpha^* x}], & x \to -\infty \\ (b_2/a_2)^{-\frac{1}{2}(b_1/b_2)}[A_2 e^{\beta x} + B_2 e^{\beta^* x}], & x \to +\infty, \end{cases} \tag{64.8}$$

respectively, where

$$\begin{matrix}\alpha^* \\ \alpha^*\end{matrix} = -\frac{1}{2}\frac{a_1}{a_2} \pm ik_1, \qquad \begin{matrix}\beta \\ \beta^*\end{matrix} = -\frac{1}{2}\frac{b_1}{b_2} \pm ik_2. \tag{64.9}$$

For any assigned pair of incoming wave amplitudes A_1, A_2 the related outgoing pair B_1, B_2 are given by the linear combinations

$$B_1 = S_{11}A_1 + S_{12}A_2, \qquad B_2 = S_{21}A_1 + S_{22}A_2 \tag{64.10}$$

which involve the 'scattering matrix' elements S_{ij} that depend exclusively on the parameters of the wave number profile (64.4).

The differential equation (64.1) possesses a contour integral solution

$$\bar{s}(x) = \frac{1}{2\pi \mathrm{i}} \int_{\Gamma} \mathrm{e}^{px} \hat{s}(p) \, \mathrm{d}p \tag{64.11}$$

if the Laplace transform, $\hat{s}(p)$, of $\bar{s}(x)$ satisfies a first order difference equation

$$[a_2 p^2 + a_1 p + a_0]\hat{s}(p) + [b_2(p-1)^2 + b_1(p-1) + b_0]\hat{s}(p-1) = 0,$$

with the alternative form

$$a_2(p-\alpha)(p-\alpha^*)\hat{s}(p) + b_2(p-1-\beta)(p-1-\beta^*)\hat{s}(p-1) = 0 \tag{64.12}$$

when the parameters (64.9) are introduced. On referring to the difference equation for the Gamma function,

$$\Gamma(z+1) - z\Gamma(z) = 0,$$

we readily verify that

$$\hat{s}(p) = C_1 \left(\frac{b_2}{a_2}\right)^p \frac{\Gamma(-p+\alpha)\Gamma(-p+\alpha^*)\Gamma(p-\beta)}{\Gamma(1-p+\beta^*)} + C_2 \left(\frac{b_2}{a_2}\right)^p \frac{\Gamma(p-\beta)\Gamma(p-\beta^*)\Gamma(-p+\alpha^*)}{\Gamma(1+p-\alpha)} \tag{64.13}$$

complies with (64.12), the constants C_1, C_2 being arbitrary.

The transform $\hat{s}(p)$ has simple poles in the complex p-plane which correspond to vanishing or negative integral arguments of the individual Gamma functions in the numerator of the separate components of (64.13); if the infinite contour Γ, whose asymptotes are parallel to the imaginary axis of the p-plane, passes between the sequences of poles at

$$p = \alpha + n, \alpha^* + n \quad \text{and} \quad p = \beta - n, \beta^* - n, \qquad n = 0, 1, \ldots$$

with the former on its right-hand side, we obtain from (64.11),

$$\bar{s}(x) = \begin{cases} -\sum_{n=0}^{\infty} [\mathrm{Res}(\mathrm{e}^{px}\hat{s}(p))_{p=\alpha+n} + \mathrm{Res}(\mathrm{e}^{px}\hat{s}(p))_{p=\alpha^*+n}], & x + \log \dfrac{b_2}{a_2} < 0 \\ \sum_{n=0}^{\infty} [\mathrm{Res}(\mathrm{e}^{px}\hat{s}(p))_{p=\beta-n} + \mathrm{Res}(\mathrm{e}^{px}\hat{s}(p))_{p=\beta^*-n}], & x + \log \dfrac{b_2}{a_2} > 0. \end{cases} \tag{64.14}$$

On identifying the specific residue contributions at $p = \alpha$ and $p = \beta^*$ with the first or incident wave terms of the asymptotic expansions (64.8) in the respective limits $x \to -\infty, +\infty$, the constants C_1, C_2 are fixed in terms of the amplitude factors A_1, A_2, viz.

$$
\begin{aligned}
C_1 &= A_1\left(\frac{b_2}{a_2}\right)^{-ik_1}\frac{\Gamma(1-\alpha+\beta^*)}{\Gamma(-\alpha+\alpha^*)\Gamma(\alpha-\beta)},\\
C_2 &= A_2\left(\frac{b_2}{a_2}\right)^{-ik_2}\frac{\Gamma(1+\beta^*-\alpha)}{\Gamma(\beta^*-\beta)\Gamma(-\beta^*+\alpha^*)}.
\end{aligned}
\tag{64.15}
$$

After matching both of the $n=0$ contributions in the separate expansions (64.14) with the corresponding pairs that appear in (64.8), we obtain explicit characterizations for the scattering matrix elements, namely

$$
\begin{aligned}
S_{11} &= \left(\frac{b_2}{a_2}\right)^{-2ik_1}\frac{\Gamma(1-\alpha+\beta^*)\Gamma(-\alpha^*+\alpha)\Gamma(\alpha^*-\beta)}{\Gamma(-\alpha+\alpha^*)\Gamma(\alpha-\beta)\Gamma(1-\alpha^*+\beta^*)},\\
S_{12} &= \left(\frac{b_2}{a_2}\right)^{-ik_1+ik_2}\frac{\Gamma(1+\beta^*-\alpha)\Gamma(\alpha^*-\beta)}{\Gamma(\beta^*-\beta)\Gamma(1+\alpha^*-\alpha)},\\
S_{21} &= \left(\frac{b_2}{a_2}\right)^{-ik_1+ik_2}\frac{\Gamma(1-\alpha+\beta^*)\Gamma(-\beta+\alpha^*)}{\Gamma(-\alpha+\alpha^*)\Gamma(1-\beta+\beta^*)},\\
S_{22} &= \left(\frac{b_2}{a_2}\right)^{2ik_2}\frac{\Gamma(1+\beta^*-\alpha)\Gamma(\beta-\beta^*)\Gamma(-\beta+\alpha^*)}{\Gamma(\beta^*-\beta)\Gamma(-\beta^*+\alpha^*)\Gamma(1+\beta-\alpha)}.
\end{aligned}
\tag{64.16}
$$

The preceding results enable a ready determination of elements for the particular profile

$$
k^2(x) = k_1^2\left(1-\frac{1-(k_2/k_1)^2}{1+\mathrm{e}^{-2x/L}}\right) = \frac{k_1^2+k_2^2\,\mathrm{e}^{2x/L}}{1+\mathrm{e}^{2x/L}}, \qquad -\infty<x<\infty \tag{64.17}
$$

which implies a smooth alteration in the wave number between the unequal asymptotic values $k_1(x\to-\infty)$ and $k_2(x\to\infty)$, with a gradient that is controlled by the magnitude of the parameter L. Thus, the assignments $a_2=b_2=1$, $a_1=b_1(K=0)$ bring (64.4) into the form (64.17) and, after employing the consequent representations

$$
\begin{matrix}\alpha\\ \alpha^*\end{matrix} = -\frac{1}{2}a_1 \pm \mathrm{i}k_1, \qquad \begin{matrix}\beta\\ \beta^*\end{matrix} = -\frac{1}{2}a_1 \pm \mathrm{i}k_2
$$

along with the scale transformation

$$
k_1 \to k_1\frac{L}{2}, \qquad k_2 \to k_2\frac{L}{2}, \tag{64.18}
$$

we deduce from (64.16) a reflection coefficient

$$
r = S_{11} = \frac{\Gamma(\mathrm{i}k_1L)\Gamma\left(-\mathrm{i}\dfrac{k_1+k_2}{2}L\right)\Gamma\left(1-\mathrm{i}\dfrac{k_1+k_2}{2}L\right)}{\Gamma(-\mathrm{i}k_1L)\Gamma\left(\mathrm{i}\dfrac{k_1-k_2}{2}L\right)\Gamma\left(1+\mathrm{i}\dfrac{k_1-k_2}{2}L\right)} \tag{64.19}
$$

corresponding to incidence on the left. A simpler expression for the absolute magnitude of r is forthcoming on the basis of (63.11) and the analogous relation,

$$|\Gamma(\mathrm{i}x)| = \left(\frac{\pi}{x \sinh x}\right)^{1/2}, \quad x \text{ real}, \tag{64.20}$$

namely,

$$|r| = \frac{\sinh\left|\pi\dfrac{k_1 - k_2}{2}L\right|}{\sinh\left(\pi\dfrac{k_1 + k_2}{2}L\right)}. \tag{64.21}$$

If $(k_1 + k_2)L \ll 1$ the principal variations of wave number are confined to a range, adjoining the coordinate origin, that is short in comparison with the limiting wave lengths $\lambda_1 = 2\pi/k_1$, $\lambda_2 = 2\pi/k_2$; and use of the power series expansion for a hyperbolic sine function underlies the estimate obtained from (64.21),

$$|r| = \left|\frac{k_1 - k_2}{k_1 + k_2}\right| (1 + 0(k_1 k_2 L^2)),$$

whose leading term is appropriate for a discontinuous change in the wave number. The inequality $|k_1 - k_2|L \gg 1$ implies that comparable changes of the wave number profile (64.17) occur on an extended part of the range $(-\infty, \infty)$ and, since the arguments of the hyperbolic function in (64.21) are much larger than unity, we find

$$|r| \doteq \mathrm{e}^{-\pi k_> L}, \qquad k_> = \max(k_1, k_2), \qquad k_< L \gg 1 \tag{64.22}$$

approximately. There is evidently a significant contrast, when short wave lengths and slowly varying local specifications prevail, between the exponentially small order of magnitude for $|r|$ in (64.22), which determination refers to a smooth form of wave number possessing continuous derivatives of any order, and the algebraic or inverse power estimate supplied by (63.13) in the circumstance that a discontinuity in first derivative of the wave number is present.

Another special wave number profile, with features duplicating those found in the family (64.4), viz.

$$k^2(x) = k^2\left(1 + \frac{\alpha}{\cosh^2(x/L)}\right) = k^2 \frac{1 + 2(1 + 2\alpha)\,\mathrm{e}^{2x/L} + \mathrm{e}^{4x/L}}{(1 + \mathrm{e}^{2x/L})^2}, \tag{64.23}$$

$$\alpha > 0, \quad -\infty < x < \infty$$

has a symmetrical character and a diminishing span of non-uniformity about the origin the smaller the magnitude of L. The selections

$$k_1 = k_2 = k, \qquad a_2 = b_2 = 1,$$

and

$$K = 4k^2\alpha,$$

which yields [cf. (64.5)]

$$a_1 - b_1 = -1 + \sqrt{(1 + 4\alpha(kL)^2)} = -1 + \delta, \text{ say,} \tag{64.24}$$

after the change of scale (64.18), are prerequisite to applying the formulas (64.16); and we thus obtain, in particular,

$$r = S_{11} = S_{22} = \frac{\Gamma(\mathrm{i}kL)\Gamma\left(\frac{1}{2} - \frac{1}{2}\delta - \mathrm{i}kL\right)\Gamma\left(\frac{1}{2} + \frac{1}{2}\delta - \mathrm{i}kL\right)}{\Gamma(-\mathrm{i}kL)\Gamma\left(\frac{1}{2} - \frac{1}{2}\delta\right)\Gamma\left(\frac{1}{2} + \frac{1}{2}\delta\right)}$$

$$= \frac{\cos\dfrac{\pi\delta}{2}}{\cos\left(\dfrac{\pi\delta}{2} - \mathrm{i}\pi kL\right)} \cdot \frac{\Gamma(\mathrm{i}kL)}{\Gamma(-\mathrm{i}kL)} \cdot \frac{\Gamma\left(\frac{1}{2} - \frac{1}{2}\delta - \mathrm{i}kL\right)}{\Gamma\left(\frac{1}{2} - \frac{1}{2}\delta + \mathrm{i}kL\right)}, \tag{64.25}$$

on taking account of the relations

$$\Gamma(z)\Gamma(1 - z) = \frac{\pi}{\sin \pi z},$$

$$\Gamma\left(\frac{1}{2} + z\right)\Gamma\left(\frac{1}{2} - z\right) = \frac{\pi}{\cos \pi z}.$$

Since the Gamma function ratios in (64.25) have unit absolute value, it follows readily that

$$|r| = \frac{\left|\cos\left(\dfrac{\pi}{2}\sqrt{(1 + 4\alpha(kL)^2)}\right)\right|}{\left\{\cos^2\left(\dfrac{\pi}{2}\sqrt{(1 + 4\alpha(kL)^2)}\right) + \sinh^2(\pi kL)\right\}^{1/2}} \tag{64.26}$$

and thence

$$|r| \simeq 2|\cos(\pi kL\sqrt{\alpha})|\,\mathrm{e}^{-\pi kL}, \quad kL \gg 1 \tag{64.27}$$

with an exponentially small magnitude of the reflection coefficient typical of short wave length excitations in smoothly varying wave number environments.

The profile (64.23) can be brought into correspondence with one suited to a load M at the midpoint of a string which has uniform density ρ, viz.

$$k^2(x) = k^2\left[1 + \frac{M}{\rho}\delta(x)\right],$$

by first enforcing the moment equality,

$$\int_{-\infty}^{\infty} \frac{M}{\rho}\delta(x)\,\mathrm{d}x = \frac{M}{\rho} = \alpha \int_{-\infty}^{\infty} \frac{\mathrm{d}x}{\cosh^2 \dfrac{x}{L}} = \alpha L \int_{-\infty}^{\infty} \mathrm{d}\tanh u = 2\alpha L,$$

and then increasing the parameter α so that

$$\lim_{L\to 0} \alpha L = M/2\rho.$$

If we simplify (64.26) in accordance with the estimates

$$\sinh \pi kL \doteq \pi kL$$

$$\cos\left(\frac{\pi}{2}\sqrt{(1+4\alpha(kL)^2)}\right) \doteq \cos\left(\frac{\pi}{2}[1+2\alpha L\cdot k^2L]\right)$$

$$\doteq \cos\left(\frac{\pi}{2}\left[1+\frac{M}{\rho}k^2L\right]\right) \doteq -\frac{\pi}{2}\frac{Mk^2L}{\rho}$$

that are available in the limit $L\to 0$, the outcome,

$$|r| = \frac{Mk/2\rho}{\sqrt{(1+(Mk/2\rho)^2)}},$$

properly reproduces the back scattering amplitude for a point load.

It may be noted that a null value of $|r|$ obtains when the argument of the cosine function in (64.26) equals a half odd integer multiple of π, or

$$(1+4\alpha k^2L^2)^{1/2} = 2n+1, \quad n = 1, 2, \ldots;$$

and for large values of n an approximate version of the latter condition

$$\sqrt{(\alpha)}kL \doteq n, \quad n \gg 1$$

bears a structural resemblance to (26.10).

§65. *Connection Formulas and Their Applications*

One or more differential equations customarily provide the analytical and physical basis for determining motions of dynamical configurations and their individual solutions define particular modes or states of excitation. The availability of solutions, together with ease of interpretation, simplifies the theory for homogeneous systems, wherein stationary and progressive wave patterns are readily identified. A more complicated situation is the rule when spatial inhomogeneity figures, and here the relationships between solutions, rather than individual attributes, acquire prominence.

Consider the variable wave number profile

$$k^2(x) = k^2(1+\beta^2\,\mathrm{e}^{-2x/L}), \qquad -\infty < x < \infty \tag{65.1}$$

for which the differential equation $s''(x) + k^2(x)s(x) = 0$ admits a regular solution

$$s(x) = J_{-\mathrm{i}kL}(\beta kL\,\mathrm{e}^{-x/L}) \tag{65.2}$$

that has the asymptotic form

$$s(x) \simeq (\beta kL/2)^{-\mathrm{i}kL}\frac{\mathrm{e}^{\mathrm{i}kx}}{\Gamma(1-\mathrm{i}kL)}, \qquad x\to\infty \tag{65.3}$$

appropriate to a progressive (time-periodic) wave advancing in the positive x-direction. It is useful, for purposes of tracing and describing the excitation, rendered by (65.2), in other parts of the coordinate range to call upon a general cylinder function connection implicit in (63.18),

$$2J_\nu(z) = H_\nu^{(1)}(z) + H_\nu^{(2)}(z); \tag{65.4}$$

the latter represents a specific embodiment of the fact that any three solutions of a linear second order (homogeneous) differential equation are linearly dependent and thus satisfy a constant coefficient interrelation. The version of (65.4) which involves the Bessel function (65.2) with wholly imaginary order,

$$2J_{-\mathrm{i}kL}(\beta kL\,\mathrm{e}^{-x/L}) = H^{(1)}_{-\mathrm{i}kL}(\beta kL\,\mathrm{e}^{-x/L}) + H^{(2)}_{-\mathrm{i}kL}(\beta kL\,\mathrm{e}^{-x/L}), \tag{65.5}$$

features a pronounced disparity in magnitudes of the last pair of terms if $kL \gg 1$ and the values of x are such that $\mathrm{e}^{x/L} \lesssim \beta kL$; more precisely, as the expression (63.12) serves to bring out, these stipulations ensure a predominance for the Hankel function of the second kind in comparison with that of the first kind. Hence, we may look upon the virtual equality between the first and third terms of (65.5) when $x \to -\infty$ in the nature of a continuation for the Bessel function from the range where the values of x are large and positive. It is consistent, moreover, to associate the individual terms of the connection formula (65.5) with the amplitude factors $s(x)$ that describe, respectively, a large primary stimulus at $x = -\infty(H^{(2)}_{-\mathrm{i}kL})$, the transmitted wave at $x = +\infty(2J_{-\mathrm{i}kL})$ and a low level return at $x = -\infty(H^{(1)}_{-\mathrm{i}kL})$.

A variable amount of 'reflection', $r(x)$, attendant on the foregoing manner of excitation, can thus be characterized by the ratio of Hankel functions,

$$r(x) = \frac{H^{(1)}_{-\mathrm{i}kL}(\beta kL\,\mathrm{e}^{-x/L})}{H^{(2)}_{-\mathrm{i}kL}(\beta kL\,\mathrm{e}^{-x/L})}, \tag{65.6}$$

and, after we rely upon their asymptotic forms,

$$H^{(1),(2)}_{-\mathrm{i}kL}(\beta kL\,\mathrm{e}^{-x/L}) \simeq \left(\frac{2}{\pi kL}\right)^{1/2}\frac{\exp\left\{\mp\frac{\pi}{2}kL \mp \mathrm{i}\frac{\pi}{4}\right\}}{(1+\beta^2\,\mathrm{e}^{-2x/L})^{1/4}}\exp\left\{\pm\mathrm{i}kL\left[(1+\beta^2\,\mathrm{e}^{-2x/L})^{1/2} - \log\left(\frac{1}{\beta}\mathrm{e}^{x/L} + \left(1+\frac{1}{\beta^2}\mathrm{e}^{2x/L}\right)^{1/2}\right)\right]\right\}, \quad \beta kL\,\mathrm{e}^{-x/L} > 1 \tag{65.7}$$

it appears that our estimated measure of reflection

$$r(x) = -\mathrm{i}\,\mathrm{e}^{-\pi kL} \exp\left\{2\,\mathrm{i}kL\left[(1+\beta^2\,\mathrm{e}^{-2x/L})^{1/2} - \log\left(\frac{1}{\beta}\,\mathrm{e}^{x/L} + \left(1+\frac{1}{\beta^2}\,\mathrm{e}^{-2x/L}\right)^{1/2}\right)\right]\right\}$$

has a uniform modulus

$$|r(x)| \doteq \mathrm{e}^{-\pi kL}, \quad kL \gg 1 \tag{65.8}$$

with the exponentially small magnitude appropriate to a smooth wave number profile and short wave lengths.

The transmission factor defined by

$$t = \lim_{x\to+\infty} \frac{2J_{-\mathrm{i}kL}(\beta kL\,\mathrm{e}^{-x/L})}{H^{(2)}_{-\mathrm{i}kL}(\beta kL)} \cdot \mathrm{e}^{-\mathrm{i}kx} \tag{65.9}$$

involves a comparison of the Bessel function at large positive values of x (where (65.3) applies) with the large Hankel function of the second kind at $x = 0$, say, (where (65.7) remains in force); applying the representation

$$\frac{1}{\Gamma(1-\mathrm{i}kL)} = \left(\frac{\sinh \pi kL}{\pi kL}\right)^2 \exp\{-\mathrm{i} \arg \Gamma(1-\mathrm{i}kL)\}$$

we obtain the estimate for its modulus

$$|t| = (1-\mathrm{e}^{-2\pi kL})^{1/2}(1+\beta^2)^{1/4}, \quad kL \gg 1 \tag{65.10}$$

and thence the balance equation

$$k(\infty)|t|^2 = k(0)(1-|r|^2).$$

Connection formulas or linear relations among solutions of the hypergeometric differential equation,

$$x(1-x)\frac{\mathrm{d}^2y}{\mathrm{d}x^2} + [\gamma - (\alpha+\beta+1)x]\frac{\mathrm{d}y}{\mathrm{d}x} - \alpha\beta y = 0, \tag{65.11}$$

find a direct and significant use in analyzing particular models of layered or inhomogeneous media, stemming from the feasibility of transforming said equation into another linear version,

$$\left(\frac{\mathrm{d}^2}{\mathrm{d}\xi^2} + k^2(\xi)\right) s(\xi) = 0 \tag{65.12}$$

where the lone coefficient function $k(\xi)$ may be regarded as a variable wave number. The hypergeometric power series,

$$F(\alpha,\beta,\gamma;x) = 1 + \frac{\alpha\beta}{1\cdot\gamma}x + \frac{\alpha(\alpha+1)\beta(\beta+1)}{1\cdot 2\cdot \gamma(\gamma+1)}x^2 + \cdots, \tag{65.13}$$

which is manifestly convergent on the range $|x| < 1$ and satisfies (65.11), plays a central role in representing fundamental systems (i.e., linearly independent pairs) of solutions relative to the (regular) singular points of (65.11), located at

$x = 0, 1, \infty$. The fundamental systems are, in fact, characterized by the expressions

$$\left.\begin{aligned} y_1(x) &= F(\alpha, \beta, \gamma; x) \\ y_2(x) &= x^{1-\gamma} F(\alpha-\gamma+1, \beta-\gamma+1, 2-\gamma; x) \end{aligned}\right\} \text{ at } x = 0,$$

$$\left.\begin{aligned} y_3(x) &= F(\alpha, \beta, \alpha+\beta-\gamma+1; 1-x) \\ y_4(x) &= (1-x)^{\gamma-\alpha-\beta} F(\gamma-\alpha, \gamma-\beta, \gamma-\alpha-\beta+1; 1-x) \end{aligned}\right\} \text{ at } x = 1,$$

$$\left.\begin{aligned} y_5 &= x^{-\alpha} F\left(\alpha, \alpha-\gamma+1, \alpha-\beta+1; \frac{1}{x}\right) \\ y_6 &= x^{-\beta} F\left(\beta, \beta-\gamma+1, \beta-\alpha+1; \frac{1}{x}\right) \end{aligned}\right\} \text{ at } x = \infty$$

with the proviso that the difference between any two of the parameters α, β, γ is non-integral.

Although the existence of a singular point at $x = 1$ limits the convergence of the expansions for y_1, y_2 and y_5, y_6 to the intervals $|x| < 1$ and $|x| > 1$, respectively, the analytic functions defined by these expansions possess continuations outside the aforesaid intervals which are likewise solutions of the differential equation. The requisite linear dependence of three solutions to a second order differential equation implies, for instance, that the continuation of y_5 onto the range $|x| < 1$ is a linear combination of the functions y_1, y_2; and the pertinent connection has the explicit form

$$y_5(x) = (-1)^{-\alpha} \frac{\Gamma(\alpha-\beta+1)\Gamma(1-\gamma)}{\Gamma(1-\beta)\Gamma(1+\alpha-\gamma)} y_1(x) + (-1)^{\gamma-1-\alpha} \frac{\Gamma(\alpha-\beta+1)\Gamma(\gamma-1)}{\Gamma(\gamma-\beta)\Gamma(\alpha)} y_2(x) \tag{65.14}$$

where the coefficients are given in terms of Gamma functions.

To relate solutions of the hypergeometric and time-reduced wave equations, (65.11) and (65.12), we observe, firstly, that the special transformations $(x, y) \to \xi(s)$ of both independent and dependent variables, viz.

$$x = x(\xi), \quad y = sp(\xi) \tag{65.15}$$

preserve the linearity insofar as the latter is concerned and, secondly, that the condition

$$2 \frac{\mathrm{d}p}{\mathrm{d}\xi} \frac{\mathrm{d}x}{\mathrm{d}\xi} - p(\xi) \left[\frac{\mathrm{d}^2 x}{\mathrm{d}\xi^2} - \frac{\gamma-(\alpha+\beta+1)}{x(1-x)} \left(\frac{\mathrm{d}x}{\mathrm{d}\xi} \right)^2 \right] = 0 \tag{65.16}$$

eliminates a term containing the first order derivative, $\mathrm{d}s/\mathrm{d}\xi$, from (65.12). Thus, leaving the transformation of independent variables, $x = x(\xi)$, unspecified and integrating the first order equation (65.16) for $p(\xi)$, we arrive at a link between solutions of (65.11), (65.12),

$$s = x^{\gamma/2}(1-x)^{1/2(\alpha+\beta-\gamma+1)}\left(\frac{dx}{d\xi}\right)^{-1/2} y, \tag{65.17}$$

which omits an inessential scale factor.

If the particular choice

$$x = -e^{\xi} \tag{65.18}$$

is made, the wave number profile in (65.12) turns out to have a form originally investigated by Epstein and previously considered (§64) without reference to the hypergeometric function, viz.

$$k^2(\xi) = C_1 + \frac{(C_2 - C_1)\,e^{\xi}}{1+e^{\xi}} + \frac{C_3\,e^{\xi}}{(1+e^{\xi})^2} \tag{65.19}$$

where the parameters or coefficients are such that

$$C_1 = -\frac{1}{4}(\gamma-1)^2, \qquad C_2 = -\frac{1}{4}(\alpha-\beta)^2, \qquad C_3 = \frac{1}{4}[(\alpha+\beta-\gamma)^2 - 1]. \tag{65.20}$$

On the basis of the power series developments for the functions $y_1(x)$, $y_2(x)$ about $x = 0$ and the coupling between small values of x and large (negative) magnitudes of ξ implicit in the transformation (65.18), we deduce the asymptotic behaviors of their counterparts $s_1(\xi)$, $s_2(\xi)$, namely

$$s_{1,2}(\xi) \simeq (-1)^{\pm(\gamma-1)2}\exp\left(\pm\frac{\gamma-1}{2}\xi\right), \quad \xi \to -\infty \tag{65.21}$$

through recourse to (65.17). Furthermore, having regard for the paired solutions $y_5(x)$ and $s_5(\xi)$, it follows as a consequence of approximation to the former at large negative values of x (or small values of $1/x$) that the latter possesses the asymptotic form

$$s_5(\xi) \simeq (-1)^{\{(\gamma-1)/2\}-\alpha}\exp\left(\frac{\beta-\alpha}{2}\xi\right), \quad \xi \to +\infty. \tag{65.22}$$

If we let

$$\frac{\gamma-1}{2} = ik_1, \qquad \frac{\beta-\alpha}{2} = ik_2,$$

the functions s_1, s_2 and s_5 can be associated, respectively, with incoming and outgoing waves at $\xi = -\infty$ and an outgoing wave at $\xi = +\infty$. Accordingly, the connection formula (65.14) provides, after directly comparing the coefficients for these wave factors and taking account of their relative normalizations (65.21), (65.22), the definite ratios of outgoing to incoming amplitudes,

$$r = \frac{s_2(\xi)}{s_1(\xi)}\,e^{-2ik_1\xi} = \frac{\Gamma(\gamma-1)\Gamma(1-\beta)\Gamma(1+\alpha-\gamma)}{\Gamma(1-\gamma)\Gamma(\gamma-\beta)\Gamma(\alpha)},$$

and

$$t = \frac{s_5(\xi)}{s_1(\xi)} = \frac{\Gamma(1-\beta)\Gamma(1+\alpha-\gamma)}{\Gamma(1-\gamma)\Gamma(\alpha-\beta+1)}. \tag{65.23}$$

§66. Approximate Solutions of a Wave-like Nature for Inhomogeneous Settings

As a consequence of the limitations on coping exactly with the variable coefficient differential equation

$$\left(\frac{\mathrm{d}^2}{\mathrm{d}x^2}+k^2(x)\right)s(x)=0, \tag{66.1}$$

substantial importance attaches to efforts which aim at securing uniformly accurate approximate solutions. It may be anticipated that conditions are favorable to the latter objective if the wave number $k(x)$ is endowed with a slowly varying behavior on the scale of dimensions set by the local wave length $\lambda(x)=2\pi/k(x)$, so that

$$\frac{1}{k}\frac{\mathrm{d}k}{\mathrm{d}x}\cdot\lambda=2\pi\frac{\mathrm{d}k/\mathrm{d}x}{(k(x))^2}\ll 1, \tag{66.2}$$

or if

$$k(x)=kf\left(\frac{x}{L}\right) \tag{66.3}$$

and there is but little change in $k(x)$ over intervals of the coordinate x whose span is a fraction of the reference length L. In terms of the variable $\xi=x/L$ the wave factor differential equation becomes

$$\left(\frac{\mathrm{d}^2}{\mathrm{d}\xi^2}+(kL)^2f^2(\xi)\right)s(\xi)=0 \tag{66.4}$$

and greater magnitudes for the parameter or scale factor L are companion to reduced local variation of $f(\xi)$. The circumstance of a large numerical parameter prompts the search for an asymptotic solution of the differential equation (66.4), with the expectation of relative simplicity and overall validity on intervals where $f(\xi)$ is bounded away from zero. To characterize such solutions in a preliminary or unsystematic manner we express the wave factor in terms of (real) amplitude and phase functions, $A(\xi)$, $\Phi(\xi)$, namely

$$s(\xi)=A(\xi)\,\mathrm{e}^{\mathrm{i}kL\Phi(\xi)} \tag{66.5}$$

and thence obtain from (66.4) the relationship

$$\{[(kL)^2\{f^2(\xi)-(\Phi'(\xi))^2\}+A''(\xi)]+\mathrm{i}kL[2\Phi'(\xi)A'(\xi)+\Phi''(\xi)A(\xi)]\}\,\mathrm{e}^{\mathrm{i}kL\Phi(\xi)}=0$$

in which primes signify differentiation. Equating to zero both real and imaginary parts of the complex multiplier for the exponential factor, a pair of coupled equations follow,

$$(\Phi'(\xi))^2=f^2(\xi)+\frac{1}{(kL)^2}A''(\xi), \tag{66.6}$$

$$\frac{A'(\xi)}{A(\xi)} = -2\frac{\Phi''(\xi)}{\Phi'(\xi)}. \tag{66.7}$$

When the term involving a small coefficient, $1/(kL)^2$, is left out of (66.6) we obtain the phase representation

$$\Phi(\xi) = \pm \int^{\xi} f(\xi)\,\mathrm{d}\xi + C_1 \tag{66.8}$$

and the subsequent integral of (66.7) specifies $A(\xi)$, viz.

$$A(\xi) = \frac{C_2}{\{f(\xi)\}^{1/2}}, \tag{66.9}$$

with two arbitrary constants C_1, C_2; the corresponding approximation for the wave factor is, according to (66.5),

$$s(\xi) = \frac{1}{\{f(\xi)\}^{1/2}}\left\{A_1 \exp\left(\mathrm{i}kL\int_{\xi_1}^{\xi} f(\xi)\,\mathrm{d}\xi\right) + B_1 \exp\left(-\mathrm{i}kL\int_{\xi_1}^{\xi} f(\xi)\,\mathrm{d}\xi\right)\right\}$$

or

$$s(x) = \frac{1}{\{k(x)\}^{1/2}}\left\{A_1 \exp\left(\mathrm{i}\int_{x_1}^{x} k(x)\,\mathrm{d}x\right) + B_1 \exp\left(-\mathrm{i}\int_{x_1}^{x} k(x)\,\mathrm{d}x\right)\right\} \tag{66.10}$$

if we revert to the original coordinate variable.

Taking into account a complex time-factor $\exp(-\mathrm{i}\omega t)$, an inspection of the complete phase functions linked to the component terms of (66.10),

$$\Phi_{\pm}(x,t) = \pm\int_{x_1}^{x} k(x)\,\mathrm{d}x - \omega t, \qquad \frac{\partial \Phi_{\pm}}{\partial x} = \pm k(x),$$

reveals their aptness for disturbances which travel along the positive and negative x-directions, respectively, with local wave number $k(x)$; in addition to its obvious role as the generator of phase changes along the route of a progressive disturbance, the variable wave number also has a direct influence on the amplitude thereof, manifest by the factor $\{k(x)\}^{-1/2}$ in (66.10). If we employ the nomenclature

$$\varphi_{\pm}(x) = \frac{1}{\{k(x)\}^{1/2}} \exp\left\{\pm \mathrm{i}\int_{x_1}^{x} k(x)\,\mathrm{d}x\right\} \tag{66.11}$$

in connection with the individual members of the representation (66.10) and note the corresponding derivatives

$$\varphi'_{\pm}(x) = \pm \mathrm{i}k(x)\left[1 \pm \frac{\mathrm{i}}{2}\frac{k'(x)}{(k(x))^2}\right]\varphi_{\pm}(x) \tag{66.12}$$

it appears that a slow variation for $k(x)$ in the sense implied by (66.2) yields

$$\varphi'_{\pm}(x) \doteq \pm \mathrm{i}k(x)\varphi_{\pm}(x)$$

approximately, which result is tantamount to neglecting the amplitude changes of $\varphi_\pm(x)$.

The independence of the functions $\varphi_\pm(x)$ also receives support from the easily demonstrable proposition that the affiliated wave-like disturbances convey a fixed time-average energy flux; specifically, when the composite representation (66.10) and the relations (66.12) are employed for an evaluation of the quantity $\operatorname{Im}(s^*(\mathrm{d}s/\mathrm{d}x))$ the outcome is simply $|A_1|^2-|B_1|^2$, without any amplitude cross product or interference contribution. The perfectly matched or reflectionless character of $\varphi_\pm(x)$ has led to their placement in the category of geometrical optics wave functions, though they are more commonly known by the names of early proponents (Liouville, Green) and later investigators (Rayleigh, Jeffreys, Kramers, Wentzel, and Brillouin, to cite but a few) who dealt with asymptotic solutions of the differential equation (66.1).

It need hardly be emphasized that definitive conclusions based on the preceding approximation, such as the existence of non-reflecting wave functions in inhomogeneous settings, cannot be expected to possess general validity for excitations which obey the equation (66.1); the functions $\varphi_\pm(x)$ are, in fact, solutions of another differential equation

$$\left(\frac{\mathrm{d}^2}{\mathrm{d}x^2}+k^2(x)+\frac{1}{2}\frac{k''(x)}{k(x)}-\frac{3}{4}\left(\frac{k'(x)}{k(x)}\right)^2\right)\varphi_\pm(x)=0, \tag{66.13}$$

save for the special circumstance wherein the wave number satisfies the equation

$$\frac{k''(x)}{k(x)}=\frac{3}{2}\left(\frac{k'(x)}{k(x)}\right)^2$$

or

$$\{k(x)\}^{1/2}\frac{\mathrm{d}}{\mathrm{d}x}\left[\{k(x)\}^{-3/2}\frac{\mathrm{d}k}{\mathrm{d}x}\right]=0$$

and is given explicitly by the two parameter form

$$k(x)=(\alpha+\beta x)^{-2}. \tag{66.14}$$

The functions $\varphi_\pm(x)$ associated with the wave number profile (66.14) thus enjoy an exact, rather than asymptotic, status and describe wholly uncoupled disturbances. We conclude our description of the functions $\varphi_\pm(x)$ with the remark that their evident singularity at a null (or turning) point of $k(x)$ undermines any claim to local adequacy as an approximate solution of (66.1), for this behavior is not shared by any actual solution thereof.

Before taking up the matter of improvements on the geometrical approximations $\varphi_\pm(x)$ in the vicinity of turning points, there is merit in detailing the relationship of such functions to a systematic development and an alternative method for obtaining integrals of (66.1).

Let us adopt the representations

$$s_+(\xi) = \exp\left\{ikL \int_{\xi_1}^{\xi} \chi(\xi)\,\mathrm{d}\xi\right\}, \qquad s_-(\xi) = \exp\left\{-ikL \int_{\xi_1}^{\xi} \chi^*(\xi)\,\mathrm{d}\xi\right\} \tag{66.15}$$

for prospective solutions of (64.4), with the complex-valued functions

$$\chi(\xi) = \psi(\xi) + \mathrm{i}\eta(\xi), \qquad \chi^*(\xi) = \psi(\xi) - \mathrm{i}\eta(\xi) \tag{66.16}$$

whose real and imaginary parts generate, separately, the phase and amplitude variations of $s_\pm(\xi)$. The function $s_+(\xi)$ acquires its intended character if $\chi(\xi)$ satisfies a first order, non-linear differential equation

$$\chi^2(\xi) - f^2(\xi) - \frac{\mathrm{i}}{kL}\chi'(\xi) = 0, \tag{66.17}$$

and, pursuant to the assumption that a solution thereof admits the development

$$\chi(\xi) = \chi_0(\xi) + \frac{1}{kL}\chi_1(\xi) + \frac{1}{(kL)^2}\chi_2(\xi) + \cdots \tag{66.18}$$

in reciprocal powers of kL, we obtain from (66.17) the relation

$$(\chi_0^2 - f^2) + \frac{1}{kL}(2\chi_0\chi_1 - \mathrm{i}\chi_0') + \frac{1}{(kL)^2}(2\chi_0\chi_2 + \chi_1^2 - \mathrm{i}\chi_1')$$
$$+ \frac{1}{(kL)^3}(2\chi_0\chi_3 + 2\chi_1\chi_2 - \mathrm{i}\chi_2') + \cdots = 0$$

after grouping terms with like powers of kL. There follows, on securing the nullity of such groups, the particular determinations

$$\chi_0 = \pm f, \qquad \chi_1 = \frac{\mathrm{i}}{2}\frac{f'}{f}, \qquad \chi_2 = \pm\frac{1}{8f}\left[3\left(\frac{f'}{f}\right)^2 - 2\frac{f''}{f}\right]$$
$$\chi_3 = -\frac{\mathrm{i}}{8f^2}\left[\frac{f'''}{f} - 6\frac{f'f''}{f^2} + 6\frac{f'^3}{f^3}\right], \ldots \tag{66.19}$$

as well as the general recurrence

$$\mathrm{i}\chi_{n-1}' = 2\chi_0\chi_n + 2\chi_1\chi_{n-1} + 2\chi_2\chi_{n-2} + \cdots + \begin{cases} \chi_{(n/2)}^2, & n \text{ even} \\ 2\chi_{(n+1/2)}\chi_{(n-1/2)}, & n \text{ odd.} \end{cases} \tag{66.20}$$

It is proper to associate the alternative choices of sign which enter into the specification of $\chi_0, \chi_2, \ldots$ with the respective functions $s_\pm(\xi)$ and thus, noting the general connection, implied by (66.17), between the real and imaginary parts of χ, namely

$$\eta(\xi) = \frac{1}{2kL}\frac{\psi'(\xi)}{\psi(\xi)}, \tag{66.21}$$

to record the consequent versions of (66.15)

$$s_{\pm}(\xi) = \frac{1}{\{\psi(\xi)\}^{1/2}} \exp\left\{\pm ikL \int_{\xi_1}^{\xi} \psi(\xi)\, d\xi\right\} \tag{66.22}$$

that involve a single real function ψ. If the latter is characterized by the initial term of the expansion

$$\begin{aligned}\psi &= \operatorname{Re} \chi \\ &= f + \frac{1}{8(kL)^2}\left[3\left(\frac{f'}{f}\right)^2 - 2\frac{f''}{f}\right] + \cdots\end{aligned}$$

and the coordinate variable $x = \xi L$ is reinstated, we obtain a prior (geometrical) approximation to the wave factor, i.e.,

$$s_{\pm}(\xi) \to \varphi_{\pm}(x), \quad kL \to \infty.$$

The utility of the development (66.18) rests upon suitable magnitudes for the ratios of individual terms and it may be surmised, on the basis of the specific pair

$$\frac{1}{kL}\frac{|\chi_1|}{\chi_0} = \frac{1}{2}\frac{k'(x)}{(k(x))^2}$$

and

$$\frac{1}{(kL)^2}\frac{\chi_2}{\chi_0} = \frac{1}{8}\left\{3\left(\frac{k'(x)}{(k(x))^2}\right)^2 - 2\frac{k''(x)}{(k(x))^3}\right\},$$

that sufficient conditions of a favorable, though impractical, kind are expressed by a sequence of inequalities

$$\frac{k'(x)}{(k(x))^2} \ll 1, \qquad \frac{k''(x)}{(k(x))^3} \ll 1, \qquad \frac{k'''(x)}{(k(x))^4} \ll 1, \ldots \tag{66.23}$$

which involve all orders of wave number differentiation.

We observe, before terminating our direct consideration of the wave-like functions $\varphi_{\pm}(x)$, that these enable us to supplant the second order differential equation (66.1) with a simultaneous pair of first order equations. Let

$$s(x) = a(x)\varphi_+(x) + b(x)\varphi_-(x) \tag{66.24}$$

and note that

$$s'(x) = a(x)\varphi_+'(x) + b(x)\varphi_-'(x) \tag{66.25}$$

if the relationship

$$a'(x)\varphi_+(x) + b'(x)\varphi_-(x) = 0 \tag{66.26}$$

is imposed on the functions $a(x)$, $b(x)$. Further to the differentiation of (66.25) and utilization of the equations (66.1), (66.13) satisfied by $s(x)$, $\varphi_{\pm}(x)$, respectively, as well as (66.24), (66.26), there obtains

$$\begin{aligned}\frac{\mathrm{d}a}{\mathrm{d}x} &= -\mathrm{i}\hat{k}(x)\left[a(x)+b(x)\exp\left\{-2\,\mathrm{i}\int_{x_1}^{x} k(\xi)\,\mathrm{d}\xi\right\}\right]\\ \frac{\mathrm{d}b}{\mathrm{d}x} &= -\exp\left\{2\,\mathrm{i}\int_{x_1}^{x} k(\xi)\,\mathrm{d}\xi\right\}\frac{\mathrm{d}a}{\mathrm{d}x}\\ &= \mathrm{i}\hat{k}(x)\left[a(x)\exp\left\{2\,\mathrm{i}\int_{x_1}^{x} k(\xi)\,\mathrm{d}\xi\right\}+b(x)\right]\end{aligned} \tag{66.27}$$

with

$$\hat{k}(x)=\frac{1}{2k(x)}\left[\frac{1}{2}\frac{k''(x)}{k(x)}-\frac{3}{4}\left(\frac{k'(x)}{k(x)}\right)^2\right]. \tag{66.28}$$

The first order system (66.27) assumes a more compact version in terms of the functions

$$c_{\pm}(x)=\binom{a(x)}{b(x)}\exp\left\{\pm\mathrm{i}\int_{x_1}^{x}\hat{k}(\xi)\,\mathrm{d}\xi\right\}, \tag{66.29}$$

namely,

$$\frac{\mathrm{d}c_{\pm}(x)}{\mathrm{d}x}=\hat{k}(x)c_{\mp}(x)\exp\left\{\mp 2\,\mathrm{i}\int_{x_1}^{x}(k(\xi)-\hat{k}(\xi))\,\mathrm{d}\xi\mp\mathrm{i}\pi/2\right\}, \tag{66.30}$$

and it follows readily therefrom that

$$|c_+|^2-|c_-|^2=|a|^2-|b|^2=\text{constant} \tag{66.31}$$

provided the wave number $k(x)$ is real.

§67. Improvements on the WKBJ or Geometrical Optics Wave Functions

We present and apply in this section another differential equation formulation that facilitates the derivation of refinements to the WKBJ wave functions and has readily interpretable consequences; the analysis rests on a suitably chosen or related second order equation with variable coefficients that involve derivatives of the inhomogeneity parameter and exact solutions of the WKBJ type in the homogeneous limit. To arrive at a related equation we transform both the independent and dependent variable of the original, say

$$\left(\frac{\mathrm{d}^2}{\mathrm{d}x^2}+k^2(x)\right)s(x)=0, \tag{67.1}$$

in accordance with the prescriptions

$$\xi=\int_0^x k(x)\,\mathrm{d}x \tag{67.2}$$

and

$$s = \frac{\psi}{\sqrt{k}}; \tag{67.3}$$

evidently, the new independent variable, ξ, has a scale which is determined by the local wave length $2\pi/k(x)$, inasmuch as $\mathrm{d}\xi = k(x)\,\mathrm{d}x$. The outcome of effecting such transformations in (67.1) is an equation

$$\frac{\mathrm{d}^2\psi}{\mathrm{d}\xi^2} + \left\{1 + \frac{1}{4}\left(\frac{\mathrm{d}k/\mathrm{d}\xi}{k(\xi)}\right)^2 - \frac{1}{2}\left(\frac{\mathrm{d}^2k/\mathrm{d}\xi^2}{k(\xi)}\right)\right\}\psi = 0 \tag{67.4}$$

to be contrasted with the original in that the variable coefficients are small when k changes gradually.

If our physical model refers to plane polarized electromagnetic excitations in a medium stratified along the x-direction, the position factors $E_y(x)$, $H_z(x)$ for a pair of non-vanishing field components

$$E_y(x)\,\mathrm{e}^{-\mathrm{i}\omega t}, \quad H_z(x)\,\mathrm{e}^{-\mathrm{i}\omega t}$$

satisfy the Maxwell equations

$$\mathrm{i}kH_z = \frac{\mathrm{d}E_y}{\mathrm{d}x}, \quad \mathrm{i}kn^2(x)E_y = \frac{\mathrm{d}H_z}{\mathrm{d}x} \tag{67.5}$$

wherein $n^2(x)$, $k = \omega/c$ characterize the dielectric constant of the medium and the uniform wave number in vacuo, respectively. It follows, through successive eliminations in the first order system (67.5), that E_y and H_z are individually specified by the same equation

$$\left(\frac{\mathrm{d}^2}{\mathrm{d}x^2} + k^2n^2(x)\right)\begin{bmatrix} E_y \\ H_z \end{bmatrix} = 0, \tag{67.6}$$

which is identical in form with (67.1). Pursuant to the adoption of an independent variable

$$\xi = k\int_0^x n(x)\,\mathrm{d}x \tag{67.7}$$

the system (67.5) becomes

$$\mathrm{i}H_z = n(\xi)\frac{\mathrm{d}E_y}{\mathrm{d}\xi}, \quad \mathrm{i}n(\xi)E_y = \frac{\mathrm{d}H_z}{\mathrm{d}\xi} \tag{67.8}$$

and thus, after introducing a single dependent variable, ψ, such that

$$H_z = \sqrt{(n)}\psi, \quad E_y = \frac{-\mathrm{i}}{n}\frac{\mathrm{d}}{\mathrm{d}\xi}(\sqrt{(n)}\psi), \tag{67.9}$$

where the latter expression stems directly from the second of the relations (67.8), we obtain the differential equation

$$\frac{\mathrm{d}^2\psi}{\mathrm{d}\xi^2} + \left\{1 + \frac{1}{2}\frac{\mathrm{d}^2n/\mathrm{d}\xi^2}{n(\xi)} - \frac{3}{4}\left(\frac{\mathrm{d}n/\mathrm{d}\xi}{n(\xi)}\right)^2\right\}\psi = 0 \tag{67.10}$$

as a consequence of substituting both of the expressions (67.9) into the first of the relations (67.8).

The related equations for an auxiliary wave function ψ, which provides a full accounting of excitations in the circumstances envisaged, are usefully recast in terms of the logarithmic derivative of k or n; thus, if

$$\nu = \frac{1}{2}\frac{\mathrm{d}}{\mathrm{d}\xi}\log\frac{1}{k}, \tag{67.11}$$

and a like measure of variation is chosen for n, it appears that

$$\frac{\mathrm{d}^2\psi}{\mathrm{d}\xi^2} + \left\{1 \pm \frac{\mathrm{d}\nu}{\mathrm{d}\xi} - \nu^2\right\}\psi = 0 \tag{67.12}$$

with the alternate ($\pm$) signs linked to the original equations (67.4) and (67.10), respectively. An especially simple related equation obtains when the wave number has a particular composite representation

$$k(x) = \begin{matrix} k_0\dfrac{L}{x}, & x > L \\ k_0, & x < L \end{matrix}, \tag{67.13}$$

which incorporates both uniform and non-uniform ranges; in keeping with the respective coordinate transformations,

$$\xi = \begin{matrix} k_0 L \log\dfrac{x}{L}, & x > L \\ k_0(x - L), & x < L \end{matrix}, \tag{67.14}$$

whose common version

$$\xi = \int_L^x k(x)\,\mathrm{d}x$$

evidently fixes an origin for ξ at the position $x = L$, we find that

$$k(\xi) = k_0\,\mathrm{e}^{-\xi/k_0 L}, \quad \xi > 0 \tag{67.15}$$

and thence

$$\nu = \begin{matrix} 1/2k_0L, & \xi > 0 \\ 0, & \xi < 0 \end{matrix}, \tag{67.16}$$

Accordingly, the related equations on both halves of the ξ-range feature constant coefficients, namely

$$\frac{\mathrm{d}^2\psi}{\mathrm{d}\xi^2} + (1 - \varepsilon^2)\psi = 0, \quad \xi > 0, \tag{67.17}$$

and

$$\frac{\mathrm{d}^2\psi}{\mathrm{d}\xi^2} + \psi = 0, \qquad \xi < 0 \tag{67.18}$$

where

$$\varepsilon = \frac{1}{2k_0L}. \tag{67.19}$$

The connections between solutions of (67.17), (67.18) at $\xi = 0$ acquire their definitive form through the requisite continuity of $s = \psi/\surd(k)$ and ds/dx; inasmuch as k is everywhere continuous, these connections prove to be

$$(\psi)_{\xi\to 0_+} = (\psi)_{\xi\to 0_-}, \tag{67.20}$$

and

$$\frac{d}{d\xi}(e^{\varepsilon\xi}\psi)_{\xi\to 0_+} = \frac{d}{d\xi}(\psi)_{\xi\to 0_-}$$

or

$$\left(\frac{d\psi}{d\xi}\right)_{\xi\to 0_+} + \varepsilon(\psi)_{\xi\to 0_+} = \left(\frac{d\psi}{d\xi}\right)_{\xi\to 0_-}. \tag{67.21}$$

If our focus is directed at the specific integrals of (67.17), (67.18),

$$\psi = \begin{cases} e^{i\xi} + r\,e^{-i\xi}, & \xi < 0 \\ t\,e^{i\surd(1-\varepsilon^2)\xi}, & \xi > 0 \end{cases} \tag{67.22}$$

which manifest an incoming disturbance $e^{i\xi} = e^{ik_0(x-L)}$, on the uniform range $\xi < 0$, along with reflected and transmitted waves of amplitude r, t, and use be made of the connections (67.20), (67.21) we secure the explicit determinations

$$\begin{aligned} r &= \frac{i}{\varepsilon}[1 - \surd(1-\varepsilon^2)], \\ t &= \frac{i}{\varepsilon}[1 - \surd(1-\varepsilon^2) - i\varepsilon]. \end{aligned} \tag{67.23}$$

There is a direct correlation between partial contributions to the reflected wave amplitude r, given by individual terms of the series development obtained from (67.23),

$$\begin{aligned} r &= i\sum_{n=1}^{\infty}(-1)^{n-1}\binom{1/2}{n}\varepsilon^{2n-1} \\ &= i\left\{\frac{\varepsilon}{2} + \frac{\varepsilon^3}{8} + \frac{\varepsilon^5}{16} + \cdots\right\}, \quad \varepsilon = \frac{1}{2k_0L} < 1 \\ &= i\left\{\frac{1}{4k_0L} + \frac{1}{64(k_0L)^3} + \frac{1}{512(k_0L)^5} + \cdots\right\}, \end{aligned} \tag{67.24}$$

and an aggregate number $(1, 3, 5, \ldots)$ of reflections on the non-uniform range which account for the magnitude of such terms; the schematic rendering (after Bremmer 1950) in Fig. 46 brings out the fixed and pairwise increase in number of internal reflections on $x > L$ that accompany the generation of successive terms in the development (67.24). It may be noted, furthermore,

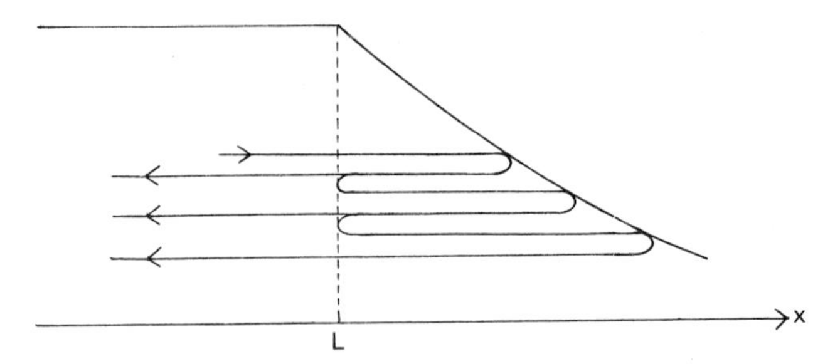

Fig. 46. Multiple internal reflections.

that the initial term of the latter, whose convergence improves with rising values of k_0L, has the alternative representation

$$r = -\frac{\mathrm{i}}{4}\frac{\left[\dfrac{\mathrm{d}k}{\mathrm{d}x}\right]_{x=L-0}^{x=L+0}}{(k(L))^2}$$

and thus conforms to a prior short wave length estimate [(63.13)] for the reflection coefficient when a discontinuity in first derivative of the wave number occurs.

An expansion in powers of ε, for the coefficient t and exponential factor of the wave function ψ in the non-uniform sector $\xi > 0$, yields

$$\psi = \mathrm{e}^{\mathrm{i}\xi} + \frac{\mathrm{i}\varepsilon}{2}\mathrm{e}^{\mathrm{i}\xi} - \frac{\mathrm{i}\varepsilon^2}{2}\xi\,\mathrm{e}^{\mathrm{i}\xi} + \varepsilon^3\left(\frac{\mathrm{i}}{8} + \frac{\xi}{4}\right)\mathrm{e}^{\mathrm{i}\xi} + \cdots, \quad \varepsilon < 1 \tag{67.25}$$

whence, after reverting to the original variables,

$$s(x) = \frac{1}{\sqrt{(k_0)}}\left(\frac{x}{L}\right)^{\mathrm{i}k_0L+1/2}\left\{1 + \frac{\mathrm{i}}{4k_0L} - \frac{\mathrm{i}}{8k_0L}\log\frac{x}{L} + \frac{\left(\mathrm{i} + 2k_0L\log\dfrac{x}{L}\right)}{64(k_0L)^3} + \cdots\right\}, \quad \begin{matrix} k_0L > 2 \\ x > L \end{matrix}. \tag{67.26}$$

The first term of (67.25) represents a literal continuation of the incident wave and its counterpart in (67.26) is the progressive WKBJ wave function

$$\frac{1}{\sqrt{(k(x))}}\exp\left\{\mathrm{i}\int_L^x k(x)\,\mathrm{d}x\right\}$$

which, predictably, furnishes a more reliable estimate for $s(x)$ the shorter the wave length $\lambda_0 = 2\pi/k_0$ in comparison with the reference length L. Subsequent terms of the transmitted wave characterization (67.26) manifest an amplitude depletion or reflection loss for those components of the wave complex with an opposite sense of propagation that are simultaneously present on the inhomogeneous range.

To trace the interconnections between forward and backward moving wave components in a setting wherein gradual parametric variations occur, let us deal with the coupled field equations (67.5) and express their solutions in the composite forms

$$E_y(x) = \frac{1}{\sqrt{(n(x))}}\left[\chi_+(x)\exp\left\{ik_0\int_0^x n(\zeta)\,\mathrm{d}\zeta\right\} - \chi_-(x)\exp\left\{-ik_0\int_0^x n(\zeta)\,\mathrm{d}\zeta\right\}\right],$$
$$H_z(x) = \sqrt{(n(x))}\left[\chi_+(x)\exp\left\{ik_0\int_0^x n(\zeta)\,\mathrm{d}\zeta\right\} + \chi_-(x)\exp\left\{-ik_0\int_0^x n(\zeta)\,\mathrm{d}\zeta\right\}\right] \tag{67.27}$$

where $\chi_\pm(x)$ require determination. It is easily shown that (67.27) and (67.5) are mutually compatible provided the functions $\chi_\pm(x)$ satisfy a first order system of equations

$$\frac{\mathrm{d}\chi_+}{\mathrm{d}x} = -\frac{1}{2}\left\{\frac{\mathrm{d}}{\mathrm{d}x}\log n(x)\right\}\chi_-(x)\exp\left\{-2\,\mathrm{i}\int_0^x n(\zeta)\,\mathrm{d}\zeta\right\},$$
$$\frac{\mathrm{d}\chi_-}{\mathrm{d}x} = -\frac{1}{2}\left\{\frac{\mathrm{d}}{\mathrm{d}x}\log n(x)\right\}\chi_+(x)\exp\left\{2\,\mathrm{i}\int_0^x n(\zeta)\,\mathrm{d}\zeta\right\} \tag{67.28}$$

which are the counterparts of (66.30) appropriate to the case at hand.

If we replace the system (67.28) by another,

$$\frac{\mathrm{d}\chi_+}{\mathrm{d}x} = \varepsilon\nu_1(x)\chi_-(x), \quad \frac{\mathrm{d}\chi_-}{\mathrm{d}x} = \varepsilon\nu_2(x)\chi_+(x), \tag{67.29}$$

which places in evidence a small order of magnitude or scale factor ε for the logarithmic derivative of $n(x)$, and postulate the developments

$$\chi_+(x) = \chi_+^{(0)} + \varepsilon\chi_+^{(1)} + \varepsilon^2\chi_+^{(2)} + \cdots,$$
$$\chi_-(x) = \chi_-^{(0)} + \varepsilon\chi_-^{(1)} + \varepsilon^2\chi_-^{(2)} + \cdots \tag{67.30}$$

there is compliance with (67.29) when terms of the latter obey the relations

$$\frac{\mathrm{d}\chi_+^{(n)}}{\mathrm{d}x} = \nu_1(x)\chi_-^{(n-1)}(x), \qquad \frac{\mathrm{d}\chi_-^{(n)}}{\mathrm{d}x} = \nu_2(x)\chi_+^{(n-1)}(x), \qquad n \ge 1. \tag{67.31}$$

Let us assign the arbitrary and constant values

$$\chi_+^{(0)} = A, \qquad \chi_-^{(0)} = B \tag{67.32}$$

to the initial terms of the respective developments (67.30) and envisage that the integrals of (67.31),

$$\chi_+^{(n)}(x) = \int_{-\infty}^x \nu_1(\zeta)\chi_-^{(n-1)}(\zeta)\,\mathrm{d}\zeta, \qquad \chi_-^{(n)}(x) = -\int_x^\infty \nu_2(\zeta)\chi_+^{(n-1)}(\zeta)\,\mathrm{d}\zeta,$$
$$n \ge 1 \tag{67.33}$$

provide, through repeated use, explicit n-fold integral representations for $\chi_+^{(n)}(x)$ and $\chi_-^{(n)}(x)$; we thus obtain, in a formal manner and subject to the convergence of (67.30), a solution of the first order system with the generality afforded by a pair of arbitrary constants.

The zero order terms, $\chi_\pm^{(0)}$, of the developments (67.30) specify, through the representations (67.27), field strengths E_y, H_z that have a WKBJ nature, with forward and backward wave motions corresponding to the subscripts $\pm$ or the phase factors

$$\exp\left\{ik_0 \int_0^x (n(\zeta)\,\mathrm{d}\zeta\right\}, \quad \exp\left\{-ik_0 \int_0^x n(\zeta)\,\mathrm{d}\zeta\right\}.$$

It appears, on the basis of the formulas (67.31) and their integrals (67.33), that the corrections to the WKBJ wave functions embodied in the first order terms $\chi_\pm^{(1)}$ of (67.30) are attributable to reflection losses from the oppositely directed zero order terms; and that successive reflections underlie the generation of higher order terms.

§68. *Integral Equation Formulations and Their Consequences*

Since there are practical complexities that beset the direct analysis of excitations based on a homogeneous, variable coefficient, second order differential equation, or a like system of first order equations, it becomes worthwhile to examine integral equation reformulations and related possibilities for successive approximations when such equations have an inhomogeneous nature. An exact integral representation for solutions of the differential equation

$$\left(\frac{\mathrm{d}^2}{\mathrm{d}x^2} + k^2(x)\right) s(x) = 0, \quad k(x) > 0 \tag{68.1}$$

is readily found with the help of the Green's function

$$G(x, x') = \frac{\mathrm{i}}{2}\varphi_+(x_>)\varphi_-(x_<), \quad \begin{matrix} x_> \\ x_< \end{matrix} = \begin{matrix} \mathrm{Max} \\ \mathrm{Min} \end{matrix}\,(x, x') \tag{68.2}$$

that contains the *WKBJ* wave factors

$$\varphi_\pm(x) = \{k(x)\}^{-1/2} \exp\left\{\pm\mathrm{i} \int_0^x k(\xi)\,\mathrm{d}\xi\right\} \tag{68.3}$$

and has the limiting form

$$G(x, x') = \frac{\mathrm{i}}{2k}\,\mathrm{e}^{ik(x_> - x_<)} = \frac{\mathrm{i}}{2k}\,\mathrm{e}^{ik|x-x'|}$$

in the case $k(x) = k$, a constant; on referring to the expressions (66.12) for $\varphi'_\pm(x)$ we establish the existence of a fixed jump discontinuity in first derivative of $G(x, x')$ [precisely as obtains when k is constant], namely

$$\frac{\mathrm{d}}{\mathrm{d}x} G(x, x') \Big|_{x=x'-0}^{x=x'+0} = -1,$$

and thence, recalling the equation (66.13) satisfied by $\varphi_{\pm}(x)$, it appears that

$$\left(\frac{\mathrm{d}^2}{\mathrm{d}x^2} + k^2(x) + \frac{1}{2}\frac{k''(x)}{k(x)} - \frac{3}{4}\left(\frac{k'(x)}{k(x)}\right)^2\right) G(x, x') = -\delta(x - x') \tag{68.4}$$

provides a full differential characterization of the source function.

If we suppose that $k(x) \to k$, $|x| \to \infty$, and consider a solution of (68.1) which features an incoming or primary wave factor at $x = -\infty$, the outcome of multiplying (68.1), (68.4) by $G(x, x')$ and $s(x)$, respectively, and subsequently integrating their difference over all x is a representation

$$s(x) = \varphi_+(x) - \int_{-\infty}^{\infty} G(x, x') \left[\frac{1}{2}\frac{k''(x')}{k(x')} - \frac{3}{4}\left(\frac{k'(x')}{k(x')}\right)^2\right] s(x')\,\mathrm{d}x' \tag{68.5}$$

with the character of an inhomogeneous Fredholm integral equation of the second kind. Let us henceforth refer, in this section, to the coordinate range $0 < x < \infty$ and make the assumptions

$$k(x) > 0, \qquad k'(0) = k'(\infty) = 0$$
$$\int_0^{\infty} |k'(x)|\,\mathrm{d}x < \infty \tag{68.6}$$

relative to the wave number; after choosing $x' = 0$ as the lower limit of the integral in (68.5) and separating the portions which arise on the intervals $(0, x)$ and (x, ∞), we find

$$s(x) = \varphi_+(x) - \frac{\mathrm{i}}{2}\varphi_+(x) \int_0^x \mathscr{K}(x)\varphi_-(x)s(x)\,\mathrm{d}x$$
$$- \frac{\mathrm{i}}{2}\varphi_-(x) \int_x^{\infty} \mathscr{K}(x)\varphi_+(x)s(x)\,\mathrm{d}x \tag{68.7}$$

where

$$\mathscr{K}(x) = \frac{1}{2}\frac{k''(x)}{k(x)} - \frac{3}{4}\left(\frac{k'(x)}{k(x)}\right)^2. \tag{68.8}$$

A transformation of both the integrals in (68.7) commences with the rearrangements

$$\begin{aligned} I &= \int_0^x \mathscr{K}(x)\varphi_-(x)s(x)\,\mathrm{d}x \\ &= \int_0^x \frac{\mathscr{K}(x)}{\{k(x)\}^{1/2}} \exp\left\{-\mathrm{i}\int_0^x k(\xi)\,\mathrm{d}\xi\right\} s(x)\,\mathrm{d}x \\ &= \int_0^x \frac{\mathrm{d}}{\mathrm{d}x}\left[\frac{k'(x)}{2\{k(x)\}^{3/2}}\right] \exp\left\{-\mathrm{i}\int_0^x k(\xi)\,\mathrm{d}\xi\right\} s(x)\,\mathrm{d}x, \end{aligned}$$

$$J = \int_x^\infty \mathcal{K}(x)\varphi_+(x)s(x)\,dx$$
$$= \int_x^\infty \frac{d}{dx}\left[\frac{k'(x)}{2\{k(x)\}^{3/2}}\right]\exp\left\{i\int_0^x k(\xi)\,d\xi\right\}s(x)\,dx,$$

following which an integration by parts yields

$$I = \frac{1}{2}\frac{k'(x)}{k(x)}\varphi_-(x)s(x) + \frac{i}{2}\int_0^x k'(x)\varphi_-(x)s(x)\,dx$$
$$-\frac{1}{2}\int_0^x \frac{k'(x)}{k(x)}\varphi_-(x)s'(x)\,dx,$$

and

$$J = -\frac{1}{2}\frac{k'(x)}{k(x)}\varphi_+(x)s(x) - \frac{i}{2}\int_x^\infty k'(x)\varphi_+(x)s(x)\,dx$$
$$-\frac{1}{2}\int_x^\infty \frac{k'(x)}{k(x)}\varphi_+(x)s'(x)\,dx.$$

Accordingly, if we introduce the decomposition

$$s(x) = s_+(x) + s_-(x) \tag{68.9}$$

and postulate the relation

$$s'(x) = ik(x)[s_+(x) - s_-(x)] \tag{68.10}$$

which imparts to $s_+(x)$, $s_-(x)$ the attributes of forward and backward moving wave factors, there obtains from (68.7) an interconnection,

$$s_+(x) + s_-(x) = \varphi_+(x) - \frac{i}{2}\varphi_+(x)I - \frac{i}{2}\varphi_-(x)J$$
$$= \varphi_+(x) + \frac{1}{2}\varphi_+(x)\int_0^x k'(x)\varphi_-(x)s_-(x)\,dx \tag{68.11}$$
$$-\frac{1}{2}\varphi_-(x)\int_x^\infty k'(x)\varphi_+(x)s_+(x)\,dx,$$

and thence a pair of coupled integral equations for $s_+(x)$, $s_-(x)$, viz.:

$$s_+(x) = \varphi_+(x) + \frac{1}{2}\varphi_+(x)\int_0^x k'(x)\varphi_-(x)s_-(x)\,dx$$
$$= \varphi_+(x) - i\int_0^x G(x, x')k'(x')s_-(x')\,dx', \tag{68.12}$$

$$s_-(x) = -\frac{1}{2}\varphi_-(x)\int_x^\infty k'(x)\varphi_+(x)s_+(x)\,dx$$
$$= i\int_x^\infty G(x, x')k'(x')s_+(x')\,dx' \tag{68.13}$$

after separately linking those parts of (68.11) with the respective directionality characteristics. The representation (68.12) evidently synthesizes the forward wave factor, $s_+(x)$, from a primary component together with reflected backward components on the antecedent range $(0, x)$, whereas (68.13) expresses the backward wave factor, $s_-(x)$, in terms of reflected forward components at points further along the coordinate range, i.e., in (x, ∞).

The mutual compatibility or equivalence of the differential and integral specifications for the wave factor receives support from a simple demonstration that the latter are consistent with both the first and second order differential equations (68.10), (68.1); thus, differentiation of (68.12), (68.13) and use of (66.12) yields the relations

$$
\begin{aligned}
s'_+(x) &= \left\{ik(x) - \frac{1}{2}\frac{k'(x)}{k(x)}\right\} s_+(x) + \frac{1}{2}\frac{k'(x)}{k(x)} s_-(x), \\
s'_-(x) &= -\left\{ik(x) + \frac{1}{2}\frac{k'(x)}{k(x)}\right\} s_-(x) + \frac{1}{2}\frac{k'(x)}{k(x)} s_+(x)
\end{aligned}
\qquad (68.14)
$$

whose sum reproduces (68.10). Furthermore, when a derivative is next applied to the first order equation (68.10) we obtain, taking account of (68.14),

$$
\begin{aligned}
s''(x) &= ik'(x)[s_+(x) - s_-(x)] + ik(x)[s'_+(x) - s'_-(x)] \\
&= -k^2(x)s(x)
\end{aligned}
$$

in agreement with (68.1). A slightly different version of the system (68.12), (68.13), originally formulated by Bellman and Kalaba (1959) on the basis of an indirect procedure, is

$$
\begin{aligned}
s_+(x) &= \varphi_+(x) + \int_0^x \varphi_+(x, x') \left\{\frac{1}{2}\frac{d}{dx'} \log k(x')\right\} s_-(x')\, dx' \\
s_-(x) &= \int_x^\infty \varphi_-(x, x') \left\{-\frac{1}{2}\frac{d}{dx'} \log k(x')\right\} s_+(x')\, dx'
\end{aligned}
\qquad x \geq 0 \qquad (68.15)
$$

where the kernels

$$
\varphi_+(x, x') = \frac{\varphi_+(x)}{\varphi_+(x')}, \qquad \varphi_-(x, x') = \frac{\varphi_-(x)}{\varphi_-(x')} \qquad (68.16)
$$

are uniformly bounded functions in $x \geq 0$ if $k(x)$ has a continuously differentiable nature and

$$
\int_0^\infty \left| \frac{d}{dx} \log k(x) \right| dx < \infty. \qquad (68.17)
$$

Predicated on the hypothesis (68.17), the inhomogeneous Volterra equations (68.15) may be shown (Atkinson, 1960) to possess unique solutions $s_+(x)$, $s_-(x)$ of continuous and uniformly bounded nature, with a sum that satisfies (68.1); this conclusion depends, in part, on ruling out the existence of non-trivial solutions for the associated homogeneous equations

$$s_+(x) = \int_0^x \varphi_+(x, x')\left[\frac{1}{2}\frac{\mathrm{d}}{\mathrm{d}x'}\log k(x')\right]s_-(x')\,\mathrm{d}x',$$
$$s_-(x) = \int_x^\infty \varphi_-(x, x')\left[-\frac{1}{2}\frac{\mathrm{d}}{\mathrm{d}x'}\log k(x')\right]s_+(x')\,\mathrm{d}x'. \qquad x \geq 0 \tag{68.18}$$

That the latter system, and its particular implications

$$s_+(0) = 0, \qquad s_-(\infty) = 0, \tag{68.19}$$

entails the consequences

$$s_+(x) = s_-(x) = 0, \quad x \geq 0$$

is readily demonstrated on applying the relation

$$k(x)\{|s_+(x)|^2 - |s_-(x)|^2\} = \text{constant} \tag{68.20}$$

which expresses the linear independence (or invariable Wronskian) of a pair of solutions $s(x) = s_+(x) + s_-(x)$, $s^*(x) = s_+^*(x) + s_-^*(x)$ to the differential equation (68.1). Substituting $x = 0$ and $x = \infty$ in (68.20) and simplifying on the basis of (68.19), we find

$$-k(0)|s_-(0)|^2 = \lim_{x\to+\infty} k(x)|s_+(x)|^2$$

with non-positive/negative magnitudes for the left-/right-hand members; thus $s_-(0)$ vanishes and, inasmuch as $s_+(0) = 0$, we infer that $s_+(x) = s_-(x) = 0$, $x \geq 0$.

A method of successive approximations can be utilized for purposes of solving the inhomogeneous system of equations (68.15), giving rise to the developments (or Neumann series)

$$s_+(x) = \varphi_+^{(0)}(x) + \varphi_+^{(2)}(x) + \cdots + \varphi_+^{(2n)}(x) + \cdots,$$
$$s_-(x) = \varphi_-^{(1)}(x) + \varphi_-^{(3)}(x) + \cdots + \varphi_-^{(2n+1)}(x) + \cdots \tag{68.21}$$

whose terms are defined in the recursive fashion

$$\varphi_+^{(0)}(x) = \varphi_+(x), \qquad \varphi_+^{(2n)}(x) = \frac{1}{2}\int_0^x \varphi_+(x, x')\left[\frac{\mathrm{d}}{\mathrm{d}x'}\log k(x')\right]\varphi_-^{(2n-1)}(x')\,\mathrm{d}x', \quad n \geq 1,$$
$$\varphi_-^{(2n-1)}(x) = -\frac{1}{2}\int_x^\infty \varphi_-(x, x')\left[\frac{\mathrm{d}}{\mathrm{d}x'}\log k(x')\right]\varphi_+^{(2n-2)}(x')\,\mathrm{d}x'. \tag{68.22}$$

The combined development,

$$s(x) = s_+(x) + s_-(x)$$
$$= \sum_{n=0}^\infty \{\varphi_+^{(2n)}(x) + \varphi_-^{(2n+1)}(x)\},$$

known as the Bremmer (1951) series in the present context, provides a solution of the differential equation (68.1) on conditions of absolute and uniform convergence, which are assured if

$$\int_0^\infty \left| \frac{\mathrm{d}}{\mathrm{d}x} \log k(x) \right| \mathrm{d}x \leq \pi$$

according to Atkinson.

§69. *Aspects of Reflection for Inhomogeneous Settings*

Stationary wave forms, with a common (instantaneous) value of the phase over their full range, constitute particular excitations that are realizable in both homogeneous and inhomogeneous settings; progressive wave forms, on the other hand, whose change after an arbitrary lapse of time is expressed by a proportionate shift of position, cannot be counted among the possible modes of excitation when inhomogeneous circumstances prevail. Hence, the identification and specification of locally reflected waves becomes meaningful to the extent that an acceptable resolution of the wave function, generally predicated on small departures from homogeneity, is available.

An overall measure of reflection or transmission, based on suitable asymptotic behaviors for the progressive wave factors in the uniform distant parts of the coordinate domain, can be characterized exactly in terms of a single integral—that involves the wave factor throughout the range of inhomogeneity—and readily estimated if the wave number changes are gradual. The integral equation formulation described in §68 lends itself to such purposes; thus, we have a uniformly valid relation which incorporates an incoming excitation from the left ($x \to -\infty$),

$$s(x) = \varphi_+(x) - \int_{-\infty}^{\infty} G(x, x') \left[\frac{1}{2} \frac{k''(x')}{k(x')} - \frac{3}{4} \left(\frac{k'(x')}{k(x')} \right)^2 \right] s(x')\, \mathrm{d}x', \qquad -\infty < x < \infty \tag{69.1}$$

where

$$\begin{aligned} G(x, x') &= \frac{\mathrm{i}}{2} \varphi_+(x_>) \varphi_-(x_<) \\ &= \frac{\mathrm{i}}{2} (k(x) k(x'))^{-1/2} \exp\left\{ \mathrm{i} \int_0^{x_>} k(\xi)\, \mathrm{d}\xi - \mathrm{i} \int_0^{x_<} k(\xi)\, \mathrm{d}\xi \right\} \\ &= \frac{\mathrm{i}}{2} (k(x) k(x'))^{-1/2} \exp\left\{ \mathrm{i}k|x - x'| + \mathrm{i} \int_{x_<}^{x_>} (k(\xi) - k)\, \mathrm{d}\xi \right\}, \end{aligned} \tag{69.2}$$

and

$$\varphi_+(x) = \frac{1}{\sqrt{(k(x))}} \exp\left\{ \mathrm{i}kx + \mathrm{i} \int_0^x (k(\xi) - k)\, \mathrm{d}\xi \right\} \tag{69.3}$$

if

$$\operatorname*{Lim}_{|x|\to\infty} k(x) = k.$$

Through recourse to (69.2) and (69.3) a pair of limiting forms of the representation (69.1) are inferred, namely

$$s(x) \sim \frac{e^{ikx}}{\sqrt{k}} \exp\left\{-i\int_{-\infty}^{0} (k(\xi)-k)\,d\xi\right\} \times$$

$$-\frac{1}{2\sqrt{k}} e^{-ikx} \int_{-\infty}^{\infty} \exp\left\{i\int_{-\infty}^{x} (k(\xi)-k)\,d\xi + ikx\right\} \left[\frac{1}{2}\frac{k''(x)}{k(x)} - \frac{3}{4}\left(\frac{k'(x)}{k(x)}\right)^2\right] \frac{s(x)}{\sqrt{(k(x))}}\,dx,$$

$$x \to -\infty, \tag{69.4}$$

$$s(x) \sim \frac{e^{ikx}}{\sqrt{k}} \exp\left\{i\int_{0}^{\infty} (k(\xi)-k)\,d\xi\right\} \times$$

$$-\frac{i}{2\sqrt{k}} e^{ikx} \int_{-\infty}^{\infty} \exp\left\{i\int_{x}^{\infty} (k(\xi)-k)\,d\xi - ikx\right\} \left[\frac{1}{2}\frac{k''(x)}{k(x)} - \frac{3}{4}\left(\frac{k'(x)}{k(x)}\right)^2\right] \frac{s(x)}{\sqrt{(k(x))}}\,dx,$$

$$x \to +\infty \tag{69.5}$$

with the proviso that

$$\int_{-\infty}^{\infty} (k(x)-k)\,dx < \infty;$$

mutual consistency between the latter and the stipulated asymptotic forms

$$s(x) \sim \begin{cases} e^{ikx} + r\,e^{-ikx}, & x \to -\infty \\ t\,e^{ikx}, & x \to +\infty \end{cases}$$

yields the specifications

$$r = -\frac{i}{2}\exp\left\{i\int_{-\infty}^{0} (k(\xi)-k)\,d\xi\right\} \int_{-\infty}^{\infty} \frac{d}{dx}\left(\frac{k'(x)}{2(k(x))^{3/2}}\right) s(x)$$

$$\times \exp\left\{i\int_{-\infty}^{x} (k(\xi)-k)\,d\xi + ikx\right\} dx$$

$$= \frac{i}{4}\exp\left\{i\int_{-\infty}^{0} (k(\xi)-k)\,d\xi\right\} \int_{-\infty}^{\infty} \frac{k'(x)}{(k(x))^{3/2}} [s'(x) + ik(x)s(x)]$$

$$\times \exp\left\{i\int_{-\infty}^{x} (k(\xi)-k)\,d\xi + ikx\right\} dx \tag{69.6}$$

and

$$t = \exp\left\{i\int_{-\infty}^{\infty} (k(\xi)-k)\,d\xi\right\} - i\frac{i}{2}\exp\left\{i\int_{-\infty}^{0} (k(\xi)-k)\,d\xi\right\} \int_{-\infty}^{\infty} \frac{d}{dx}\left(\frac{k'(x)}{2(k(x))^{3/2}}\right)$$

$$\times\ s(x)\exp\left\{i\int_{x}^{\infty} (k(\xi)-k)\,d\xi - ikx\right\} dx$$

$$= \exp\left\{\mathrm{i}\int_{-\infty}^{\infty}(k(\xi)-k)\,\mathrm{d}\xi\right\} + \frac{\mathrm{i}}{4}\exp\left\{\mathrm{i}\int_{-\infty}^{0}(k(\xi)-k)\,\mathrm{d}\xi\right\}\int_{-\infty}^{\infty}\frac{k'(x)}{(k(x))^{3/2}}$$

$$\times [s'(x)-\mathrm{i}k(x)s(x)] \times \exp\left\{\mathrm{i}\int_{x}^{\infty}(k(\xi)-k)\,\mathrm{d}\xi - \mathrm{i}kx\right\}\mathrm{d}x$$

appropriate to the reflection and transmission coefficients, respectively.

If the first and second derivatives of the wave number have small magnitudes we may rely upon a process of iteration to generate a series development for the solution of the integral equation (69.1). The leading terms are given, explicitly, by

$$s(x) = \varphi_+(x) - \int_{-\infty}^{\infty} G(x, x')\left[\frac{1}{2}\frac{k''(x')}{k(x')} - \frac{3}{4}\left(\frac{k'(x')}{k(x')}\right)^2\right]\varphi_+(x')\,\mathrm{d}x' + \cdots \tag{69.8}$$

and, after transformation of the integral herein, it appears that

$$s(x) = \varphi_+(x) - \frac{\mathrm{i}}{8}\varphi_+(x)\int_{-\infty}^{\infty}\frac{(k'(x))^2}{(k(x))^3}\,\mathrm{d}x$$

$$-\frac{\mathrm{i}}{8}\varphi_-(x)\int_{x}^{\infty}\left[\frac{(k'(x))^2}{(k(x))^3} - 4\,\mathrm{i}\frac{k'(x)}{k(x)}\right]\exp\left\{2\,\mathrm{i}\int_0^x k(\xi)\,\mathrm{d}\xi\right\}\mathrm{d}x + \cdots$$

whence

$$s'(x) + \mathrm{i}k(x)s(x) = \varphi_+(x)\left[2\,\mathrm{i}k(x) + \left(\frac{k(x)}{4} + \frac{\mathrm{i}}{16}\frac{k'(x)}{k(x)}\right)\int_{x}^{\infty}\frac{(k'(x))^2}{(k(x))^3}\,\mathrm{d}x\right]$$

$$+\varphi_-(x)\left[\frac{\mathrm{i}}{16}\frac{k'(x)}{k(x)}\int_{x}^{\infty}\left(\frac{(k'(x))^2}{(k(x))^3} - 4\,\mathrm{i}\frac{k'(x)}{k(x)}\right)\exp\left\{2\,\mathrm{i}\int_0^x k(\xi)\,\mathrm{d}\xi\right\}\right]$$

$$\doteq 2\,\mathrm{i}k(x)\varphi_+(x)\left\{1 - \frac{\mathrm{i}}{8}\int_{x}^{\infty}\frac{(k'(x))^2}{(k(x))^3}\,\mathrm{d}x\right\} \tag{69.9}$$

and

$$s'(x) - \mathrm{i}k(x)s(x)$$

$$= \varphi_+(x)\left[\frac{\mathrm{i}}{16}\frac{k'(x)}{k(x)}\int_{-\infty}^{x}\frac{(k'(x))}{(k(x))^3}\,\mathrm{d}x\right]$$

$$+\varphi_-(x)\left[\left(-\frac{k(x)}{4} + \frac{\mathrm{i}}{16}\frac{k'(x)}{k(x)}\right)\int_{x}^{\infty}\left[\frac{(k'(x))^2}{(k(x))^3} - 4\,\mathrm{i}\frac{k'(x)}{k(x)}\right]\exp\left\{2\,\mathrm{i}\int_0^x k(\xi)\,\mathrm{d}\xi\right\}\mathrm{d}x\right]$$

$$\doteq \mathrm{i}k(x)\varphi_-(x)\left\{1 - \frac{\mathrm{i}}{4}\frac{k'(x)}{(k(x))^2}\right\}\int_{x}^{\infty}\frac{k'(x)}{k(x)}\exp\left\{2\,\mathrm{i}\int_0^x k(\xi)\,\mathrm{d}\xi\right\}\mathrm{d}x. \tag{69.10}$$

On employing the principal part or initial term of the approximation (69.9) we obtain from (69.6) an estimate for the reflection coefficient,

$$r=-\frac{1}{2}\int_{-\infty}^{\infty}\frac{k'(x)}{k(x)}\exp\left\{2\,\mathrm{i}\int_{-\infty}^{x}(k(\xi)-k)\,\mathrm{d}\xi+2\,\mathrm{i}kx\right\}\mathrm{d}x, \tag{69.11}$$

which is accurate to the first order in the (small) measure of inhomogeneity, $(\mathrm{d}/\mathrm{d}x)\log k(x)=k'(x)/k(x)$. The analogous transmission coefficient estimate, deduced from (69.7) and the principal part of (69.10), has the form

$$t=\exp\left\{\mathrm{i}\int_{-\infty}^{\infty}(k(\xi)-k)\,\mathrm{d}\xi\right\}\left[1-\frac{1}{4}\int_{-\infty}^{\infty}\mathrm{d}x\,\frac{k'(x)}{k(x)}\exp\left\{-2\,\mathrm{i}\int_{0}^{x}k(\xi)\,\mathrm{d}\xi\right\}\right.$$
$$\left.\times\int_{x}^{\infty}\mathrm{d}x'\,\frac{k'(x')}{k(x')}\exp\left\{2\,\mathrm{i}\int_{0}^{x'}k(\xi)\,\mathrm{d}\xi\right\}\right] \tag{69.12}$$

inclusive of second order terms in the aforesaid measure. We find, on the basis of (69.12) and the suppression of higher order terms, that

$$\begin{aligned}|t|^2=1&-\frac{1}{4}\int_{-\infty}^{\infty}\mathrm{d}x\frac{k'(x)}{k(x)}\exp\left\{-2\,\mathrm{i}\int_{0}^{x}k(\xi)\,\mathrm{d}\xi\right\}\int_{x}^{\infty}\mathrm{d}x'\frac{k'(x')}{k(x')}\exp\left\{2\,\mathrm{i}\int_{0}^{x'}k(\xi)\,\mathrm{d}\xi\right\}\\&-\frac{1}{4}\int_{-\infty}^{\infty}\mathrm{d}x\frac{k'(x)}{k(x)}\exp\left\{2\,\mathrm{i}\int_{0}^{x}k(\xi)\,\mathrm{d}\xi\right\}\int_{x}^{\infty}\mathrm{d}x'\frac{k'(x')}{k(x')}\exp\left\{-2\,\mathrm{i}\int_{0}^{x'}k(\xi)\,\mathrm{d}\xi\right\}\\=1&-\frac{1}{4}\int_{-\infty}^{\infty}\mathrm{d}x\frac{k'(x)}{k(x)}\exp\left\{2\,\mathrm{i}\int_{0}^{x}k(\xi)\,\mathrm{d}\xi\right\}\int_{-\infty}^{\infty}\mathrm{d}x'\frac{k'(x')}{k(x')}\exp\left\{-2\,\mathrm{i}\int_{0}^{x'}k(\xi)\,\mathrm{d}\xi\right\}\end{aligned}$$

consequent to reversing the integration orders in the first double integral. Since

$$r=-\frac{1}{2}\exp\left\{2\,\mathrm{i}\int_{-\infty}^{0}(k(\xi)-k)\,\mathrm{d}\xi\right\}\int_{-\infty}^{\infty}\frac{k'(x)}{k(x)}\exp\left\{2\,\mathrm{i}\int_{0}^{x}k(\xi)\,\mathrm{d}\xi\right\}\mathrm{d}x \tag{69.13}$$

the estimates (69.11), (69.12) comply with the energy conservation relation

$$|r|^2+|t|^2=1.$$

The representation (66.24), with its component wave functions $\varphi_{\pm}(x)$ and coefficients $a(x)$, $b(x)$ that vary slowly if $k(x)$ does likewise, affords a basis for specifying and estimating local reflection in slightly inhomogeneous circumstances. Let us envisage an incoming wave motion at one extremity of the x-interval, say $x=-\infty$, and effect a sequential integration of the coupled equations (66.30) consonant with the asymptotic conditions

$$C_-(x)\to 0,\qquad C_+(x)\to C,\quad x\to\infty;$$

this yields, further to the choice $x_1=0$,

$$\begin{aligned}C_-(x)&=C\{I_1(x)+I_3(x)+\cdots\},\\C_+(x)&=C\{1+I_2(x)+I_4(x)+\cdots\}\end{aligned} \tag{69.14}$$

where

$$I_1(x)=\int_\infty^x \hat{k}(x)\exp\left\{2\,\mathrm{i}\int_0^x (k(\xi)-\hat{k}(\xi))\,\mathrm{d}\xi+\mathrm{i}\pi/2\right\}\mathrm{d}x, \tag{69.15}$$

$$\begin{aligned} I_{2n+1}(x)&=\int_\infty^x \hat{k}(x)\exp\left\{2\,\mathrm{i}\int_0^x (k(\xi)-\hat{k}(\xi))\,\mathrm{d}\xi+\mathrm{i}\pi/2\right\}I_{2n}(x)\,\mathrm{d}x, \quad n\geq 1\\ I_{2n}(x)&=\int_\infty^x \hat{k}(x)\exp\left\{-2\,\mathrm{i}\int_0^x (k(\xi)-\hat{k}(\xi))\,\mathrm{d}\xi-\mathrm{i}\pi/2\right\}I_{2n-1}(x)\,\mathrm{d}x. \end{aligned} \tag{69.16}$$

Thus, when we define a reflecting factor in terms of the ratio of components making up the representation (66.24), and utilize the developments (69.14), it follows that

$$\begin{aligned} r(x)&=\frac{b(x)\varphi_-(x)}{a(x)\varphi_+(x)}=\frac{C_-(x)}{C_+(x)}\exp\left\{-2\,\mathrm{i}\int_0^x (k(\xi)-\hat{k}(\xi))\,\mathrm{d}\xi\right\}\\ &=\frac{I_1(x)+I_3(x)+\cdots}{1+I_2(x)+I_4(x)+\cdots}\exp\left\{-2\,\mathrm{i}\int_0^x (k(\xi)-\hat{k}(\xi))\,\mathrm{d}\xi\right\}. \end{aligned} \tag{69.17}$$

Restricted changes in the wave number promote rapid convergence for the appertaining series and thus enhance the practicality of the latter expression; a particular deduction therefrom, viz.,

$$\begin{aligned} r&=\lim_{x\to-\infty}[r(x)\,\mathrm{e}^{2ikx}]\doteq\lim_{x\to-\infty}\left[I_1(x)\exp\left\{-2\,\mathrm{i}\int_0^x k(\xi)\,\mathrm{d}\xi+2\,\mathrm{i}kx\right\}\right]\\ &=I_1(-\infty)\exp\left\{2\,\mathrm{i}\int_{-\infty}^0 (k(\xi)-k)\,\mathrm{d}\xi\right\}, \end{aligned}$$

agrees with the first order reflection coefficient estimate (69.13), inasmuch as

$$\begin{aligned} I_1(-\infty)&=-\frac{\mathrm{i}}{2}\int_{-\infty}^\infty \frac{\mathrm{d}}{\mathrm{d}x}\left(\frac{k'(x)}{2(k(x))^{3/2}}\right)\exp\left\{2\,\mathrm{i}\int_0^x (k(\xi)-\hat{k}(\xi))\,\mathrm{d}\xi\right\}\frac{\mathrm{d}x}{(k(x))^{1/2}}\\ &=\frac{\mathrm{i}}{4}\int_{-\infty}^\infty \frac{k'(x)}{(k(x))^{3/2}}\left[2\,\mathrm{i}\frac{k(x)-\hat{k}(x)}{(k(x))^{1/2}}\right.\\ &\qquad\left.-\frac{1}{2}\frac{k'(x)}{(k(x))^{3/2}}\right]\exp\left\{2\,\mathrm{i}\int_0^x (k(\xi)-\hat{k}(\xi))\,\mathrm{d}\xi\right\}\mathrm{d}x\\ &\doteq-\frac{1}{2}\int_{-\infty}^\infty \frac{k'(x)}{k(x)}\exp\left\{2\,\mathrm{i}\int_0^x k(\xi)\,\mathrm{d}\xi\right\}\mathrm{d}x \end{aligned}$$

if $k(x)\to k$, $|x|\to\infty$ and only the principal contributions to the magnitude and phase of the integral are retained.

A direct scheme of successive approximations to the reflection factor $r(x)$ is readily devised after the first order differential equation (58.14) and its lone supplementary condition, $r(x)\to 0$, $x\to\infty$, are jointly incorporated in the (non-linear) integral equation

$$r(x)=\exp\{-\mathrm{i}\mu(x)\}\int_x^\infty \nu(\xi)[1-r^2(\xi)]\exp\{\mathrm{i}\mu(\xi)\}\,\mathrm{d}\xi \tag{69.18}$$

where

$$\mu(x) = 2 \int_{x_1}^{x} k(\xi)\,\mathrm{d}\xi \tag{69.19}$$

and

$$\nu(x) = \frac{1}{2}\frac{\mathrm{d}}{\mathrm{d}x}\log\frac{1}{k(x)} = -\frac{k'(x)}{2k(x)}. \tag{69.20}$$

Consider a sequence of functions

$$\begin{aligned} r^{(0)}(x) &= 0, \\ r^{(1)}(x) &= \exp\{-\mathrm{i}\mu(x)\} \int_{x}^{\infty} \nu(\xi)\exp\{\mathrm{i}\mu(\xi)\}\,\mathrm{d}\xi, \\ &\vdots \\ r^{(n)}(x) &= \exp\{-\mathrm{i}\mu(x)\} \int_{x}^{\infty} \nu(\xi)[1-(r^{(n-1)}(\xi))^2]\exp\{\mathrm{i}\mu(\xi)\}\,\mathrm{d}\xi, \\ &\vdots \end{aligned} \tag{69.21}$$

whose initial member is found on assuming a null value for $\nu(\xi)$ in the integrand of (69.18) and whose succeeding members are recursively coupled via the non-linear term which occurs therein. That a sequence defined in such fashion converges to the solution of (69.18) is assured when the measure of inhomogeneity, $\nu(x)$, has a bounded magnitude everywhere; evidently, the convergence becomes more rapid as the magnitudes of both $\nu(x)$ and the iterates $r^{(n)}$ diminish.

It can be inferred, by simulating the postulated inhomogeneity with a model of locally uniform sections, that the successive approximations $r^{(1)}(x)$, $r^{(2)}(x), \ldots$ pertain to corresponding orders of multiply reflected components. Let us examine, in particular, the first approximation

$$r^{(1)}(x_1) = \int_{x_1}^{\infty} \nu(x)\exp\left\{2\,\mathrm{i}\int_{x_1}^{\infty} k(\xi)\,\mathrm{d}\xi\right\}\mathrm{d}x \tag{69.22}$$

and, presuming a near equality in values $k_n = k(x_n)$, $k_{n+1} = k(x_{n+1})$ of the wave number at closely separated points x_n, x_{n+1}, adopt the estimate

$$\frac{k_n - k_{n+1}}{k_n + k_{n+1}}\exp\left\{2\,\mathrm{i}\int_{x_1}^{x_n} k(\xi)\,\mathrm{d}\xi\right\}\Delta x_n \tag{69.23}$$

for the contribution to the integral (69.22) from the range $\Delta x_n = x_{n+1} - x_n$; since the phase factor in (69.23) is characteristic of geometrical (or short wave) propagation from x_1 to x_n and back, while the first factor describes a reflection coefficient at x_n consonant with an abrupt change in the wave number from a uniform value k_n on one side to a different uniform value k_{n+1} on the other side, we shall regard (69.22) as a synthesis of single reflections at points $x > x_1$, leaving out any variation in the primary wave amplitude. In the subsequent stages of the approximating sequence (69.21) for the reflection factor we encounter integrals of increasing multiplicity and, by a similar estimation of their phase and magnitude, it appears that account is then taken of a depleted primary wave amplitude along with multiple internal reflection in the sectioned model.

If the variable part of the wave number is explicitly isolated through a decomposition

$$k^2(x) = k^2(1 + \eta(x)), \tag{69.24}$$

where

$$\int_{-\infty}^{\infty} |\eta(x)| \, \mathrm{d}x < \infty,$$

and reference be made to the reflection coefficient estimate (69.11), expressed in the form

$$r = -\frac{1}{2}\int_{-\infty}^{\infty} \left(\frac{\mathrm{d}}{\mathrm{d}x} \log \frac{k(x)}{k}\right) \exp\left\{2\,\mathrm{i} \int_{-\infty}^{\infty} (k(\xi) - k)\, \mathrm{d}\xi + 2\,\mathrm{i}kx\right\} \mathrm{d}x$$

$$= \frac{\mathrm{i}}{2}\int_{-\infty}^{\infty} (1 + \eta(x))^{1/2} \log(1 + \eta(x)) \exp\left\{2\,\mathrm{i}k \int_{-\infty}^{\infty} [(1 + \eta(\xi))^{1/2} - 1]\, \mathrm{d}\xi + 2\,\mathrm{i}kx\right\} \mathrm{d}x$$

which lends itself to approximation when $|\eta(x)| \ll 1$, there follows a first order characterization

$$r = \frac{\mathrm{i}k}{2}\int_{-\infty}^{\infty} \eta(x) \exp\left\{2\,\mathrm{i}kx + \mathrm{i}k \int_{-\infty}^{x} \eta(\xi)\, \mathrm{d}\xi\right\} \mathrm{d}x \tag{69.25}$$

in terms of $\eta(x)$. An identical result obtains from the approximate reflection factor $r^{(1)}(x)$ of the sequence (69.21), in accordance with the prescription

$$r = \lim_{x \to -\infty} \{\mathrm{e}^{2\,\mathrm{i}kx} r^{(1)}(x)\}$$

that fixes the origin as a point of comparison.

§70. *Phase Calculations and Turning Points*

The phase difference between oppositely directed wave trains conveys information, and exclusively so in circumstances of total reflection, about the nature of the environment which supports and couples them. Measurements have been taken, in particular, of the phase difference between upgoing waves at a transmitter and downgoing waves at a receiver for purposes of ascertaining the distribution of ionized atmospheric layers and their effect on electromagnetic wave transmission. Since the presence of inhomogeneity generally precludes the feasibility of exact calculation, there is an incentive to develop practical phase approximations and to make comparison with such definitive results as may be readily available.

Let us consider, at the outset, one-dimensional excitations in a setting which includes homogeneous ($x<0$) and inhomogeneous ($x>0$) parts and assume, moreover, that $k^2(x)<0$ at sufficiently large values of x, thus ensuring the totality of reflection for waves incident on the latter range. If

$$s(x) = A\,\mathrm{e}^{ikx} + B\,\mathrm{e}^{-ikx}, \quad x \le 0 \tag{70.1}$$

describes the resultant wave factor in the homogeneous range (where $k(x) = k$), the phase difference at $x=0$ between its components reflected from and incident on the inhomogeneous range,

$$\varphi = \arg \frac{B}{A}, \tag{70.2}$$

can be expressed in the form

$$\varphi = \arg\left\{-\frac{\mathrm{d}s/\mathrm{d}x - iks}{\mathrm{d}s/\mathrm{d}x + iks}\right\}_{x=0}. \tag{70.3}$$

When the respective functions $k(x)$, $s(x)$, $s'(x)$ are continuous throughout and $s_+(x)$ denotes a real solution of the differential equation

$$\left(\frac{\mathrm{d}^2}{\mathrm{d}x^2} + k^2(x)\right) s(x) = 0, \quad x>0 \tag{70.4}$$

that approaches zero as $x \to \infty$, an explicit determination of φ [modulo 2π] is afforded by (70.3), namely

$$\varphi = 2\tan^{-1}\left(\frac{\mathrm{d}s_+/\mathrm{d}x}{ks_+(x)}\right)_{x=0}. \tag{70.5}$$

The special choice

$$k^2(x) = k^2\left(1-\frac{x}{L}\right), \quad x \ge 0 \tag{70.6}$$

enables us to draw upon extensive knowledge about solutions of the equation (70.4), which acquires the form

$$\left(\frac{\mathrm{d}^2}{\mathrm{d}\xi^2} - \xi\right) s(\xi) = 0 \tag{70.7}$$

in terms of a new independent variable

$$\xi = -(kL)^{2/3}\left(1-\frac{x}{L}\right) \tag{70.8}$$

ranging over the interval $-(kL)^{2/3} < \xi < \infty$. A particular integral of (70.7), regular and real for all real ξ, is given by the Airy function $\mathrm{Ai}(\xi)$ [cf. §37], with the complementary cylinder function representations

$$\mathrm{Ai}(\xi)=\begin{cases}\frac{1}{3}\xi^{1/2}\left[J_{1/3}\left(\frac{2}{3}\xi^{3/2}\right)+J_{-1/3}\left(\frac{2}{3}\xi^{3/2}\xi^{3/2}\right)\right], & \xi>0\\ \frac{1}{3}(-\xi)^{1/2}\left[-I_{1/3}\left(\frac{2}{3}(-\xi)^{3/2}\right)+I_{-1/3}\left(\frac{2}{3}(-\xi)^{3/2}\right)\right], & \xi<0\end{cases} \tag{70.9}$$

and the asymptotic behavior

$$\mathrm{Ai}(\xi)\sim\frac{1}{2}\pi^{-1/2}\xi^{-1/4}\exp\left(-\frac{2}{3}\xi^{3/2}\right),\quad \xi\gg 1.$$

The latter contrasts with the exponential growth of a linearly independent solution, $B\mathrm{i}(\xi)$, of (70.7) and uniquely qualifies the function $\mathrm{Ai}(\xi)$ for a role in the phase determination (70.5), with the result that

$$\varphi=2\tan^{-1}\left\{(kL)^{-1/3}\frac{\mathrm{Ai}'(-(kL)^{2/3})}{\mathrm{Ai}(-(kL)^{2/3})}\right\} \tag{70.10}$$

where the prime signifies an argument derivative.

If the positive values of $k^2(x)$ are distributed over an interval $0<x<L$ whose span greatly exceeds the reference wave length $\lambda=2\pi/k$ on the homogeneous sector, and the Airy function estimate

$$\mathrm{Ai}(\xi)\sim\pi^{-1/2}(-\xi)^{-1/4}\cos\left\{\frac{2}{3}(-\xi)^{3/2}-\frac{\pi}{4}\right\},\quad \xi\ll -1 \tag{70.11}$$

is applied in connection with the rigorous formula (70.10), it appears that

$$\varphi\sim\frac{4}{3}kL-\frac{\pi}{2},\quad kL\gg 1. \tag{70.12}$$

A direct phase evaluation, based on the local wave number and an appropriate 'optical path' (in the ray terminology) recommends itself when $k(x)$ varies slowly; thus, on integrating $k(x)$ over a path which extends from $x=0$ to the turning point $x=L$ and back, we obtain a phase shift (relative to the starting point) of magnitude

$$2\int_0^L k(x)\,\mathrm{d}x=2k\int_0^L\left\{1-\frac{x}{L}\right\}^{1/2}\mathrm{d}x=\frac{4}{3}kL,$$

in accordance with the leading term of (70.12). The second term of (70.12) traces its origin to separate contributions, $-\pi$ and $+\pi/2$, that represent a phase shift characteristic of total (or sharp, $L\to 0$) reflection and account for the range $x>L$ on which $k^2(x)<0$, respectively.

The existence of a null or turning point for a function $k^2(x)$ which varies throughout the coordinate interval imparts distinctive features to solutions of (70.4). We note, in particular, that each of the linearly independent solutions of a specific differential equation whose variable coefficient vanishes at the origin,

$$\left(\frac{d^2}{dx^2}+x\right)s(x)=0, \quad -\infty<x<\infty, \tag{70.13}$$

namely

$$s(x)=\sqrt{(x)}H_{1/3}^{(i)}\left(\frac{2}{3}x^{3/2}\right), \quad i=1,2 \tag{70.14}$$

exhibit a branch point at the same location. The resolution of (70.14) into what may be termed incident and reflected wave factors depends on whether or not, in the course of a passage for x between ∞ and $-\infty$, the secondary variable

$$z=\frac{2}{3}x^{3/2}$$

traverses a branch cut of the Hankel functions $H_{1/3}^{(1),(2)}(z)$ issuing from the origin of the z-plane; such is the case when a cut extends along the negative imaginary axis and x traces a curve passing below the origin of the x-plane, so that the values $\arg z=0,\ 3\pi/4,\ -3\pi/2$ correspond to $x=\infty, 0, -\infty$, respectively.

Let us imagine a source at one extremity ($x=\infty$) of the coordinate range and, in keeping with the fact that propagation is ruled out for the sector where $k^2(x)=x<0$, impose the requirement $s(x)\to 0,\ x\to-\infty$ on the appropriate solution of (70.13). Since $H_{1/3}^{(1)}(|z|\,e^{i\pi/2})\to 0$ as $|z|\to\infty$ and $\arg z=-3\pi/2$ in the limit $x\to-\infty$, the representation

$$s(x)=x^{1/2}H_{1/3}^{(1)}\left(\frac{2}{3}x^{3/2}\,e^{2\pi i}\right)$$

conforms to our stipulated asymptotic behavior of the wave factor; on taking note of the relation

$$H_{1/3}^{(1)}\left(\frac{2}{3}x^{3/2}\,e^{2\pi i}\right)=-H_{1/3}^{(1)}\left(\frac{2}{3}x^{3/2}\right)-e^{-i\pi/3}H_{1/3}^{(2)}\left(\frac{2}{3}x^{3/2}\right), \quad x>0,$$

which owes its existence to a branch point singularity for the Hankel functions at $x=0$; it may be deduced that the amplitudes of the incident and reflected wave factors, $H_{1/3}^{(1)}(\frac{2}{3}x^{3/2})$ and $H_{1/3}^{(2)}(\frac{2}{3}x^{3/2})$, have a ratio whose modulus equals unity.

A wave number profile with more significance in terms of a physical model is specified by the real function $k^2(x)$ which tends to the fixed asymptotic value k^2 as $x\to-\infty$, varies in linear fashion near the turning point $x=L$, and assumes negative values when $x>L$. We are obliged to seek a composite approximation for the solution of the differential equation (70.4), in the absence of an explicit overall version, or else to utilize a different formulation of the problem. A preliminary stage in converting to an integral equation formulation involves the selection of a suitable Green's function and here, for reasons of efficiency and simplicity, our aim is to endow the latter with proper local behaviors near the turning point and far away therefrom.

On the basis of developments which hold in the neighborhood of $x = L$, namely

$$k^2(x) = (L - x)[a_0 + a_1(L - x) + \cdots], \quad a_0 > 0,$$

and

$$k(x) = (L - x)^{1/2}[b_0 + b_1(L - x) + \cdots], \quad b_0 > 0, \tag{70.15}$$

we find through termwise integration

$$\begin{aligned}\Phi(x) &= \int_L^x k(\xi)\,\mathrm{d}\xi \\ &= -(L - x)^{3/2}\left[\frac{2}{3}b_0 + \frac{2}{5}b_1(L - x) + \cdots\right];\end{aligned} \tag{70.16}$$

thus, given the determinations

$$\left.\begin{aligned}\arg(L - x) &= \pi,\\ \arg \Phi(x) &= \pi/2,\end{aligned}\right. \quad x > L \tag{70.17}$$

it follows, by proceeding in a clockwise sense below the point $x = L$ in a complex plane, that

$$\left.\begin{aligned}\arg(L - x) &= 0,\\ \arg \Phi(x) &= -\pi,\end{aligned}\right. \quad x < L. \tag{70.18}$$

Consider next a pair of functions,

$$F_+(x) = \mathrm{e}^{2\pi\,\mathrm{i}/3}\left(\frac{\Phi(x)}{k(x)}\right)^{1/2} H_{1/3}^{(1)}(\Phi(x)) \tag{70.19}$$

and

$$F_-(x) = \mathrm{e}^{\pi\,\mathrm{i}/6}\left(\frac{\Phi(x)}{k(x)}\right)^{1/2} \{H_{1/3}^{(1)}(\Phi(x)) + 2\mathrm{e}^{-\pi\,\mathrm{i}/3} H_{1/3}^{(2)}(\Phi(x))\}, \tag{70.20}$$

which constitute exact solutions of the differential equation (70.4) in case $k^2(x)$ maintains a linear variation throughout the coordinate range, and note the constancy of their Wronskian, viz.

$$\begin{aligned}F_+F_-' - F_+'F_- &= 2\,\mathrm{i}\Phi(x)\{H_{1/3}^{(1)}(\Phi(x))H_{1/3}^{(2)\prime}(\Phi(x)) - H_{1/3}^{(1)\prime}(\Phi(x))H_{1/3}^{(2)}(\Phi(x))\} \\ &= 8/\pi.\end{aligned} \tag{70.21}$$

Pursuant to the phase specifications (70.18) and the formulas

$$\begin{aligned}H_{1/3}^{(1)}(|\Phi(x)|\,\mathrm{e}^{-\mathrm{i}\pi}) &= H_{1/3}^{(1)}(|\Phi(x)|) + \mathrm{e}^{-\pi\,\mathrm{i}/3} H_{1/3}^{(2)}(|\Phi(x)|)\\ H_{1/3}^{(2)}(|\Phi(x)|\,\mathrm{e}^{-\mathrm{i}\pi}) &= -\,\mathrm{e}^{\mathrm{i}\pi/3} H_{1/3}^{(1)}(|\Phi(x)|)\end{aligned}$$

we obtain

$$\left.\begin{aligned} F_+(x) &= e^{i\pi/6}\left(\frac{|\Phi(x)|}{k(x)}\right)^{1/2}\{H^{(1)}_{1/3}(|\Phi(x)|) + e^{-\pi i/3}H^{(2)}_{1/3}(|\Phi(x)|)\}, \\ F_-(x) &= e^{-i\pi/3}\left(\frac{|\Phi(x)|}{k(x)}\right)^{1/2}\{-H^{(1)}_{1/3}(|\Phi(x)|) + e^{-\pi i/3}H^{(2)}_{1/3}(|\Phi(x)|)\}, \end{aligned}\right\} \quad x < L$$

and thence, making use of the asymptotic estimates (63.20), it turns out that

$$\left.\begin{aligned} F_+(x) &\sim 2\sqrt{(2/\pi k)}\cos\left\{|\Phi(x)| - \frac{\pi}{4}\right\}, \\ F_-(x) &\sim -2\sqrt{(2/\pi k)}\sin\left\{|\Phi(x)| - \frac{\pi}{4}\right\}, \end{aligned}\right\} \quad x \to -\infty. \tag{70.22}$$

In accordance with (70.17) and the representations

$$H^{(1)}_{1/3}(|\Phi(x)|\, e^{i\pi/2}) = \frac{2}{\pi\, i}\, e^{-\pi i/6}K_{1/3}(|\Phi(x)|)$$

$$H^{(2)}_{1/3}(|\Phi(x)|\, e^{i\pi/2}) = -\frac{2}{\pi\, i}\, e^{-\pi i/6}K_{1/3}(|\Phi(x)|) + 2\, e^{i\pi/6}I_{1/3}(|\Phi(x)|)$$

we secure the expressions

$$\left.\begin{aligned} F_+(x) &= \frac{2}{\pi}\, e^{-i\pi/4}\left(\left|\frac{\Phi(x)}{k(x)}\right|\right)^{1/2}K_{1/3}(|\Phi(x)|), \\ F_-(x) &= 4e^{-i\pi/4}\left(\left|\frac{\Phi(x)}{k(x)}\right|\right)^{1/2}\{I_{1/3}(|\Phi(x)|) + \sqrt{(3)}K_{1/3}(|\Phi(x)|)\}, \end{aligned}\right\} \quad x > L$$

and the consequent asymptotic forms

$$\left.\begin{aligned} F_+(x) &\sim e^{-i\pi/4}\left(\frac{2}{\pi|k(x)|}\right)^{1/2} e^{-|\Phi(x)|}, \\ F_-(x) &\sim 2e^{-i\pi/4}\left(\frac{2}{\pi|k(x)|}\right)^{1/2} e^{|\Phi(x)|}, \end{aligned}\right\} \quad x \to \infty. \tag{70.23}$$

Since the Hankel functions satisfy a common differential equation

$$\left(\frac{d^2}{dz^2} + \frac{1}{z}\frac{d}{dz} + 1 - \frac{1}{9z^2}\right)H^{(1),(2)}_{1/3}(z) = 0,$$

the same is true of $F_\pm(x)$ and, in point of fact, the outcome of a straightforward calculation yields

$$\left[\frac{d^2}{dx^2} + k^2(x) + \frac{1}{2}\frac{k''(x)}{k(x)} - \frac{3}{4}\left(\frac{k'(x)}{k(x)}\right)^2 + \frac{5}{36}\left(\frac{k(x)}{\Phi(x)}\right)^2\right]F_\pm(x) = 0$$

or

$$\left[\frac{d^2}{dx^2} + k^2(x) - \mathscr{K}(x)\right]F_\pm(x) = 0 \tag{70.24}$$

with the coefficient function

$$\hat{\mathscr{K}}(x) = (\Phi(x))^{-1/6}(k(x))^{1/2}\frac{d^2}{dx^2}[(\Phi(x))^{1/6}(k(x))^{-1/2}] \tag{70.25}$$

that vanishes when $k^2(x)$ has a strictly linear behavior. There is a qualitative similarity between the desired wave factor $s(x)$ and the function $F_+(x)$ insofar as their standing wave nature in the limits $x \to -\infty, k^2(x) \to k^2$ and their exponentially small magnitudes for $x \to +\infty$ are concerned; moreover, from the convergence of the respective equations for $s(x)$ and $F_\pm(x)$ as the turning point is approached, we may anticipate a useful role for the latter in fashioning an overall characterization of the wave factor. To implement this scheme let us, following Saxon (1959), define the Green's function

$$G(x, x') = \frac{\pi}{8}F_+(x_>)F_-(x_<) \tag{70.26}$$

whose differential specification is, in view of (70.21), (70.24),

$$\left(\frac{d^2}{dx^2} + k^2(x) - \hat{\mathscr{K}}(x)\right)G(x, x') = -\delta(x - x') \tag{70.27}$$

and subsequently, by combining (70.27) with the equation

$$\left(\frac{d^2}{dx^2} + k^2(x)\right)s(x) = 0,$$

establish an integral equation

$$s(x) = F_+(x) + \int_{-\infty}^{\infty} G(x, x')\hat{\mathscr{K}}(x')s(x')\,dx' \tag{70.28}$$

for a wave factor of the intended class. An immediate deduction from (70.28) is that

$$s(x) \sim F_+(x) + \frac{\pi}{8}F_-(x)\int_{-\infty}^{\infty} F_+(x)\hat{\mathscr{K}}(x)s(x)\,dx, \quad x \to -\infty \tag{70.29}$$

or, when the asymptotic forms (70.22) of $F_\pm(x)$ are introduced herein,

$$s(x) \sim 2\left(\frac{2}{\pi k}\right)^{1/2}\frac{\cos\left\{\int_x^L k(\xi)\,d\xi - \frac{\pi}{4} + \delta\right\}}{\cos\delta}, \quad x \to -\infty \tag{70.30}$$

with

$$\tan\delta = \frac{\pi}{8}\int_{-\infty}^{\infty} F_+(x)\hat{\mathscr{K}}(x)s(x)\,dx. \tag{70.31}$$

If $\hat{\mathscr{K}}(x)$ assumes small values everywhere and we rely on the approximate solution $s(x) \doteq F_+(x)$ of (70.28), it appears that

$$\tan\delta = \frac{\pi}{8}\int_{-\infty}^{\infty} \hat{\mathcal{K}}(x) F_+^2(x)\,\mathrm{d}x$$
$$\doteq \frac{1}{2}\int_{-\infty}^{\infty} \frac{\hat{\mathcal{K}}(x)}{k(x)}\,\mathrm{d}x$$

provides a first order estimate for the phase shift δ.

§71. *Related Equations and Improved Asymptotic Solutions*

Approximate solutions are available, in explicit and general fashion, for the differential equation

$$\frac{\mathrm{d}^2 s}{\mathrm{d}x^2} + k_0^2 f(x) s = 0, \quad -\infty < x < \infty \tag{71.1}$$

when the coefficient function $f(x)$ remains bounded away from zero and has only minor variation on intervals whose scale $\lambda_0 = 2\pi/k_0$ is determined by a large coefficient factor k_0; if $f(x)$ vanishes at one or more so-called turning points, the construction of solutions which retain accuracy in their vicinity presents a more complicated task. The original efforts aimed at securing uniformly accurate representations provide connection formulas or rules for matching locally suitable solutions on the respective sides of any turning point(s). We owe to Langer (1931) another method which obviates the need for such connections, this based on the introduction of a related or comparison differential equation with known solutions that are asymptotically equivalent to those of (71.1).

Transformations of both the dependent and independent variables are a prerequisite to selecting related equations; with the particular choices (§67)

$$s = f^{-1/4}\psi, \quad \xi = k_0 \int^x (f(x'))^{1/2}\,\mathrm{d}x' \tag{71.2}$$

the given equation (71.1) becomes

$$\frac{\mathrm{d}^2\psi}{\mathrm{d}\xi^2} + \psi - \left\{ f^{-3/4} \frac{\mathrm{d}^2}{\mathrm{d}(k_0 x)^2} f^{-1/4} \right\}\psi = 0, \tag{71.3}$$

and

$$\frac{\mathrm{d}^2\psi}{\mathrm{d}\xi^2} + \psi = 0 \tag{71.4}$$

meets the criteria for a related equation if f be slowly varying. The solutions of (71.4) evidently furnish us with WKBJ type wave factors whose singularity (and consequent inadmissibility) at a turning point is introduced by way of the original transformation.

To achieve a non-singular transformation we replace the factor $f^{-1/4}$ in the first of the relations (71.2) by another, $g^{-1/4}$, and incorporate a disposable function, $F(\xi)$, in the second or independent variable relation, so that

$$s = g^{-1/4}\psi, \quad F(\xi) = \int^{x} (f(x'))^{1/2}\,\mathrm{d}x'; \tag{71.5}$$

on converting the equation (71.1) by means of the latter transformation pair, one finds

$$\frac{\mathrm{d}^2\psi}{\mathrm{d}\xi^2} + \frac{\mathrm{d}F/\mathrm{d}\xi}{2f^{1/2}}\left[\frac{\mathrm{d}}{\mathrm{d}x}\log\frac{f}{\left(\dfrac{\mathrm{d}F}{\mathrm{d}\xi}\right)^2 g}\right]\frac{\mathrm{d}\psi}{\mathrm{d}\xi} + k_0^2\left(\frac{\mathrm{d}F}{\mathrm{d}\xi}\right)^2\psi + \left[\frac{(\mathrm{d}F/\mathrm{d}\xi)^2}{f}g^{1/4}\frac{\mathrm{d}^2}{\mathrm{d}x^2}g^{-1/4}\right]\psi = 0$$

and a simpler version follows, namely

$$\frac{\mathrm{d}^2\psi}{\mathrm{d}\xi^2} + k_0^2\left(\frac{\mathrm{d}F}{\mathrm{d}\xi}\right)^2\psi - \left[g^{-1/4}\frac{\mathrm{d}^2}{\mathrm{d}\xi^2}g^{1/4}\right]\psi = 0 \tag{71.6}$$

if

$$g = \frac{f}{\left(\dfrac{\mathrm{d}F}{\mathrm{d}\xi}\right)^2} \tag{71.7}$$

affords one condition relating to the presently unspecified functions g, F. The regularity of g, and thence of s, requires that $(\mathrm{d}F/\mathrm{d}\xi)^2$ shall possess zeros (and singularities) to match those of f.

In the special case that the coefficient function $f(x)$ has a single and simple zero at $x = 0$, we set $(\mathrm{d}F/\mathrm{d}\xi)^2 = \xi$ and adopt the transformation relation

$$F(\xi) = \frac{2}{3}\xi^{3/2} = \int_0^{x} (f(x'))^{1/2}\,\mathrm{d}x' \tag{71.8}$$

which establishes a one-to-one correspondence between x and ξ if suitable branch choices for $f^{1/2}$ and $\xi^{3/2}$ are made; a mapping of positive/negative x-ranges onto their ξ-counterparts is implied by (71.8), along with the local version

$$\xi \doteq a_1^{1/3}x, \quad x \to 0$$

when (71.9)

$$f(x) = a_1x + a_2x^2 + \cdots, \quad a_1 > 0.$$

The amplitude factor $g = f/\xi$ that enters into the second transformation relation,

$$s = g^{-1/4}\psi = \left(\frac{f}{\xi}\right)^{-1/4}\psi \tag{71.10}$$

proves to be a non-vanishing positive function and the equation (71.6) for ψ becomes

$$\frac{d^2\psi}{d\xi^2} + k_0^2 \xi \psi - \left[g^{-1/4} \frac{d^2}{d\xi^2} g^{1/4} \right] \psi = 0. \tag{71.11}$$

Solutions of the latter can thus be approximated, when k_0 is large, by Airy functions which satisfy the related equation

$$\frac{d^2\psi}{d\xi^2} + k_0^2 \xi \psi = 0 \tag{71.12}$$

and their asymptotic estimates for $|k_0 \xi^{3/2}| \gg 1$ are, in fact, recognizable as the WKBJ solutions of (71.12).

Let us select the coefficient function

$$f(x) = \tanh \frac{x}{L} \tag{71.13}$$

for purposes of illustrating the related equation method and seek, in particular, to characterize a solution of (71.1) that has the asymptotic behaviors

$$s(x) \sim \begin{cases} e^{-ik_0 x} + r\, e^{ik_0 x}, & x \to +\infty \\ t\, e^{k_0 x}, & x \to -\infty \end{cases} \tag{71.14}$$

which are consistent with the limit values $f \to \pm 1, x \to \pm \infty$. Since $f(x)$ possesses a simple zero at $x = 0$ and $f'(0) = 1/L > 0$, we take over the coordinate transformation (71.8) and, anticipating their proximate use, record the asymptotic versions

$$\begin{aligned} \frac{2}{3} \xi^{3/2} &\simeq x - x_0, & x \to +\infty, \\ \frac{2}{3} (-\xi)^{3/2} &\simeq -(x + x_0), & x \to -\infty, \end{aligned} \tag{71.15}$$

wherein

$$x_0 = L \int_0^\infty (1 - \tanh^{1/2} \zeta)\, d\zeta. \tag{71.16}$$

It is the Airy function $\mathrm{Ai}(-k_0^{2/3}\xi)$, with exponentially small magnitude for $-k_0^{2/3}\, \xi \gg 1$, which constitutes an appropriate solution of the related equation (71.12); hence, referring to the transformation (71.10), we obtain as an approximate solution of (71.1),

$$s = C \left(\frac{\tanh \dfrac{x}{L}}{\xi} \right)^{-1/4} \mathrm{Ai}(-k_0^{2/3}\xi), \tag{71.17}$$

where the scale factor C allows for normalization in the manner given by the first of the expressions (71.14). When x and $\xi \to +\infty$ it follows from (71.15), (71.17) that

$$s \simeq e^{-ik_0 x} + e^{ik_0 x_0 - i\pi/2} \cdot e^{2ik_0 x_0 - i\pi/2}, \quad x \to \infty$$

if

$$C = 2\pi^{1/2} k_0^{1/6}\, e^{-i(k_0 x_0 + \pi/4)}, \tag{71.18}$$

and direct comparison with (71.14) provides an explicit reflection coefficient,

$$r = -i \exp\{-2\, ik_0 x_0\} = -i \exp\left\{2\, ik_0 L \int_0^\infty (1 - \tanh^{1/2} \zeta)\, d\zeta\right\} \tag{71.19}$$

of unit absolute value. In the opposite limit, $x \to -\infty$, where the representation (71.17) implies that

$$s \simeq \frac{1}{2} C\pi^{-1/2} k_0^{-1/6}\, e^{k_0(x + x_0)}, \quad x \to -\infty$$

we verify a concordance with the asymptotic form (71.14) and, additionally, secure an estimate for the transmitted wave factor amplitude, viz.

$$\begin{aligned} t &= \frac{1}{2} C\pi^{-1/2} k_0^{-1/6}\, e^{k_0 x_0} \\ &= e^{k_0 x_0}\, e^{-i(k_0 x_0 + \pi/4)}. \end{aligned} \tag{71.20}$$

To assess the magnitude of the term

$$P = -g^{-1/4} \frac{d^2}{d\xi^2} g^{1/4} = g^{-3/4} \frac{d^2}{dx^2} g^{-1/4} \tag{71.21}$$

which appears in the exact differential equation (71.11) and is absent from the related equation (71.12), let us note that differentiation of (71.8) yields

$$\left(\frac{d\xi}{dx}\right)^2 \equiv (\xi')^2 = \frac{f}{\xi} = g,$$

whence

$$P = (\xi')^{-3/2} \frac{d^2}{dx^2} (\xi')^{-1/2} = \frac{3}{4} (\xi')^{-4} (\xi'')^2 - \frac{1}{2} (\xi')^{-3} \xi'''.$$

Extracting from (71.15) the less precise, though sufficiently accurate, estimate $\zeta \sim x^{2/3}$, $x \to +\infty$, we obtain

$$P \sim x^{-4/3}, \quad x \to +\infty$$

and the same order of magnitude for P applies in the limit $x \to -\infty$. Thus, it may be held that an accurate description for the dominant part of $s(x)$ results, on utilizing a solution of the related equation, when $|x|$ is sufficiently large; and that our expressions (71.19), (71.20) for r, t possess lowest order validity whatever the value of the scale factor L in the coefficient function $f(x)$. To support the latter contention, we remark on the agreement between the limiting forms of (71.19), (71.20),

$$r \to -\mathrm{i}, \qquad t \to \mathrm{e}^{-\mathrm{i}\pi/4}, \qquad L \to 0$$

and the independently verified determination of the secondary wave amplitudes in the case of a step function coefficient profile, that is, $f = \pm 1, x \gtrless 0$.

Near the turning point, $x = 0$, ξ and its derivatives have power series expansions in x which are individually determinable from a consequence of the transformation formula (71.8), e.g.,

$$\frac{2}{3}\xi^{3/2} = \int_0^x \left(\tanh\frac{x'}{L}\right)^{1/2} \mathrm{d}x' = \frac{2}{3}L^{-1/2}x^{3/2} - \frac{1}{21}L^{-5/2}x^{7/2} + \cdots, \quad x > 0$$

and there follows

$$P = \frac{1}{7}L^{-4/3}\left\{1 + \frac{6}{7}\left(\frac{x}{L}\right)^2 + \cdots\right\}, \quad x > 0.$$

The affiliated representation for the full coefficient of ψ in the equation (71.11),

$$k_0^2\xi + P = k_0^2 L^{2/3}\left\{\frac{x}{L} + \frac{1}{7(k_0L)^2} + 0(x/L)^2\right\}, \quad x \to 0_+$$

whose explicit terms reveal a displacement in the null or turning point from the origin to $x = -1/7\,k_0^2L$, suggests that our prior characterization (71.17) for $s(x)$ retains overall accuracy provided $k_0L \gg 1$.

If it be desired to refine the conclusions regarding the differential equation

$$\frac{\mathrm{d}^2\psi}{\mathrm{d}\xi^2} + (k_0^2\xi - P(\xi))\psi = 0 \tag{71.22}$$

which are based on solution of a related equation with $P = 0$, we may first of all identify the null point, ξ^*, of its coefficient function, i.e.,

$$k_0^2\xi^* - P(\xi^*) = 0, \tag{71.23}$$

and introduce the new independent variable $\eta = \xi - \xi^*$; then, assuming a development

$$P(\xi) = P(\xi^*) + P'(\xi^*)\eta + \frac{1}{2}P''(\xi^*)\eta^2 + \cdots \tag{71.24}$$

and recasting (71.22) in the form

$$\frac{\mathrm{d}^2\psi}{\mathrm{d}\eta^2} + k_0^2 h(\eta)\psi = 0, \tag{71.25}$$

where

$$h(\eta) = \left(\frac{P'(\xi^*)}{k_0^2}\right)\eta + \frac{1}{k_0^2}Q(\eta), \quad Q(\eta) = -\sum_{n=2}^{\infty} \frac{\eta^n}{n!} P^{(n)}(\xi^*), \tag{71.26}$$

we are in a position to re-employ the transformations that connect the original equation (71.1) with (71.22). Specifically, the second transformation set,

$$H(\zeta) = \frac{2}{3}\zeta^{3/2} = \int_0^\eta (h(\eta'))^{1/2}\, d\eta', \qquad \psi = j^{-1/4}\varphi, \qquad j = \frac{h}{\zeta}, \tag{71.27}$$

converts (71.25) to

$$\frac{d^2\varphi}{d\zeta^2} + (k_0^2\zeta - R(\zeta))\varphi = 0 \tag{71.28}$$

where

$$R(\zeta) = j^{-1/4}\frac{d^2}{d\zeta^2}j^{1/4} = -j^{-3/4}\frac{d^2}{d\eta^2}j^{-1/4}. \tag{71.29}$$

That $R(\zeta)$ has a lesser order of magnitude, for large values of the parameter k_0, than its counterpart, $P(\zeta) = 0(1)$, is inferred from the development

$$\begin{aligned} H &= \alpha^{1/2}\int_0^\eta (\eta')^{1/2}\left(1 + \frac{Q(\eta')}{\eta'\alpha k_0^2}\right)^{1/2} d\eta' \\ &= \frac{2}{3}\alpha^{1/2}\eta^{3/2} + \frac{1}{2}\alpha^{-1/2}k_0^{-2}\int_0^\eta \frac{Q(\eta)}{\eta}\, d\eta + 0\left(\frac{1}{k_0^4}\right), \quad \alpha = 1 - \frac{P'(\xi^*)}{k_0^2} \end{aligned}$$

which yields

$$\zeta = \alpha^{1/3}\eta + 0\left(\frac{1}{k_0^2}\right), \qquad j = \alpha^{-2/3} + 0\left(\frac{1}{k_0^2}\right)$$

and hence an estimate $R(\zeta) = 0(1/k_0^2)$. Thanks to the turning point correction which precedes the establishment of (71.28), a related equation therefor,

$$\frac{d^2\varphi}{d\zeta^2} + k_0^2\zeta\varphi = 0,$$

surpasses a prior equation, with the identical form, as regards closeness to the originals; and thus permits an improved approximation when k_0 is large. Successive turning point corrections of the same nature make available higher order approximations.

§72. Perturbation Calculations in Cases of Non-uniformity

If the deviations from regularity or homogeneity are slight, we can adopt the techniques of perturbation theory to establish approximate results in convenient and general fashion. As a first example of their use in connection with equations that bear on wave phenomena, let us refer to a class of tidal motions along a channel whose section is nearly uniform. For one-dimensional or longitudinal movements gauged by a mass flux, $\Psi(x, t)$, equal to the product of density, velocity and sectional area $A(x)$, classic dynamical arguments provide an equation

$$\frac{\partial^2 \Psi}{\partial t^2} = gA(x)\frac{\partial}{\partial x}\left\{\frac{1}{b(x)}\frac{\partial \Psi}{\partial x}\right\}, \tag{72.1}$$

where $b(x)$ designates the (equilibrium) free surface breadth and g the gravitational constant. Following the introduction of a new independent variable,

$$\xi(x) = \frac{\int_0^x b(x)\,\mathrm{d}x}{\int_0^L b(x)\,\mathrm{d}x} \tag{72.2}$$

with a range $0 \le \xi \le 1$ that corresponds to the full extent $0 \le x \le L$ of the channel, (72.1) is turned into a variable coefficient wave equation

$$\frac{\partial^2 \Psi}{\partial \xi^2} = \frac{1}{f(\xi)}\frac{\partial^2 \Psi}{\partial t^2} \tag{72.3}$$

where

$$f(\xi) = \frac{gb(x(\xi))A(x(\xi))}{\left(\int_0^L b(x)\,\mathrm{d}x\right)^2} \tag{72.4}$$

and $x(\xi)$ represents the inverse of the function $\xi(x)$.

In the case of simply-periodic motions we set

$$\Psi = \psi(\xi)\,\mathrm{e}^{-\mathrm{i}\omega t} \tag{72.5}$$

and the time-independent factor $\psi(\xi)$ satisfies an ordinary differential equation,

$$\frac{\mathrm{d}^2\psi}{\mathrm{d}\xi^2} + \frac{\omega^2}{f(\xi)}\psi = 0, \tag{72.6}$$

together with the boundary conditions

$$\psi(0) = \psi(1) = 0 \tag{72.7}$$

if the channel is closed off at the ends by fixed walls. The solution of (72.6) which complies with (72.7) is readily forthcoming when the coefficient function $f(\xi)$ has a fixed magnitude

$$\bar{f} = \frac{g\bar{A}}{L^2\bar{b}} \tag{72.8}$$

expressed in terms of mean values for the shape factors $A(x)$, $b(x)$; and the normalized eigenfunctions prove to be

$$\psi_n^{(0)} = \sqrt{(2)} \sin n\pi\xi, \qquad \int_0^1 (\psi_n^{(0)})^2 \, \mathrm{d}\xi = 1, \qquad n = 1, \ldots \tag{72.9}$$

while the proper frequencies of the tidal motions are given by

$$\omega_n^{(0)} = n\pi\sqrt{\bar{f}} = \frac{n\pi}{L}\sqrt{g\bar{h}} \tag{72.10}$$

where $\bar{h} = \bar{A}/\bar{b}$ is the mean depth of the channel.

To characterize a particular solution of (72.6), namely the perturbed eigenfunction

$$\psi_n(\xi) = \psi_n^{(0)}(\xi) + \psi_n^{(1)}(\xi) \tag{72.11}$$

we introduce, concurrently, the representations

$$\begin{aligned} \omega_n^2 &= (\omega_n^{(0)})^2[1 + \delta_n] \\ f(\xi) &= \bar{f} + \Delta f(\xi), \qquad \int_0^1 \Delta f \, \mathrm{d}\xi = 0 \end{aligned} \tag{72.12}$$

and assume that $\psi_n^{(1)}$, δ and Δf are small enough for the disregard of all related products. On substituting (72.11), along with the expression

$$\frac{\omega_n^2}{f(\xi)} = \frac{(\omega_n^{(0)})^2[1 + \delta]}{\bar{f} + \Delta f} \doteq \frac{(\omega_n^{(0)})^2}{\bar{f}} + (\omega_n^{(0)})^2 \left\{ \frac{\delta_n}{\bar{f}} - \frac{\Delta f}{(\bar{f})^2} \right\}$$

into (72.6), and retaining only terms of a comparable (that is, first order magnitude in the perturbed quantities), an equation for $\psi_n^{(1)}$ results, viz.

$$\left(\frac{\mathrm{d}^2}{\mathrm{d}\xi^2} + \frac{(\omega_n^{(0)})^2}{\bar{f}}\right)\psi_n^{(1)} = \frac{(\omega_n^{(0)})^2}{\bar{f}}\left\{\frac{\Delta f}{\bar{f}} - \delta_n\right\}\psi_n^{(0)}, \tag{72.13}$$

with a known form of the inhomogeneous member. Since

$$\left(\frac{\mathrm{d}^2}{\mathrm{d}\xi^2} + \frac{(\omega_n^{(0)})^2}{\bar{f}}\right)\psi_n^{(0)} = 0$$

and $\psi_n^{(0)}$, $\psi_n^{(1)}$ individually satisfy the boundary conditions (72.7), it is easy to confirm that

$$\int_0^1 \psi_n^{(0)} \left(\frac{\mathrm{d}^2}{\mathrm{d}\xi^2} + \frac{(\omega_n^{(0)})^2}{\bar{f}}\right)\psi_n^{(1)} \, \mathrm{d}\xi = 0$$

and therefore (72.13) implies

$$\int_0^1 \left\{\frac{\Delta f}{\bar{f}} - \delta_n\right\}(\psi_n^{(0)})^2 \, \mathrm{d}\xi = 0 \tag{72.14}$$

or

$$\delta_n = \frac{1}{\bar{f}} \int_0^1 \Delta f(\xi)(\psi_n^{(0)}(\xi))^2 \, \mathrm{d}\xi. \tag{72.15}$$

Hence, first order perturbation theory yields the proper frequency estimates

$$\omega_n^2 = (\omega_n^{(0)})^2 \left\{1 + \frac{1}{\bar{f}} \int_0^1 \Delta f(\xi)(\psi_n^{(0)}(\xi))^2 \, \mathrm{d}\xi\right\},$$

and (72.16)

$$\omega_n^2 = (\omega_n^{(0)})^2 \left\{1 + 2\int_0^L \frac{\Delta f(x)}{\bar{f}} \sin^2\left(n\pi \frac{\int_0^x b(\zeta)\,\mathrm{d}\zeta}{\int_0^L b(\zeta)\,\mathrm{d}\zeta}\right) \frac{b(x)\,\mathrm{d}x}{\int_0^L b(\zeta)\,\mathrm{d}\zeta}\right\},$$

according as the transformed or original coordinate variables appear therein. Particulars of the eigenfunctions in this and higher order approximations can be supplied without difficulty, though we refrain from so doing here.

Our next example of a perturbation calculation has relevance to progressive motions in a fluid or elastic material whose dynamical response is controlled by two parameters, namely the compressibility κ and the density ρ. We shall suppose that the latter vary along the x-direction and admit the decompositions

$$\kappa(x) = \kappa_0 + \kappa_1(x)$$
$$\rho(x) = \rho_0 + \rho_1(x)$$

where κ_1, ρ_1 account for the small and localized departures from homogeneity. In states of excitation which feature an exclusive displacement $\xi(x, t)$ along the aforesaid direction, a single equation of motion figures, namely

$$\frac{\partial}{\partial x}\left\{\frac{1}{\kappa(x)} \frac{\partial \xi}{\partial x}\right\} - \rho(x)\frac{\partial^2 \xi}{\partial t^2} = 0. \tag{72.18}$$

Let us resolve ξ in the fashion

$$\xi(x, t) = \xi^{(0)}(x, t) + \xi^{(1)}(x, t) \tag{72.19}$$

and identify one part,

$$\xi^{(0)}(x, t) = A \exp\left\{-\mathrm{i}\omega\left(t - \frac{x}{c}\right)\right\}, \quad c^2 = \frac{1}{\kappa_0 \rho_0} \tag{72.20}$$

as a primary (and progressive) wave function befitting uniform values κ_0, ρ_0 of the material parameters; we associate the other part, $\xi^{(1)}(x, t)$, with secondary waves attributable to the small irregularities κ_1, ρ_1 of these parameters. After recourse to the approximation

$$\frac{1}{\kappa(x)}=\frac{1}{\kappa_0+\kappa_1(x)}\doteq\frac{1}{\kappa_0}-\frac{\kappa_1(x)}{\kappa_0^2}$$

and joint manipulation of (72.17)–(72.19), we obtain an inhomogeneous equation for $\xi^{(1)}$,

$$\left(\frac{\partial}{\partial x^2}-\frac{1}{c^2}\frac{\partial^2}{\partial t^2}\right)\xi^{(1)}=\frac{\partial}{\partial x}\left\{\frac{\kappa_1(x)}{\kappa_0}\frac{\partial\xi^{(0)}}{\partial x}\right\}+\rho_1(x)\kappa_0\frac{\partial^2\xi^{(0)}}{\partial t^2}, \tag{72.21}$$

with source terms that are parently linear in κ_1, ρ_1 and thus appropriate to a first order stage of perturbation theory.

Having regard for the explicit and time-periodic representation (72.20) of $\xi^{(0)}$, a synchronous solution of (72.21) is rendered by

$$\xi^{(1)}(x,t)=\frac{\mathrm{d}\Pi_1}{\mathrm{d}x}\,\mathrm{e}^{-\mathrm{i}\omega t}+\Pi_2(x)\,\mathrm{e}^{-\mathrm{i}\omega t} \tag{72.22}$$

where

$$\begin{aligned}\left(\frac{\mathrm{d}^2}{\mathrm{d}x^2}+\frac{\omega^2}{c^2}\right)\Pi_1&=A\frac{\kappa_1(x)}{\kappa_0}\frac{\mathrm{d}}{\mathrm{d}x}\,\mathrm{e}^{\mathrm{i}\omega x/c}=\mathrm{i}A\frac{\omega}{c}\frac{\kappa_1(x)}{\kappa_0}\,\mathrm{e}^{\mathrm{i}\omega x/c},\\ \left(\frac{\mathrm{d}^2}{\mathrm{d}x^2}+\frac{\omega^2}{c^2}\right)\Pi_2&=-A\omega^2\kappa_0\rho_1(x)\,\mathrm{e}^{\mathrm{i}\omega x/c}.\end{aligned} \tag{72.23}$$

If the measures of non-uniformity in compressibility and density, κ_1, ρ_1, are confined to a range Δ, the particular integrals of (72.23),

$$\begin{aligned}\Pi_1(x)&=-\frac{\mathrm{i}c}{2\omega}\int_\Delta\exp\left\{\mathrm{i}\frac{\omega}{c}|x-x'|\right\}\left[\mathrm{i}A\frac{\omega}{c}\frac{\kappa_1(x')}{\kappa_0}\,\mathrm{e}^{\mathrm{i}\omega x'/c}\right]\mathrm{d}x',\\ \Pi_2(x)&=-\frac{\mathrm{i}c}{2\omega}\int_\Delta\exp\left\{\mathrm{i}\frac{\omega}{c}|x-x'|\right\}\left[-A\omega^2\kappa_0\rho_1(x')\,\mathrm{e}^{\mathrm{i}\omega x'/c}\right]\mathrm{d}x'\end{aligned} \tag{72.24}$$

secure for $\xi^{(1)}$ an outgoing wave character in the exterior of Δ. When Δ overlaps the coordinate origin and its extension is small enough for us to regard κ_1, ρ_1 as constants, the approximate versions of (72.24) become

$$\Pi_1(x)=\frac{A\Delta}{2}\frac{\kappa_1}{\kappa_0}\,\mathrm{e}^{\mathrm{i}\omega|x|c},$$

$$\Pi_2(x)=\mathrm{i}A\frac{\omega c}{2}\kappa_0\rho_1\,\mathrm{e}^{\mathrm{i}\omega|x|c},$$

and

$$\xi^{(1)}(x,t)=\mathrm{i}\frac{A\Delta}{2}\frac{\omega}{c}\left[\frac{\kappa_1}{\kappa_0}\operatorname{sgn}x+\frac{\rho_1}{\rho_0}\right]\exp\left\{-\mathrm{i}\omega\left(t-\frac{|x|}{c}\right)\right\} \tag{72.25}$$

furnishes the secondary or scattered wave function. Assuming a direct proportionality between density and compressibility changes, viz.

$$\frac{\kappa_1}{\kappa_0} = \alpha \frac{\rho_1}{\rho_0},$$

there follows

$$\xi^{(1)}(x, t) = \mathrm{i}\frac{A\Delta}{2}\frac{\omega\rho_1}{c\rho_0}[1 + \alpha \operatorname{sgn} x] \exp\left\{-\mathrm{i}\omega\left(t - \frac{|x|}{c}\right)\right\}. \tag{72.26}$$

Given that

$$S = -\frac{1}{\kappa_0}\frac{\partial \xi}{\partial x}\frac{\partial \xi}{\partial t} \tag{72.27}$$

specifies the instantaneous energy flux along the positive x-direction, computations with (72.26) yield

$$\bar{S}_+ = \frac{1}{2}\frac{\omega^4}{\kappa_0 c^3}\left(\frac{A\Delta}{2} \cdot \frac{\rho_1}{\rho_0}\right)^2 (1 + \alpha)^2,$$

$$\bar{S}_- = \frac{1}{2}\frac{\omega^4}{\kappa_0 c^3}\left(\frac{A\Delta}{2} \cdot \frac{\rho_1}{\rho_0}\right)^2 (1 - \alpha)^2$$

for the time-average amounts in the secondary waves that travel outwards on the respective sides of the source region Δ. A measure of the scattering efficiency for localized material parameter changes is expressed by the ratio of the net flux in secondary waves, $\bar{S}_+ + \bar{S}_-$, to the primary wave flux, $(1/2)(\omega^2/\kappa_0 c)A^2$; and our first order perturbation theory estimate for this comparative power ratio takes the form

$$\frac{1}{2}\left(\frac{\omega\Delta}{c} \cdot \frac{\rho_1}{\rho_0}\right)^2 (1 + \alpha^2) = \frac{1}{2}\left(\frac{\omega\Delta}{c}\right)^2 \left[\left(\frac{\rho_1}{\rho_0}\right)^2 + \left(\frac{\kappa_1}{\kappa_0}\right)^2\right].$$

§73. A Short Wave Length Expansion Technique

If the equation appropriate to harmonic excitations of an inhomogeneous setting,

$$\frac{\mathrm{d}^2 s}{\mathrm{d}x^2} + k^2(1 + \eta(x))s = 0, \quad -\infty < x < \infty \tag{73.1}$$

has a smooth (and absolutely integrable) function $\eta(x)$ with an effective range characterized by the length L, we anticipate but little reflection for a primary disturbance

$$s^{inc} = \mathrm{e}^{ikx}, \quad x \to -\infty \tag{73.2}$$

whose wave length $\lambda = 2\pi/k$ is small compared to L. When the effect of inhomogeneity on the primary wave function (73.2) is sought after in terms of a multiplicative factor, $\psi(x)$, such that

$$s(x) = \mathrm{e}^{ikx}\psi(x), \quad \psi(-\infty) = 1 \tag{73.3}$$

the equation for $\psi(x)$ which follows from (73.1) takes the form

$$\left(\frac{d^2}{dx^2}+2\,ik\frac{d}{dx}\right)\psi=-k^2\eta(x)\psi. \tag{73.4}$$

On the supposition that the scale for significant changes in ψ is proportional to L rather than λ, the first term on the left-hand side of (73.4) possesses an order of magnitude smaller than that of the second by a factor $1/kL$; and by ignoring the former we readily deduce an approximate solution of (73.4), with the correct asymptotic behavior in the limit $x\to-\infty$, viz.

$$\psi^{(0)}(x)=\exp\left\{\frac{ik}{2}\int_{-\infty}^{x}\eta(\xi)\,d\xi\right\} \tag{73.5}$$

which yields, in turn, the estimate

$$s^{(0)}(x)=\exp\left\{ikx+\frac{ik}{2}\int_{-\infty}^{x}\eta(\xi)\,d\xi\right\}. \tag{73.6}$$

An expression with the same features is arrived at from the affiliated progressive wave function of the WKBJ representation (66.10) by substituting k and $k(1+\frac{1}{2}\eta(x))$ for $k(x)$ in the amplitude and phase factors, respectively. The invariable amplitude or absolute value of the wave function (73.6) makes it evident that the consequences of reflection have not been incorporated therein; to confirm this omission in another manner, let us employ an integral equation,

$$s(x)=e^{ikx}+\frac{ik}{2}\int_{-\infty}^{\infty}e^{ik|x-x'|}\eta(x')s(x')\,dx',\quad -\infty<x<\infty \tag{73.7}$$

which subsumes the differential and incoming wave specifications of $s(x)$ and is readily deduced from (73.3) with the help of the Green's function (17.11). If the corresponding relation for $\psi(x)$ is displayed in the form

$$\psi(x)=1+\frac{ik}{2}\int_{-\infty}^{x}\eta(x')\psi(x')\,dx'+\frac{ik}{2}\,e^{-2\,ikx}\int_{x}^{\infty}e^{2\,ikx'}\eta(x')\psi(x')\,dx'$$

the second integral describes a regressive component of the excitation at the point x, inasmuch as its support is located farther along the primary wave direction ($x'>x$). On discarding this component (with a magnitude smaller than the preceding one by virtue of the oscillatory integrand factor $e^{2\,ikx'}$) the resultant integral equation for $\psi(x)$ can be turned into a first order differential equation

$$\frac{d\psi}{dx}=\frac{ik}{2}\eta(x)\psi(x),$$

whose solution (that assumes the stipulated unit value at $x=-\infty$) is given by (73.5).

A two component representation for the wave function,

$$s(x) = e^{ikx}\psi_+(x) + e^{-ikx}\psi_-(x), \tag{73.8}$$

where both multipliers $\psi_\pm(x)$ have small variation on intervals of length λ, enables us to correct the exposed deficiency in our prior estimate. When we express the outcome of differentiating $s(x)$ in the form

$$\frac{ds}{dx} = ik[e^{ikx}\psi_+(x) - e^{-ikx}\psi_-(x)] \tag{73.9}$$

a first condition relative to the determination of $\psi_\pm(x)$ is implied, namely,

$$e^{ikx}\psi'_+(x) + e^{-ikx}\psi'_-(x) = 0; \tag{73.10}$$

and a second condition obtains after the derivative of (73.9), along with (73.8), is substituted in (73.3), viz.

$$e^{ikx}\psi'_+(x) - e^{-ikx}\psi'_-(x) = ik\eta(x)[e^{ikx}\psi_+(x) + e^{-ikx}\psi_-(x)]. \tag{73.11}$$

The solution of (73.10), (73.11) for $\psi'_\pm(x)$ yields a pair of coupled first-order differential equations

$$\begin{aligned} \frac{d\psi_+}{dx} &= \frac{ik}{2}\eta(x)\psi_+(x) + \frac{ik}{2}\eta(x)\, e^{-2ikx}\psi_-(x), \\ \frac{d\psi_-}{dx} &= -\frac{ik}{2}\eta(x)\psi_-(x) - \frac{ik}{2}\eta(x)\, e^{2ikx}\psi_+(x), \end{aligned} \tag{73.12}$$

which are, in association with (73.8), fully equivalent to the original second-order equation (73.3).

Bearing in mind the assumption that only a single incoming wave (at $x = -\infty$) is present, the requisite integrals of the system (73.12) are selected on the basis of the conditions

$$\psi_+(-\infty) = 1, \quad \psi_-(\infty) = 0. \tag{73.13}$$

For purposes of an iterative solution, we adopt the expansions

$$\begin{aligned} \psi_+(x) &= \sum_{j=0}^{\infty} \psi_+^{(2j)}(x), \\ \psi_-(x) &= \sum_{j=1}^{\infty} \psi_-^{(2j-1)}(x) \end{aligned} \tag{73.14}$$

whose terms are successively and alternately defined by the equations

$$\begin{aligned} \frac{d\psi_-^{(2j-1)}}{dx} + \frac{ik}{2}\eta(x)\psi_-^{(2j-1)}(x) &= -\frac{ik}{2}\eta(x)\, e^{2ikx}\psi_+^{(2j-2)}(x), \\ \frac{d\psi_+^{(2j)}}{dx} - \frac{ik}{2}\eta(x)\psi_+^{(2j)}(x) &= \frac{ik}{2}\eta(x)\, e^{-2ikx}\psi_-^{(2j-1)}(x), \end{aligned} \quad j \geq 0 \tag{73.15}$$

together with the supplemental prescriptions

$$\psi_-^{(2j-1)}(x) \equiv 0, \quad j < 1$$
$$\psi_+^{(2j)}(-\infty) = 0, \quad \psi_-^{(2j-1)}(+\infty) = 0, \quad j \geq 1. \tag{73.16}$$

In particular,

$$\psi_+^{(0)}(x) = \exp\left\{\frac{ik}{2}\int_{-\infty}^{x} \eta(\xi)\,d\xi\right\} \tag{73.17}$$

$$\psi_-^{(1)}(x) = \frac{ik}{2}\exp\left\{-\frac{ik}{2}\int_{-\infty}^{x} \eta(\xi)\,d\xi\right\}\int_{x}^{\infty} \eta(x')\,e^{2ikx'}\exp\left\{ik\int_{-\infty}^{x'} \eta(\xi)\,d\xi\right\}dx', \tag{73.18}$$

$$\psi_+^{(2)}(x) = -\frac{k^2}{4}\exp\left\{\frac{ik}{2}\int_{-\infty}^{x} \eta(\xi)\,d\xi\right\}\int_{-\infty}^{x} \eta(x')\,e^{-2ikx'}\exp\left\{-ik\int_{-\infty}^{x'} \eta(\xi)\,d\xi\right\}$$
$$\times \int_{x'}^{\infty} \eta(x'')\,e^{2ikx''}\exp\left\{ik\int_{-\infty}^{x''} \eta(\xi)\,d\xi\right\}dx'\,dx'', \tag{73.19}$$

etc.

A determination of the reflection coefficient follows directly from the (complex) amplitude, $\psi_-(-\infty)$, of the outgoing wave at $x = -\infty$; and the first order estimate therefore,

$$r^{(1)} = \lim_{x\to-\infty} \psi_-^{(1)}(x)$$
$$= \frac{ik}{2}\int_{-\infty}^{\infty} \eta(x)\,e^{2ikx}\exp\left\{ik\int_{-\infty}^{x} \eta(\xi)\,d\xi\right\}dx$$

agrees in all respects with the comparable expression (69.25). Since the wave factor $\exp(ikx)\psi_+^{(0)}(x)$ takes no account of reflection it is fitting that the related transmission coefficient,

$$t^{(0)} = \lim_{x\to+\infty} \psi_+^{(0)}(x) = \exp\left\{\frac{ik}{2}\int_{-\infty}^{\infty} \eta(\xi)\,d\xi\right\},$$

has unit absolute magnitude for any real function $\eta(x)$. Less than complete transmission is manifest in the next stage of our outgoing wave representation at $x = +\infty$, and the result

$$t^{(2)} = \lim_{x\to+\infty} (\psi_+^{(0)}(x) + \psi_+^{(2)}(x))$$
$$= \exp\left\{\frac{ik}{2}\int_{-\infty}^{\infty} \eta(\xi)\,d\xi\right\}\left[1 - \frac{k^2}{4}\int_{-\infty}^{\infty} \eta(x)\,e^{-2ikx}\exp\left(-ik\int_{-\infty}^{x} \eta(\xi)\,d\xi\right)\right. \tag{73.20}$$
$$\left.\times \int_{x}^{\infty} \eta(x')\,e^{2ikx'}\exp\left(ik\int_{-\infty}^{x'} \eta(\xi)\,d\xi\right)dx\,dx'\right]$$

is consistent with a previous estimate (69.12), stated in terms of the full wave number $k(x) = k(1+\eta(x))^{1/2}$, when $\eta(x) \ll 1$; conservation of energy (for real $\eta(x)$) dictates the balance equation

$$|t^{(2)}|^2 = 1 - |r^{(1)}|^2,$$

which can be verified along the lines employed in §69.

To acquire a different perspective of the links between earlier prototypes of approximate solutions for the variable coefficient differential equation

$$\left(\frac{\mathrm{d}^2}{\mathrm{d}x^2} + k^2 n^2(x)\right) s(x) = 0, \tag{73.21}$$

assuming that the parameter k is large and that

$$n(x) = (1+\eta(x))^{1/2} > 0, \tag{73.22}$$

let us define the vector [cf. van Kampen, 1967]

$$\varphi(x) = \begin{pmatrix} \varphi_1(x) \\ \varphi_2(x) \end{pmatrix} = \begin{pmatrix} s(x) \\ \dfrac{1}{k}\dfrac{\mathrm{d}s}{\mathrm{d}x} \end{pmatrix} \tag{73.23}$$

and replace (73.21) by a matrix relation

$$\frac{\mathrm{d}}{\mathrm{d}x}\varphi(x) = kM(x)\varphi(x), \quad M(x) = \begin{pmatrix} 0 & 1 \\ -n^2(x) & 0 \end{pmatrix}. \tag{73.24}$$

On diagonalization of the matrix M in accordance with the scheme

$$(A(x))^{-1} M(x) A(x) = B(x), \tag{73.25}$$

where

$$A(x) = \begin{pmatrix} 1 & 1 \\ in(x) & -in(x) \end{pmatrix}, \quad B(x) = \begin{pmatrix} in(x) & 0 \\ 0 & -in(x) \end{pmatrix},$$
$$(A(x))^{-1} = \begin{pmatrix} \dfrac{1}{2} & -\dfrac{\mathrm{i}}{2n(x)} \\ \dfrac{1}{2} & \dfrac{\mathrm{i}}{2n(x)} \end{pmatrix} \tag{73.26}$$

it follows that the vector

$$\Phi(x) = A(x)\varphi(x) \tag{73.27}$$

satisfies a first-order matrix differential equation

$$\frac{\mathrm{d}\Phi}{\mathrm{d}x} = \mathrm{i}kn(x)\begin{pmatrix} 1 & 0 \\ 0 & -1 \end{pmatrix}\Phi - \frac{n'(x)}{2n(x)}\begin{pmatrix} 1 & -1 \\ -1 & 1 \end{pmatrix}\Phi. \tag{73.28}$$

If we omit the second part of the right-hand side in (73.28), which lacks the large coefficient k present in the first part, the resulting specifications are

$$\begin{pmatrix} \Phi_1(x) \\ \Phi_2(x) \end{pmatrix} = \exp\left\{\pm \mathrm{i}k \int_0^x \eta(\xi)\,\mathrm{d}\xi\right\}\begin{pmatrix} \Phi_1(0) \\ \Phi_2(0) \end{pmatrix}$$

and the sole effect of inhomogeneity is registered by a cumulative phase factor, akin to that appearing in the estimate (73.6). When the diagonal components of the second part are included, Φ_1, Φ_2 remain independent and their versions

$$\begin{pmatrix}\Phi_1(x)\\ \Phi_2(x)\end{pmatrix} = \frac{1}{\sqrt{(n(x))}} \exp\left\{\pm ik \int_0^x \eta(\xi)\,\mathrm{d}\xi\right\}\begin{pmatrix}\Phi_1(0)\\ \Phi_2(0)\end{pmatrix} \tag{73.29}$$

have the character of WKBJ functions. The non-diagonal elements in (73.28) provide a coupling or interaction between Φ_1 and Φ_2 which is explicitly revealed in the component equations

$$\begin{aligned}\frac{\mathrm{d}\Phi_1}{\mathrm{d}x} - \left(ikn(x) - \frac{n'(x)}{2n(x)}\right)\Phi_1 &= \frac{n'(x)}{2n(x)}\Phi_2,\\ \frac{\mathrm{d}\Phi_2}{\mathrm{d}x} + \left(ikn(x) + \frac{n'(x)}{2n(x)}\right)\Phi_2 &= \frac{n'(x)}{2n(x)}\Phi_1.\end{aligned} \tag{73.30}$$

We may ascertain the initial corrections to the WKBJ forms (73.29) by substituting the latter into the second members of (73.30) and thereafter determining particular solutions of the resultant inhomogeneous differential equations; it appears that, as regards Φ_1, said correction term (with a non-diagonal nature) is expressed through the integral

$$\int_0^x \frac{n'(x)}{2n(x)} \exp\left(-2ik \int_0^x n(\xi)\,\mathrm{d}\xi\right)\mathrm{d}x$$

whose order of magnitude, $1/k$, is smaller than one suggested by the explicitly appearing factors k in (73.28).

§74. *Variational and Other Efficient Characterizations of Scattering Coefficients*

The use of formally exact integral expressions for scattering (i.e., reflection/transmission) coefficients is contingent on having particulars of the wave function; and a distinction may be drawn between different expressions of such nature as regards the accuracy of their predictions on the basis of approximate forms of the wave function. Let us refer, by way of example, to an expression for the reflection coefficient deduced from (73.7),

$$\begin{aligned} r &= \lim_{x\to-\infty} \mathrm{e}^{ikx}[s(x) - s^{inc}(x)]\\ &= \frac{ik}{2}\int_{-\infty}^{\infty} \mathrm{e}^{ikx}\eta(x)s(x)\,\mathrm{d}x,\end{aligned} \tag{74.1}$$

and note that the first order reflection coefficient [cf. (69.25)]

$$r^{(1)} = \frac{ik}{2}\int_{-\infty}^{\infty} \eta(x)\,e^{2ikx}\exp\left\{ik\int_{-\infty}^{x}\eta(\xi)\,d\xi\right\}dx \tag{74.2}$$

does not follow therefrom on substituting the zeroth approximation

$$s^{(0)}(x) = \exp\left\{ikx + \frac{ik}{2}\int_{-\infty}^{x}\eta(\xi)\,d\xi\right\} \tag{74.3}$$

to the wave factor. However, the latter distribution suffices for confirming the prediction (74.2) if the calculation is made with another version of the reflection coefficient, namely one distinguished by a stationary property relative to small variations of $s(x)$ about its exact values. To construct the efficient representation of r we observe, firstly, that the result of multiplying the equation (73.7) by $(ik/2)\eta(x)s(x)$ and then integrating is an identity

$$\frac{ik}{2}\int_{-\infty}^{\infty}\eta(x)s^2(x)\,dx = \frac{ik}{2}\int_{-\infty}^{\infty}e^{ikx}\eta(x)s(x)\,dx$$
$$+\left(\frac{ik}{2}\right)^2\int_{-\infty}^{\infty}\eta(x)s(x)\,e^{ik|x-x'|}\eta(x')s(x')\,dx\,dx';$$

and, secondly, that this identity enables us to express r in the fashion

$$r = ik\int_{-\infty}^{\infty}e^{ikx}\eta(x)s(x)\,dx - \frac{ik}{2}\int_{-\infty}^{\infty}\eta(x)s^2(x)\,dx$$
$$+\left(\frac{ik}{2}\right)^2\int_{-\infty}^{\infty}\eta(x)s(x)\,e^{ik|x-x'|}\eta(x')s(x')\,dx\,dx'. \tag{74.4}$$

Thanks to the evident symmetry between x and x' in the double integral of (74.4), the first variation of the functional $r[s(x)]$, with $\eta(x)$ fixed, becomes

$$\delta r = ik\int_{-\infty}^{\infty}\eta(x)\delta s(x)\left[e^{ikx} - s(x) + \frac{ik}{2}\int_{-\infty}^{\infty}e^{ik|x-x'|}\eta(x')s(x')\,dx'\right]dx$$

and thus the necessary condition for a vanishing value thereof, when $\delta s(x)$ is arbitrary, takes the form of the integral equation (73.7) that specifies $s(x)$. Accordingly, we may conclude that an inaccuracy of the small magnitude ε in a trial function $s(x)$ generates one of the lesser magnitude ε^2 in the corresponding estimate of the reflection coefficient obtained from (74.4), whereas the original and non-variational expression for r supplies an estimate whose inaccuracy is comparable in magnitude to that of the trial function.

Let us, for the purpose of corroborating the latter aspect, introduce the trial function $s^{(0)}(x)$ in (74.4) and consider the relevant double integral

$$\begin{aligned}
I &= \left(\frac{ik}{2}\right)^2 \int_{-\infty}^{\infty} \eta(x) s^{(0)}(x)\, \mathrm{e}^{ik|x-x'|} \eta(x') s^{(0)}(x')\, \mathrm{d}x\, \mathrm{d}x' \\
&= \left(\frac{ik}{2}\right)^2 \int_{-\infty}^{\infty} \eta(x) s^{(0)}(x)\, \mathrm{e}^{ikx} \left(\int_{-\infty}^{x} \mathrm{e}^{-ikx'} \eta(x') s^{(0)}(x')\, \mathrm{d}x' \right) \mathrm{d}x \\
&\quad + \left(\frac{ik}{2}\right)^2 \int_{-\infty}^{\infty} \eta(x) s^{(0)}(x)\, \mathrm{e}^{-ikx} \left(\int_{x}^{\infty} \mathrm{e}^{ikx'} \eta(x') s^{(0)}(x')\, \mathrm{d}x' \right) \mathrm{d}x \\
&= I_1 + I_2,
\end{aligned}$$

say. We have

$$\begin{aligned}
I_1 &= \frac{ik}{2} \int_{-\infty}^{\infty} \eta(x) s^{(0)}(x)\, \mathrm{e}^{ikx} \left(\int_{-\infty}^{x} \mathrm{d}_{x'} \exp\left\{ \frac{ik}{2} \int_{-\infty}^{x'} \eta(x'')\, \mathrm{d}x'' \right\} \right) \mathrm{d}x \\
&= \frac{ik}{2} \int_{-\infty}^{\infty} \eta(x) s^{(0)}(x)\, \mathrm{e}^{ikx} \left[\exp\left\{ \frac{ik}{2} \int_{-\infty}^{x} \eta(x')\, \mathrm{d}x' \right\} - 1 \right] \mathrm{d}x \\
&= \frac{ik}{2} \int_{-\infty}^{\infty} \eta(x) [s^{(0)}(x)]^2\, \mathrm{d}x - \frac{ik}{2} \int_{-\infty}^{\infty} \eta(x) s^{(0)}(x)\, \mathrm{e}^{ikx}\, \mathrm{d}x,
\end{aligned}$$

and, using the temporary designation

$$F(x) = \int_{x}^{\infty} \mathrm{e}^{ikx'} \eta(x') s^{(0)}(x')\, \mathrm{d}x', \quad F(\infty) = 0,$$

it appears that

$$\begin{aligned}
I_2 &= \frac{ik}{2} \int_{-\infty}^{\infty} F(x)\, \mathrm{d}_x \exp\left\{ \frac{ik}{2} \int_{-\infty}^{x} \eta(x'')\, \mathrm{d}x'' \right\} \\
&= -\frac{ik}{2} F(-\infty) - \frac{ik}{2} \int_{-\infty}^{\infty} \exp\left\{ \frac{ik}{2} \int_{-\infty}^{x} \eta(x'')\, \mathrm{d}x'' \right\} \frac{\mathrm{d}F}{\mathrm{d}x}\, \mathrm{d}x \\
&= -\frac{ik}{2} \int_{-\infty}^{\infty} \mathrm{e}^{ikx} \eta(x) s^{(0)}(x)\, \mathrm{d}x \\
&\quad + \frac{ik}{2} \int_{-\infty}^{\infty} \mathrm{e}^{ikx} \eta(x) s^{(0)}(x) \exp\left\{ \frac{ik}{2} \int_{-\infty}^{x} \eta(x')\, \mathrm{d}x' \right\} \mathrm{d}x;
\end{aligned}$$

hence, the result yielded by the variational or stationary expression (74.4) with a zero order wave function turns out to be

$$r = \frac{ik}{2} \int_{-\infty}^{\infty} \eta(x)\, \mathrm{e}^{ikx} \exp\left\{ ik \int_{-\infty}^{x} \eta(\xi)\, \mathrm{d}\xi \right\} \mathrm{d}x$$

and possesses the anticipated first order accuracy.

A formal characterization of the transmission coefficient implied by (73.7), namely

$$\begin{aligned}
t &= \operatorname*{Lim}_{x \to +\infty} [\mathrm{e}^{-ikx} s(x)] \\
&= 1 + \frac{ik}{2} \int_{-\infty}^{\infty} \mathrm{e}^{-ikx} \eta(x) s(x)\, \mathrm{d}x, \qquad (74.5)
\end{aligned}$$

features an integral whose exponential factor, e^{-ikx}, contrasts with another, e^{ikx}, that enters into the corresponding expression (74.1) for the reflection coefficient. As a consequence of this alteration, the integral equation (73.7) for $s(x)$ no longer suffices to establish a stationary representation for t, though such a version may be derived on joint utilization of the independent pair of equations that specify the functions $s_\pm(x)$ appropriate to the oppositely directed incident waves $e^{\pm ikx}$. Thus, if we multiply the equation for $s_+(x)$ by $\eta(x)s_-(x)$ and the equation for $s_-(x)$ by $\eta(x)s_+(x)$ it follows, after integrating over all values of x, that

$$\int_{-\infty}^{\infty} \eta(x)s_+(x)s_-(x)\,dx \\ = \int_{-\infty}^{\infty} e^{ikx}\eta(x)s_-(x)\,dx + \frac{ik}{2}\int_{-\infty}^{\infty} \eta(x)s_-(x)\,e^{ik|x-x'|}\eta(x')s_+(x')\,dx\,dx', \quad (74.6)$$

$$\int_{-\infty}^{\infty} \eta(x)s_+(x)s_-(x)\,dx \\ = \int_{-\infty}^{\infty} e^{-ikx}\eta(x)s_+(x)\,dx + \frac{ik}{2}\int_{-\infty}^{\infty} \eta(x)s_+(x)\,e^{ik|x-x'|}\eta(x')s_-(x')\,dx\,dx'. \quad (74.7)$$

Comparison of the latter equations furnishes an identity

$$\int_{-\infty}^{\infty} e^{-ikx}\eta(x)s_+(x)\,dx = \int_{-\infty}^{\infty} e^{ikx}\eta(x)s_-(x)\,dx \quad (74.8)$$

which, in concert with expressions of the type (74.5), reveals the equality of the transmission coefficients for either direction of incidence. Moreover, when this identity is employed together with any one of the antecedent relations (74.6), (74.7), it appears that we may display the transmission coefficient in the form

$$t = 1 + \frac{ik}{2}\int_{-\infty}^{\infty} e^{-ikx}\eta(x)s_+(x)\,dx + \frac{ik}{2}\int_{-\infty}^{\infty} e^{ikx}\eta(x)s_-(x)\,dx \\ -\frac{ik}{2}\int_{-\infty}^{\infty} \eta(x)s_+(x)s_-(x)\,dx + \left(\frac{ik}{2}\right)^2\int_{-\infty}^{\infty} \eta(x)s_-(x)\,e^{ik|x-x'|}\eta(x')s_+(x')\,dx\,dx'. \quad (74.9)$$

The claim to significance on behalf of the representation (74.9) lies in the readily verified fact that the functional $t[s_+(x), s_-(x)]$ is stationary for arbitrary and independent small variations of $s_+(x)$ and $s_-(x)$ about their proper determinations.

If the zeroth order approximations

$$s_+^{(0)}(x) = \exp\left\{ikx + i\frac{k}{2}\int_{-\infty}^{x} \eta(\xi)\,d\xi\right\}, \quad s_-^{(0)}(x) = \exp\left\{-ikx + \frac{ik}{2}\int_{x}^{\infty} \eta(\xi)\,d\xi\right\}$$

are substituted into (74.9), the outcome is

$$t=\exp\left\{\mathrm{i}\frac{k}{2}\int_{-\infty}^{\infty}\eta(\xi)\,\mathrm{d}\xi\right\}-\frac{k^2}{4}\int_{-\infty}^{\infty}\eta(x)s_-^{(0)}(x)\,\mathrm{e}^{-\mathrm{i}kx}\int_x^{\infty}\mathrm{e}^{\mathrm{i}kx'}\eta(x')s_+^{(0)}(x')\,\mathrm{d}x'\,\mathrm{d}x$$

$$=\exp\left\{\mathrm{i}\frac{k}{2}\int_{-\infty}^{\infty}\eta(\xi)\,\mathrm{d}\xi\right\}\left[1-\frac{k^2}{4}\int_{-\infty}^{\infty}\eta(x)\,\mathrm{e}^{-2\mathrm{i}kx}\exp\left\{-\frac{\mathrm{i}k}{2}\int_{-\infty}^{x}\eta(\xi)\,\mathrm{d}\xi\right\}\right.$$

$$\left.\times\int_x^{\infty}\eta(x')\,\mathrm{e}^{2\mathrm{i}kx'}\exp\left\{\frac{\mathrm{i}k}{2}\int_{-\infty}^{x'}\eta(\xi)\,\mathrm{d}\xi\right\}\mathrm{d}x'\,\mathrm{d}x\right],$$

and we observe that, save for a numerical factor of 1/2 in the exponential arguments, $\mp(\mathrm{i}k/2)\int_{-\infty}^{x}\eta(\xi)\,\mathrm{d}\xi$, there is full agreement with the estimate (73.20) which accurately describes the initial pair of terms in the short wave development of $t^{(2)}$. On utilizing the first order trial functions

$$s_+(x)=s_+^{(0)}(x)+s_+^{(1)}(x),\quad s_-(x)=s_-^{(0)}(x)+s_-^{(1)}(x),$$

where

$$s_+^{(1)}(x)=\frac{\mathrm{i}k}{2}\exp\left(-\mathrm{i}kx-\frac{\mathrm{i}k}{2}\int_{-\infty}^{x}\eta(\xi)\,\mathrm{d}\xi\right)\int_x^{\infty}\eta(x')\,\mathrm{e}^{-2\mathrm{i}kx'}\exp\left(\mathrm{i}k\int_{-\infty}^{x'}\eta(\xi)\,\mathrm{d}\xi\right)\mathrm{d}x'$$

and

$$s_-^{(1)}(x)=\frac{\mathrm{i}k}{2}\exp\left(\mathrm{i}kx-\frac{\mathrm{i}k}{2}\int_{x}^{\infty}\eta(\xi)\,\mathrm{d}\xi\right)\int_{-\infty}^{x}\eta(x')\,\mathrm{e}^{-2\mathrm{i}kx'}\exp\left(\mathrm{i}k\int_{x}^{\infty}\eta(\xi)\,\mathrm{d}\xi\right)\mathrm{d}x',$$

there follows from (74.9), after a more elaborate reduction procedure,

$$t=t^{(2)}+\left(\frac{k}{2}\right)^4\mathrm{e}^{\mathrm{i}k\delta/2}\int_{-\infty}^{\infty}\eta(x)\,\mathrm{e}^{2\mathrm{i}kx}\exp\left(\frac{\mathrm{i}k}{2}\int_{-\infty}^{x}\eta(\xi)\,\mathrm{d}\xi\right)P(x)$$

$$\times\int_{-\infty}^{x}\eta(x')\,\mathrm{e}^{-2\mathrm{i}kx'}\exp\left(-\frac{\mathrm{i}k}{2}\int_{-\infty}^{x'}\eta(\xi)\,\mathrm{d}\xi\right)Q(x')\,\mathrm{d}x'\,\mathrm{d}x$$

with

$$\delta=\int_{-\infty}^{+\infty}\eta(\xi)\,\mathrm{d}\xi,\qquad P(x)=\int_{-\infty}^{x}\eta(x')\,\mathrm{e}^{-2\mathrm{i}kx'}\exp\left(-\mathrm{i}k\int_{-\infty}^{x'}\eta(\xi)\,\mathrm{d}\xi\right)\mathrm{d}x',$$

and

$$Q(x)=\int_x^{\infty}\eta(x')\,\mathrm{e}^{2\mathrm{i}kx'}\exp\left(\mathrm{i}k\int_{-\infty}^{x'}\eta(\xi)\,\mathrm{d}\xi\right)\mathrm{d}x'.\tag{74.11}$$

Since the final term if (74.10) has a lesser magnitude than those entering into $t^{(2)}$, for slowly varying or small scale functions $\eta(x)$, the efficiency of a variational characterization is confirmed again.

A realization of practical advantages through the establishment of alternative, including non-stationary, formulations of the scattering coefficients merits attention. If we account for an inhomogeneous presence by use of the Green's function (69.2) and adopt the integral equation [cf. (69.1)]

$$s(x)=\varphi_+(x)+\int_{-\infty}^{\infty}G(x,x')K(x')s(x')\,\mathrm{d}x',\quad -\infty<x<\infty\tag{74.12}$$

where

$$K(x)=-\mathcal{K}(x)=\frac{3}{4}\left(\frac{k'(x)}{k(x)}\right)^2-\frac{1}{2}\frac{k''(x)}{k(x)},\quad k(x)\to k,\quad |x|\to\infty,$$

the reflection coefficient characterization

$$r = \frac{i}{2} e^{2 i\alpha} \int_{-\infty}^{\infty} \varphi_+(x) K(x) s(x) \, dx \tag{74.13}$$

with

$$\alpha = \int_{-\infty}^{0} (k(\xi) - k) \, d\xi$$

is a direct consequence of the asymptotic form

$$s(x) \sim k^{-1/2} e^{ikx - i\alpha} + \frac{i}{2} k^{-1/2} e^{-ikx + i\alpha} \int_{-\infty}^{\infty} \varphi_+(x) K(x) s(x) \, dx, \quad x \to -\infty.$$

When (74.12) is multiplied by $\varphi_+(x)K(x)$ prior to integration over the coordinate range, and we take note of an equivalent version for the integral in (74.13), it appears that the representation

$$\begin{aligned} r = \frac{i}{2} e^{2 i\alpha} \int_{-\infty}^{\infty} \varphi_+^2(x) K(x) \, dx - \frac{1}{4} e^{2 i\alpha} \Big[\int_{-\infty}^{\infty} \varphi_+^2(x) K(x) \int_{-\infty}^{x} \varphi_-(x') K(x') s(x') \, dx' \, dx \\ + \int_{-\infty}^{\infty} \frac{K(x)}{k(x)} \int_{x}^{\infty} \varphi_+(x') K(x') s(x') \, dx' \, dx \Big] \end{aligned} \tag{74.14}$$

can be employed in place of (74.13).

A comparison of (74.13) and (74.14) suggests that the latter provides a superior basis for estimating the reflection coefficient in the circumstance that gradual changes of $k(x)$ imply small values for $K(x)$; it is, moreover, a simple matter to confirm that the initial (and explicit) term of (74.14) accords with the first order estimate (74.2). Inasmuch as the Green's function remains unaltered on exchange of its argument variables, the further characterization

$$\begin{aligned} r = \frac{i}{2} e^{2 i\alpha} \Big[2 \int_{-\infty}^{\infty} \varphi_+(x) K(x) s(x) \, dx - \int_{-\infty}^{\infty} K(x) s^2(x) \, dx \\ + \int_{-\infty}^{\infty} K(x) s(x) G(x, x') K(x') s(x') \, dx \, dx' \Big] \end{aligned} \tag{74.15}$$

possesses a stationary nature in respect of variations for $s(x)$ relative to the proper values which satisfy (74.12). If the trial function $s(x) = \varphi_+(x)$ be introduced into (74.14) and (74.15) we arrive at identical estimates.

§75. *A Scattering Matrix and its Variational Characterization*

The comprehensive variational description of outgoing or secondary wave amplitudes in the exterior of inhomogeneous ranges, encompassing those given earlier (§74) for the reflection and transmission coefficients appropriate to a particular choice of incident wave direction, is naturally presented in terms of a scattering matrix. Let us assume that the combinations

$$s(x) \simeq \begin{cases} A_1\,\mathrm{e}^{ikx} + B_1\,\mathrm{e}^{-ikx}, & x \to -\infty \\ A_2\,\mathrm{e}^{-ikx} + B_2\,\mathrm{e}^{ikx}, & x \to +\infty \end{cases}, \tag{75.1}$$

wherein the coefficients A_1, A_2 specify incoming or primary wave amplitudes, describe the asymptotic forms of a solution to the differential equation

$$\begin{gathered} \left(\frac{\mathrm{d}^2}{\mathrm{d}x^2} + k^2\right) s(x) = -k\eta(x)s(x), \quad -\infty < x < \infty \\ \eta(x) \to 0, \quad |x| \to \infty. \end{gathered} \tag{75.2}$$

From an integral of (75.2) which incorporates these primary wave specifications and contains the response or secondary excitation in its final member, viz.:

$$s(x) = A_1\,\mathrm{e}^{ikx} + A_2\,\mathrm{e}^{-ikx} + \frac{\mathrm{i}}{2}\int_{-\infty}^{\infty} \mathrm{e}^{\mathrm{i}k|x-x'|}\eta(x')s(x')\,\mathrm{d}x', \tag{75.3}$$

we find

$$s(x) \simeq \begin{matrix} A_1\,\mathrm{e}^{ikx} + A_2\,\mathrm{e}^{-ikx} + \dfrac{\mathrm{i}}{2}\,\mathrm{e}^{-ikx}\displaystyle\int_{-\infty}^{\infty} \mathrm{e}^{ikx}\eta(x)s(x)\,\mathrm{d}x, & x \to -\infty \\ A_1\,\mathrm{e}^{ikx} + A_2\,\mathrm{e}^{-ikx} + \dfrac{\mathrm{i}}{2}\,\mathrm{e}^{ikx}\displaystyle\int_{-\infty}^{\infty} \mathrm{e}^{-ikx}\eta(x)s(x)\,\mathrm{d}x, & x \to +\infty \end{matrix}$$

and thence the representations of outgoing wave amplitudes

$$\begin{aligned} B_1 &= A_2 + \frac{\mathrm{i}}{2}\int_{-\infty}^{\infty} \mathrm{e}^{\mathrm{i}kx}\eta(x)s(x)\,\mathrm{d}x \\ B_2 &= A_2 + \frac{\mathrm{i}}{2}\int_{-\infty}^{\infty} \mathrm{e}^{-\mathrm{i}kx}\eta(x)s(x)\,\mathrm{d}x \end{aligned} \quad .$$

On introducing the resolution

$$s(x) = A_1\psi_1(x) + A_2\psi_2(x) \tag{75.5}$$

into (75.3) and exploiting the arbitrariness which is inherent in a selection of the coefficients A_1, A_2, we obtain an independent pair of integral equations for ψ_1, ψ_2, namely

$$\psi_{\binom{1}{2}}(x) = \mathrm{e}^{\pm ikx} + \frac{\mathrm{i}}{2}\int_{-\infty}^{\infty} \mathrm{e}^{\mathrm{i}k|x-x'|}\eta(x')\psi_{\binom{1}{2}}(x')\,\mathrm{d}x', \quad -\infty < x < \infty. \tag{75.6}$$

Linear relations between the coefficients B_1, B_2 and A_1, A_2 are found after substituting (75.5) into (75.4) and, in their collective form

$$\begin{pmatrix} B_1 \\ B_2 \end{pmatrix} = R \begin{pmatrix} A_2 \\ A_1 \end{pmatrix} = S \begin{pmatrix} A_1 \\ A_2 \end{pmatrix}, \tag{75.7}$$

we encounter the matrices

$$R = I + \mathrm{i}T = \begin{pmatrix} 1 & 0 \\ 0 & 1 \end{pmatrix} + \mathrm{i}\begin{pmatrix} T_{11} & T_{12} \\ T_{21} & T_{22} \end{pmatrix} = \begin{pmatrix} 1+\mathrm{i}T_{11} & \mathrm{i}T_{12} \\ \mathrm{i}T_{21} & 1+\mathrm{i}T_{22} \end{pmatrix},$$

and

$$S = R\begin{pmatrix} 0 & 1 \\ 1 & 0 \end{pmatrix} = \begin{pmatrix} \mathrm{i}T_{12} & 1+\mathrm{i}T_{11} \\ 1+\mathrm{i}T_{22} & \mathrm{i}T_{21} \end{pmatrix} \tag{75.8}$$

whose elements depend on the integrals

$$T_{11} = \frac{1}{2}\int_{-\infty}^{\infty} \mathrm{e}^{ikx}\eta(x)\psi_2(x)\,\mathrm{d}x = \frac{1}{2}\int_{-\infty}^{\infty} \mathrm{e}^{-ikx}\eta(x)\psi_1(x)\,\mathrm{d}x = T_{22},$$

$$T_{12} = \frac{1}{2}\int_{-\infty}^{\infty} \mathrm{e}^{ikx}\eta(x)\psi_1(x)\,\mathrm{d}x, \qquad T_{21} = \frac{1}{2}\int_{-\infty}^{\infty} \mathrm{e}^{-ikx}\eta(x)\psi_2(x)\,\mathrm{d}x; \tag{75.9}$$

the fact that $T_{11} = T_{22}$ follows from an evident symmetry of the complex kernel function $\mathrm{e}^{ik|x-x'|}$ in (75.6) with regard to the exchange $x \leftrightarrow x'$ of its argument variables. There is no essential distinction between S, the previously defined scattering matrix, and

$$R = S\begin{pmatrix} 0 & 1 \\ 1 & 0 \end{pmatrix} = \begin{pmatrix} S_{12} & S_{11} \\ S_{22} & S_{21} \end{pmatrix}; \tag{75.10}$$

both possess a unitary character when $\eta(x)$ is real and energy conservation requires that $|A_1|^2 + |A_2|^2 = |B_1|^2 + |B_2|^2$. It is fitting to look on R, rather than S, as a direct scattering or "collision" matrix, since the former tends to the unit matrix when $\eta \to 0$ and secondary waves are absent.

The three distinct elements T_{12}, T_{21}, T_{11} of the T matrix can be exhibited in different stationary forms; we have, for instance, the version

$$T_{12} = \frac{2}{\mathrm{i}} S_{11} = \frac{1}{2}\left\{2\int_{-\infty}^{\infty} \mathrm{e}^{ikx}\eta(x)\psi_1(x)\,\mathrm{d}x - \int_{-\infty}^{\infty} \eta(x)\psi_1^2(x)\,\mathrm{d}x \right.$$
$$\left. + \frac{\mathrm{i}}{2}\int_{-\infty}^{\infty} \eta(x)\psi_1(x)\,\mathrm{e}^{ik|x-x'|}\eta(x')\psi_1(x')\,\mathrm{d}x\,\mathrm{d}x'\right\}, \tag{75.11}$$

akin to the prior reflection coefficient characterization (74.4), and another

$$\frac{1}{T_{12}} = \frac{2\displaystyle\int_{-\infty}^{\infty} \eta(x)\psi_1^2(x)\,\mathrm{d}x - \mathrm{i}\int_{-\infty}^{\infty} \eta(x)\psi_1(x)\,\mathrm{e}^{ik|x-x'|}\eta(x')\psi_1(x')\,\mathrm{d}x\,\mathrm{d}x'}{\left(\displaystyle\int_{-\infty}^{\infty} \mathrm{e}^{ikx}\eta(x)\psi_1(x)\,\mathrm{d}x\right)^2} \tag{75.12}$$

which evidently furnishes the correct value of T_{12} if ψ_1 satisfies the integral equation (75.6). That (75.12) is stationary for small variations $\delta\psi_1$, relative to the proper function ψ_1, follows from a calculation of the concomitant variation δT_{12}, in accordance with (75.6), (75.9), viz.

$$\begin{aligned}
\delta T_{12}&\left[2\int_{-\infty}^{\infty}\eta(x)\psi_1^2(x)\,\mathrm{d}x-\mathrm{i}\int_{-\infty}^{\infty}\eta(x)\,\mathrm{e}^{\mathrm{i}k|x-x'|}\eta(x')\psi_1(x')\,\mathrm{d}x\,\mathrm{d}x'\right]\\
&=\int_{-\infty}^{\infty}\eta(x)\delta\psi_1(x)\Bigg\{2\,\mathrm{e}^{ikx}\int_{-\infty}^{\infty}\mathrm{e}^{ikx}\eta(x)\psi_1(x)\,\mathrm{d}x\\
&\quad-T_{12}\left[4\psi_1(x)-2\,\mathrm{i}\int_{-\infty}^{\infty}\mathrm{e}^{\mathrm{i}k|x-x'|}\eta(x')\psi_1(x')\,\mathrm{d}x\right]\Bigg\}\,\mathrm{d}x\\
&=2\int_{-\infty}^{\infty}\eta(x)\delta\psi_1(x)\left\{\mathrm{e}^{ikx}\int_{-\infty}^{\infty}\mathrm{e}^{ikx}\eta(x)\psi_1(x)\,\mathrm{d}x-2\,\mathrm{e}^{ikx}T_{12}\right\}\mathrm{d}x\\
&=0.
\end{aligned}$$

A noteworthy contrast between the respective variational expressions (75.11), (75.12) stems from the obvious fact that the latter functional, $T_{12}\{\psi_1\}$, has a homogeneous, or scale independent, nature as regards the argument function ψ_1.

The corresponding homogeneous characterizations for T_{21} and $T_{11}=T_{22}$, namely

$$\frac{1}{T_{21}}=\frac{2\int_{-\infty}^{\infty}\eta(x)\psi_2^2(x)\,\mathrm{d}x-\mathrm{i}\int_{-\infty}^{\infty}\eta(x)\psi_2(x)\,\mathrm{e}^{\mathrm{i}k|x-x'|}\eta(x')\psi_2(x')\,\mathrm{d}x\,\mathrm{d}x'}{\left(\int_{-\infty}^{\infty}\mathrm{e}^{-ikx}\eta(x)\psi_2(x)\,\mathrm{d}x\right)^2},\tag{75.13}$$

$$\frac{1}{T_{11}}=\frac{2\int_{-\infty}^{\infty}\eta(x)\psi_1(x)\psi_2(x)\,\mathrm{d}x-\mathrm{i}\int_{-\infty}^{\infty}\eta(x)\psi_1(x)\,\mathrm{e}^{\mathrm{i}k|x-x'|}\eta(x')\psi_2(x')\,\mathrm{d}x\,\mathrm{d}x'}{\int_{-\infty}^{\infty}\mathrm{e}^{ikx}\eta(x)\psi_1(x)\,\mathrm{d}x\int_{-\infty}^{\infty}\mathrm{e}^{-ikx}\eta(x)\psi_2(x)\,\mathrm{d}x},\tag{75.14}$$

are stationary for arbitrary variations of ψ_2, and of ψ_1, ψ_2 independently, relative to solutions of the integral equations (75.6).

Estimates for elements of the T matrix secured from the preceding variational formulas through the use of specific trial functions ψ_1, ψ_2 do not generally conform with the relations implied by unitarity of the actual matrix $R=I+\mathrm{i}T$. It is possible, however, to acquire variational estimates of the R matrix which meet the unitary requirement if those of the T matrix are obtained in a special manner. With this objective in mind, let us first rewrite the integral equations (75.6) for ψ_1, ψ_2, based on a separation of the kernel function $\mathrm{e}^{\mathrm{i}k|x-x'|}$ into real and imaginary parts and a decomposition of the former into exponential terms; this yields, in keeping with the specifications (75.9) for elements of the T matrix

$$\begin{aligned}\psi_1(x)+\frac{1}{2}\int_{-\infty}^{\infty}\sin k|x-x'|\eta(x')\psi_1(x')\,\mathrm{d}x' &= \left(1+\frac{\mathrm{i}T_{11}}{2}\right)\mathrm{e}^{\mathrm{i}kx}+\frac{\mathrm{i}T_{12}}{2}\mathrm{e}^{-\mathrm{i}kx},\\ \psi_2(x)+\frac{1}{2}\int_{-\infty}^{\infty}\sin k|x-x'|\eta(x')\psi_2(x')\,\mathrm{d}x' &= \frac{\mathrm{i}T_{21}}{2}\mathrm{e}^{\mathrm{i}kx}+\left(1+\frac{\mathrm{i}T_{11}}{2}\right)\mathrm{e}^{-\mathrm{i}kx}.\end{aligned}\tag{75.15}$$

If we define a pair of functions $\varphi_1(x)$, $\varphi_2(x)$ that satisfy the equations

$$\varphi_{\binom{1}{2}}(x)+\frac{1}{2}\int_{-\infty}^{\infty}\sin k|x-x'|\eta(x')\varphi_{\binom{1}{2}}(x')\,\mathrm{d}x' = \mathrm{e}^{\pm\mathrm{i}kx},\quad -\infty<x<\infty \tag{75.16}$$

the linear connections between $\psi_{1,2}(x)$ and $\varphi_{1,2}(x)$ implied by (75.15) can be expressed in the matrix form

$$\begin{pmatrix}\psi_1(x)\\ \psi_2(x)\end{pmatrix}=\left(I+\frac{\mathrm{i}T}{2}\right)\begin{pmatrix}\varphi_1(x)\\ \varphi_2(x)\end{pmatrix}. \tag{75.17}$$

Since the integral equations (75.16) possess a real and symmetric kernel function, $\sin k|x-x'|$, it appears that

$$\varphi_1^*(x)=\varphi_2(x) \tag{75.18}$$

and, moreover, that the matrix U with elements

$$\begin{aligned}U_{11}&=\frac{1}{2}\int_{-\infty}^{\infty}\mathrm{e}^{\mathrm{i}kx}\eta(x)\varphi_2(x)\,\mathrm{d}x, & U_{22}&=\frac{1}{2}\int_{-\infty}^{\infty}\mathrm{e}^{-\mathrm{i}kx}\eta(x)\varphi_1(x)\,\mathrm{d}x\\ U_{12}&=\frac{1}{2}\int_{-\infty}^{\infty}\mathrm{e}^{\mathrm{i}kx}\eta(x)\varphi_1(x)\,\mathrm{d}x, & U_{21}&=\frac{1}{2}\int_{-\infty}^{\infty}\mathrm{e}^{-\mathrm{i}kx}\eta(x)\varphi_2(x)\,\mathrm{d}x\end{aligned}\tag{75.19}$$

has a Hermitian character, namely

$$\tilde{U}^*=U, \tag{75.20}$$

inasmuch as

$$U_{11}=U_{22}\,(\text{real}),\qquad U_{12}^*=U_{21}. \tag{75.21}$$

On utilizing the relation

$$T=U\left(I+\frac{\mathrm{i}T}{2}\right), \tag{75.22}$$

which is directly inferred from (75.17), we are next able to express T and thence R in terms of U, viz.

$$T=\left(I-\frac{\mathrm{i}}{2}U\right)^{-1}U=\frac{U}{I-\frac{\mathrm{i}}{2}U}, \tag{75.23}$$

$$R = I + \mathrm{i}T = \frac{I + \frac{\mathrm{i}}{2}U}{I - \frac{\mathrm{i}}{2}U}. \tag{75.24}$$

A unitary character of R is assured by the latter representation, as

$$\tilde{R}^* = \frac{I - \frac{\mathrm{i}}{2}\tilde{U}^*}{I + \frac{\mathrm{i}}{2}\tilde{U}^*} = \frac{I - \frac{\mathrm{i}}{2}U}{I + \frac{\mathrm{i}}{2}U} = R^{-1},$$

and this consequently provides an advantageous means of basing estimates of the scattering matrix upon those of U. We can employ, in particular, the stationary expressions

$$\frac{1}{U_{\mathrm{ij}}} = \frac{2\int_{-\infty}^{\infty} \eta(x)\varphi_{\mathrm{i}}(x)\varphi_{\mathrm{j}}^*(x)\,\mathrm{d}x + \int_{-\infty}^{\infty} \eta(x)\varphi_{\mathrm{i}}(x)\sin k|x-x'|\eta(x')\varphi_{\mathrm{j}}^*(x')\,\mathrm{d}x\,\mathrm{d}x'}{\int_{-\infty}^{\infty} \zeta_{\mathrm{j}}^*(x)\eta(x)\varphi_{\mathrm{i}}(x)\,\mathrm{d}x \int_{-\infty}^{\infty} \zeta_{\mathrm{i}}(x)\eta(x)\varphi_{\mathrm{j}}^*(x)\,\mathrm{d}x} \tag{75.25}$$

where $\mathrm{i}, j = 1, 2$ and

$$\begin{pmatrix} \zeta_1(x) \\ \zeta_2(x) \end{pmatrix} = \begin{pmatrix} \mathrm{e}^{ikx} \\ e^{-ikx} \end{pmatrix}$$

to approximate the elements of a Hermitian matrix for use in (75.24).

§76. *Formalities of a Variational Calculation*

Our direct application of variational principles has thus far been effected with trial functions of fully determinate form, embodying one or two initial terms from a development of the actual wave function(s) (cf. §74). It is useful, as regards their systematic and practical employment, to indicate how more flexible trial functions are utilized, and how successive variational estimates may be contrasted. Suppose we consider the functional

$$I\{s(x)\} = 2\int f(x)g(x)s(x)\,\mathrm{d}x - \int g(x)s^2(x)\,\mathrm{d}x + \int g(x)s(x)K(x,x')g(x')s(x')\,\mathrm{d}x\,\mathrm{d}x' \tag{76.1}$$

wherein $f(x)$, $g(x)$, and $K(x, x')$ are known, and the integrals extend over a common and arbitrary range L; the so-called Euler-Lagrange equation for $s(x)$ predicated on a vanishing first variation of this functional, namely, of the difference $I\{s(x)+\delta s(x)\}-I\{s(x)\}$ to the first order in, and for arbitrary (small) values of, $\delta s(x)$ is an integral equation with the form

$$s(x)-\int K(x,x')g(x')s(x')\,\mathrm{d}x' = f(x), \quad x \text{ in } L \tag{76.2}$$

provided that $K(x,x')=K(x',x)$. When f, g, K, and L are given the specifications found in (73.7) the stationary (and complex) measure of I,

$$I=\int f(x)g(x)s(x)\,\mathrm{d}x, \tag{76.3}$$

represents the actual value of an outgoing wave amplitude or complex reflection coefficient.

For purposes of estimating said value, let us introduce a trial function

$$s(x)=\sum_{\mathrm{i}=1}^{n} A_{\mathrm{i}}\psi_{\mathrm{i}}(x) \tag{76.4}$$

with arbitrary coefficients $A_1,\ldots,A_n$ and individual functions $\psi_1(x),\ldots,\psi_n(x)$ chosen in a manner that befits the particular circumstances envisaged ($\psi_1=1, \psi_2=x,\ldots,$ say), if the incident wave length has a magnitude large compared with the range of inhomogeneity, L, or support of the function $g(x)$). On substituting the representation (76.4) into (76.1) we obtain a quadratic form in the A_{i}, viz.

$$I(A_1,\ldots,A_n)=2\sum_{\mathrm{i}=1}^{n} A_{\mathrm{i}}B_{\mathrm{i}}-\sum_{\mathrm{i},j=1}^{n} A_{\mathrm{i}}D_{\mathrm{i}j}A_j \tag{76.5}$$

with

$$B_{\mathrm{i}}=\int f(x)g(x)\psi_{\mathrm{i}}(x)\,\mathrm{d}x, \tag{76.6}$$

$$D_{\mathrm{ij}}=D_{\mathrm{ji}}=\int g(x)\psi_{\mathrm{i}}(x)\psi_j(x)\,\mathrm{d}x-\int g(x)\psi_{\mathrm{i}}(x)K(x,x')g(x')\psi_j(x')\,\mathrm{d}x\,\mathrm{d}x'.$$

If the coefficients A_{i} are determined by a system of linear equations,

$$\sum_{j=1}^{n} D_{\mathrm{i}j}A_j=B_{\mathrm{i}}, \quad \mathrm{i}=1,2,\ldots,n, \tag{76.7}$$

which stem from the conditions

$$\frac{\partial I}{\partial A_{\mathrm{i}}}=0, \quad \mathrm{i}=1,2,\ldots,n \tag{76.8}$$

the resultant or stationary value of the quadratic form (76.5)

$$I^{(n)} = \sum_{i=1}^{n} A_i B_i \tag{76.9}$$

provides an explicit variational estimate for I itself.

Succeeding stages of such a procedure involve different numbers of coefficients without simple interrelations; however, it is possible to exhibit a compact and useful expression for the concomitant change, $\Delta I = I^{(n)} - I^{(n-1)}$, in stationary value of I. We begin with the linear equations of the $n, n-1$ stages,

$$\sum_{j=1}^{n} D_{ij} A_j = B_i, \quad i = 1, \dots, n \tag{76.7}$$

$$\sum_{j=1}^{n-1} D_{ij} A'_j = B_i, \quad i = 1, \dots, n-1 \tag{76.10}$$

and, pursuant to the definitions

$$C_i = A_i - A'_i, \quad i = 1, \dots, n-1; \quad C_n = A_n, \tag{76.11}$$

obtain from them a system of equations for the C_i,

$$\sum_{j=1}^{n} D_{ij} C_j = B'_i, \quad i = 1, \dots, n \tag{76.12}$$

where

$$B'_i = \begin{cases} 0, & i = 1, \dots, n-1 \\ B_n - \sum_{j=1}^{n-1} D_{nj} A'_j, & i = n. \end{cases} \tag{76.13}$$

After multiplying in (76.7) and (76.12) with C_i and A_i, respectively, and summing over i it follows, on account of the symmetry relation $D_{ij} = D_{ji}$, that

$$\sum_{i,j=1}^{n} C_i D_{ij} A_j = \sum_{i,j=1}^{n} A_i D_{ij} C_j = \sum_{i=1}^{n} B_i C_i = \sum_{i=1}^{n} A_i B'_i. \tag{76.14}$$

Thus

$$\begin{aligned} \Delta = I^{(n)} - I^{(n-1)} &= \sum_{i=1}^{n} A_i B_i - \sum_{i=1}^{n-1} A'_i B_i = \sum_{i=1}^{n} B_i C_i \\ &= A_n B'_n = C_n B'_n = (B'_n)^2 \frac{\mathscr{D}_{n-1}}{\mathscr{D}_n} \end{aligned} \tag{76.15}$$

where $\mathscr{D}_n$, $\mathscr{D}_{n-1}$ are notations for the determinants

$$|D_n| = \begin{vmatrix} D_{11} & & D_{1n} \\ & \ddots & \\ D_{n1} & & D_{nn} \end{vmatrix}, \qquad |D_{n-1}| = \begin{vmatrix} D_{11} & & D_{1\,n-1} \\ & \ddots & \\ D_{n-1\,1} & & D_{n-1\,n-1} \end{vmatrix} \tag{76.16}$$

with given elements. To express B'_n (and thence Δ) in a fully definite manner we write $A'_j = -\alpha_j/\alpha_n$, $j = 1, \dots, n-1$, and observe that the equations (76.10),

(76.13) acquire a homogeneous form in terms of the quantities α_i, i – 1, . . . , n, namely

$$\sum_{j=1}^{n-1} D_{ij}\alpha_j + B_i\alpha_n = 0, \quad i = 1, \ldots, n-1$$
$$\sum_{j=1}^{n-1} D_{nj}\alpha_j + (B_n - B'_n)\alpha_n = 0.$$

The condition which assures non-trivial determinations of the α_i,

$$\begin{vmatrix} D_{11} \ldots D_{1\,n-1} & B_1 \\ \ddots & \\ D_{n1} \ldots D_{n\,n-1} & B_n - B_{n'} \end{vmatrix} = 0$$

allows us to infer that

$$B'_n = \frac{\mathscr{D}_n^{(B)}}{\mathscr{D}_{n-1}} \tag{76.17}$$

where $\mathscr{D}_n^{(B)}$ is the determinant

$$|D_n^{(B)}| = \begin{vmatrix} D_{11} \ldots D_{1\,n-1} & B_1 \\ \ddots & \\ D_{n1} \ldots D_{n\,n-1} & B_n \end{vmatrix}, \tag{76.18}$$

arrived at by substituting $B_1, \ldots, B_n$ for the corresponding elements $D_{1n}, \ldots, D_{nn}$ in the last column of $\mathscr{D}_n$. Combining (76.15) and (76.17) we secure, finally, the requisite characterization of a first difference in stationary values of I,

$$\Delta = I^{(n)} - I^{(n-1)} = \frac{(\mathscr{D}_n^{(B)})^2}{\mathscr{D}_{n-1}\mathscr{D}_n}; \tag{76.19}$$

since

$$I^{(1)} = B_1^2/D_{11}$$

it appears, moreover, that

$$I^{(n)} = I^{(1)} + \sum_{j=1}^{n} (I^{(j)} - I^{(j-1)}) = \frac{B_1^2}{D_{11}} + \sum_{j=2}^{n} \frac{(\mathscr{D}_j^{(B)})^2}{\mathscr{D}_{j-1}\,\mathscr{D}_j}. \tag{76.20}$$

Two separate phases figure in the preceding (Ritz-Galerkin) variational procedure for achieving estimates of quantities that are descriptive of wave excitations (or other physical phenomena); the first involves construction of trial functions, or specification of the set $\psi_1(x), \ldots, \psi_n(x)$ in (76.4), and the second involves an evaluation, based on said functions, of formulas with the above nature. There is another, or so-called variation-iteration, approach which incorporates a systematic generation of improved trial functions and requires but a single choice at the outset; we refer again, for descriptive purposes, to the functional (76.1) and suppose that the function satisfies (by appropriate selection of an internal parameter, say) the relation

$$\int g(x)\psi_1^2(x)\,dx - \int g(x)\psi_1(x)K(x,x')g(x')\psi_1(x')\,dx\,dx' = \int f(x)g(x)\psi_1(x)\,dx. \tag{76.21}$$

The value of I which follows from the substitution $s_1(x)=\psi_1(x)$ therein, and likewise from the stipulation that the scale factor A of the trial function $s_1(x)=A\psi_1(x)$, selected by the condition $dI/dA=0$, shall be equal to unity, is

$$I^{(1)} = \int f(x)g(x)\psi_1(x)\,dx. \tag{76.22}$$

We next define the function

$$\psi_2(x) = \psi_1(x) - \int K(x,x')g(x')\psi_1(x')\,dx' - f(x) \tag{76.23}$$

that would vanish if $\psi_1(x)$ were a solution of the Euler-Lagrange equation (76.2) for the given functional, and note the orthogonality of ψ_1, ψ_2 with respect to the weight factor $g(x)$, viz.

$$\int g(x)\psi_1(x)\psi_2(x)\,dx = 0, \tag{76.24}$$

in consequence of (76.21). When the second trial function, represented as a linear combination of ψ_1, ψ_2,

$$s_2(x) = A_1\psi_1(x) + A_2\psi_2(x), \tag{76.25}$$

is introduced into (76.1) and the conditions $\partial I/\partial A_1 = \partial I/\partial A_2 = 0$ are applied, it appears that the coefficients A_1, A_2 satisfy the linear equations

$$\begin{aligned} A_1D_{11} + A_2D_{12} &= B_1 \\ A_1D_{12} + A_2D_{22} &= B_2 \end{aligned} \tag{76.26}$$

with

$$B_{1,2} = \int f(x)g(x)\psi_{1,2}(x)\,dx,$$

$$D_{ii} = \int g(x)\psi_i^2(x)\,dx - \int g(x)\psi_i(x)K(x,x')g(x')\psi_i(x')\,dx\,dx', \quad i=1,2$$

$$D_{12} = D_{21} = -\int g(x)\psi_1(x)K(x,x')g(x')\psi_2(x')\,dx\,dx' \tag{76.27}$$

if $K(x, x') = K(x', x)$. Since $B_1 = D_{11}$ we may, utilizing (76.23) and (76.24) in addition, bring the equations (76.26) to a form

$$(A_1 - 1)D_{11} + A_2 D_{12} = 0$$

$$(A_1 - 1)D_{12} + A_2 D_{22} = \int g(x)\psi_2(x)\left\{f(x) - \int K(x, x')g(x')\psi_1(x')\,dx'\right\}dx$$

$$= -\int g(x)\psi_2^2(x)\,dx$$

that readily yields

$$A_1 = 1 + \frac{C_2 D_{12}}{D_{11}D_{22} - D_{12}^2}, \qquad A_2 = -\frac{C_2 D_{11}}{D_{11}D_{22} - D_{12}^2} \tag{76.28}$$

where

$$C_2 = \int g(x)\psi_2^2(x)\,dx. \tag{76.29}$$

The stationary value

$$I^{(2)} = A_1 B_1 + A_2 B_2, \tag{76.30}$$

achieved with the determinations (76.28), differs from an original estimate of the functional I by the amount

$$I^{(2)} - I^{(1)} = (A_1 - 1)B_1 + A_2 B_2$$

$$= (A_1 - 1)D_{11} + A_2 \int g(x)\psi_2(x)\left[\psi_1(x) - \psi_2(x) - \int K(x, x')g(x')\psi_1(x')\,dx'\right]dx$$

$$= (A_1 - 1)D_{11} + A_2 D_{12} - A_2 C_2$$

$$= -A_2 C_2 = \frac{(C_2)^2 D_{11}}{D_{11}D_{22} - D_{12}^2}. \tag{76.31}$$

To carry on with this self-contained scheme, we define a function

$$\psi_3(x) = \psi_2(x) - \int K(x, x')g(x')s_2(x')\,dx' - f(x) \tag{76.32}$$

such that

$$\int g(x)\psi_i(x)\psi_3(x)\,dx = 0, \quad i = 1, 2$$

and thence a third approximation trial function

$$s_3(x) = A_1\psi_1(x) + A_2\psi_2(x) + A_3\psi_3(x) \tag{76.33}$$

whose coefficients A_1, A_2, A_3 are fixed by the conditions $\partial I/\partial A_1 = \partial I/\partial A_2 = \partial I/\partial A_3 = 0$; it follows, in particular, that

$$I^{(3)} - I^{(2)} = \frac{(C_3)^2 \mathscr{D}_2}{\mathscr{D}_3} \tag{76.34}$$

where

$$C_3 = \int g(x)\psi_3^2(x)\,\mathrm{d}x \tag{76.35}$$

and $\mathscr{D}_2, \mathscr{D}_3$ represent $2\times 2, 3\times 3$ versions of the determinantal forms (76.16), with elements D_{ij} specified as in (76.27). More generally, on advancing from the $n-1^{\text{st}}$ to the nth approximation stage of the variation-iteration procedure, there is a change

$$I^{(n)} - I^{(n)} = \frac{(C_n)^2 \mathscr{D}_{n-1}}{\mathscr{D}_n} \tag{76.36}$$

in stationary value of the functional I.

§77. *A Green's Function Representation and its Variational Characterization*

The specific variational formulations described earlier, and introduced for practical reasons, furnish efficient means of estimating secondary (or outgoing) wave amplitudes when the source of excitation is infinitely remote. To bring out their adaptability and general usefulness, we shall next present a variational characterization for the Green's function, or fundamental solution of a differential equation with an arbitrarily located source point, and obtain therefrom a basic representation of said function. It is appropriate to deal with the respective unit source functions in a pair of distinct configurations, and, specifically, to utilize an expression for their difference which has a self-reciprocal nature as regards the determination of one from the other.

Let the functions $G(x, \xi)$, $\mathfrak{G}(x, \eta)$, that refer to unit sources at arbitrary points ξ, η in separate configurations, satisfy the equations

$$LG(x, \xi) = -\delta(x - \xi), \qquad \mathscr{L}\mathfrak{G}(x, \eta) = -\delta(x - \eta), \quad -\infty < x, \xi, \eta < \infty \tag{77.1}$$

involving the differential operators

$$L = \frac{\mathrm{d}^2}{\mathrm{d}x^2} + k^2(p(x) + q(x)), \qquad \mathscr{L} = \frac{\mathrm{d}^2}{\mathrm{d}x^2} + k^2 p(x) \tag{77.2}$$

whose difference,

$$L - \mathscr{L} = k^2 q(x),$$

is determined by a function $q(x)$ with support Δ. On the basis of the preceding equations and a collateral stipulation that the pertinent solutions manifest an outward sense of wave propagation for $|x|\to\infty$ (and $p(x)\to 1$), we deduce the individual symmetries

$$G(x,\xi)=G(\xi,x),\qquad \mathfrak{G}(x,\eta)=\mathfrak{G}(\eta,x) \tag{77.3}$$

and the relations

$$G(x,x')-\mathfrak{G}(x,x')=k^2\int_\Delta q(\zeta)\mathfrak{G}(x,\zeta)G(\zeta,x')\,\mathrm{d}\zeta \tag{77.4}$$

$$=k^2\int_\Delta q(\zeta)G(x,\zeta)\mathfrak{G}(\zeta,x')\,\mathrm{d}\zeta \tag{77.5}$$

involving both source functions. The relation (77.4) acquires a deterministic role, namely that of an integral equation for G, if the source function $\mathfrak{G}$ is given and the argument variable x is restricted to Δ; likewise, (77.4) constitutes an integral equation for $\mathfrak{G}$ if the source function G is known and the variables x, ζ share a common range.

Another relationship that contains the difference of fundamental solutions,

$$\begin{aligned}G(x,x')-\mathfrak{G}(x,x')&=k^2\int_\Delta q(\xi)\mathfrak{G}(x,\xi)G(x',\xi)\,\mathrm{d}\xi+k^2\int_\Delta q(\xi)\mathfrak{G}(x',\xi)G(x,\xi)\,\mathrm{d}\xi\\&\quad-k^2\int_\Delta q(\xi)\mathfrak{G}(x,\xi)\mathfrak{G}(x',\xi)\,\mathrm{d}\xi\\&\quad-k^4\int_\Delta q(\xi)\mathfrak{G}(x,\xi)G(\xi,\eta)q(\eta)\mathfrak{G}(x',\eta)\,\mathrm{d}\xi\,\mathrm{d}\eta,\end{aligned} \tag{77.6}$$

whose justification rests on integral identities obtained from (77.4), (77.5), namely

$$\begin{aligned}&\int_\Delta q(\xi)\mathfrak{G}(x,\xi)\mathfrak{G}(x',\xi)\,\mathrm{d}\xi+k^2\int_\Delta q(\xi)\mathfrak{G}(x,\xi)G(\xi,\eta)q(\eta)\mathfrak{G}(x',\eta)\,\mathrm{d}\xi\,\mathrm{d}\eta\\&\qquad=\int_\Delta q(\xi)\mathfrak{G}(x',\xi)G(x,\xi)\,\mathrm{d}\xi=\int_\Delta q(\xi)\mathfrak{G}(x,\xi)G(x',\xi)\,\mathrm{d}\xi,\end{aligned}$$

has a stationary behavior relative to small independent variations of $q(\xi)\mathfrak{G}(x,\xi)$ and $q(\xi)\mathfrak{G}(x',\xi)$, since the concomitant Euler-Lagrange equations coincide with the integral equations (77.4) for $G(x',\xi)$ and $G(x,\xi)$.

The integrals in (77.4)–(77.6), which offer a formal characterization of $G-\mathfrak{G}$, can be identified with secondary source distributions over the whole range Δ; in the absence of explicit details bearing on the latter distributions (or a knowledge of G and $\mathfrak{G}$ individually) an indirect approach becomes necessary when particulars of $G-\mathfrak{G}$ are sought after by means of these expressions. To establish a symmetrical expansion of the difference between a pair of fundamental solutions, for instance, we may (as suggested by Filippi, 1973) introduce auxiliary source distributions on Δ that produce a common response in each configuration and thence designate a function basis suitable for the desired representation. Specifically, if the functions $\varphi_n(x)$, $n = 1, \ldots$ provide an orthonormal, square integrable basis on Δ, and if a second basis with functions of like character, $\psi_n(x)$, $n = 1, \ldots,$ is thereafter defined in terms of the first by the pairwise relations

$$\psi_n(x) = \varphi_n(x) - k^2 \int_\Delta \mathfrak{G}(x, \xi) q(\xi) \varphi_n(\xi)\, \mathrm{d}\xi \tag{77.7}$$

it follows that $\varphi_n(x)$ and $\psi_n(x)$ are equivalent, as source distribution functions, for the respective configurations with fundamental solutions G, $\mathfrak{G}$, inasmuch as

$$k^2 \int_\Delta \mathfrak{G}(x, \xi) q(\xi) \varphi_n(\xi)\, \mathrm{d}\xi = k^2 \int_\Delta G(x, \xi) q(\xi) \psi_n(\xi)\, \mathrm{d}\xi = \chi_n(x), \quad \text{say.} \tag{77.8}$$

It is the latter functions, $\chi_n(x)$, with alternative realizations in terms of G, ψ or g, φ, which are appropriate for characterizing the source function difference, $G(x, x') - \mathfrak{G}(x, x')$; to develop this role in a constructive manner, we utilize the variational expression (77.6) and assume that

$$g(x, \xi) = \sum_{n=1}^{N} c_n(x) \psi_n(\xi), \quad \mathfrak{G}(x', \xi) = \sum_{n=1}^{N} c_n(x') \psi_n(\xi) \tag{77.9}$$

where the factors $c_n(x)$, $c_n(x')$ are presently unspecified. On substituting the trial functions (77.9) into (77.6) and applying (77.8), it follows that

$$\begin{aligned} G(x, x') - \mathfrak{G}(x, x') = &\sum_{n=1}^{N} \{c_n(x) \chi_n(x') + c_n(x') \chi_n(x)\} \\ &- k^2 \sum_{m,n=1}^{N} \{A_{mn} + B_{mn}\} c_m(x) c_n(x') \end{aligned} \tag{77.10}$$

with

$$\begin{aligned} A_{mn} &= \int_\Delta q(\xi) \psi_m(\xi) \psi_n(\xi)\, \mathrm{d}\xi = A_{nm}, \\ B_{mn} &= \int_\Delta q(\xi) \psi_m(\xi) G(\xi, \eta) q(\eta) \psi_n(\eta)\, \mathrm{d}\xi\, \mathrm{d}\eta \\ &= \int_\Delta q(\xi) \psi_m(\xi) \chi_n(\xi)\, \mathrm{d}\xi = \int_\Delta q(\xi) \psi_n(\xi) \chi_m(\xi)\, \mathrm{d}\xi = B_{nm}. \end{aligned}$$

The stationary conditions which express a vanishing derivative of $G - \mathfrak{G}$ with respect to the individual factors $c_n(x)$ comprise a linear system for the factors

$c_n(x')$, namely

$$\chi_n(x') = k^2 \sum_{m=1}^{N} \{A_{mn} + B_{mn}\} c_m(x'), \quad n = 1, \ldots, N \tag{77.12}$$

and there is an identical system for the $c_n(x)$. If we employ a ready consequence of (77.12), arrived at by multiplying the respective equations with $c_n(x')$ and forming their sum, the representation (77.10) acquires a simpler (and stationary) version,

$$G(x, x') - \mathfrak{G}(x, x') = \sum_{n=1}^{N} c_n(x')\chi_n(x), \tag{77.13}$$

which is rendered explicit, in connection with the chosen trial functions, once the $c_n(x')$ are found from (77.12). When we have recourse to the equality

$$\begin{aligned} S_N(x, x') &= \sum_{n=1}^{N} c_n(x')\chi_n(x) \\ &= (S_N - S_{N-1}) + (S_{N-1} - S_{N-2}) + \cdots + (S_2 - S_1) + S_1 \end{aligned} \tag{77.14}$$

and the characterization (of a type described in §76),

$$S_N - S_{N-1} = \frac{\mathcal{D}_N\{\chi_i(x)\}\mathcal{D}_N\{\chi_i(x')\}}{D_{N-1}D_N}, \quad N \geq 2 \tag{77.15}$$

where D_N is the determinant formed with elements

$$D_{mn} = k^2\{A_{mn} + B_{mn}\} = k^2 \int_{\Delta} q(\xi)\psi_m(\xi)\varphi_n(\xi)\, d\xi = D_{nm}, \quad m, n = 1, \ldots, N,$$

and (77.16)

$$\mathcal{D}_N\{\chi_i(x)\} = \begin{vmatrix} D_{11} & & D_{1\,N-1} & \chi_1(x) \\ & \ddots & \vdots & \vdots \\ D_{N1} & & D_{N\,N-1} & \chi_N(x) \end{vmatrix}$$

it follows, after permitting N to increase without limit, that

$$G(x, x') - \mathfrak{G}(x, x') = \frac{\chi_1(x)\chi_1(x')}{D_{11}} + \sum_{n=2}^{\infty} \frac{\mathcal{D}_n\{\chi_i(x)\}\mathcal{D}_n\{\chi_i(x')\}}{D_{n-1}D_n}. \tag{77.18}$$

This constitutes the sought for representation, with a manifest symmetry relative to the argument variables x, x' and fully interchangeable links to the pair of configurations involved; as is the circumstance for analogous Ritz-Galerkin developments pertaining to other stationary and quadratic functionals, the original selection of particular basis functions, $\varphi_n(x), n = 1, 2, \ldots,$ merely affects the rate (though not the fact) of convergence in (77.18).

The variational characterization of a difference between source functions given by (77.6) also holds when the contrasting range, Δ, for the respective configurations has an infinite extent. Let us, with a view to assessing variational estimates in a case where precise results are available, choose the coefficient functions

$$p(x) = 1, \quad -\infty < x < \infty, \quad q(x) = x/L, \quad x > 0$$

in (77.2); if $x' < 0$, a straightforward calculation which fits together appropriate solutions of the homogeneous equations (77.1) in $x < x', x' < x < 0$ and $x > 0$ yields

$$G(x, x') - \mathfrak{G}(x, x') = \frac{\mathrm{i}}{2k} \mathrm{e}^{-ik(x+x')} \frac{H^{(1)}_{1/3}(\zeta_0) + \mathrm{i}H^{(1)}_{-2/3}(\zeta_0)}{H^{(1)}_{1/3}(\zeta_0) - \mathrm{i}H^{(1)}_{-2/3}(\zeta_0)}, \quad x, x' < 0$$

$$G(x, x') = \frac{\mathrm{i}}{k}\left(\frac{\zeta(x)}{\zeta_0}\right)^{1/3} \frac{\mathrm{e}^{-ikx'} H^{(1)}_{1/3}(\zeta(x))}{H^{(1)}_{1/3}(\zeta_0) - \mathrm{i}H^{(1)}_{-2/3}(\zeta_0)}, \quad x > 0, \quad x' < 0$$

with

$$\mathfrak{G}(x, x') = \frac{\mathrm{i}}{2k} \mathrm{e}^{\mathrm{i}k|x-x'|}, \tag{77.20}$$

$$\zeta(x) = \frac{2}{3} kL\left(1 + \frac{x}{L}\right)^{3/2}, \quad \zeta(0) = \zeta_0 = \frac{2}{3} kL. \tag{77.21}$$

When there is but little change in $q(x)$ over an interval with the span $\lambda = 2\pi/k$, or $kL \gg 1$ and $\zeta_0 \gg 1$, use of the asymptotic developments

$$\left.\begin{aligned} H^{(1)}_{1/3}(\zeta_0) &\sim (2/\pi\zeta_0)^{1/2} \exp\left\{\mathrm{i}\left(\zeta_0 - \frac{5\pi}{12}\right)\right\}\left[1 + \frac{5}{72\,\mathrm{i}\zeta_0} + \cdots\right], \\ H^{(1)}_{-2/3}(\zeta_0) &\sim (2/\pi\zeta_0)^{1/2} \exp\left\{\mathrm{i}\left(\zeta_0 + \frac{\pi}{12}\right)\right\}\left[1 - \frac{7}{72\,\mathrm{i}\zeta_0} + \cdots\right] \end{aligned}\right\} \zeta_0 \gg 1 \tag{77.22}$$

reveals that

$$G(x, x') - \mathfrak{G}(x, x') \sim \frac{1}{16k^2L} \mathrm{e}^{-ik(x+x')}, \quad \begin{matrix} kL \gg 1 \\ x, x' < 0 \end{matrix}; \tag{77.23}$$

contrariwise, if kL and ζ_0 are both much smaller than unity,

$$G(x, x0) - \mathfrak{G}(x, x') \sim -\frac{\mathrm{i}}{2k} \mathrm{e}^{-ik(x+x')}, \quad \begin{matrix} kL \ll 1 \\ x, x' < 0 \end{matrix} \tag{77.24}$$

since $H^{(1)}_{-2/3}(\zeta_0)/H^{(1)}_{1/3}(\zeta_0) = 0(\zeta_0^{-1}),\ \zeta_0 \to 0$.

To secure a single term estimate for $G - \mathfrak{G}$ along the lines indicated above, viz.

$$G(x, x') - \mathfrak{G}(x, x') \doteq \frac{\chi_1(x)\chi_1(x')}{D_{11}}, \tag{77.25}$$

we first assume that

$$\varphi_1(x) = \mathrm{e}^{ikx - x/l} \tag{77.26}$$

and thence obtain, employing (77.7), (77.8), and (77.20), the concomitant expressions

$$\chi_1(x) = \psi_1(x) - \varphi_1(x) \begin{cases} = -\dfrac{ikl^2}{2L} \dfrac{e^{-ikx}}{(1-2\,ikl)^2}, & x<0 \\ = -\dfrac{ikl^2}{2L} e^{ikx}\left[1 + e^{-x/l}\left\{\dfrac{4\,ikl(1-ikl)}{(1-2\,ikl)^2} + \dfrac{2\,ikx}{1-2\,ikl}\right\}\right], & x>0, \end{cases}$$

and

$$\begin{aligned} D_{11} &= \frac{k^2}{L}\int_0^\infty x\psi_1(x)\varphi_1(x)\,dx \\ &= \frac{(kl)^2}{4L}\left[\frac{1}{1-ikl} - \frac{2\,ikl^2}{L}\left\{\frac{1}{1-2\,ikl} + \frac{ikl}{2(1-ikl)^2}\right\}\right]\frac{1}{1-ikl}. \end{aligned} \tag{77.28}$$

It may be directly verified, on referring to the approximate forms

$$D_{11} \doteq -\frac{1}{4L}, \qquad kL \gg (kl)^2 \gg 1,$$

$$D_{11} \doteq -\frac{i\,k^3 l^4}{2\,L^2}, \qquad kL \ll (kl)^2 \ll 1,$$

that the variational estimate stemming from (77.25), (77.27), and (77.28) accords with the prior characterizations (77.23), (77.24).

§78. *Other Inhomogeneous Realizations*

The equations descriptive of inhomogeneous systems may involve several coefficient functions with a variable nature, rather than the lone function envisaged thus far. We consider in this section two prototypes of systems whose inhomogeneity is specified by a pair of configurational variables and describe some analytical approaches, along previous lines, suitable for particular excitations therein.

Let us first refer to a compressible medium with the local equilibrium density $\rho(x)$ and sound speed $c(x)$; and assume that the coupled equations [of momentum and mass conservation, respectively]

$$\rho(x)\frac{\partial v}{\partial t} = -\frac{\partial p}{\partial x}, \quad \rho(x)\frac{\partial v}{\partial t} = -\frac{1}{c^2(x)}\frac{\partial p}{\partial t} \tag{78.1}$$

correlate the deviations in pressure, $p(x, t)$, and velocity, $v(x, t)$, from their (given) static values $\bar{p}(x), 0$. We shall suppose, moreover, that both p and v remain continuous even if ρ and/or c have unequal limits (and thence a discontinuity) at one or more points of the medium. After the successive elimination of v and p from the first order system (78.1) uncoupled second order equations for these quantities result, namely

$$\frac{\partial^2 p}{\partial t^2} = \rho(x)c^2(x)\frac{\partial}{\partial x}\left(\frac{1}{\rho(x)}\frac{\partial p}{\partial x}\right), \qquad \rho(x)\frac{\partial^2 v}{\partial t^2} = \frac{\partial}{\partial x}\left(\rho(x)c^2(x)\frac{\partial v}{\partial x}\right), \tag{78.2}$$

and it appears that the density and compressibility

$$\kappa(x) = \frac{1}{\rho(x)c^2(x)} \tag{78.3}$$

characterize separate measures of inhomogeneity.

When the disturbances in the medium are time-periodic and the representations

$$p(x, t) = \mathrm{Re}\{P(x)\,\mathrm{e}^{-\mathrm{i}\omega t}\}, \quad v(x, t) = \mathrm{Re}\{V(x)\,\mathrm{e}^{-\mathrm{i}\omega t}\} \tag{78.4}$$

apply, the simultaneous equations for $P(x)$, $V(x)$ become

$$\frac{\mathrm{d}P}{\mathrm{d}x} = \mathrm{i}\omega\rho(x)V, \quad \frac{\mathrm{d}V}{\mathrm{d}x} = \mathrm{i}\omega\kappa(x)P \tag{78.5}$$

while those of individual type assume the forms

$$\frac{\mathrm{d}^2P}{\mathrm{d}x^2} - \left(\frac{\mathrm{d}}{\mathrm{d}x}\log\rho(x)\right)\frac{\mathrm{d}P}{\mathrm{d}x} + \frac{\omega^2}{c^2(x)}P = 0, \tag{78.6}$$

$$\frac{\mathrm{d}^2V}{\mathrm{d}x^2} - \left(\frac{\mathrm{d}}{\mathrm{d}x}\log\kappa(x)\right)\frac{\mathrm{d}V}{\mathrm{d}x} + \frac{\omega^2}{c^2(x)}V = 0. \tag{78.7}$$

In this circumstance it suffices, as a procedural matter, to determine either P or V on the basis of (78.6) or (78.7), since the other can be readily deduced from one of the coupled equations (78.5). If ρ and c possess constant values ρ_0, c_0 a single function $\varphi(x)$ that satisfies the equation

$$\frac{\mathrm{d}^2\varphi}{\mathrm{d}x^2} + k_0^2\varphi = 0, \quad k_0 = \frac{\omega}{c_0} \tag{78.8}$$

yields, in accordance with the scheme

$$V = -\mathrm{d}\varphi/\mathrm{d}z, \quad P = -\mathrm{i}\omega\rho_0\varphi, \tag{78.9}$$

joint determinations of V and P; there is, however, no useful role served by the velocity potential φ when ρ and c have a variable nature.

It is easy to eliminate the first derivative term from the equations that describe P and V, and thereby secure versions which resemble those already dealt with in cases of inhomogeneity; thus, if we assume that the distant parts of the medium are uniform, viz.

$$\rho(x) \to \rho_0, \quad c(x) \to c_0, \quad |x| \to \infty$$

and set

$$P(x) = \left(\frac{\rho(x)}{\rho_0}\right)^{1/2}\psi(x), \quad V(x) = \sqrt{(\rho_0 c_0^2\kappa(x))}\,\chi(x) \tag{78.10}$$

the equations for $\psi(x)$, $\chi(x)$ which follow from (78.6), (78.7) are

$$\frac{d^2\psi}{dx^2} + \left\{\frac{1}{2}\frac{\rho''}{\rho} - \frac{3}{4}\left(\frac{\rho'}{\rho}\right)^2 + \frac{\omega^2}{c^2(x)}\right\}\psi = 0, \tag{78.11}$$

$$\frac{d^2\chi}{dx^2} + \left\{\frac{1}{2}\frac{\kappa''}{\kappa} - \frac{3}{4}\left(\frac{\kappa'}{\kappa}\right)^2 + \frac{\omega^2}{c^2(x)}\right\}\chi = 0. \tag{78.12}$$

A rearrangement of (78.11), namely

$$\frac{d^2\psi}{dx^2} + k_0^2\psi = \left\{k_0^2(1 - n^2(x)) + \frac{3}{4}\left(\frac{\rho'}{\rho}\right)^2 - \frac{1}{2}\frac{\rho''}{\rho}\right\}\psi \tag{78.13}$$

with

$$n(x) = \frac{c_0}{c(x)} \tag{78.14}$$

and a collection of terms on the right-hand side that owe their existence to nonuniformity in $\rho(x)$ and $c(x)$, facilitates the derivation of an integral equation for $\psi(x)$,

$$\psi(x) = e^{ik_0x} + \frac{i}{2k_0}\int_{-\infty}^{\infty} e^{ik_0|x-\xi|}\left\{k_0^2(n^2(\xi) - 1) + \frac{1}{2}\frac{\rho''(\xi)}{\rho(\xi)} - \frac{3}{4}\left(\frac{\rho'(\xi)}{\rho(\xi)}\right)^2\right\}\psi(\xi)\,d\xi, \tag{78.15}$$

whose inhomogeneous member is a solution of (78.13) when the material parameters are invariable ($n = 1, \rho = \rho_0$). The corresponding representation for $P(x)$ has one component, $P^{inc}(x) = \sqrt{(\rho(x)/\rho_0)}\, e^{ik_0x}$, that describes an incoming wave of unit amplitude at $x = -\infty$, and another

$$P^{sec}(x) = \frac{i}{2k_0}\int_{-\infty}^{\infty} e^{ik_0|x-\xi|}\left\{k_0^2(n^2(\xi) - 1) + \frac{1}{2}\frac{\rho''(\xi)}{\rho(\xi)} - \frac{3}{4}\left(\frac{\rho'(\xi)}{\rho(\xi)}\right)^2\right\}\left(\frac{\rho(x)}{\rho_0}\right)^{1/2}\psi(\xi)\,d\xi \tag{78.16}$$

which formally characterizes the secondary disturbance resulting from variations of these parameters. It may be inferred that conditions are favorable for an estimation of $P^{sec}(x)$, on the basis of (78.16) and the integral equation (78.15), when the density undergoes gradual changes and has a continuous nature. To accomodate a discontinuous behavior of the density, let us first express the differential equation for ψ in another form

$$\frac{d^2\psi}{d\xi^2} + \left\{\eta^2(\xi) + \frac{1}{2}\frac{\eta''(\xi)}{\eta(\xi)} - \frac{3}{4}\left(\frac{\eta'(\xi)}{\eta(\xi)}\right)^2\right\}\psi = (\eta^2(\xi) - n^2(\xi))\psi, \tag{78.17}$$

where

$$\xi = k_0x,$$

$$\rho(x)/\rho_0 = \eta(\xi), \quad n(\xi) = \frac{c_0}{c(\xi/k_0)}; \tag{78.18}$$

then, utilizing a Green's function $G(\xi, \xi')$ which satisfies the equation

$$\left(\frac{d^2}{d\xi^2} + \eta^2(\xi) + \frac{1}{2}\frac{\eta''(\xi)}{\eta(\xi)} - \frac{3}{4}\left(\frac{\eta'(\xi)}{\eta(\xi)}\right)^2\right) G(\xi, \xi') = -\delta(\xi - \xi')$$

and has the explicit representation [cf. §69]

$$G(\xi, \xi') = \frac{i}{2}\varphi_+(\xi_>)\varphi_-(\xi_<), \quad \varphi_\pm(\xi) = (\eta(\xi))^{-1/2}\exp\left\{\pm i\int_0^\xi \eta(\tau)\, d\tau\right\} \tag{78.19}$$

we construct an integral of (78.17),

$$\psi(\xi) = \varphi_+(\xi) + \int_{-\infty}^{\infty} G(\xi, \xi')[n^2(\xi') - \eta^2(\xi')]\psi(\xi')\, d\xi', \tag{78.20}$$

that incorporates, within the term $\varphi_+(\xi)$, an inward directed component of unit amplitude at $\xi = -\infty$. Since derivatives of the density or sound speed are absent from the second term of (78.20), this formulation is suitable when the material parameters undergo discontinuous change.

A different formulation with unimpaired validity in the latter circumstance rests on a second order system of equations linking the pressure and velocity, viz.,

$$\frac{d^2P}{d\xi^2} + P = (1 - n^2(\xi))P + i\rho_0 c_0 \eta'(\xi) V, \tag{78.21}$$

$$\frac{d^2V}{d\xi^2} + V = (1 - n^2(\xi))V + ic_0\kappa'(\xi)P, \tag{78.22}$$

which are direct consequences of (78.5)–(78.7) and the definitions (78.18). An integral of (78.21), making provision for an incoming wave at $\xi = -\infty$, is expressed by

$$P(\xi) = e^{i\xi} + \frac{i}{2}\int_{-\infty}^{\infty} e^{i|\xi-\xi'|}\left[(n^2(\xi') - 1)P(\xi') - i\rho_0 c_0 \frac{d\eta(\xi')}{d\xi'} V(\xi')\right] d\xi'$$

or, consequent to integration by parts in the term that involves V, by

$$P(\xi) = e^{i\xi} + \int_{-\infty}^{\infty} G(\xi, \xi') f(\xi') P(\xi')\, d\xi' - i\frac{d}{d\xi}\int_{-\infty}^{\infty} G(\xi, \xi') g(\xi') V(\xi')\, d\xi' \tag{78.23}$$

where

$$f(\xi) = \frac{n^2(\xi)}{\eta(\xi)} - 1, \quad g(\xi) = \rho_0 c_0(\eta(\xi) - 1),$$

$$G(\xi, \xi') = \frac{i}{2} e^{i|\xi-\xi'|}; \tag{78.24}$$

the corresponding versions which pertain to the differential equation (78.22) take the forms

$$V(\xi) = \frac{1}{\rho_0 c_0} e^{i\xi} + \frac{i}{2}\int_{-\infty}^{\infty} e^{i|\xi-\xi'|}\left[(n^2(\xi')-1)V(\xi') - ic_0\frac{d\kappa(\xi')}{d\xi'}P(\xi)\right] d\xi',$$

and

$$\rho_0 c_0 V(\xi) = e^{i\xi} + \int_{-\infty}^{\infty} G(\xi,\xi')g(\xi')V(\xi')\,d\xi' - i\frac{d}{d\xi}\int_{-\infty}^{\infty} G(\xi,\xi')f(\xi')P(\xi')\,d\xi', \tag{78.25}$$

respectively.

The asymptotic representations yielded by (78.23), (78.25),

$$P(\xi) \simeq e^{i\xi} + r\,e^{-i\xi}, \quad V(\xi) \simeq \frac{1}{\rho_0 c_0}[e^{i\xi} - re^{-i\xi}], \quad \xi \to -\infty \tag{78.26}$$

conform to the interrelations

$$\frac{dP}{d\xi} = i\rho_0 c_0 V, \quad \frac{dV}{d\xi} = i\frac{P}{\rho_0 c_0}, \quad \xi \to -\infty$$

which are consistent with (78.5) and, moreover, acquire their definitive characterization in terms of the common reflection coefficient,

$$r = \frac{i}{2}\int_{-\infty}^{\infty} e^{i\xi}[f(\xi)P(\xi) - g(\xi)V(\xi)]\,d\xi. \tag{78.27}$$

Determinations of both $P(\xi)$ and $V(\xi)$ are a prerequisite, in the operational sense, to calculating r from (78.27) and may be sought after on the basis of the coupled integro-differential equations (78.23) and (78.25); it is appropriate, in this regard, to note that the expression

$$\begin{aligned} r = \frac{i}{2}\Big\{ & 2\int_{-\infty}^{\infty} e^{i\xi}[f(\xi)P(\xi) - g(\xi)V(\xi)]\,d\xi - \int_{-\infty}^{\infty} f(\xi)P^2(\xi)\,d\xi + \rho_0 c_0\int_{-\infty}^{\infty} g(\xi)V^2(\xi)\,d\xi \\ & + \int_{-\infty}^{\infty} f(\xi)P(\xi)G(\xi,\xi')f(\xi')P(\xi')\,d\xi\,d\xi' \\ & - \int_{-\infty}^{\infty} g(\xi)V(\xi)G(\xi,\xi')g(\xi')V(\xi')\,d\xi\,d\xi' \\ & - 2i\int_{-\infty}^{\infty} f(\xi)P(\xi)\frac{d}{d\xi}G(\xi,\xi')g(\xi')V(\xi')\,d\xi\,d\xi' \Big\} \end{aligned} \tag{78.28}$$

provides r with a stationary character insofar as $P(\xi)$ and $V(\xi)$ undergo first order variations about their proper values.

Our second prototype of an inhomogeneous model whose specification involves a pair of configuration variables relates to fluid motions in a lake or channel with continuously altering breadth and depth. Let the x-axis point along the lake and suppose that $b(x)$, $A(x)$ designate the breadth of its free surface and cross-sectional area, respectively, in the equilibrium state; then, for motions described by a longitudinal fluid velocity, $u(x, t)$, and a vertical free surface displacement relative to the undisturbed level, $\zeta(x, t)$, the equations of balance for momentum and mass in small scale local movements are

$$\frac{\partial u}{\partial t} = -g\frac{\partial \zeta}{\partial x}, \quad \frac{\partial}{\partial x}(A(x)u) = -b(x)\frac{\partial \zeta}{\partial t} \tag{78.29}$$

if compressibility is ignored and the gravitational force alone is taken into account. To develop the analysis in terms of a fundamental equation, we may advantageously recast the system (78.29) before eliminating one or the other of the dependent variables; thus, on introducing the mass flux $S(x, t)$, or product of the fluid density ρ, velocity u and cross-sectional area A, the equations (78.29) become

$$\frac{\partial S}{\partial t} = -\rho g A(x)\frac{\partial \zeta}{\partial x}, \quad \frac{\partial S}{\partial x} = -\rho b(x)\frac{\partial \zeta}{\partial t}, \tag{78.30}$$

and simple manipulation furnishes the individual second order equations

$$\frac{\partial^2 S}{\partial t^2} = gA(x)\frac{\partial}{\partial x}\left(b(x)\frac{\partial S}{\partial x}\right), \tag{78.31}$$

$$b(x)\frac{\partial^2 \zeta}{\partial t^2} = g\frac{\partial}{\partial x}\left(A(x)\frac{\partial \zeta}{\partial x}\right). \tag{78.32}$$

When the motions are time-periodic and it is assumed that

$$\begin{gathered} S(x, t) = \mathrm{Re}\{M(x)\,\mathrm{e}^{-\mathrm{i}\omega t}\} \\ A(x) = b(x)h(x), \quad c^2(x) = gh(x) \end{gathered} \tag{78.33}$$

where h characterizes the mean depth over a cross-section and c the local velocity for long waves in a fluid of such depth (§45), we deduce from (78.31) an equation

$$\frac{\mathrm{d}^2 M}{\mathrm{d}x^2} - \left(\frac{\mathrm{d}}{\mathrm{d}x}\log b(x)\right)\frac{\mathrm{d}M}{\mathrm{d}x} + \frac{\omega^2}{c^2(x)}M = 0 \tag{78.34}$$

whose form duplicates that of an earlier pair, namely (78.6), (78.7). A vanishing mass flux at the respective ends ($x = 0, L$) of a lake is implied by the terminal conditions

$$M(0) = M(L) = 0 \tag{78.35}$$

which thus play a selective role in determining the proper solutions and eigenvalues ω of the equation (78.34).

The boundary value problem linked with the ensemble or trio of relations (78.34), (78.35) can be reformulated in terms of a homogeneous integral equation; let us, for such purpose, introduce the real Green's function

$$G(x, x') = -\cot\{\Phi(L)\}\varphi_1(x_<)\varphi_2(x_>) \tag{78.36}$$

where

$$\Phi(L) = \int_0^L b(x)\,\mathrm{d}x,$$

and

$$\begin{aligned}\varphi_1(x) &= \{b(x)\}^{-1/2} \sin\left\{\int_0^x b(\xi)\,\mathrm{d}\xi\right\} = \{b(x)\}^{-1/2} \sin \Phi(x) \\ \varphi_2(x) &= \varphi_1(x) - \tan\{\Phi(L)\} \cdot \{b(x)\}^{-1/2} \cos \Phi(x)\end{aligned} \qquad (78.37)$$

constitute a linearly independent pair of solutions to the differential equation

$$\left(\frac{\mathrm{d}^2}{\mathrm{d}x^2} + b^2(x) + \frac{1}{2}\frac{b''(x)}{b(x)} - \frac{3}{4}\left(\frac{b'(x)}{b(x)}\right)^2\right)\varphi_{1,2}(x) = 0. \qquad (78.38)$$

It is easy to verify that $G(x, x')$ also satisfies the equation (78.38) when $x \neq x'$, has a unit jump or discontinuity in the first derivative at $x = x'$, viz.

$$\frac{\mathrm{d}}{\mathrm{d}x} G(x, x')\Bigg|_{x=x'-0}^{x=x'+0} = -1 \qquad (78.39)$$

and vanishes at $x = 0, L$ since $\varphi_1(0) = \varphi_2(L) = 0$. If we transform (78.34) in accordance with the substitution $M(x) = \{b(x)\}^{1/2}\sigma(x)$ and refer to a particular version of the equation for $\sigma(x)$, viz.

$$\left(\frac{\mathrm{d}^2}{\mathrm{d}x^2} + b^2(x) + \frac{1}{2}\frac{b''(x)}{b(x)} - \frac{3}{4}\left(\frac{b'(x)}{b(x)}\right)^2\right)\sigma(x) = \left(b^2(x) - \frac{\omega^2}{c^2(x)}\right)\sigma(x), \qquad (78.40)$$

the enumerated properties of G support our contention that

$$\sigma(x) = \int_0^L G(x, x')\left[\frac{\omega^2}{c^2(x')} - b^2(x')\right]\sigma(x')\,\mathrm{d}x', \quad 0 < x < L \qquad (78.41)$$

characterizes a solution of (78.40) with null boundary values, i.e., $\sigma(0) = \sigma(L) = 0$. The homogeneous integral equation (78.41) thus unites the various specifications demanded of $\sigma(x)$ and determinate values of the parameter ω therein, or proper frequencies of the tidal motions, are correlated with its non-trivial solutions.

Pursuant to multiplication in (78.41) by $((\omega^2/c^2(x)) - b^2(x))\sigma(x)$ and integration over $0 < x < L$, we obtain the relationship

$$\int_0^L \left(\frac{\omega^2}{c^2(x)} - b^2(x)\right)\sigma^2(x)\,\mathrm{d}x = \int_0^L \left(\frac{\omega^2}{c^2(x)} - b^2(x)\right)\sigma(x) \times G(x, x')\left(\frac{\omega^2}{c^2(x')} - b^2(x')\right)\sigma(x')\,\mathrm{d}x\,\mathrm{d}x'; \qquad (78.42)$$

inasmuch as $G(x, x')$ is a symmetrical function of its argument variables, it follows from (78.42) that variations of ω and $\sigma(x)$ satisfy the equation

$$\omega\delta\omega\left\{\int_0^L \frac{\sigma^2(x)}{c^2(x)}\,\mathrm{d}x - 2\int_0^L \frac{\sigma(x)}{c^2(x)}\,G(x,x')\left(\frac{\omega^2}{c^2(x')} - b^2(x')\right)\sigma(x')\,\mathrm{d}x\,\mathrm{d}x'\right\}$$
$$= \int_0^L \delta\sigma(x)\left(\frac{\omega^2}{c^2(x)} - b^2(x)\right)\left\{\int_0^L G(x,x')\left(\frac{\omega^2}{c^2(x')} - b^2(x')\right)\sigma(x')\,\mathrm{d}x' - \sigma(x)\right\}\mathrm{d}x.$$

Hence, if $\sigma(x)$ is a solution of (78.41) and $\delta\sigma(x)$ designates the concomitant variation, we obtain $\delta\omega = 0$, which signifies that (78.42) affords a stationary means of estimating ω.

§79. *The Born Approximation and Reflection at Short Wave Lengths*

A widely used procedural technique at short wave lengths, termed the Born approximation, traces its inception to the replacement, in expressions with a formally exact nature, of the actual wave function by the corresponding primary or incident wave function; thus, the first Born approximation to a function $\psi(x)$ which satisfies the differential equation

$$\left(\frac{\mathrm{d}^2}{\mathrm{d}x^2} + k^2 + V(x)\right)\psi(x) = 0 \tag{79.1}$$

or the integral equation

$$\psi(x) = \psi^{inc}(x) + \frac{\mathrm{i}}{2k}\int_{-\infty}^{\infty} \mathrm{e}^{\mathrm{i}k|x-x'|}V(x')\psi(x')\,\mathrm{d}x', \quad -\infty < x < \infty \tag{79.2}$$

has the explicit characterization

$$\psi_1(x) = \psi^{inc}(x) + \frac{\mathrm{i}}{2k}\int_{-\infty}^{\infty} \mathrm{e}^{\mathrm{i}k|x-x'|}V(x')\psi^{inc}(x')\,\mathrm{d}x'. \tag{79.3}$$

If

$$\psi^{inc}(x) = \mathrm{e}^{\mathrm{i}kx}$$

a collateral approximation to the reflection coefficient

$$r = \lim_{x\to-\infty} \mathrm{e}^{\mathrm{i}kx}(\psi(x) - \mathrm{e}^{\mathrm{i}kx})$$
$$= \frac{\mathrm{i}}{2k}\int_{-\infty}^{\infty} \mathrm{e}^{\mathrm{i}kx}V(x)\psi(x)\,\mathrm{d}x, \tag{79.4}$$

namely,

$$r_1 = \frac{\mathrm{i}}{2k}\int_{-\infty}^{\infty} \mathrm{e}^{2\mathrm{i}kx}V(x)\,\mathrm{d}x, \tag{79.5}$$

depends in linear fashion on the measure of inhomogeneity, $V(x)$. Subsequent approximations can be devised with wave functions that are obtained from the integral equation (79.2) through a process of iteration and connected by relations of the type

$$\psi_N(x) = \psi^{inc}(x) + \frac{\mathrm{i}}{2k}\int_{-\infty}^{\infty} \mathrm{e}^{\mathrm{i}k|x-x'|}V(x')\psi_{N-1}(x')\,\mathrm{d}x', \quad N \geq 1 \tag{79.6}$$

though these are seldom pursued for reasons of analytical complexity. To infer a general criterion for the convergence of such expansions and to reach some conclusions regarding the accuracy of the first term, we rely on details of individual cases associated with specific forms of $V(x)$.

If $V(x)$ is a delta function or distribution, say

$$V(x) = \frac{Mk^2}{\rho}\delta(x), \tag{79.7}$$

where the normalization suits the circumstance of mass loading on a string (§17), the first Born approximation to the reflection coefficient,

$$r_1 = \frac{\mathrm{i}Mk}{2\rho},$$

serves as the generator of all successive contributions, since

$$r_n = (r_1)^n, \quad n > 1;$$

and the Born expansion

$$r = \sum_{n=1}^{\infty} r^n = \frac{\mathrm{i}Mk/2\rho}{1 - \mathrm{i}\dfrac{Mk}{2\rho}} \tag{79.8}$$

converges to the exact result (17.19), provided that

$$\frac{M}{2\rho} < \frac{1}{k}.$$

When $V(x)$ has a uniform magnitude, V_0, on the range $0 < x < L$ and vanishes elsewhere, it follows from (26.8) that

$$r = \frac{2\mathrm{i}V_0}{k^2}\frac{\sin nkL}{(n+1)^2\,\mathrm{e}^{-\mathrm{i}nkL} - (n-1)^2\,\mathrm{e}^{\mathrm{i}nkL}}, \quad n^2 = 1 + \frac{V_0}{k^2}$$

and hence

$$r \simeq \frac{\mathrm{i}V_0}{2k^2}\mathrm{e}^{\mathrm{i}kL}\sin kL, \quad (kL)^2 \gg V_0L^2$$

in agreement with the first Born approximation.

A deficiency in the latter, or discrepancy between r and r_1, emerges on consideration of a smooth and infinitely extended profile, viz.

$$V(x) = \frac{\alpha k^2}{\cosh^2 \dfrac{x}{L}}, \quad -\infty < x < \infty; \tag{79.9}$$

here the first Born approximation to the reflection coefficient is given by

$$r_1 = \mathrm{i}\alpha kL \int_{-\infty}^{\infty} \cos(2kL\zeta)\,\mathrm{sech}^2\,\zeta\,\mathrm{d}\zeta = \mathrm{i}\alpha kL \cdot 2\pi kL\,\mathrm{cosech}\,\pi kL$$

and thus

$$|r_1| \simeq 4\pi\alpha(kL)^2\,\mathrm{e}^{-\pi kL}, \quad kL \gg 1 \tag{79.10}$$

whereas the comparable estimate supplied by the exact representation (64.26) is

$$|r| \simeq 2|\cos(\pi\sqrt{(\alpha)}kL)|\,\mathrm{e}^{-\pi kL}, \quad kL \gg 1. \tag{79.11}$$

The dissimilarity between expressions, for large kL, of r_1 and r is further accentuated in the case of a Gaussian profile,

$$V(x) = -V\,\mathrm{e}^{-(x/L)^2}, \quad -\infty < x < \infty.$$

From an assessment of these and other specific examples, it may be surmised that the Born expansions converge at sufficiently large values of the wave number k (or small values of the wave length $\lambda = 2\pi/k$) for absolutely integrable functions $V(x)$; and, furthermore, that the accuracy of the first Born approximation depends on the analytical nature of said functions. The conditions are most favorable for an asymptotic validity of the first Born approximation when $V(x)$ lacks a continuation from the real axis into the complex x-plane and, contrariwise, substantial differences between r and r_1 appear when $V(x)$ is an entire function of x. For purposes of a more precise statement (Wu, 1966), let us distinguish between functions $V(x)$ which are

1) analytic and integrable on closed real intervals whose union comprises the real axis, or
2) analytic throughout the strip $|\mathrm{Im}\, x| \leq x_0$ and integrable along any line parallel to the real axis therein, or
3) analytic in the strip $|\mathrm{Im}\, x| < x_0$, integrable along any line parallel to the real axis therein, with a finite number of singularities on the boundary lines $|\mathrm{Im}\, x| = x_0$, or
4) entire functions of the complex variable x.

The conclusions respecting individual and joint asymptotic properties of the reflection coefficient, r, and its first Born approximation, r_1, for large values of k are these: $r_1 \sim r$ in class 1; $r\exp(2kx_0) \to 0$, $r_1 \exp(2kx_0) \to 0$ in class 2; r, r_1 share a common exponential dependence, $\exp(-2kx_0)$, in class 3 and have no particular likeness in class 4.

That a singularity of $V(x)$ with the least positive imaginary part has a decisive role in shaping the behavior of the reflection coefficient when $k \to \infty$ is implied by the representation (79.4) of r as a Fourier transform of $\psi(x)$ with the weight factor $V(x)$; for, if we grant the existence of a regular continuation of $\psi(x)$, it falls upon $V(x)$ to offset, through singular behavior, such reduction in magnitude of the transform integral as is brought about by the exponential factor e^{ikx} when the contour recedes farther from the real axis in the upper half of the x-plane. This observation also suggests that an asymptotic estimate for the reflection coefficient in the limit $k \to \infty$ may be deduced by means of an approximation to the wave function $\psi(x)$ which is accurate near the relevant singularity of $V(x)$.

To illustrate the latter approach we consider the profile (79.9), whose analytic continuation possesses double poles at

$$x = \pm\left(n - \frac{1}{2}\right)\pi\, \mathrm{i}L, \quad n = 0, \pm 1, \ldots,$$

and set

$$x = \mathrm{i}x_0 + \xi, \qquad x_0 = \pi L/2 \tag{79.12}$$

which places x near the pole with the least positive imaginary part if ξ is small. Since

$$\cosh^2 \frac{x}{L} \doteq -(\xi/L)^2, \quad \xi \to 0,$$

an approximate differential equation for $\psi(\xi)$ in this neighborhood takes the form

$$\frac{\mathrm{d}^2\psi}{\mathrm{d}\xi^2} + k^2\left(1 - \alpha\frac{L^2}{\xi^2}\right)\psi = 0 \tag{79.13}$$

with all of its solutions comprised in the basic representation

$$\psi(\xi) = \xi^{1/2} Z_\nu(k\xi) \tag{79.14}$$

where Z_ν denotes a cylinder function of the order

$$\nu = \left\{\frac{1}{4} + \alpha(kL)^2\right\}^{1/2}. \tag{79.15}$$

Bearing in mind, firstly, that the general solution of (79.13) reduces to a linear combination of the exponential functions $\exp(\pm \mathrm{i}k\xi)$ when $|\xi| \to \infty$ and, secondly, that the Hankel functions of the first and second kinds have the asymptotic forms

$$\left.\begin{aligned} H_\nu^{(1)}(z) &\simeq \left(\frac{\pi}{2}z\right)^{-1/2} \exp\left\{\mathrm{i}\left(z - \frac{\nu\pi}{2} - \frac{\pi}{4}\right)\right\}, && -\pi < \arg z < 2\pi \\ H_\nu^{(2)}(z) &\simeq \left(\frac{\pi}{2}z\right)^{-1/2} \exp\left\{-\mathrm{i}\left(z - \frac{\nu\pi}{2} - \frac{\pi}{4}\right)\right\}, && -2\pi < \arg z < \pi \end{aligned}\right\} \quad |z| \to \infty \tag{79.16}$$

we adopt the normalized version of (79.14),

$$\hat{\psi}(\xi) = \left(\frac{\pi}{2}\xi\right)^{-1/2} e^{-\pi kL/2} \exp\left\{i\left(\frac{\nu\pi}{2}+\frac{\pi}{4}\right)\right\} H_\nu^{(1)}(k\xi),$$

$$\simeq e^{-\pi kL/2}\, e^{ik\xi} = e^{ikx}, \qquad \xi \to \infty, \qquad \arg \xi = 0 \tag{79.17}$$

which characterizes an outgoing wave on the right or incidence from the left.

That approximations to $V(x)$ and $\psi(x)$ near the singular point of the former at $x = ix_0 = i\pi L/2$ yield an accurate estimate of the reflection coefficient at short wave lengths ($kL \gg 1$) can be substantiated by utilizing (79.4) or the alternative characterizations

$$r = \operatorname*{Lim}_{x\to-\infty} e^{ikx}(\hat{\psi}(x) - e^{ikx})$$

$$= \operatorname*{Lim}_{\xi\to\infty\exp(-i\pi)} \{e^{-\pi kL/2}\, e^{ik\xi}(\hat{\psi}(\xi) - e^{-\pi kL/2}\, e^{ik\xi})\}. \tag{79.18}$$

Inasmuch as the asymptotic representation (79.16) for the Hankel function of the first kind does not apply at the (Stokes) line whereupon $\arg \xi = -\pi$, we evaluate the limit in (79.18) after a preliminary transformation of (79.17) based on the cylinder function identity

$$H_\nu^{(1)}(z\, e^{-i\pi}) = 2\cos \pi\nu H_\nu^{(1)}(z) + e^{-i\pi\nu} H_\nu^{(2)}(z)$$

and subsequent application of (79.16); thus

$$r = \operatorname*{Lim}_{\xi\to\infty\exp(-i\pi)} \left\{2\cos\pi\nu\, e^{-\pi kL}\, e^{ik\xi}\left(\frac{\pi}{2}k\xi\right)^{1/2} \exp\left\{i\left(\frac{\nu\pi}{2}+\frac{\pi}{4}\right)\right\} H_\nu^{(1)}(-k\xi)\right\}$$

$$= -2i\cos\pi\nu\, e^{-\pi kL}$$

$$\simeq -2i\cos(\pi\sqrt{(\alpha)}kL)e^{-\pi kL}, \quad kL \gg 1$$

in agreement with (64.27).

Let us observe, prior to concluding this section, that when $V(x)$ possesses an infinitesimal magnitude and $\psi(x)$ is identified with the incident (or unperturbed) wave function $\psi^{inc} = e^{ikx}$, the Fourier transform representation for the reflection coefficient implied by (79.4), namely

$$r(k) = \frac{i}{2k}\int_{-\infty}^{\infty} V(x)\, e^{2ikx}\, dx, \quad V(x) \to 0$$

has a reciprocal form

$$V(x) = \frac{2}{\pi i}\int_{-\infty}^{\infty} kr(k)\, e^{-2ikx}\, dk. \tag{79.19}$$

The latter relation finds a direct use in the so-called inverse scattering problem, that is, the reconstruction of the configurational profile $V(x)$ from a knowledge of the reflection coefficient $r(k)$ at all wave numbers k. For example, on designating the scale factor of the profile (79.9) by $V_0 = \alpha k^2$ and employing the exact determination (64.25) to secure the estimate

$$r(k) = \frac{\mathrm{i}\pi V_0 L^2}{\sinh \pi k L}, \quad V_0 \to 0$$

which incorporates the correct linear dependence on V_0, it follows from (79.19) that

$$\begin{aligned} V(x) &= 2V_0 L^2 \int_{-\infty}^{\infty} \frac{k}{\sinh \pi k L} \mathrm{e}^{-2\mathrm{i}kx}\, \mathrm{d}k \\ &= 2V_0 L^2 \frac{\mathrm{d}}{\mathrm{d}x} \int_0^{\infty} \frac{\sin 2kx}{\sinh \pi k L}\, \mathrm{d}k \\ &= V_0 L \frac{\mathrm{d}}{\mathrm{d}x} \tanh \frac{x}{L} = \frac{V_0}{\cosh^2 \dfrac{x}{L}}, \end{aligned}$$

as expected.

§80. *A Long Wave Approximation*

There exists another opportunity for constructing solutions of the respective equations,

$$\frac{\mathrm{d}^2\psi}{\mathrm{d}x^2} + k^2 n^2(x)\psi = 0, \tag{80.1}$$

$$-\infty < x < \infty$$

$$\frac{\mathrm{d}^2\psi}{\mathrm{d}x^2} + \frac{2m}{\hbar^2}(E - V(x))\psi = 0, \tag{80.2}$$

which encompasses a whole class of variable coefficients $n(x)$ and $V(x)$; and this is conditional, in the first instance, on small magnitudes of either kL or $(\sqrt{(2mE)}/\hbar)L$, where L is any length that characterizes the scale of variation for $n(x)$ or $V(x)$ separately. The solutions of (80.1), for example, are expressed by means of power series in k whose individual terms have explicit, multiple integral, representations that involve the coefficient function $n^2(x)$. Since powers of the integration variable also enter, convergence of the series warrants a further stipulation regarding the coefficient function, viz., if $n^2(x) \to 1$ as $|x| \to \infty$, then

$$\int_{-\infty}^{\infty} x^l \eta(x)\, \mathrm{d}x = \int_{-\infty}^{\infty} x^l (n^2(x) - 1)\, \mathrm{d}x < \infty \tag{80.3}$$

for all (positive) integer values of l.

The juxtaposition of k and x in the asymptotic forms

$$\psi \sim e^{\pm ikx}, \quad |x| \to \infty$$

of independent solutions to (80.1) implies that a small k—or long wave—expansion of the corresponding wave function ψ does not converge uniformly for all x. Uniformity of convergence on a semi-infinite range is an expected attribute of like expansions for the ratios $\psi/e^{\pm ikx}$ which possess definite (numerical) limiting values when $x \to +\infty$ or $-\infty$; and we may thus endeavor to realize, through an aggregate of linked expansions, an overall description of wave functions that have given asymptotic properties in the respective limits $x \to \pm\infty$. To exemplify the latter procedure, let us introduce a trio of normalized functions $\psi_i(x)$, $\psi_r(x)$, and $\psi_t(x)$, $-\infty < x < \infty$, such that

$$\psi_i \sim e^{ikx}, \qquad \psi_r \sim e^{-ikx}, \quad x \to -\infty, \tag{80.4}$$

$$\psi_t \sim e^{ikx}, \qquad x \to +\infty; \tag{80.5}$$

their designations evidently suggest (self-consistent) relative pairings, namely (ψ_i, ψ_r) with incident/reflected waves and (ψ_i, ψ_t) with incident/transmitted waves.

If we postulate an expansion of the form

$$\chi_t(x) = \psi_t(x)\, e^{-ikx} = \sum_{n=0}^{\infty} (ik)^n f_n(x), \quad x \geq 0 \tag{80.6}$$

substitute in (80.1) and collect terms which contain identical powers of k, it follows that the functions $f_n(x)$ satisfy a recursive system of differential equations

$$\begin{aligned} &f_0'' = 0, \qquad f_1'' = -2f_0' \\ &f_n'' = -2f_{n-1}' + \eta(x) f_{n-2}, \quad n \geq 2. \end{aligned} \tag{80.7}$$

Individual determinations of the f_n, based on (80.7) and the conditions

$$\begin{aligned} &f_0(\infty) = 1, \qquad f_0'(\infty) = 0, \\ &f_n(\infty) = f_n'(\infty) = 0, \quad n \geq 1 \end{aligned} \tag{80.8}$$

prompted by (80.5), are readily forthcoming, i.e.,

$$\begin{aligned} &f_0 = 1, \qquad f_1 = 0 \\ &f_2 = \int_\infty^x d\xi \int_\infty^\xi \eta(\xi')\, d\xi' \\ &f_3 = -2 \int_\infty^x d\xi \int_\infty^\xi d\xi' \int_\infty^{\xi'} \eta(\xi'')\, d\xi'' \\ &\ldots.; \end{aligned} \tag{80.9}$$

and we then secure a partial or half-range representation

$$\psi_t(x) = e^{ikx} \sum_{n=0}^{\infty} (ik)^n f_n(x), \quad x \geq 0 \tag{80.10}$$

of the wave function (80.5). The analogous representations

$$\left.\begin{aligned}\psi_i(x) &= e^{ikx}\sum_{n=0}^{\infty}(ik)^n g_n(x),\\ \psi_r(x) &= e^{-ikx}\sum_{n=0}^{\infty}(-ik)^n g_n(x)\end{aligned}\right\}\quad x\le 0 \tag{80.11}$$

are appropriate for the wave functions (80.4) if the coefficient function in (80.1) is real; and it suffices merely to replace ∞ by $-\infty$ in the expressions (80.9) for the purpose of specifying the individual functions $g_n(x)$. Absolute and uniform convergence of the series in (80.10), (80.11) may be proved (Drazin, 1962) when k is suitably bounded and $|\eta(x)|\le A\,e^{-a|x|}$, $-\infty<x<\infty$.

Linear interdependence of the functions ψ_i, ψ_r, and ψ_t is a consequence of the fact that these all satisfy the same differential equation. If we multiply the respective functions by constants A, B, C and assert the equality of a resultant pair, namely

$$\begin{aligned}\psi_-(x) &= A\psi_i(x)+B\psi_r(x),\\ \psi_+(x) &= C\psi_t(x),\end{aligned} \tag{80.12}$$

two of the constant factors are uniquely fixed in terms of the other one through the matching conditions

$$\psi_-(0)=\psi_+(0),\qquad \psi_-'(0)=\psi_+'(0). \tag{80.13}$$

When the incoming wave amplitude A is assigned, it follows from (80.12), (80.13) that the outgoing (or reflected and transmitted) wave amplitudes are given by

$$\frac{B}{A}=\frac{W(\psi_i,\psi_t)}{W(\psi_t,\psi_r)},\qquad \frac{C}{A}=\frac{W(\psi_i,\psi_r)}{W(\psi_t,\psi_r)} \tag{80.14}$$

where the Wronskian forms,

$$W(\varphi_1,\varphi_2)=\varphi_1\varphi_2'-\varphi_1'\varphi_2,$$

are evaluated at $x=0$. To obtain the developments, in ascending powers of k, for the reflection and transmission coefficients $r=B/A$, $t=C/A$, we first generate those of the Wronskians by utilizing (80.9)–(80.11); straightforward calculations reveal, in particular, that

$$\begin{aligned}W(\psi_i,\psi_t) &= \chi_i(0)[\chi_t'(0)+ik\chi_t(0)]-\chi_t(0)[\chi_i'(0)+ik\chi_i(0)]\\ &= -k^2(f_2'(0)-g_2'(0))-ik^3(f_3'(0)-g_3'(0))+0(k^4)\\ &= k^2\int_{-\infty}^{\infty}\eta(\xi)\,d\xi\\ &\quad +2ik^3\left\{\int_0^{\infty}d\xi\int_{\xi}^{\infty}\eta(\xi')\,d\xi'-\int_{-\infty}^{0}d\xi\int_{-\infty}^{\xi}\eta(\xi')\,d\xi'\right\}+0(k^4),\end{aligned}$$

$$W(\psi_t,\psi_r)=-2ik-k^2\int_{-\infty}^{\infty}\eta(\xi)+0(k^4),$$

whence

$$r = \frac{ik}{2}\int_{-\infty}^{\infty} \eta(\xi)\,d\xi - \frac{k^2}{4}\left\{\left(\int_{-\infty}^{\infty} \eta(\xi)\,d\xi\right)^2 + 4\int_0^{\infty} d\xi \int_{\xi}^{\infty} \eta(\xi')\,d\xi' - 4\int_{-\infty}^{0} d\xi \int_{-\infty}^{\xi} \eta(\xi')\,d\xi'\right\} + 0(k^4) \tag{80.15}$$

§81. *The Schrödinger Wave Equation and Potential Barrier Problems*

Schrödinger's formulation of quantum mechanics, interwoven with the classical theory of short wave propagation through non-uniform media, provides the one-dimensional wave equation

$$i\hbar\frac{\partial \Psi}{\partial t} = -\frac{\hbar^2}{2m}\frac{\partial^2 \Psi}{\partial x^2} + V(x)\Psi \tag{81.1}$$

for a particle of mass m and potential energy, $V(x)$, derived from external sources; here $\hbar = h/2\pi$ designates a quantum constant, named after Planck, with a universal magnitude and the dimensions of action or energy multiplied by time. If the (probabilistic) wave function $\Psi(x, t)$ is expressed in the product form

$$\Psi(x, t) = \psi(x)\,e^{-iEt/\hbar} \tag{81.2}$$

the coordinate factor $\psi(x)$ satisfies a second order differential equation

$$\frac{d^2\psi}{dx^2} + \frac{2m}{\hbar^2}(E - V(x))\psi = 0 \tag{81.3}$$

that manifestly resembles the equation

$$\frac{d^2\psi}{dx^2} + k^2 n^2(x)\psi = 0 \tag{81.4}$$

relating to excitations of an inhomogeneous continuum wherein $k(x) = kn(x)$ specifies the local wave number. Classical mechanics supplies an interpretation for the coefficient function in (81.3) (apart from the scale factor $\hbar^{-2}$), namely

$$2m(E - V(x)) = p^2,$$

where p designates the momentum of a particle having the total energy E; and thus, when $V(x) = 0$, the progressive wave functions representing free particles,

$$\Psi(x, t) = e^{i(\pm kx - \omega t)} \tag{81.5}$$

have the parameter assignments

$$k = \frac{\sqrt{(2mE)}}{\hbar} = \frac{p}{\hbar}, \qquad \omega = \frac{E}{\hbar} \tag{81.6}$$

which relate wave number and frequency to momentum and energy, respectively, in accordance with de Broglie's hypothesis.

Stationary or bound states for any individual potential energy distribution $V(x)$ are associated with square integrable solutions of the time-independent Schrödinger equation (81.3); and their energies are directly obtained from the corresponding determinations of the eigenvalue E. In the case of a linear oscillator, or mass acted on by a restoring force proportional to the distance x from a fixed point ($x = 0$), the potential energy is a quadratic function, say

$$V(x) = \frac{1}{2}Kx^2 = \frac{1}{2}m\omega^2 x^2, \quad -\infty < x < \infty \tag{81.7}$$

where the second version contains a parameter $\omega = \sqrt{(K/m)}$ that specifies (uniquely) the classical magnitude of the oscillator frequency in terms of the mass m and force constant K. It is a well documented result that the physically acceptable or bounded solutions of the relevant Schrödinger equation,

$$\frac{d^2\psi}{dx^2} + \left(\frac{2mE}{\hbar^2} - \frac{m^2\omega^2 x^2}{\hbar^2}\right)\psi = 0, \quad -\infty < x < \infty \tag{81.8}$$

take the form

$$\psi_n^{(1)}(x) = \exp\left\{-\frac{m\omega}{2\hbar}x^2\right\} P_n\left(\left(\frac{m\omega}{\hbar}\right)^{1/2} x\right) \tag{81.9}$$

where P_n is an nth order (Hermite) polynomial; and that the bound state energies of the oscillator are given by the expression

$$E_n = \left(n + \frac{1}{2}\right)\hbar\omega, \quad n = 0, 1, \ldots . \tag{81.10}$$

The asymptotic behavior of the wave function (81.9), viz.

$$\psi_n^{(1)}(x) \simeq x^n \exp\left(-\frac{m\omega}{2\hbar}x^2\right), \quad |x| \to \infty \tag{81.11}$$

contrasts with that of a linearly independent and unbounded solution to the differential equation (81.8) for the same energy eigenvalue E_n, viz.

$$\psi_n^{(2)}(x) \simeq x^{-(n+1)} \exp\left(\frac{m\omega}{2\hbar}x^2\right), \quad |x| \to \infty. \tag{81.12}$$

Wave functions of a different type are encountered when $V(x) \neq 0$ on a limited range Δ and vanishes elsewhere; these include composite forms indicative of free particle reflection from and transmission through Δ, as in the scattering of waves on a string by an inhomogeneous segment. According to classical mechanics, particles cannot penetrate (and are thus totally reflected from) regions where the potential energy $V(x)$ has a magnitude in excess of their assigned total energy E, since negative values of the kinetic energy $E - V(x)$ are ruled out; in wave mechanics, however, transmission or tunnelling through potential barriers is feasible because solutions of the Schrödinger equation (81.3) are equally well defined in regions where $E - V(x) \gtrless 0$, and can be appropriately connected at their separation points to achieve an acceptable overall determination.

Let us detail various aspects of scattering by a barrier with a truncated oscillator potential, i.e.,

$$V(x) = \begin{array}{ll} \frac{1}{2}Kx^2 = \frac{1}{2}m\omega^2x^2, & |x| \leq L \\ 0, & |x| > L \end{array} \tag{81.13}$$

when a uniform beam of free particles is incident from the left-hand side (see Fig. 47). We shall assume, firstly, that the energy E, common to the incident, reflected ($x < -L$) and transmitted ($x > L$) beams, satisfies the condition

$$E \ll V_{\max} = \frac{1}{2}KL^2 \tag{81.14}$$

and, secondly, that the asymptotic representations for all solutions of the Schrödinger equation inside the barrier are applicable at $x = \pm L$; the inequality

$$L \gg \surd(\hbar/m\omega), \tag{81.15}$$

which may be relied on in support of the latter assumption, attributes to the potential (81.13) a breadth substantially greater than the turning point separation for a classical oscillator whose energy $E_0 = \hbar\omega/2$ has the least magnitude in the wave mechanical spectrum (81.10).

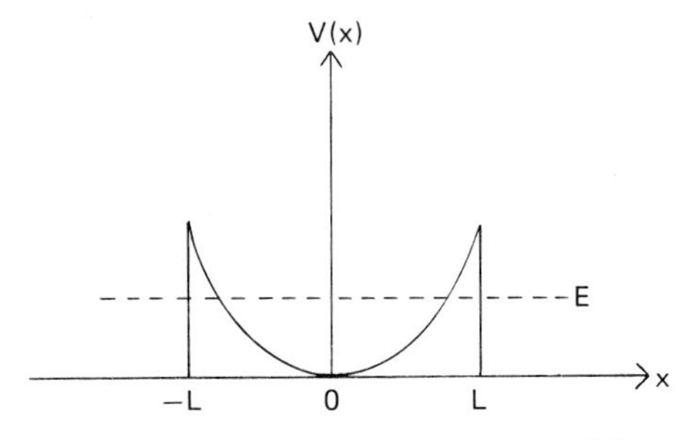

Fig. 47. A truncated parabolic potential.

Outside the barrier the appropriate free particle wave functions are

$$\psi(x) = \begin{matrix} e^{ikx} + r\,e^{-ikx}, & x < -L \\ t\,e^{ikx}, & x > L \end{matrix}, \tag{81.16}$$

incorporating a given wave number

$$k = \frac{2\pi}{\lambda} = \frac{\sqrt{(2mE)}}{\hbar}$$

along with the unspecified reflection and transmission coefficients r, t. For the purpose of determining r and t we are obliged to join the respective functions in (81.16), at the boundaries $x = \pm L$ of the barrier, with Schrödinger wave functions characteristic of the inner potential; it may be anticipated, and subsequently verified, that the procedure yields significantly different results according as the incident beam energy E coincides with one of the explicit values E_n (81.10) for bound states of a complete oscillator, or not.

If the oscillator equation (81.8) is recast in terms of the parameter

$$\delta = \frac{2E}{\hbar\omega} \tag{81.17}$$

and a new independent variable, $\xi = (m\omega/\hbar)^{1/2}x$, we find

$$\frac{d^2\psi}{d\xi^2} + (\delta - \xi^2)\psi = 0 \tag{81.18}$$

and thus, introducing the representations

$$\psi(\xi) = \varphi_{\mp}(\xi)\exp(\mp\xi^2/2), \tag{81.19}$$

it follows that

$$\frac{d^2\varphi_-}{d\xi^2} - 2\xi\frac{d\varphi}{d\xi} - (1-\delta)\varphi_- = 0, \tag{81.20}$$

$$\frac{d^2\varphi_+}{d\xi^2} + 2\xi\frac{d\varphi_+}{d\xi} + (1+\delta)\varphi_+ = 0. \tag{81.21}$$

Taking note of the particular specifications

$$\varphi_-(\xi) \sim \xi^{(\delta-1)/2}, \quad \xi \to +\infty, \qquad \varphi_+(\xi) \sim |\xi|^{-(\delta+1)/2}, \quad \xi \to -\infty,$$

consistent with (81.20) and (81.21), respectively, we may therefore define a pair of real and linearly independent solutions to the Schrödinger equation, $\psi^{(1)}(x)$ and $\psi^{(2)}(x) = \psi^{(1)}(-x)$, which have the asymptotic behaviors

$$\psi^{(1)}(x) \simeq \begin{cases} (x/L)^{(\delta-1)/2}\exp\left(-\dfrac{m\omega}{2\hbar}x^2\right), & x \to +\infty, \\ C|x/L|^{-(\delta+1)/2}\exp\left(\dfrac{m\omega}{2\hbar}x^2\right), & x \to -\infty, \end{cases} \tag{81.22}$$

$$\psi^{(2)}(x) \simeq \begin{cases} C(x/L)^{-(\delta+1)/2} \exp\left(\dfrac{m\omega}{2\hbar}x^2\right), & x \to +\infty, \\ |x/L|^{(\delta-1)/2} \exp\left(-\dfrac{m\omega}{2\hbar}x^2\right), & x \to -\infty \end{cases} \tag{81.23}$$

where C denotes a constant whose actual value has no special significance in what follows. If $\psi^{(1)}(x)$ and $\psi^{(2)}(x)$ are chosen as basis functions, a general solution of the Schrödinger equation within the potential barrier is provided by the combination

$$\psi(x) = A_1\psi^{(1)}(x) + A_2\psi^{(2)}(x) \tag{81.24}$$

that involves arbitrary coefficients.

A continuous variation for both ψ and its derivative ψ' (whose product relates to the particle current density) accords with the physical interpretation of Schrödinger wave functions in the present circumstances; and is assured everywhere, once the complementary representations (81.16), (81.24) are joined together in the requisite manner at the points of division, $x = \pm L$. The four relations yielded by the latter procedure enable us, moreover, to determine r, t, A_1, A_2 and thence an overall wave function which describes the scattering of a monoenergetic beam of particles by the potential barrier. After we utilize expressions for $\psi^{(1)}(\pm L \mp 0)$, $\psi^{(2)}(\pm L \mp 0)$ taken directly from (81.22), (81.23), and also avail ourselves of the concurrent estimates

$$\left.\begin{aligned} \psi^{(1)\prime}(\pm L \mp 0) &\doteq -\frac{m\omega}{\hbar}L\psi^{(1)}(\pm L \mp 0) \\ \psi^{(2)\prime}(\pm L \mp 0) &\doteq \frac{m\omega}{\hbar}L\psi^{(2)}(\pm L \mp 0) \end{aligned}\right\}, \quad \frac{m\omega}{\hbar}L^2 \gg 1 \tag{81.25}$$

there results, in particular, a transmission coefficient

$$t = \frac{4}{iC}\left(\frac{\hbar k}{m\omega L}\right)\frac{e^{-2ikL}}{\left(1 - i\dfrac{\hbar k}{m\omega L}\right)^2}\exp\left(-\frac{m\omega}{\hbar}L^2\right) \tag{81.26}$$

with an exponentially small magnitude, since

$$\frac{m\omega}{\hbar}L^2 \gg 1, \qquad \frac{\hbar k}{m\omega L} = \sqrt{(E/V_{max})} \ll 1. \tag{81.27}$$

The preceding conclusion and its corollary, viz. that the reflection coefficient differs from unity by an exponentially small amount, is significantly altered when the incident beam has an energy equal to that of any bound state for the infinitely extended wave mechanical oscillator. If this coincidence or resonance occurs, the parameter defined by (81.17) assumes an integral value, namely

$$\delta_n = 2n+1, \qquad n = 0, 1, \dots \tag{81.28}$$

and the differential equation (81.18) admits a pair of linearly independent solutions such that

$$\begin{aligned} \psi_n^{(1)}(x) &\simeq (x/L)^n \exp\left(\frac{m\omega}{2\hbar}x^2\right), \\ \psi_n^{(2)}(x) &\simeq (x/L)^{-(n+1)} \exp\left(\frac{m\omega}{2\hbar}x^2\right), \end{aligned} \qquad |x| \to \infty, \quad n = 0, 1, \dots \tag{81.29}$$

where $\psi_n^{(1)}$ is evidently an asymptotic version of the oscillator wave function (81.9). On adopting the representation

$$\psi_n(x) = \bar{A}_1 \psi_n^{(1)}(x) + \bar{A}_2 \psi_n^{(2)}(x), \quad |x| < L \tag{81.30}$$

in place of (81.24) and matching this function with those outside the barrier, as before, it follows that

$$\begin{aligned} t_n &= 2\,\mathrm{i}(-1)^n \frac{(\hbar k_n/m\omega L)}{1+\left(\dfrac{\hbar k_n}{m\omega L}\right)^2} \mathrm{e}^{-2\,\mathrm{i}k_n L} \\ &= 2\,\mathrm{i}(-1)^n \frac{(E_n/V_{\max})^{1/2}}{1+\left(\dfrac{E_n}{V_{\max}}\right)} \mathrm{e}^{-2\,\mathrm{i}k_n L}. \end{aligned} \qquad \begin{aligned} (\hbar k_n)^2 &= 2mE_n \\ &= (2n+1)m\hbar\omega \end{aligned} \tag{81.31}$$

Although t_n possesses a small absolute value, in consequence of the prevailing inequality $E_n/V_{\max} \ll 1$, this exceeds – by a large exponential factor – the same measure of the corresponding transmission coefficient estimate (81.26) when $E \neq E_n$; and we may thus infer a comparatively stronger interaction between the incident beam and the limited oscillator in the resonance case.

More generally, the analysis of barrier problems requires solution of the differential equation (81.4) in the circumstance that the coefficient function (or carrier) $k^2n^2(x)$ assumes real and negative values on one or more distinct ranges, $x_1 < x < x_2, x_3 < x < x_4$, and is real and positive elsewhere. Approximation procedures, which are obligatory for all but a limited number of such problems, apart from whether or not $n^2(x)$ changes sign, meet with particular technical complications when it comes to specifying and then linking solutions of different (viz., exponential and oscillatory) character on opposite sides of the individual transition points where $n^2(x)$ vanishes. If we refer to a pair of approximate solutions inside the barrier range(s), with exponentially large and small magnitudes, respectively, and observe that corrections to the former can exceed the latter, a need for subtlety in handling and interpreting their linear combination becomes evident. The literature (methodically surveyed by Heading, 1962) documents analyses and results based on solutions of WKBJ type and also includes an account of the related equation approach (Fröman and Fröman, 1965). Heading (1972) demonstrates the advantage of utilizing a formal representation (cf. §60), that features the ratio of independent solutions with disparate magnitudes, in order to confirm a reflection coefficient estimate

$$|r| \doteq 1 - \frac{1}{2}\exp\left(-2k\int_{x_1}^{x_2} \surd(-n^2(\xi))\,\mathrm{d}\xi\right), \quad k(x_2 - x_1) \gg 1$$

for the single barrier located on the range (x_1, x_2).

§82. *Inverse Scattering Theory*

The direct scattering problems hitherto dealt with, which aim at a description of the secondary waves resulting from given systemic irregularities (such as loads or inhomogeneous segments), have counterparts or inverses wherein the object is that of deducing properties of the scatterer when the secondary wave features are known. An important class of inverse problems acquire their formulation in terms of the one-dimensional and stationary Schrödinger equation with a velocity dependent potential V, viz.:

$$\frac{\mathrm{d}^2\psi}{\mathrm{d}\zeta^2} + \omega^2\psi = V(\zeta, \omega)\psi, \tag{82.1}$$

where $\psi = \psi(\zeta, \omega)$ denotes the wave function and the chosen independent variable $\zeta = x/c$ measures the travel time over a segment x at the speed c. Since a pair of complex numbers, namely the reflection and transmission coefficients corresponding to the incident wave $e^{i\omega\zeta}$, comprise the full particulars of the scattering at any frequency in a one-dimensional abstraction, there is no likelihood of determining the potential unless these coefficients are given for all frequencies. Additional information must be supplied, as a matter of fact, because it is in general impossible to characterize $V(\zeta, \omega)$, a function of two variables, by specifying two functions of one variable each. Thus, a realistic attack on the inverse scattering problem awaits simple initial assumptions about the form of the potential; and the respective choices

i) $V(\zeta, \omega) = V_1(\zeta)$ [i.e., frequency independence]

or

ii) $V(\zeta, \omega) = V_1(\zeta) - i\omega V_2(\zeta)$ [i.e., linear dependence on the frequency]

are exclusively featured in the existing literature.

A lossy electrical medium provides another model for the exploration of configurational details through scattering data, and a suitable one-dimensional wave equation is

$$\left[\frac{\partial^2}{\partial x^2}-\frac{1}{c^2}\varepsilon_r(x)\frac{\partial^2}{\partial t^2}-\frac{\sigma(x)}{c^2\varepsilon_0}\frac{\partial}{\partial t}\right]E(x,t)=0; \tag{82.2}$$

here $E(x, t)$ defines a (real) electric field component normal to the direction of propagation (x), $\varepsilon_r(x)$, and $\sigma(x)$ denote the relative dielectric constant and conductivity of the medium, respectively, while ε_0 and c register the values of the permittivity and wave speed in free space. For a reason akin to that advanced in circumscribing the behavior of the Schrödinger potential function $V(\zeta, \omega)$, an assumption which implicitly shapes the explicit form of (82.2) is that retardation effects (i.e., frequency variation) are without influence on the medium parameters $\varepsilon_r(x)$ and $\sigma(x)$.

If $\bar{E}(x, \omega)$ represents the Fourier transform of the electric field, namely

$$\bar{E}(x,\omega)=\frac{1}{2\pi}\int_{-\infty}^{\infty}E(x,t)\,\mathrm{e}^{\mathrm{i}\omega t}\,\mathrm{d}t, \tag{82.3}$$

it follows from (82.2) that

$$\frac{\mathrm{d}^2}{\mathrm{d}x^2}\bar{E}(x,\omega)+\frac{\omega^2}{c^2}\left[\varepsilon_r(x)+\frac{\mathrm{i}\sigma(x)}{\omega\varepsilon_0}\right]\bar{E}(x,\omega)=0; \tag{82.4}$$

and, consequent to the joint (Liouville) transformations of both the independent and dependent variables

$$\frac{\mathrm{d}\zeta}{\mathrm{d}x}=\eta^2(x)=\frac{1}{c}\varepsilon_r(x) \tag{82.5}$$

$$\psi(\zeta,\omega)=\eta(x)\bar{E}(x,\omega), \tag{82.6}$$

the differential equation (82.4) becomes

$$\frac{\mathrm{d}^2}{\mathrm{d}\zeta^2}\psi(\zeta,\omega)+\omega^2\psi(\zeta,\omega)=[V_1(\zeta)-\mathrm{i}\omega V_2(\zeta)]\psi(\zeta,\omega) \tag{82.7}$$

where

$$V_1(\zeta)=\frac{1}{\eta}\frac{\mathrm{d}^2\eta}{\mathrm{d}\zeta^2}=\varepsilon_r^{-1/4}\frac{\mathrm{d}^2}{\mathrm{d}\zeta^2}\varepsilon_r^{1/4}$$

and (82.8)

$$V_2(\zeta)=\frac{\sigma}{\varepsilon_0\varepsilon_r}.$$

Accordingly, in its latter version (which involves a pair of real functions $V_{1,2}(\zeta)$ and another travel time variable, ζ, reckoned on the basis of a non-uniform wave speed $c/\varepsilon_r(x)$), the lossy electromagnetic and wave mechanical or Schrödinger problem with a two-part potential (case ii) have a common mathematical expression. The linkage of both problems with a one-dimensional setting reflects a deliberate bias, inasmuch as very little is known about inverse scattering in higher dimensions. Spherically symmetric potentials invariably enter into three-dimensional inverse scattering theory and the concomitant Schrödinger equation again takes the form (82.1), though ζ assumes only non-negative values.

To gain some appreciation of how structural details can be inferred from external wave measurements rather than internal probes, consider a stratified medium formed by individually uniform layers. Suppose that $\sigma(x) \equiv 0$ and

$$\varepsilon_r(x) = \begin{matrix} \varepsilon_0, & x < x_1, & x > x_N \\ \varepsilon_i, & x_i < x < x_{i+1}, & i = 1, 2, \ldots, N-1 \end{matrix} \tag{82.9}$$

where ε_i specifies the relative dielectric constant of the lossless i^{th} layer; then there exists a smooth or continuously differentiable solution of the equation

$$\frac{d^2}{dx^2}\bar{E} + \frac{\omega^2}{c^2}\varepsilon_r(x)\bar{E} = 0$$

whose value at $x = x_1 = 0$, say, is rendered by the series

$$\begin{aligned} \bar{E}(0, \omega) = 1 + r_1 + r_2(1 - r_1^2)\,\mathrm{e}^{2\,\mathrm{i}\varphi_1} + r_3(1 - r_1^2)(1 - r_2^2)\,\mathrm{e}^{2\,\mathrm{i}(\varphi_1 + \varphi_2)} + \cdots \\ - (1 - r_1^2) r_1 r_2^2\,\mathrm{e}^{4\,\mathrm{i}\varphi_1} - \cdots \\ - r_1(1 - r_1^2)(1 - r_2^2) r_3^2\,\mathrm{e}^{4\,\mathrm{i}(\varphi_1 + \varphi_2)} - \cdots \end{aligned} \tag{82.10}$$

containing single interface reflection coefficients

$$r_{i+1} = \frac{1 - \sqrt{(\varepsilon_{i+1}/\varepsilon_i)}}{1 + \sqrt{(\varepsilon_{i+1}/\varepsilon_i)}} \tag{82.11}$$

and the 'electrical' lengths of individual layers,

$$\varphi_i = \frac{\omega}{c}\sqrt{(\varepsilon_i)}(x_{i+1} - x_i). \tag{82.12}$$

The first term in (82.10) befits an incoming (from $x = -\infty$) or incident wave of unit amplitude, namely $\mathrm{e}^{\mathrm{i}\omega x/c}$, and the others identify outgoing (towards $x = -\infty$) wave contributions whose amplitudes and phases are consistent with definite sequences of partial reflection inside the layered medium. From the time-dependent affiliate of (82.10), which obtains after inverse Fourier transformation, i.e.,

$$E(0, t) = \delta(t) + r_1\delta(t) + r_2(1 - r_1^2)\delta\left(t - 2\frac{\sqrt{(\varepsilon_1)}}{c}(x_2 - x_1)\right) + \cdots, \tag{82.13}$$

it appears that the electric field at the (outside) surface of the medium exhibits a recurring series of localized pulses, with arrival times linked to the prior number of internal reflections and the appertaining scale or range of the disturbance. Thus, observation of the reflected pulses and reference to the individual terms of the series (82.13) enables a resolution of the inverse scattering problem for the layered structure; in particular, the amplitudes of the earliest and succeeding reflected pulses determine, respectively, the dielectric constants ϵ_1 and ϵ_2 of the first and second layers, while the time interval between their reception permits a calculation of the layer breadth $x_2 - x_1$. Properties of the medium, from its boundary as far as some interior layer, are describable after a finite and specific period of observation, initiated when the incident pulse reaches the surface and concluded with the arrival thereat of a disturbance which has travelled to this layer and back. It is also significant that the structure resolution can be effected on the basis of information about a single element of the scattering matrix, namely a reflection coefficient for incidence on one side thereof; reciprocity and unitarity (in the lossless case) of said matrix permit the deduction of other elements.

Early contributions to inverse scattering theory (e.g., Levinson, 1947) dwelt upon the matters of uniqueness and construction of potentials, given the phase shifts or an asymptotic form of the Schrödinger wave function; Gelfand and Levitan (1957) ushered in a more active phase of the subject with a paper, on three-dimensional wave mechanical scattering by a spherically symmetric and velocity (frequency) independent potential (choice i, above), which relates the solution of the inverse scattering problem to that of a single linear integral equation. Several years afterwards, Faddeev (1967) improved on the latter reformulation and gave a version suitable for one-dimensional problems wherein the variable ζ takes on all positive and negative values. Schmidt and Wu (1974) have recently shown how to bring dissipation, or loss attributable to a frequency proportional component of the potential function (choice ii, above), within the scope of a Gelfand-Levitan theory. Since a considerable amount of complex analysis underlies relations of the Gelfand–Levitan type (cf. Roseau, 1976), their presentation – rather than derivation – is offered here.

Let us assume that the Schrödinger equation (82.1) admits, when the potential $V(\zeta, \omega)$ is non-zero over a finite range of the variable ζ, an independent pair of solutions, $f_1(\zeta, \omega)$ and $f_2(\zeta, \omega)$, possessing the distinct asymptotic behaviors

$$f_1(\zeta, \omega) \sim \begin{cases} e^{i\omega\zeta} + \hat{S}_{12}(\omega)\, e^{-i\omega\zeta}, & \zeta \to -\infty, \\ \hat{S}_{11}(\omega)\, e^{i\omega\zeta}, & \zeta \to \infty, \end{cases} \tag{82.14}$$

$$f_2(\zeta, \omega) \sim \begin{cases} \hat{S}_{22}(\omega)\, e^{i\omega\zeta}, & \zeta \to -\infty, \\ e^{-i\omega\zeta} + \hat{S}_{21}(\omega)\, e^{i\omega\zeta}, & \zeta \to \infty. \end{cases} \tag{82.15}$$

The oppositely directed incident wave terms $e^{i\omega\zeta}$ and $e^{-i\omega\zeta}$ are identifiable as such in the excitations characterized by the preceding solutions, and the respective pairs of factors $\hat{S}_{12}$, $\hat{S}_{21}$ and $\hat{S}_{11}$, $\hat{S}_{22}$ embody reflection and transmission coefficients on the left and right; these coefficients enter, as an ensemble, into a 2×2 matrix, $\hat{S}(\omega)$, and correspond, in the individual manners

$$S_{11} = \hat{S}_{12}, \qquad S_{12} = \hat{S}_{22}, \qquad S_{21} = \hat{S}_{11}, \qquad S_{22} = \hat{S}_{21}$$

with elements of a previously defined scattering matrix, S [§48]. Analytic continuation of the matrix elements, from real to complex values of ω, is feasible and vital for inverse scattering theory. Causality, which holds that secondary waves appear after the primary ones reach the site of configurational changes, or come within the range of the potential, directly affects the behavior of said continuations; and its particular implication is that $\hat{S}_{12}(\omega)$ possesses a bounded and analytic character in the upper half of the complex ω-plane if $V(\zeta, \omega) = 0$, $\zeta < 0$, arranged for by a variable translation. [Poles of the reflection coefficient may occur, in conjunction with bound states of wave mechanical systems.]

The representation

$$\hat{S}_{12}(\omega) = \int_0^\infty \hat{R}(t)\, e^{2\, i\omega t}\, dt \tag{82.16}$$

secures the cited behavior of the reflection coefficient and, moreover, introduces the concomitant transform, $\hat{R}(t)$, which has a conspicuous place in the Gelfand-Levitan theory. Given $\hat{R}(t)$, the latter furnishes an integral equation

$$\hat{R}(\zeta + y) + \int_{-\zeta - y}^{0} \hat{R}(\zeta + y + z) B(\zeta, z)\, dz + B(\zeta, y) = 0, \quad y < 0 \tag{82.17}$$

whose solution, $B(\zeta, y)$, specifies the potential function $V_1(\zeta)$ in accordance with the explicit formula

$$V_1(\zeta) = \frac{d}{d\zeta} B(\zeta, 0). \tag{82.18}$$

When two components, $V_1(\zeta)$ and $V_2(\zeta)$, of a potential function that is linear in ω (choice ii) are sought after, it no longer suffices to employ a single reflection coefficient or the transform, $R(t)$, thereof; a substitute for, and generalization of, the Gelfand–Levitan equation (82.17) in this situation necessitates the deployment of a second reflection coefficient on the same side of the scattering range. Since the originally conceived problem affords no such quantity, the construction of another (and possibly fictitious) problem becomes indicated, bearing in mind the guidelines that 1) a proper transition to the dissipationless limit ($V_2(\zeta) = 0$) obtains, and that 2) the second reflection coefficient, with suitable analytic behavior, can be expressed directly in terms of the matrix $\hat{S}$. Schmidt and Wu suggest, in this regard, the use of an adjoint problem, featuring the differential equation [compare with (82.7)]

$$\frac{d^2}{d\zeta^2} \tilde{\psi}(\zeta, \omega) + \omega^2 \tilde{\psi}(\zeta, \omega) = [V_1(\zeta) + i\omega V_2(\zeta)] \tilde{\psi}(\zeta, \omega) \tag{82.19}$$

and the relation (where † signifies the Hermitian adjoint)

$$\hat{S}(\omega) \tilde{S}^\dagger(\omega) = \tilde{S}^\dagger(\omega) \hat{S}(\omega) = 1 \tag{82.20}$$

between matrices of the separate problems. The element $\tilde{S}_{12}(\omega)$ is analytic in the upper half of the ω-plane, provided that the loss remains moderate, and thus admits a representation analogous to (82.16), i.e.,

$$\tilde{S}_{12}(\omega) = \int_0^\infty \tilde{R}(t)\, e^{2i\omega t}\, dt. \tag{82.21}$$

A knowledge of both $\hat{R}$ and $\tilde{R}$, jointly gained from the scattering matrix $\tilde{S}$, imparts to the Gelfand–Levitan system of integral and differential equations,

$$\begin{aligned}
&[1+B_0(\zeta)]\hat{R}(\zeta+y)+\int_{-\zeta-y}^{0}\hat{R}(\zeta+y+z)B(\zeta,z)\,dz+C(\zeta,y)=0, \quad y<o,\\
&\frac{\tilde{R}(\zeta+y)}{1+B_0(\zeta)}+\int_{-\zeta-y}^{0}\tilde{R}(\zeta+y+z)C(\zeta,z)\,dz+B(\zeta,y)=0, \quad y<0,
\end{aligned} \tag{82.22}$$

$$2B_0'(\zeta)+B(\zeta,0)-C(\zeta,0)[1+B_0(\zeta)]^2=0,$$

a self-determining character as regards the three unknowns, $B(\zeta, y)$, $C(\zeta, y)$, and $B_0(\zeta)$. Consequent to a resolution of the system, that of the inverse scattering problem is reached through the expressions

$$\begin{aligned}
V_1(\zeta) &= \frac{B_0''(\zeta)}{1+B_0(\zeta)}+\frac{d}{d\zeta}\left[\frac{B(\zeta,0)}{1+B_0(\zeta)}\right],\\
V_2(\zeta) &= \frac{2B_0'(\zeta)}{1+B_0(\zeta)}.
\end{aligned} \tag{82.23}$$

§83. *Variational Theory and its Implications*

To establish, in their primary or differential form, the basic equations governing particular movements of mass continua we have relied on a direct mathematical transcription of physical concepts regarding the indestructibility of matter and the role of forces; and afterwards derived equations with a secondary rank, such as those pertaining to energy and momentum, through algebraic manipulations. A considerable significance attaches, therefore, to the fact that all these dynamical equations may be individually characterized in variational terms, as is also the case when electromagnetic fields and forces are present.

Let us recall that a mechanical system with a finite number of degrees of freedom is completely described by a Lagrange function L which renders the value of $\int_{t_1}^{t_2} L\, dt$ an extremum (or merely stationary) for the actual motion of the system, in comparison with all neighboring states that meet the (generalized) coordinate assignments at the times t_1, t_2; hence, a compact expression for the Lagrange equations of motion of the system, which comprise a set of ordinary differential equations that determine the temporal change of its coordinates, is afforded by Hamilton's principle,

$$\delta \int_{t_1}^{t_2} L \, \mathrm{d}t = 0, \tag{83.1}$$

where the symbol δ embraces the totality of independent coordinate variations.

If matter is continuously distributed in one, two, or three dimensions, it becomes appropriate to regard L as an integral of the Lagranian density, $\mathcal{L}$, with the requisite number of coordinate (and integration) variables. Assuming that $\mathcal{L}$ depends on but one state variable, $\varphi(x, t)$, together with its first partial derivatives, viz.

$$\mathcal{L} = \mathcal{L}\left(\varphi, \frac{\partial\varphi}{\partial x}, \frac{\partial\varphi}{\partial t}, x, t\right) = \mathcal{L}(\varphi, \varphi_x, \varphi_t, x, t),$$

variations of the Lagrange function resulting from those of φ within a fixed coordinate interval $x_1 < x < x_2$ are given by

$$\begin{aligned}\delta L = \delta \int_{x_1}^{x_2} \mathcal{L} \, \mathrm{d}x &= \int_{x_1}^{x_2} \left\{ \frac{\partial\mathcal{L}}{\partial\varphi}\delta\varphi + \frac{\partial\mathcal{L}}{\partial\varphi_x}\delta\varphi_x + \frac{\partial\mathcal{L}}{\partial\varphi_t}\delta\varphi_t \right\} \mathrm{d}x \\ &= \int_{x_1}^{x_2} \left\{ \frac{\partial\mathcal{L}}{\partial\varphi} - \frac{\partial}{\partial x}\left(\frac{\partial\mathcal{L}}{\partial\varphi_x}\right) - \frac{\partial}{\partial t}\left(\frac{\partial\mathcal{L}}{\partial\varphi_t}\right) \right\} \delta\varphi \, \mathrm{d}x + \frac{\mathrm{d}}{\mathrm{d}t}\int_{x_1}^{x_2} \frac{\partial\mathcal{L}}{\partial\varphi_t}\delta\varphi \, \mathrm{d}x\end{aligned} \tag{83.2}$$

where the latter version is based on the operational equivalences

$$\delta\varphi_x = \frac{\partial}{\partial x}\delta\varphi, \qquad \delta\varphi_t = \frac{\partial}{\partial t}\delta\varphi$$

and the hypothesis that

$$\delta\varphi = 0, \qquad x = x_1, x_2.$$

Thus, if φ is also prescribed at the instants of time t_1, t_2, or

$$\delta\varphi = 0, \qquad t = t_1, t_2$$

the single condition

$$\frac{\partial}{\partial x}\left(\frac{\partial\mathcal{L}}{\partial\varphi_x}\right) + \frac{\partial}{\partial t}\left(\frac{\partial\mathcal{L}}{\partial\varphi_t}\right) - \frac{\partial\mathcal{L}}{\partial\varphi} = 0 \tag{83.3}$$

expresses a stationary character of the integral

$$\int_{t_1}^{t_2}\int_{x_1}^{x_2} \mathcal{L}\left(\varphi, \frac{\partial\varphi}{\partial x}, \frac{\partial\varphi}{\partial t}, x, t\right) \mathrm{d}x \, \mathrm{d}t \tag{83.4}$$

for arbitrary variations of φ on the open domain $x_1 < x < x_2, t_1 < t < t_2$; accordingly, the variational or Hamilton's principle supplies, in this instance, a partial differential equation, (83.3), which determines the state variable $\varphi(x, t)$.

We may identify (83.3) as the inhomogeneous linear wave equation for small transverse displacements $y(x, t)$ of a string with the density $\rho(x)$ and tension $P(t)$, acted on by a given external force (per unit length) $\mathfrak{F}(x, t)$, that is,

$$\rho(x)\frac{\partial^2 y}{\partial t^2} - P(t)\frac{\partial^2 y}{\partial x^2} = \mathfrak{F}(x, t), \tag{83.5}$$

if the Lagrangian density has the form

$$\mathscr{L}\left(y, \frac{\partial y}{\partial x}, \frac{\partial y}{\partial t}, x, t\right) = \frac{1}{2}\rho(x)\left(\frac{\partial y}{\partial t}\right)^2 - \frac{1}{2}P(t)\left(\frac{\partial y}{\partial x}\right)^2 + \mathfrak{F}(x, t)y \tag{83.6}$$

whose individual terms register kinetic or potential energy contributions.

It is noteworthy that stationary principles also afford a description of dissipative systems; for example, the equation which relates to damped motions of a (uniform) string, namely

$$\frac{\partial^2 y}{\partial t^2} + \gamma\frac{\partial y}{\partial t} + (hc)^2 y = c^2\frac{\partial^2 y}{\partial x^2}; \qquad c^2 = \frac{P}{\rho}, \quad \gamma > 0, \tag{83.7}$$

expresses a stationary property of the double integral

$$I = \int\left\{\frac{1}{2}\rho\left(\frac{\partial y}{\partial t}\right)^2 - \frac{1}{2}P\left(\frac{\partial y}{\partial x}\right)^2 - \frac{1}{2}Ph^2y^2\right\}e^{\gamma t}\,dx\,dt. \tag{83.8}$$

The integrand in (83.8), or Lagranian density, contains terms without a ready physical interpretation and manifests a direct exponential dependence on the time.

Variational characterizations are not conditional on a small amplitude (or slope) approximation in the case of string or other dynamical motions and they encompass, more generally, systems of equations. Thus, if we consider the planar modes of a homogeneous flexible string whose length is fixed and locate material points thereupon by the pair of coordinates $x(s, t)$, $y(s, t)$, where s denotes the arc length, Hamilton's principle asserts that (in the absence of external forces) the kinetic energy integral

$$\int\frac{1}{2}\rho\left\{\left(\frac{\partial x}{\partial t}\right)^2 + \left(\frac{\partial y}{\partial t}\right)^2\right\}ds\,dt$$

is stationary for variations of both x and y, subject to the equation of constraint

$$\left(\frac{\partial x}{\partial s}\right)^2 + \left(\frac{\partial y}{\partial s}\right)^2 = 1. \tag{83.9}$$

On utilizing a Lagrange multiplier, $P/2$, in connection with the latter, an alternative version of the stationary principle obtains, namely

$$\delta\int\left[\frac{1}{2}\rho\left\{\left(\frac{\partial x}{\partial t}\right)^2 + \left(\frac{\partial y}{\partial t}\right)^2\right\} - \frac{1}{2}P\left\{\left(\frac{\partial x}{\partial s}\right)^2 + \left(\frac{\partial y}{\partial s}\right)^2 - 1\right\}\right]ds\,dt = 0, \tag{83.10}$$

and the resultant trio of equations, reflecting independent variations of x, y, and P between the fixed limits of integration, include the dynamical pair

$$\rho\frac{\partial^2 x}{\partial t^2}=\frac{\partial}{\partial s}\left(P\frac{\partial x}{\partial s}\right),\qquad \rho\frac{\partial^2 y}{\partial t^2}=\frac{\partial}{\partial s}\left(P\frac{\partial y}{\partial s}\right) \tag{83.11}$$

together with the kinematical constraint. These equations make up a fundamental system which lends itself to specifying x, y, and P, and thence self-consistent motions of the string.

As intimated earlier, differential conservation relations (for energy, momentum, ...) can also be fitted into the variational scheme; such relations are, in fact, consequences of the invariance of Hamilton's principle under specific transformations like temporal displacements, coordinate translations, ... Let us first bring out some of the pertinent aspects, with a minimum of formal preliminaries, by referring to a Lagrangian density in which the time does not explicitly enter. Since a change of the time origin is then without effect on the integrated value of L between any fixed limits we have, symbolically,

$$\int_{t_1-\delta t}^{t_2-\delta t} L(t+\delta t)\,\mathrm{d}t-\int_{t_1}^{t_2} L(t)\,\mathrm{d}t=0 \tag{83.12}$$

where δt is an arbitrary constant; and it follows that the approximation to (83.12) which is of first order in δt, i.e.,

$$\delta t\frac{\mathrm{d}}{\mathrm{d}\varepsilon}\left[\int_{t_1-\varepsilon}^{t_2-\varepsilon} L(t+\varepsilon)\,\mathrm{d}t\right]_{\varepsilon=0}=\delta t\left[-L(t_2)+L(t_1)+\int_{t_1}^{t_2}(\mathrm{d}L/\mathrm{d}t)\,\mathrm{d}t\right]=0$$

warrants the status of an identity, for

$$\int_{t_1}^{t_2}\frac{\mathrm{d}L}{\mathrm{d}t}\,\mathrm{d}t=L(t_2)-L(t_1).$$

When we associate with the time displacement δt a variation of the Lagrangian,

$$\delta L=\frac{\mathrm{d}L}{\mathrm{d}t}\delta t,$$

that is evidently a sequel to replacing the actual value of each state variable (or generalized coordinate) at time t by the actual value assumed at the time $t+\delta t$, there obtains a statement of invariance,

$$\int_{t_1}^{t_2}\delta L\,\mathrm{d}t=\delta t[L(t_2)-L(t_1)]=\delta t\int_{t_1}^{t_2}\frac{\mathrm{d}L}{\mathrm{d}t}\,\mathrm{d}t, \tag{83.13}$$

whose application awaits only the specification of δL. If the Lagrangian density has the form

$$\mathcal{L}=\mathcal{L}(\varphi,\varphi_x,\varphi_t,x),$$

the effect of the concomitant variations

$$\delta\varphi = \varphi_t \delta t, \qquad \delta\varphi_t = \varphi_{tt}\delta t$$

is to induce in L a variation

$$\begin{aligned}\delta L &= \int \left\{ \left[\frac{\partial \mathscr{L}}{\partial \varphi} - \frac{\partial}{\partial x}\left(\frac{\partial \mathscr{L}}{\partial \varphi_x} \right) \right] \delta\varphi + \frac{\partial \mathscr{L}}{\partial \varphi_t} \delta\varphi_t \right\} \mathrm{d}x \\ &= \delta t \int \left\{ \left[\frac{\partial \mathscr{L}}{\partial \varphi} - \frac{\partial}{\partial x}\left(\frac{\partial \mathscr{L}}{\partial \varphi_x} \right) \right] \varphi_t + \frac{\partial \mathscr{L}}{\partial \varphi_t} \varphi_{tt} \right\} \mathrm{d}x.\end{aligned} \tag{83.14}$$

The first of the expressions in (83.14), adapted from (83.2), cannot now be justified by citing a null variation for φ at the endpoints of an arbitrary integration range, owing to the restrictive type of variation that is contemplated; we shall assume, therefore, that the x-integral extends over an infinite range and that the magnitude of φ is negligible at the outer reaches during the entire time interval under consideration.

After utilizing the equation of motion (83.3), we find

$$\delta L = \delta t \frac{\mathrm{d}}{\mathrm{d}t}\left[\int_{-\infty}^{\infty} \frac{\partial \mathscr{L}}{\partial \varphi_t} \varphi_t \, \mathrm{d}x \right] \tag{83.15}$$

whence the invariance relation (83.13) becomes

$$\delta t \int_{t_1}^{t_2} \frac{\mathrm{d}}{\mathrm{d}t}\left[\int_{-\infty}^{\infty} \frac{\partial \mathscr{L}}{\partial \varphi_t} \varphi_t \, \mathrm{d}x - L \right] \mathrm{d}t = 0$$

and implies at once that

$$E = \int_{-\infty}^{\infty} \left(\frac{\partial \mathscr{L}}{\partial \varphi_t} \varphi_t - \mathscr{L} \right) \mathrm{d}x \tag{83.16}$$

remains unchanged in time. When $\varphi(x, t)$ is identified with the displacement, $y(x, t)$, of a homogeneous string in which a uniform tensile force acts, substitution of the Lagrangian density [cf. (83.6)]

$$\mathscr{L} = \frac{1}{2}\rho\left(\frac{\partial y}{\partial t}\right)^2 - \frac{1}{2}P\left(\frac{\partial y}{\partial x}\right)^2 \tag{83.17}$$

in (83.16) reveals that

$$E = \int_{-\infty}^{\infty} \left\{ \frac{1}{2}\rho\left(\frac{\partial y}{\partial t}\right)^2 + \frac{1}{2}P\left(\frac{\partial y}{\partial x}\right)^2 \right\} \mathrm{d}x \tag{83.18}$$

is the total energy of the system.

To arrive at and extend the energy conservation result in a different manner, let us next calculate

$$\delta I = \delta \int_{t_1}^{t_2} \int_{-\infty}^{\infty} \mathscr{L}(\varphi, \varphi_x, \varphi_t, x, t) \, \mathrm{d}t \, \mathrm{d}x \tag{83.19}$$

on the hypothesis of exclusive variation in t, with the proviso that $\delta t = 0$ at $t = t_1, t_2$; this yields

$$\delta I = \int_{t_1}^{t_2} \int_{-\infty}^{\infty} \left(\frac{\partial \mathscr{L}}{\partial t} \delta t + \frac{\partial \mathscr{L}}{\partial \varphi_t} \Delta\varphi_t + \mathscr{L} \frac{\delta \, \mathrm{d}t}{\mathrm{d}t} \right) \mathrm{d}t \, \mathrm{d}x \tag{83.20}$$

where

$$\Delta\varphi_t = -\varphi_t \frac{\delta \, \mathrm{d}t}{\mathrm{d}t}$$

expresses the change of $\partial\varphi/\partial t$ that arises for the indicated reason. After integrating by parts in the last two terms of (83.20), we reach the desired form of variation, viz.

$$\delta I = \int_{t_1}^{t_2} \delta t \left[\frac{\mathrm{d}}{\mathrm{d}t} \int_{-\infty}^{\infty} \left(\frac{\partial \mathscr{L}}{\partial \varphi_t} \varphi_t - \mathscr{L} \right) \mathrm{d}x + \int_{-\infty}^{\infty} \frac{\partial \mathscr{L}}{\partial t} \mathrm{d}x \right] \mathrm{d}t. \tag{83.21}$$

When use is made of the partial differential equation (82.3) for φ, there obtains

$$\frac{\mathrm{d}}{\mathrm{d}t} \int_{-\infty}^{\infty} \frac{\partial \mathscr{L}}{\partial \varphi_t} \varphi_t \, \mathrm{d}x = \int_{-\infty}^{\infty} \left\{ \frac{\partial \mathscr{L}}{\partial \varphi_t} \varphi_{tt} + \frac{\partial \mathscr{L}}{\partial \varphi} \varphi_t - \frac{\partial}{\partial x} \left(\frac{\partial \mathscr{L}}{\partial \varphi_x} \right) \varphi_t \right\} \mathrm{d}x$$

and, since

$$\frac{\mathrm{d}}{\mathrm{d}t} \int_{-\infty}^{\infty} \mathscr{L} \, \mathrm{d}x = \int_{-\infty}^{\infty} \left\{ \frac{\partial \mathscr{L}}{\partial t} + \frac{\partial \mathscr{L}}{\partial \varphi} \varphi_t + \frac{\partial \mathscr{L}}{\partial \varphi_x} \varphi_{tt} + \frac{\partial \mathscr{L}}{\partial \varphi_x} \varphi_{xt} \right\} \mathrm{d}x,$$

it follows that

$$\frac{\mathrm{d}E}{\mathrm{d}t} + \int_{-\infty}^{\infty} \frac{\partial \mathscr{L}}{\partial t} \mathrm{d}x = -\int_{-\infty}^{\infty} \frac{\partial}{\partial x} \left(\frac{\partial \mathscr{L}}{\partial \varphi_x} \varphi_t \right) \mathrm{d}x = -\frac{\partial \mathscr{L}}{\partial \varphi_x} \varphi_t \Bigg|_{x=-\infty}^{x=+\infty}; \tag{83.22}$$

accordingly, if the state function φ pertains to a localized excitation and the integrated terms on the right-hand side of (83.22) vanish, we have

$$\frac{\mathrm{d}E}{\mathrm{d}t} + \int_{-\infty}^{\infty} \frac{\partial \mathscr{L}}{\partial t} \mathrm{d}x = 0 \tag{83.23}$$

and

$$\delta I = 0.$$

Reciprocally, (83.23) expresses the necessary condition that arbitrary time variations between t_1 and t_2 are without effect in (83.19). It may be observed that the prior conclusion, $E = \text{const.}$, is evidently recovered when $\mathscr{L}$ has no explicit time dependence and, furthermore, that (83.23) correlates changes of the local energy density with those of the tension in a string, since

$$\frac{\partial \mathscr{L}}{\partial t} = -\frac{1}{2} \frac{\mathrm{d}P}{\mathrm{d}t} \left(\frac{\partial y}{\partial x} \right)^2$$

according to (83.6). A particular form of the relation (83.22) which is appropriate to string motions on the arbitrary interval $x_1 < x < x_2$,

$$\frac{\partial}{\partial t}\int_{x_1}^{x_2}\left\{\frac{1}{2}\rho(x)\left(\frac{\partial y}{\partial t}\right)^2+\frac{1}{2}P(t)\left(\frac{\partial y}{\partial x}\right)^2\right\}\mathrm{d}x+\left(-P(t)\frac{\partial y}{\partial x}\frac{\partial y}{\partial t}\right)\Bigg|_{x=x_1}^{x=x_2}$$
$$-\int_{x_1}^{x_2}\frac{1}{2}\frac{\mathrm{d}P}{\mathrm{d}t}\left(\frac{\partial y}{\partial x}\right)^2\mathrm{d}x=0, \tag{83.24}$$

balances, in a self-consistent fashion, the overall energy content with fluxes at the endpoints and an interaction between the perturbed ($y \neq 0$) and mean ($y \equiv 0$) states of the system that stems from a changing tension.

There is need for a more general variation of functionals defined by multiple (coordinate-time) integrals if the complete forms of conservation relations are to be derived therefrom; namely, one which also reckons on variation in the domain itself as a consequence of infinitesimal transformations of both integration variables. As regards the particular functional

$$I=\int_D \mathscr{L}(\varphi, \varphi_x, \varphi_t, x, t)\,\mathrm{d}x\,\mathrm{d}t \tag{83.25}$$

associated with an arbitrary domain D of the two variables x, t, this leads us to define a family of transformations

$$x' = X(x, t, \varepsilon), \qquad t' = T(x, t, \varepsilon) \tag{83.26}$$

in which a special parametric value ($\varepsilon = 0$, say) permits retrieval of the identity transformation $x' = x$, $t' = t$, and to assume otherwise ($\varepsilon \neq 0$) the existence of distinct domains D, D' for the corresponding ranges of the variables. Given a function $\varphi'(x', t', \varepsilon)$ in D', with ε attached thereto as a reminder of the parametrization which identifies individual domains, we need only call upon the formulas (83.26) to effect its transplantation into D, or expression in terms of the variables x, t, viz.

$$\varphi'(x', t', \varepsilon) = \varphi'(X, T, \varepsilon) \overset{\text{def}}{=} \Phi(x, t, \varepsilon);$$

since the domains D' and D are identical if ε vanishes, the requirement

$$\lim_{\varepsilon\to 0} \varphi(x, t, \varepsilon) = \Phi(x, t)$$

ensures that no distinction then exists between the functions φ' and φ.

In accordance with these premises, a first variation of the functional I is defined by

$$\delta I = \varepsilon\left(\frac{\mathrm{d}I(\varepsilon)}{\mathrm{d}\varepsilon}\right)_{\varepsilon=0} \tag{83.28}$$

where

$$I(\varepsilon) = \int_{D'} \mathscr{L}\left(\varphi', \frac{\partial \varphi'}{\partial x'}, \frac{\partial \varphi'}{\partial t'}, x', t'\right) \mathrm{d}x'\,\mathrm{d}t'$$
$$= \int_{D} \mathscr{L}\left(\Phi, \frac{\partial}{\partial x'}\Phi, \frac{\partial}{\partial t'}\Phi, X, T\right) \frac{\partial(X, T)}{\partial(x, t)} \,\mathrm{d}x\,\mathrm{d}t, \tag{83.29}$$

$$J = \frac{\partial(X, T)}{\partial(x, t)} = \frac{\partial X}{\partial x}\frac{\partial T}{\partial t} - \frac{\partial X}{\partial t}\frac{\partial T}{\partial x} \tag{83.30}$$

represents the Jacobian of the transformation (83.26). Introducing the coordinate and time variations

$$\delta x = \varepsilon\left(\frac{\partial X}{\partial \varepsilon}\right)_{\varepsilon=0}, \qquad \delta t = \varepsilon\left(\frac{\partial T}{\partial \varepsilon}\right)_{\varepsilon=0} \tag{83.31}$$

that of the Jacobian becomes

$$\delta J = \varepsilon\left(\frac{\partial}{\partial \varepsilon}J\right)_{\varepsilon=0} = \frac{\partial}{\partial x}\delta x + \frac{\partial}{\partial t}\delta t; \tag{83.32}$$

moreover, the variation

$$\delta\varphi = \varepsilon\left(\frac{\partial \Phi(x, t, \varepsilon)}{\partial \varepsilon}\right)_{\varepsilon=0} \tag{83.33}$$

of the dependent function $\varphi(x, t)$ is expressible in the composite form

$$\delta\varphi = \overline{\delta\varphi} + \frac{\partial\varphi}{\partial x}\delta x + \frac{\partial\varphi}{\partial t}\delta t \tag{83.34}$$

which involves a 'local' contribution (at the fixed point x, t), namely,

$$\overline{\delta\varphi} = \varepsilon\left(\frac{\partial}{\partial \varepsilon}\varphi'(x, t, \varepsilon)\right)_{\varepsilon=0}, \tag{83.35}$$

and others attributable to the independent variable transformations. When (83.32)–(83.34) and the analogous forms that render the variations of $\partial\varphi/\partial x$ and $\partial\varphi/\partial t$ are applied for purposes of representing the derivative in (83.28), we obtain the sought-after result

$$\delta I = \int_D \left\{[\mathscr{L}]\overline{\delta\varphi} + \frac{\partial}{\partial x}\left(\mathscr{L}\delta x + \frac{\partial \mathscr{L}}{\partial \varphi_x}\overline{\delta\varphi}\right) + \frac{\partial}{\partial t}\left(\mathscr{L}\delta t + \frac{\partial \mathscr{L}}{\partial \varphi_t}\overline{\delta\varphi}\right)\right\} \mathrm{d}x\,\mathrm{d}t \tag{83.36}$$

where

$$[\mathscr{L}] \overset{\text{def}}{=} \frac{\partial \mathscr{L}}{\partial \varphi} - \frac{\partial}{\partial x}\left(\frac{\partial \mathscr{L}}{\partial \varphi_x}\right) - \frac{\partial}{\partial t}\left(\frac{\partial \mathscr{L}}{\partial \varphi_t}\right). \tag{83.37}$$

It will be perceived that $[\mathscr{L}] = 0$ describes a condition of stationarity relative to the functional I if the coordinate variations vanish and φ is prescribed at the endpoints of D (thus removing the middle and last terms of (83.36)), and (83.37) confirms that this is identical with the previously given version (83.3). Covariance of the latter (or so-called Euler-Lagrange equation) with respect to transformation of the variables implies that a similar property is manifest by the functional in the associated variational principle; conversely, and of especial importance there, is the fact that invariance of the functional under continuous groups of transformations permits the systematic deduction of explicit differential identities with a conservation nature, that is, which connect temporal changes of a density measure and the gradient (derivative) of its flux, for arbitrary coordinate intervals.

Invariance of the functional I with respect to general transformations of the independent and dependent variables, say,

$$x' = \mathscr{X}(x, t, \varphi, \varphi_x, \varphi_t), \qquad t' = \mathscr{T}(x, t, \varphi, \varphi_x, \varphi_t),$$
$$\varphi'(x', t') = \Phi(x, t, \varphi, \varphi_x, \varphi_t)$$

is understood to mean that

$$\int_{D'} \mathscr{L}(\varphi', \varphi'_{x'}, \varphi'_{t'}, x', t')\,\mathrm{d}x'\,\mathrm{d}t' = \int_D \mathscr{L}(\varphi, \varphi_x, \varphi_t, x, t)\,\mathrm{d}x\,\mathrm{d}t \qquad (83.38)$$

for arbitrary pairs of domains D, D'. The transformation relations above may contain a finite number of parameters, $\varepsilon_1, \ldots, \varepsilon_m$, viz.

$$x' = \mathscr{X}(x, t, \varphi, \varphi_x, \varphi_t, \varepsilon_1, \ldots, \varepsilon_m) = X(x, t, \varepsilon_1, \ldots, \varepsilon_m)$$
$$t' = \mathscr{T}(x, t, \varphi, \varphi_x, \varphi_t, \varepsilon_1, \ldots, \varepsilon_m) = T(x, t, \varepsilon_1, \ldots, \varepsilon_m)$$

and the latter are termed essential if their number cannot be reduced in consequence of relations such as

$$\sum_{n=1}^{m} a_n(\varepsilon_1, \ldots, \varepsilon_m)\frac{\partial X}{\partial \varepsilon_n} = 0,$$
$$\sum_{n=1}^{m} b_n(\varepsilon_1, \ldots, \varepsilon_m)\frac{\partial T}{\partial \varepsilon_n} = 0.$$

We regard the ensemble of these transformations as constituting a finite, m parameter, continuous group if any two sequential representatives $x \to x' \to x''$, $t \to t' \to t''$, with the respective parameters $\varepsilon_n, \varepsilon'_n$, are equivalent to a single transformation $x \to x''$, $t \to t''$ whose parameters are $\varepsilon''_n = \varepsilon''_n(\varepsilon_n, \varepsilon'_n)$; and confer the designation of an infinite group on the analogous transformations which involve one or more arbitrary functions rather than numerical parameters.

If the identity transformation $x' = x$, $t' = t$, $\varphi'(x', t') = \varphi(x, t)$ corresponds to the null parameter values $\varepsilon_1 = \varepsilon_2 = \cdots = \varepsilon_m = 0$, there exist group transformations which change the variables by arbitrarily small amounts. On making the expansions

$$x' = x + \delta x + \cdots, \qquad \delta x = \sum_{n=1}^{m} \varepsilon_n \left(\frac{\partial X}{\partial \varepsilon_n} \right)_{\varepsilon_1 = \varepsilon_2 = \cdots = \varepsilon_m = 0},$$
$$t' = t + \delta t + \cdots, \qquad \delta t = \sum_{n=1}^{m} \varepsilon_n \left(\frac{\partial T}{\partial \varepsilon_n} \right)_{\varepsilon_1 = \varepsilon_2 = \cdots = \varepsilon_m = 0}, \tag{83.39}$$

$$\varphi'(x', t') = \varphi(x, t) + \delta\varphi(x, t) + \cdots, \tag{83.40}$$

where $\delta\varphi$ is also of the first degree in the parameters ε_n, and thereafter disregarding all terms of higher degree, we characterize an infinitesimal (finite) Lie group of transformations; likewise, an infinitesimal (infinite) Lie group receives its specification by means of equations that are linear in the arbitrary functions which enter into the transformation formulas.

Let us next observe that if the statement of invariance (83.38) for the integral I covers a continuous group of transformations, it is evidently applicable to the infinitesimal transformations comprised therein and then takes the form

$$\delta I = 0.$$

From the constructive version of the first variation given in (83.36), (83.37), the foregoing statement becomes

$$0 = \int_D \left\{ \overline{\delta \mathscr{L}} + \frac{\partial}{\partial x}(\mathscr{L}\delta x) + \frac{\partial}{\partial t}(\mathscr{L}\delta t) \right\} \mathrm{d}x\, \mathrm{d}t \tag{83.41}$$

where $\overline{\delta \mathscr{L}}$ unites those contributions that stem from the local variation of the dependent variable, $\overline{\delta\varphi}$, namely,

$$\overline{\delta \mathscr{L}} = [\mathscr{L}]\overline{\delta\varphi} + \frac{\partial}{\partial x}\left(\frac{\partial \mathscr{L}}{\partial \varphi_x} \overline{\delta\varphi} \right) + \frac{\partial}{\partial t}\left(\frac{\partial \mathscr{L}}{\partial \varphi_t} \overline{\delta\varphi} \right). \tag{83.42}$$

Since (83.41) holds for an arbitrary domain D the integrand must vanish identically and we thus arrive at the identity

$$[\mathscr{L}]\overline{\delta\varphi} = -\frac{\partial}{\partial x}\left(\mathscr{L}\delta x + \frac{\partial \mathscr{L}}{\partial \varphi_x} \overline{\delta\varphi} \right) - \frac{\partial}{\partial t}\left(\mathscr{L}\delta t + \frac{\partial \mathscr{L}}{\partial \varphi_t} \overline{\delta\varphi} \right) \tag{83.43}$$

which can be superseded by a like set of m independent ones,

$$[\mathscr{L}]\overline{\delta\varphi}^{(n)} = -\frac{\partial}{\partial x}\left(\mathscr{L}\delta x^{(n)} + \frac{\partial \mathscr{L}}{\partial \varphi_x} \overline{\delta\varphi}^{(n)} \right) - \frac{\partial}{\partial t}\left(\mathscr{L}\delta t^{(n)} + \frac{\partial \mathscr{L}}{\partial \varphi_t} \overline{\delta\varphi}^{(n)} \right)$$
$$n = 1, 2, \ldots, m \tag{83.44}$$

in consequence of the representations

$$\delta x = \sum_{n=1}^{m} \varepsilon_n \delta x^{(n)}, \qquad \delta t = \sum_{n=1}^{m} \varepsilon_n \delta t^{(n)}$$
$$\overline{\delta\varphi} = \sum_{n=1}^{m} \varepsilon_n \overline{\delta\varphi}^{(n)} \tag{83.45}$$

and the presumed irreducibility of the m arbitrary transformation parameters. After simplification of the foregoing set of identities on the basis of the Euler-Lagrange differential equation, $[\mathscr{L}]=0$, which expresses the stationary nature of the functional I for variations of φ in a fixed domain, they provide us with the secondary differential or conservation type relations

$$\frac{\partial}{\partial x}\left(\mathscr{L}\delta x^{(n)}+\frac{\partial \mathscr{L}}{\partial \varphi_x}\overline{\delta\varphi}^{(n)}\right)+\frac{\partial}{\partial t}\left(\mathscr{L}\delta t^{(n)}+\frac{\partial \mathscr{L}}{\partial \varphi_t}\overline{\delta\varphi}^{(n)}\right)=0, \qquad n=1,2,\ldots,m \tag{83.46}$$

that stem from an invariant aspect of the functional (Noether, 1918).

For illustrative purposes, let us consider the integral (83.8), whose Euler-Lagrange equation (83.7) describes damped string vibrations, and observe that the expression

$$\left\{\frac{1}{2}\rho\left(\frac{\partial y}{\partial t}\right)^2-\frac{1}{2}P\left(\frac{\partial y}{\partial x}\right)^2-\frac{1}{2}Ph^2y^2\right\}\mathrm{e}^{\gamma t}\,\mathrm{d}x\,\mathrm{d}t$$

preserves its form under the infinitesimal transformations

$$x'=x+\varepsilon_1, \qquad t'=t+\varepsilon_2, \qquad y'=y-\frac{1}{2}\gamma\varepsilon_2 y$$

of a two-parameter group. The relation (83.46) which is associated with coordinate translations and the particular variations

$$\delta x^{(1)}=\varepsilon_1, \qquad \delta t^{(1)}=0, \qquad \overline{\delta y}^{(1)}=-\varepsilon_1\frac{\partial y}{\partial x}$$

turns out to be

$$\frac{\partial}{\partial x}\left\{\left(\frac{1}{2}\rho\left(\frac{\partial y}{\partial t}\right)^2+\frac{1}{2}P\left(\frac{\partial y}{\partial x}\right)^2-\frac{1}{2}Ph^2y^2\right)\mathrm{e}^{\gamma t}\right\}-\frac{\partial}{\partial t}\left\{\rho\frac{\partial y}{\partial t}\frac{\partial y}{\partial x}\mathrm{e}^{\gamma t}\right\}=0$$

or

$$\frac{\partial}{\partial x}\left\{\frac{1}{2}\rho\left(\frac{\partial y}{\partial t}\right)^2+\frac{1}{2}P\left(\frac{\partial y}{\partial x}\right)^2-\frac{1}{2}Ph^2y^2\right\}-\left(\gamma+\frac{\partial}{\partial t}\right)\left\{\rho\frac{\partial y}{\partial y}\frac{\partial y}{\partial x}\right\}=0. \tag{83.47}$$

It is easy to confirm that the primary equation of motion, (83.7), i.e.,

$$P\frac{\partial^2 y}{\partial x^2}=\rho\frac{\partial^2 y}{\partial t^2}+\rho\gamma\frac{\partial y}{\partial t}+Ph^2y, \tag{83.48}$$

follows from (83.47) after differentiation and cancellation of like terms; moreover, if we integrate (83.48) with respect to x between x_1 and x_2, the resulting identity

$$\frac{\mathrm{d}}{\mathrm{d}t}\int_{x_1}^{x_2}\rho\frac{\partial y}{\partial t}\,\mathrm{d}x+\int_{x_1}^{x_2}\rho\gamma\frac{\partial y}{\partial t}\,\mathrm{d}x+\int_{x_1}^{x_2}Ph^2y\,\mathrm{d}x=P\frac{\partial y}{\partial x}\bigg|_{x_1}^{x_2}$$

correlates the overall change of transverse momentum and the working of various forces on the string.

Applying (83.46) together with the variations

$$\delta x^{(2)} = 0, \qquad \delta t^{(2)} = \varepsilon_2, \qquad \overline{\delta y}^{(2)} = -\varepsilon_2\left(\frac{\partial y}{\partial t} + \frac{1}{2}\gamma y\right)$$

the outcome, after integrating over x, is

$$\left(\frac{\mathrm{d}}{\mathrm{d}t} + \gamma\right)\int_{x_1}^{x_2}\left\{\frac{1}{2}\rho\left(\frac{\partial y}{\partial t}\right)^2 + \frac{1}{2}P\left(\frac{\partial y}{\partial x}\right)^2 + \frac{1}{2}\rho\gamma y\frac{\partial y}{\partial t} + \frac{1}{2}Ph^2y^2\right\}\mathrm{d}x$$
$$-\left[\frac{1}{2}P\gamma y\frac{\partial y}{\partial x} + P\frac{\partial y}{\partial x}\frac{\partial y}{\partial t}\right]_{x_2}^{x_1} = 0$$

which constitutes an equation that involves recognizable energy components. If the string has fixed ends and the boundary terms vanish, the latter equation becomes

$$\frac{\mathrm{d}\mathscr{E}}{\mathrm{d}t} + \gamma\mathscr{E} = 0$$

where

$$\mathscr{E}(t) = \int_{x_1}^{x_2}\left\{\frac{1}{2}\rho\left(\frac{\partial y}{\partial t}\right)^2 + \frac{1}{2}P\left(\frac{\partial y}{\partial x}\right)^2 + \frac{1}{2}\rho\gamma y\frac{\partial y}{\partial t} + \frac{1}{2}Ph^2y^2\right\}\mathrm{d}x$$
$$= E(t) + \frac{1}{2}\int_{x_1}^{x_2}\rho\gamma y\frac{\partial y}{\partial t}\,\mathrm{d}x,$$

$$E(t) = \int_{x_1}^{x_2}\left\{\frac{1}{2}\rho\left(\frac{\partial y}{\partial t}\right)^2 + \frac{1}{2}P\left(\frac{\partial y}{\partial x}\right)^2 + \frac{1}{2}Ph^2y^2\right\}\mathrm{d}x$$

represents the total energy of the string movements. Time differentiation yields

$$\frac{\mathrm{d}E}{\mathrm{d}t} = \frac{\mathrm{d}\mathscr{E}}{\mathrm{d}t} - \frac{1}{2}\frac{\mathrm{d}}{\mathrm{d}t}\int_{x_1}^{x_2}\rho\gamma y\frac{\partial y}{\partial t}\,\mathrm{d}x$$
$$= -\gamma\mathscr{E} - \frac{1}{2}\int_{x_1}^{x_2}\rho\gamma\left(\frac{\partial y}{\partial t}\right)^2\mathrm{d}x - \frac{1}{2}\int_{x_1}^{x_2}\rho\gamma y\frac{\partial^2 y}{\partial t^2}\,\mathrm{d}x$$
$$= -\int_{x_1}^{x_2}\rho\gamma\left(\frac{\partial y}{\partial t}\right)^2\mathrm{d}x \tag{83.49}$$

following the use of (83.48) and an integration by parts; both the dissipation of energy, registered by a negative magnitude for $\mathrm{d}E/\mathrm{d}t$, and its quantitative measure are apparent in (83.49).

Variatonal formulations and their ramifications make up the analytical framework for recent studies (Whitham, 1975) of linear and non-linear dispersive wave problems.

§84. *Progressive Waves in Variable Configurations*

The designation of progressive waves is conferred on special solutions of the second order partial differential equation

$$c^2\frac{\partial^2 y}{\partial x^2} = \frac{\partial^2 y}{\partial t^2}, \tag{84.1}$$

namely

$$y_1 = f\left(t - \frac{x}{c}\right), \qquad y_2 = g\left(t + \frac{x}{c}\right) \tag{84.2}$$

where c has any fixed (positive) magnitude and the dimensions of velocity; it is a familiar fact (§1), moreover, that they satisfy the individual first order equations

$$c\frac{\partial f}{\partial x} = -\frac{\partial f}{\partial t}, \qquad c\frac{\partial g}{\partial x} = \frac{\partial g}{\partial t} \tag{84.3}$$

and assume constant values along the respective characteristics $ct \mp x =$ const. of (84.1). There are sufficient reasons, of theoretical and applied nature, for a wider study of progressive waves that have an arbitrary or irregular time dependence; specifically, this may be aimed at the enumeration of such forms which involve two or three coordinate variables, or the analysis of excitations in settings with a continuously variable spatial inhomogeneity (extending the prior focus on time-periodic states).

Let us refer at the outset to a linear wave equation regulating small amplitude displacements of a non-uniform (variable density) string,

$$\left(c^2(x)\frac{\partial^2}{\partial x^2} - \frac{\partial^2}{\partial t^2}\right) y = 0, \tag{84.4}$$

wherein $c(x)$ specifies the local propagation or wave speed. On noting the alternative forms of the partial differential operator in (84.1), viz.

$$\left(c(x)\frac{\partial}{\partial x} - \frac{\partial}{\partial t}\right)\left(c(x)\frac{\partial}{\partial x} + \frac{\partial}{\partial t}\right) - c(x)c'(x)\frac{\partial}{\partial x},$$

$$\left(c(x)\frac{\partial}{\partial x} + \frac{\partial}{\partial t}\right)\left(c(x)\frac{\partial}{\partial x} - \frac{\partial}{\partial t}\right) - c(x)c'(x)\frac{\partial}{\partial x'}$$

it follows that

$$y(x, t) = f(x, t) + g(x, t) \tag{84.5}$$

satisfies (84.4) if

$$\left(c(x)\frac{\partial}{\partial x}-\frac{\partial}{\partial t}\right)\left(c(x)\frac{\partial}{\partial x}+\frac{\partial}{\partial t}\right)f - c(x)c'(x)\frac{\partial f}{\partial x}$$
$$+\left(c(x)\frac{\partial}{\partial x}+\frac{\partial}{\partial t}\right)\left(c(x)\frac{\partial}{\partial x}-\frac{\partial}{\partial t}\right)g - c(x)c'(x)\frac{\partial g}{\partial x}=0.$$

Fulfillment of the latter relation is assured when the component functions f, g obey the system of first order differential equations

$$\left(c(x)\frac{\partial}{\partial x}+\frac{\partial}{\partial t}\right)f = -\left(c(x)\frac{\partial}{\partial x}-\frac{\partial}{\partial t}\right)g = \frac{1}{2}c'(x)(f-g) \tag{84.6}$$

or the equivalent pair

$$\left(c(x)\frac{\partial}{\partial x}\pm\frac{\partial}{\partial t}-\frac{1}{2}c'(x)\right)\begin{bmatrix} f \\ g \end{bmatrix} = -\frac{1}{2}c'(x)\begin{bmatrix} g \\ f \end{bmatrix} \tag{84.7}$$

which reduce, as anticipated, to the independent versions (84.3) in the circumstance (or homogeneous limit) $c(x) = \text{const}$. Accordingly, the coupled equations (84.7) suggest that, in regions where the magnitude of $c'(x)$ remains small, the complete wave function (84.5) is given by a sum of two progressive waves whose features undergo gradual change.

Particulars of individual progressive waves are obtained by recourse to the original (i.e., second-order) dynamical equations such as (84.4) or

$$\rho(x)\frac{\partial^2 y}{\partial t^2} = P(t)\frac{\partial^2 y}{\partial x^2}, \tag{84.8}$$

and

$$\frac{\partial}{\partial x}\left(E(x)\frac{\partial u}{\partial x}\right) = \rho(x)\frac{\partial^2 u}{\partial t^2}, \tag{84.9}$$

$$\frac{\partial}{\partial x}\left(\frac{1}{\rho(x)}\frac{\partial \sigma}{\partial x}\right) = \frac{1}{E(x)}\frac{\partial^2 \sigma}{\partial t^2}, \tag{84.10}$$

where the latter pair characterize one-dimensional velocity and stress distributions, u, σ, within an elastic continuum that has a variable density $\rho(x)$ and elastic modulus $E(x)$. A preliminary transformation of (84.9) and (84.10), which introduces derivatives of the material parameters, offers patent advantages in dealing with nearly homogeneous configurations and enables us to realize concomitant approximations for wave functions of the progressive type; both the independent/dependent variables, x and u/σ, are altered for said purpose, rivalling the manner of an earlier rearrangement (§67). Thus, if we express the transformation equations in the general fashion

$$u = v(\xi)w(\xi, t), \qquad \mathrm{d}x = q(\xi)\,\mathrm{d}\xi, \tag{84.11}$$

where ξ, w are the variables intended as replacements for x, u, the differential equation (84.9) may be converted to

$$\frac{\partial}{\partial \xi}\left(\frac{E}{q} v^2 \frac{\partial w}{\partial \xi}\right) + v \frac{\partial}{\partial \xi}\left(\frac{E}{q}\frac{\partial v}{\partial \xi}\right) w = \rho q v^2 \frac{\partial^2 w}{\partial t^2}. \tag{84.12}$$

On specifying v and q through the particular relations

$$\frac{E}{q} v^2 = 1, \qquad \rho q v^2 = 1$$

which imply that

$$v = (E\rho)^{-1/4}, \qquad q = \sqrt{(E/\rho)} \tag{84.13}$$

a simplified version of (84.12) obtains, namely

$$\frac{\partial^2 w}{\partial \xi^2} - \frac{\partial^2 w}{\partial t^2} = -(E\rho)^{-1/4} \frac{\mathrm{d}}{\mathrm{d}\xi}\left((E\rho)^{1/2} \frac{\mathrm{d}}{\mathrm{d}\xi}(E\rho)^{-1/4}\right) w. \tag{84.14}$$

When the material parameters E, ρ are slowly varying functions the coefficient of w on the right-hand side in (84.14) possesses a small magnitude; and, if this member be omitted altogether, the progressive wave integrals of the resulting equation,

$$w_1 = f(t - \xi), \qquad w_2 = g(t + \xi),$$

lead to an approximate form of the general solution of (84.9) in nearly homogeneous circumstances, viz.

$$u(x, t) = (E\rho)^{-1/4}\left[f\left(t - \int_{x_0}^{x} (\rho/E)^{1/2}\,\mathrm{d}\xi\right) + g\left(t + \int_{x_0}^{x} (\rho/E)^{1/2}\,\mathrm{d}\xi\right)\right] \tag{84.15}$$

which is an evident counterpart of the time-periodic *JWKB* wave functions (§66). The particulars of an analogous representation which suits the variable coefficient equation (84.4) are readily found once we employ the integral

$$y(x, t) = A(x) F(t - \varphi(x)) \tag{84.16}$$

which contains the arbitrary function F and a disposable pair $A(x), \varphi(x)$. Substitution in (84.4) yields [with primes designating argument derivatives]

$$\begin{aligned}[(c(x)\varphi'(x))^2 - 1] A(x) F'' - (c(x))^2 [2A'(x)\varphi'(x) + A(x)\varphi''(x)] F' \\ + (c(x))^2 A''(x) F = 0\end{aligned}$$

and thence, equating (separately) the coefficients of F'' and F' to zero, while ignoring the coefficient of F (inasmuch as $A'' \ll A' \ll A$ if $A(x)$ undergoes gradual change), we obtain the determinations

$$q(x) = \pm \int^{x} \frac{\mathrm{d}\xi}{c(\xi)}, \qquad A(x) = \frac{1}{\sqrt{(\varphi'(x))}} = \sqrt{(c(x))} \tag{84.17}$$

that underlie the approximate solutions of (84.4),

$$y(x, t) = \sqrt{(c(x))}\left[f\left(t - \int_{x_0}^{x} \frac{d\xi}{c(\xi)}\right) + g\left(t + \int_{x_0}^{x} \frac{d\xi}{c(\xi)}\right)\right]. \tag{84.18}$$

When the objective is to achieve a more systematic resolution of (84.9), say, a superposition of the product (or progressive wave) forms (84.16) may be utilized, as figures in the development

$$u(x, t) = \sum_{n=0}^{\infty} A_n(x) F_n(t - \varphi(x)); \tag{84.19}$$

and, since neither of the paired factors A_n, F_n enter with prior constraints, we are entitled to characterize the sequence $\{F_n\}$ through the simple interrelations

$$F_n' = F_{n-1}, \quad n \geq 1, \quad F_0 \text{ arbitrary} \tag{84.20}$$

where the prime signifies an argument derivative. After calling upon (84.9), (84.19), and (84.20), it follows that

$$\begin{aligned}\sum_{n=0}^{\infty} [\{E(x)(\varphi'(x))^2 - \rho(x)\}A_n(x)F_{n-2} \\ - \{2E(x)\varphi'(x)A_n'(x) + (E(x)\varphi''(x) + E'(x)\varphi'(x))A_n(x)\}F_{n-1} \\ + \{E(x)A_n''(x) + E'(x)A_n'(x)\}F_n] = 0\end{aligned}$$

and thence, consequent to adjusting the summation index in the first and last group of terms, we obtain

$$\begin{aligned}\{E(x)(\varphi'(x))^2 - \rho(x)\}A_{n+1}(x) - 2E(x)\varphi'(x)A_n'(x) \\ - \{E(x)\varphi''(x) + E'(x)\varphi'(x)\}A_n(x) + E(x)A_{n-1}''(x) + E'(x)A_{n-1}'(x) = 0.\end{aligned} \tag{84.21}$$

A null value for the coefficient of $A_{n+1}(x)$ herein is implied jointly by the choice $n = -1$ and the stipulations $A_0(x) \not\equiv 0$, $A_n(x) = 0$, $n < 0$, which accord with (84.19); this yields

$$E(x)(\varphi'(x))^2 = \rho(x) \tag{84.22}$$

or

$$\varphi(x) = \varphi^{(0)} \pm \int_{x_0}^{x} \frac{d\xi}{c(\xi)}, \qquad c(x) = \left(\frac{E(x)}{\rho(x)}\right)^{1/2}, \qquad \varphi^{(0)} = \varphi(x_0). \tag{84.23}$$

From the duly simplified version of (84.21),

$$A_n'(x) + \left[\frac{1}{2}\frac{d}{dx}\log\{E(x)\varphi'(x)\}\right]A_n(x) = \frac{1}{2E(x)\varphi'(x)}\frac{d}{dx}\{E(x)A_{n-1}'(x)\} \tag{84.24}$$

or

$$A_n'(x) + \left[\frac{1}{2}\frac{d}{dx}\log\{\rho(x)c(x)\}\right]A_n(x) = \pm\frac{1}{2\rho(x)c(x)}\frac{d}{dx}\{\rho(x)(c(x))^2 A_{n-1}'(x)\} \tag{84.25}$$

a recursive scheme that determines the functions $A_n(x)$ becomes feasible; this is initiated with the readily accessible solution of the ordinary, first order and homogeneous, differential equation

$$A_0'(x) + \left[\frac{1}{2}\frac{\mathrm{d}}{\mathrm{d}x}\log\{\rho(x)c(x)\}\right]A_0(x) = 0 \tag{84.26}$$

and makes subsequent use of the representation (i.e., integral of (84.25))

$$A_n(x) = A_n^{(0)}\left(\frac{\rho(x_0)c(x_0)}{\rho(x)c(x)}\right)^{1/2} \pm \frac{1}{2}\int_{x_0}^{x} \frac{\frac{\mathrm{d}}{\mathrm{d}\xi}\{\rho(\xi)(c(\xi))^2 A_{n-1}'(\xi)\}}{\sqrt{(\rho(x)\rho(\xi)c(x)c(\xi))}}\,\mathrm{d}\xi, \quad n \geq 1 \tag{84.27}$$

wherein $A_n^{(0)} = A_n(x_0)$.

If the arbitrary function F_0 in (84.19), which may be termed the "wave form," assumes a discontinuous Heaviside step profile, namely

$$F_0(t - \varphi(x)) = H(t - \varphi(x)),$$

then

$$F_n(t - \varphi(x)) = \frac{(t - \varphi(x))^n}{n!}H(t - \varphi(x))$$

and the corresponding development

$$u(x, t) = \sum_{n=0}^{\infty} A_n(x)\frac{(t - \varphi(x))^n}{n!}H(t - \varphi(x)) \tag{84.29}$$

pertains to a disturbance with the sharp front

$$t = \varphi(x). \tag{84.30}$$

Since $u(x, t)$ vanishes for $t < \varphi(x)$, and thus registers no excitation ahead of the moving front, the coefficients $A_n(x)$ in its expansion (84.29) are direct measures of the respective discontinuities (or jumps) in the t-derivatives of the wave function at the front, viz.,

$$\left[\frac{\partial^n u}{\partial t^n}\right]_{t=\varphi(x)-0}^{t=\varphi(x)+0} = A_n(x); \tag{84.31}$$

prompted by the latter association, the formulas (84.26) and (84.27) which determine the coefficients A_n are known as the transport equations.

The form of the wave functions expansion (84.29) favors convergence near the front (where the magnitude of $t - \varphi(x)$ is small) and, reciprocally, the significance of terms far along in the expansion rises in proportion to the lapse of time from the arrival of the front at a particular location. Inasmuch as a distortion or change in profile of the progressive wave is brought about by the presence of higher order terms in the aforesaid expansion, it may be presumed that the latter lends itself principally to the description of travelling discontinuities and lacks effectiveness for tracing their large scale evolutionary changes.

Another type of progressive wave, with features in marked contrast to those of the front or discontinuity, is realized by a long train which has a locally periodic nature and exhibits relatively gradual internal changes of wave number and amplitude. Such attributes acquire a formal expression through the inequalities,

$$\frac{\Delta A}{A} \ll 1, \quad \frac{\Delta k}{k} \ll 1, \quad \frac{\Delta \omega}{\omega} \ll 1 \quad \text{when} \quad \begin{matrix} \Delta x = 0(1/k) \quad \text{or} \\ \Delta t = 0(1/\omega) \end{matrix}$$

that characterize different components of the master (overall) wave function

$$y(x, t) = A(x, t) \exp\{\mathrm{i}\Psi(x, t)\}, \tag{84.32}$$

i.e., the amplitude factor $A(x, t)$ and the local values of the wave number and frequency,

$$k = \frac{\partial \Psi}{\partial x}, \qquad \omega = -\frac{\partial \Psi}{\partial t} \tag{84.33}$$

which are derivable from the phase function.

To construct the wave function of a slowly varying train as befits, say, motions on a string with variable density and tension (cf. (84.8)), we employ the small parameter, ε, for a manifest distinction between the scales of change in amplitude and phase function; thus, we postulate a development of the form

$$y(x, t) = [A_0(x, t) + \varepsilon A_1(x, t) + \varepsilon^2 A_2(x, t) + \cdots] \exp\left\{\frac{\mathrm{i}}{\varepsilon} \Psi(x, t)\right\} \tag{84.34}$$

which makes plain that the local wave length and period, proportional to $\varepsilon/(\partial\Psi/\partial x)$ and $\varepsilon/(\partial\Psi/\partial t)$, respectively, are short in comparison with the related scales of argument change in A_0. Equations for the component functions $A_n(x, t)$, $\Psi(x, t)$ become available after the development (84.34) is substituted into the wave equation (84.8) and terms are grouped according to powers of ε. A pair of reciprocal powers occur, namely the second and first, giving rise to the relations

$$\rho(x)\left(\frac{\partial \Psi}{\partial t}\right)^2 = P(t)\left(\frac{\partial \Psi}{\partial x}\right)^2, \tag{84.35}$$

and

$$2\,\mathrm{i}\left[\rho(x)\frac{\partial \Psi}{\partial t}\frac{\partial A_0}{\partial t} - P(t)\frac{\partial \Psi}{\partial x}\frac{\partial A_0}{\partial x}\right] + \mathrm{i}\left[\rho(x)\frac{\partial^2 \Psi}{\partial t^2} - P(t)\frac{\partial^2 \Psi}{\partial x^2}\right]A_0 = \\ = \left[\rho(x)\left(\frac{\partial \Psi}{\partial t}\right)^2 - P(t)\left(\frac{\partial \Psi}{\partial x}\right)^2\right]A_1. \tag{84.36}$$

If we eliminate (derivatives of) the phase function Ψ in favor of k and ω, by recourse to (84.33), there obtains from (84.35), (84.36) the local dispersion relation

$$\omega = \omega(k, x, t) = \left(\frac{P(t)}{\rho(x)}\right)^{1/2} |k| > 0 \tag{84.37}$$

and a homogeneous transport equation for the initial function, A_0, of the amplitude sequence,

$$\rho\omega \frac{\partial A_0}{\partial t} + Pk \frac{\partial A_0}{\partial x} = -\frac{1}{2}\left[\rho \frac{\partial \omega}{\partial t} + P \frac{\partial k}{\partial x}\right] A_0,$$

which acquires the version

$$\frac{\partial A_0}{\partial t} + u \frac{\partial A_0}{\partial x} = -\frac{1}{2\omega}\left[\frac{\partial \omega}{\partial t} + \frac{P}{\rho}\frac{\partial k}{\partial x}\right] A_0 \tag{84.38}$$

on introducing the group velocity $u = \partial\omega/\partial k = Pk/\rho\omega = \sqrt{(P/\rho)}$. Subsequent members $A_1, A_2, \ldots$ of the amplitude sequence may be recursively found through a linear system of inhomogeneous transport equations,

$$\left(\frac{\partial}{\partial t} + u \frac{\partial}{\partial x}\right) A_n = \mathfrak{F}_n(A_n, A_{n-1}, \ldots, A_0, x, t), \tag{84.39}$$

whose explicit forms originate in the (respective) terms of order $\varepsilon^0, \varepsilon, \varepsilon^2, \ldots$ generated by the primitive dynamical equation (84.8) following the substitution of the wave function development (84.34) therein. The family of transport equations are customarily linked with a determination of the amplitude functions along the rays, namely trajectories in the x, t plane which correspond to local values of the group velocity and represent characteristic curves of the partial differential equation

$$\frac{\partial \Psi}{\partial t} = -\omega\left(\frac{\partial \Psi}{\partial x}, x, t\right)$$

secured from the dispersion relation (84.37).

The wave function expansion (84.34) has only a claim to asymptotic (rather than convergent) status, since the absence of a reflected component (bearing the opposite sign for $\partial\Psi/\partial x$) runs counter to the expected attribute of an excitation in inhomogeneous circumstances.

§85. *A Wave Speed Determination*

Experimental measurements of wave speeds in different physical realizations (optical, acoustical, seismic) are important for interpretative and/or fundamental reasons; and those relating to light have gained an especially prominent historical place. The essential component of Fizeau's scheme for ascertaining a light (or electromagnetic wave) propagation speed is a circular disc (wheel) with uniformly spaced and identical teeth along its rim; in bare outline, when the disc rotates and interrupts a light beam directed towards the rim there obtains a variable intensity light distribution, with periodic dependence on the speed of rotation. Once the (least) angular velocity that corresponds to the minimum intensity level, say, and other configurational parameters are known, the direct calculation of a light velocity becomes feasible. If we ignore practical details and concentrate on bringing out the significance of the calculated magnitude (Cantor, 1907), various analytical simplifications are permissible. Thus, it suffices to utilize a single scalar wave function, rather than the vector wave functions of the electromagnetic field, and to simulate the discontinuous transmission of the incident beam between the teeth of the rotating disc by a time-periodic modulation of the concomitant wave at a fixed point.

Let us, furthermore, rely on one-dimensional wave functions, associating the fixed frequency progressive form,

$$\psi^{(i)}(x, t) = A \sin \omega\left(t - \frac{x}{c}\right), \tag{85.1}$$

with an incident monochromatic light beam of small cross-section; the perturbation of same, initiated at $t = 0$ and represented by a wave function $\psi^{(m)}(x, t)$ which satisfies the homogeneous equation

$$\left(\frac{\partial^2}{\partial x^2} - \frac{1}{c^2}\frac{\partial^2}{\partial t^2}\right)\psi^{(m)}(x, t) = 0, \quad x, t > 0 \tag{85.2}$$

is assumed to be a given multiple, $\cos^2(\omega_0 t/2)$, of $\psi^{(i)}(x, t)$ at the modulation point $x = 0$, or site of the rotating disc, i.e.,

$$\begin{aligned} \psi^{(m)}(0, t) &= A \cos^2 \frac{\omega_0 t}{2} \sin \omega t \\ &= \frac{A}{2}(1 + \cos \omega_0 t) \sin \omega t \\ &= \frac{A}{2} \sin \omega t + \frac{A}{4}[\sin(\omega + \omega_0)t + \sin(\omega - \omega_0)t]. \end{aligned} \tag{85.3}$$

The consequent excitation registered elsewhere ($x > 0$) after an appropriate lapse of time ($t > c$),

$$\begin{aligned}\psi^{(m)}(x,t) &= \frac{A}{2}\sin\omega\left(t-\frac{x}{c}\right)+\frac{A}{4}\left[\sin(\omega+\omega_0)\left(t-\frac{x}{c}\right)+\sin(\omega-\omega_0)\left(t-\frac{x}{c}\right)\right]\\ &= \frac{A}{2}\sin\omega\left(t-\frac{x}{c}\right)\left[1+\cos\omega_0\left(t-\frac{x}{c}\right)\right]\\ &= A\cos^2\frac{\omega_0}{2}\left(t-\frac{x}{c}\right)\sin\omega\left(t-\frac{x}{c}\right)\\ &= \cos^2\frac{\omega_0}{2}\left(t-\frac{x}{c}\right)\cdot\psi^{(i)}(x,t),\end{aligned} \tag{85.4}$$

thus comprises a progressive wave triplet, whose frequencies ω, $\omega+\omega_0$ and $\omega-\omega_0$ are narrowly separated if $\omega_0 \ll \omega$.

Since the modulated light beam undergoes reflection from a fixed mirror in the Fizeau arrangement and then passes once again between the teeth of the rotating disc before its reception at the eye, another modulation factor $\cos^2(\omega_0 t/2)$ is indicated for the perturbation model wave function, yielding an expression

$$\bar{\psi}^{(m)}(x,t) = \bar{A}(x,t)\sin\omega\left(t-\frac{x}{c}\right) \tag{85.5}$$

with the variable amplitude

$$\bar{A}(x,t) = A\cos^2\frac{\omega_0}{2}\left(t-\frac{x}{c}\right)\cos^2\frac{\omega_0 t}{2}. \tag{85.6}$$

An observational measure of the latter function is rendered by the average, during a modulation cycle $\tau_0 = 2\pi/\omega_0$, of the amplitude squared, namely

$$\begin{aligned} I &= \frac{\omega_0}{2\pi}\int_0^{2\pi/\omega_0}\bar{A}^2(x,t)\,\mathrm{d}t\\ &= \frac{3A^2}{64}\left[1+4\cos^2\left(\frac{\omega_0 x}{2c}\right)+\frac{4}{3}\cos^4\left(\frac{\omega_0 x}{2c}\right)\right],\end{aligned} \tag{85.7}$$

which manifests, at any individual location x, a periodic dependence on the frequency ω_0; thus, a common (and arbitrary) value of I corresponds to the set of frequencies $\omega_0(1+(2cn\pi/x))$, $n = 0, 1, \ldots$. If ω_0^* denotes the smallest magnitude of the set associated with the minimum value of I, it appears that

$$\omega_0^* x/2c = \pi/2$$

or

$$c = \frac{2\omega_0^* l}{\pi} = 4\nu_0^* l \tag{85.8}$$

when $x = 2l$ and $\omega_0^* = 2\pi\nu_0^*$; evidently this relation specifies the carrier wave speed c in terms of two specific observables, a frequency ν_0^* and a length l, whose counterparts in a Fizeau arrangement are the (least) rotational frequency of the disc at minimum transmitted light intensity and the separation between disc and reflecting mirror. It is clear, however, that claims for the practical application of results suggested by the above analysis ought not to be pressed before assessing a presumptive role of wave dispersion.

When c, c_+, c_- designate the phase velocities at frequencies $\omega, \omega+\omega_0$, $\omega-\omega_0$, respectively, the concomitant version of the progressive wave triplet (85.4) becomes

$$\psi^{(m)}(x,t)=\frac{A}{2}\sin\omega\left(t-\frac{x}{c}\right)+\frac{A}{4}\left[\sin(\omega+\omega_0)\left(t-\frac{x}{c_+}\right)+\sin(\omega-\omega_0)\left(t-\frac{x}{c_-}\right)\right].$$

In accordance with the hypothesis that $\omega_0 \ll \omega$, the approximate characterizations

$$c_+=c+\frac{\partial c}{\partial\omega}\omega_0=c(1+\varepsilon),$$

$$c_-=c-\frac{\partial c}{\partial\omega}\omega_0=c(1-\varepsilon)$$

where

$$\varepsilon=\frac{\omega_0}{c}\frac{\partial c}{\partial\omega} \tag{85.9}$$

are suitable; and thence, retaining only terms of first order in ε,

$$\begin{aligned}\psi^{(m)}(x,t)&=\frac{A}{2}\sin\omega\left(t-\frac{x}{c}\right)+\frac{A}{4}\left[\sin\omega\left(t-\frac{x}{c}+\frac{\delta}{c}\right)+\sin\omega\left(t-\frac{x}{c}-\frac{\delta}{c}\right)\right]\\&=A\cos^2\frac{\omega\delta}{2c}\sin\omega\left(t-\frac{x}{c}\right)\\&=A\cos^2\frac{\omega_0}{2}\left(t-\frac{\alpha x}{c}\right)\sin\omega\left(t-\frac{x}{c}\right)\end{aligned} \tag{85.10}$$

if

$$\delta/c=\frac{\omega_0}{\omega}\left(t-\frac{\alpha x}{c}\right),$$

and

$$\alpha=1-\varepsilon\frac{\omega}{\omega_0}=1-\frac{\omega}{c}\frac{\partial c}{\partial\omega}. \tag{85.11}$$

The sole difference between (85.4) and (85.10) is attributable to the numerical parameter α, which effectively alters the coordinate scale in the modulation frequency amplitude factor. It suffices to replace x with αx for modifying the expression (85.8) and, accordingly, the selective condition at the minimum intensity level takes the form

$$\omega_0^*\alpha x/2c=\pi/2$$

or

$$4\nu_0^* l = \frac{c}{\alpha} = \frac{c^2}{c - \omega \dfrac{\partial c}{\partial \omega}}. \tag{85.12}$$

When dispersion exists, therefore, it is the group velocity c/α [cf. (§33)], rather than the phase velocity c, which the twin measures ν_0^*, l of the schematic arrangement determine; and the reality of light dispersion prompts the requisite interpretation of measurements in a Fizeau apparatus.

§86. *Waves in a Random Setting*

Inhomogeneity is typified by a random character in a growing number of contemporary problems, whence it becomes necessary to alter the objectives and style of our previous analytical procedures; precise solutions are, evidently, no longer relevant when the wave setting is specified in probabilistic terms! The nature of a general and systematic approach to wave propagation analysis, assuming small random variations of the configurational parameter(s) about a mean value, can be expeditiously detailed in connection with the transverse displacements $y(x, t)$ of a taut string whose mass density is a random function of position; thus, let us suppose that the density $\rho(x) = \rho_0 + \rho'(x)$ combines a mean component ρ_0 and another, $\rho'(x)$, which varies randomly along the string and measures fluctuations about the mean. For a compatible and operational description of the string movements we shall follow Howe (1971) in advocating a synthesis of the displacement amplitude from its mean and fluctuating parts, respectively, say, $\bar{y}(x, t)$ and $y'(x, t)$; the former identifies an (ensemble) average, pertaining to representatives of a family of strings that have well defined individual values of $\rho'(x)$ at every point and a definite realization probability, of the simultaneous displacements which evolve from a given initial choice, while the latter registers the amplitude difference, $y - \bar{y}$, in any particular experimental realization.

To determine $\bar{y}$ and y' we first associate their sum

$$y(x, t) = \bar{y}(x, t) + y'(x, t) \tag{86.1}$$

with a linear wave equation,

$$\frac{\partial^2 y}{\partial t^2} - c^2 \frac{\partial^2 y}{\partial x^2} = 0 \tag{86.2}$$

and next assume that the square of the local wave speed is expressible in the fashion

$$c^2 = c_0^2(1+\eta(x)) = \frac{P}{\rho_0 + \rho'(x)} \tag{86.3}$$

where

$$c_0^2 = \overline{\{P/(\rho_0 + \rho'(x))\}} \tag{86.4}$$

represents an ensemble average of the tension/density ratio, or mean square wave speed, and $\eta(x)$ designates a random function of small magnitude whose average value is zero. On eliminating c^2 from (86.2) through the use of (86.3), it follows that

$$\frac{\partial^2 y}{\partial t^2} - c_0^2 \frac{\partial^2 y}{\partial x^2} = c_0^2 \eta \frac{\partial^2 y}{\partial x^2} \tag{86.5}$$

and thence, after performing an ensemble average, there obtains an exact relation

$$\frac{\partial^2 \bar{y}}{\partial t^2} - c_0^2 \frac{\partial^2 \bar{y}}{\partial x^2} = c_0^2 \overline{\eta \frac{\partial^2 y'}{\partial x^2}} \tag{86.6}$$

which ascribes the change in mean displacement $\bar{y}$ to a (second order) 'interaction' between correlated values of the pair of random quantities, η and y', categorizing the wave speed and profile. Another equation of exact status is arrived at by subtracting (86.6) from (86.5), namely

$$\frac{\partial^2 y'}{\partial t^2} - c_0^2 \frac{\partial^2 y'}{\partial x^2} = c_0^2 \eta \frac{\partial^2 \bar{y}}{\partial x^2} + c_0^2 \left[\eta \frac{\partial^2 y'}{\partial x^2} - \overline{\eta \frac{\partial^2 y'}{\partial x^2}} \right]; \tag{86.7}$$

here the first and second entries in the right-hand member represent a source and modifier, respectively, of the random displacement. A well posed problem, in particular, calls for the solution of the coupled differential equations (86.6), (86.7) consistent with an assigned initial displacement (or wave profile) devoid of any random component. That the coupled equations are preferable to the original one, (86.5), stems from a facility with which the particulars and interdependence of the component parts of the total displacement may be ascertained.

A systematic procedure for resolving (86.6), (86.7), suitable when η is small and y' vanishes initially, utilizes a sequence of functions $y_1', y_2', \ldots$ which obey the equations

$$\begin{aligned} \frac{\partial^2 y_1'}{\partial t^2} - c_0^2 \frac{\partial^2 y_1'}{\partial x^2} &= c_0^2 \eta \frac{\partial^2 \bar{y}}{\partial x^2} \\ \frac{\partial^2 y_n'}{\partial t^2} - c_0^2 \frac{\partial^2 y_n'}{\partial x^2} &= c_0^2 \eta \frac{\partial^2 \bar{y}}{\partial x^2} + c_0^2 \left[\eta \frac{\partial^2 y_{n-1}'}{\partial x^2} - \overline{\eta \frac{\partial^2 y_{n-1}'}{\partial x^2}} \right], \quad n > 1 \end{aligned} \tag{86.8}$$

and entails their formal construction in terms of η and $\bar{y}$; successive approximations to the mean displacement, $\bar{y}_1, \bar{y}_2, \ldots$, are next defined by substituting the aforesaid versions of $y_1', y_2', \ldots$ into the right-hand side of (86.6).

Employing the causal Green's function of a one-dimensional wave equation, namely

$$\left(\frac{\partial^2}{\partial t^2} - c_0^2 \frac{\partial^2}{\partial x^2}\right) \mathfrak{G}(x, t) = \delta(x)\delta(t), \qquad \mathfrak{G} = 0, \quad t < 0 \tag{86.9}$$

with the explicit characterization (cf. §23)

$$\mathfrak{G}(x, t) = \frac{1}{2c_0} H\left(t - \frac{|x|}{c_0}\right) \tag{86.10}$$

it appears that

$$y_1'(x, t) = c_0^2 \int_{-\infty}^{\infty} \mathfrak{G}(x - \xi, t - \tau)\eta(\xi) \frac{\partial^2 \bar{y}(\xi, \tau)}{\partial \xi^2} \, d\xi \, d\tau \tag{86.11}$$

provides the requisite integral of the first equations (86.8); and that the corresponding approximation $\bar{y}_1$, jointly shaped by (86.6), (86.11), satisfies the homogeneous integro-differential equation

$$\frac{\partial^2 \bar{y}_1}{\partial t^2} - c_0^2 \frac{\partial^2 \bar{y}_1}{\partial x^2} = c_0^4 \int_{-\infty}^{\infty} \overline{\eta(x)\eta(\xi)} \frac{\partial^2 \mathfrak{G}(x - \xi, t - \tau)}{\partial x^2} \frac{\partial^2 \bar{y}_1(\xi, \tau)}{\partial \xi^2} \, d\xi \, d\tau. \tag{86.12}$$

The statistical imprint on the latter equation enters through the correlation product

$$R(x - \xi) = \overline{\eta(x)\eta(\xi)}, \qquad R(0) = \overline{\eta^2} = \text{const.} \tag{86.13}$$

which constitutes an even function of $x - \xi$ alone if η is a stationary random function of its argument variable. Adopting the integration variable $\zeta = x - \xi$ in place of ξ, and making use of the relationships

$$\frac{\partial^2 \mathfrak{G}(x - \xi, t - \tau)}{\partial x^2} = \frac{\partial^2 \mathfrak{G}(\zeta, t - \tau)}{\partial \zeta^2} = -\frac{\delta(\zeta)\delta(t - \tau)}{c_0^2} + \frac{1}{2c_0^3} \delta'\left(t - \tau - \frac{|\zeta|}{c_0}\right)$$

[where $\delta'(\alpha) = (d/d\alpha)\delta(\alpha) = (d^2/d\alpha^2)H(\alpha)$] the equation (86.12) for the mean displacement may be recast in the simpler form

$$\frac{\partial^2 \bar{y}_1}{\partial t^2} - c_0^2(1 - \overline{\eta^2}) \frac{\partial^2 \bar{y}_1}{\partial x^2} = \frac{1}{2} c_0 \int_{-\infty}^{\infty} R(\zeta) \frac{\partial^3 \bar{y}_1\left(x - \zeta, t - \frac{|\zeta|}{c_0}\right)}{\partial t \partial x^2} \, d\zeta. \tag{86.14}$$

To further the pursuit of specific inferences from (86.14), a different version proves convenient, namely

$$\frac{\partial^2 \bar{y}_1}{\partial t^2} - c_0^2(1 - \overline{\eta^2}) \frac{\partial^2 \bar{y}_1}{\partial x^2} = \frac{c_0 L}{2} \int_{-\infty}^{\infty} R(\mu) \frac{\partial^3 \bar{y}_1(x - L\mu, t - L|\mu|/c_0)}{\partial t \partial x^2} \, d\mu, \tag{86.15}$$

wherein the integration variable

$$\mu = \zeta / L$$

is scaled on the correlation length, L, of the density fluctuations. When the magnitude of the latter remains small compared with a typical wave length in the mean displacement spectrum, the conditions are favorable to a Taylor expansion of the integrand factor $\partial^3\bar{y}_1(x - L\mu, t - L|\mu|/c_0)/\partial t \partial x^2$ involving both argument variables. If we neglect all terms of the expansion beyond the first order and recall that $R(\mu)$ is a symmetric function, the non-local or integro-differential equation (86.15) reduces to a local or fourth order partial differential equation, viz.

$$\frac{\partial^2\bar{y}_1}{\partial t^2} - c_0^2(1 - \overline{\eta^2})\frac{\partial^2\bar{y}_1}{\partial x^2} = a_0 c_0 L \frac{\partial^3\bar{y}_1}{\partial t \partial x^2} - a_1 L^2 \frac{\partial^4\bar{y}_1}{\partial t^2 \partial x^2} \tag{86.16}$$

with coefficients

$$a_0 = \int_0^\infty R(\mu)\, \mathrm{d}\mu, \qquad a_1 = \int_0^\infty \mu R(\mu)\, \mathrm{d}\mu$$

which do not depend on L. Noteworthy among the implications of (86.16) is a lesser speed for advance of the mean wave profile, $c_0(1 - \overline{\eta^2})^{1/2}$, than that appropriate to progressive disturbances on the uniform string; and also an independence of roles played by the respective terms in the right-hand member. The first term, with a single time derivative and an irreversible nature, accounts for energy transfer from the mean to the random component of displacement, whereas the second term, with a double time derivative, is responsible for dispersion among the harmonics comprised in a mean displacement wave packet. To give these assertions substance, we represent the latter by a Fourier synthesis,

$$\bar{y}_1(x, t) = \int_{-\infty}^{\infty} \bar{y}_1(k, \omega) \exp\{\mathrm{i}(kx - \omega t)\}\, \mathrm{d}k\, \mathrm{d}\omega \tag{86.17}$$

and examine the dispersion relation forthcoming from (86.16), namely

$$c_0^2(1 - \overline{\eta^2})k^2 - \omega^2 = \mathrm{i} a_0 c_0 L \omega k^2 - a_1 L^2 \omega^2 k^2, \tag{86.18}$$

which yields the expected frequency/wave number proportionality

$$\omega = \pm c_0 k$$

when the random density fluctuations are absent and their correlation length vanishes.

It is a simple matter to obtain, on the basis of (86.18) and the long wave length inequalities $kL \ll 1$, $\omega L/c_0 \ll 1$, an expansion for ω in powers of $\overline{\eta^2}$; utilizing a Gaussian form of the correlation function

$$R(\zeta) = \overline{\eta^2} \exp(-\zeta^2/L^2) \tag{86.19}$$

and the resultant coefficient determinations

$$a_0 = \tfrac{1}{2}\sqrt{(\pi)}\overline{\eta^2}, \qquad a_1 = \frac{1}{2}\overline{\eta^2},$$

the outcome, inclusive of first degree terms in $\overline{\eta^2}$, is

$$\omega = \pm c_0 k\left\{1 - \frac{1}{2}\overline{\eta^2} + \frac{1}{2}\overline{\eta^2}k^2L^2\right\} - \mathrm{i}\frac{\sqrt{(k)}}{4}\overline{\eta^2}c_0Lk^2. \tag{86.20}$$

Both a reduction in wave speed and frequency dispersion are manifest by the real part of ω, and the existence of a negative imaginary part for ω betokens, in consort with (86.17), a damping of the mean displacement in the course of time.

Quantitative aspects of the energy transfer from mean to random excitations of the string, implicit in a subsidence of the former, may be developed through recourse to a pair of balance equations which contain the time derivative of the respective energy densities. The derivation of said energy relations is served by incorporating and isolating, within the fundamental or dynamical equation of motion,

$$(\rho_0 + \rho'(x))\frac{\partial^2 y}{\partial t^2} - P\frac{\partial^2 y}{\partial x^2} = 0, \tag{86.21}$$

the small random component of density, $\rho'(x)$, whose mean value vanishes; thus, one balance equation, fashioned on multiplying the ensemble average of (86.21) with $\partial\bar{y}/\partial t$ and rearranging, takes the form

$$\frac{\partial}{\partial t}\left[\frac{1}{2}\rho_0\left(\frac{\partial\bar{y}}{\partial t}\right)^2 + \frac{1}{2}P\left(\frac{\partial\bar{y}}{\partial x}\right)^2\right] + \frac{\partial}{\partial x}\left[-P\frac{\partial\bar{y}}{\partial x}\frac{\partial\bar{y}}{\partial t}\right] = -\frac{\partial\bar{y}}{\partial t}\cdot\overline{\rho'(x)\frac{\partial^2 y'}{\partial t^2}}, \tag{86.22}$$

while another

$$\frac{\partial}{\partial t}\left[\overline{\frac{1}{2}(\rho_0 + \rho'(x))\left(\frac{\partial y'}{\partial t}\right)^2} + \frac{1}{2}P\overline{\left(\frac{\partial y'}{\partial x}\right)^2}\right] + \frac{\partial}{\partial x}\left[-P\overline{\frac{\partial y'}{\partial x}\frac{\partial y'}{\partial t}}\right] = -\frac{\partial^2\bar{y}}{\partial t^2}\cdot\overline{\rho'(x)\frac{\partial y'}{\partial t}}, \tag{86.23}$$

obtains after multiplying the difference between (86.21) and its ensemble average with $\partial y'/\partial t$, rearranging and performing a further ensemble average. Energy densities for the mean and random displacements and their rates of change are identifiable, along with flux terms, in the left-hand members of (86.22), (86.23); adding the latter relations paves the way, moreover, for exhibiting the ensemble average of an energy conservation equation which involves the resultant (or total) displacement. The fact that the right-hand terms,

$$\bar{\mathfrak{P}} = -\frac{\partial\bar{y}}{\partial t}\cdot\overline{\rho'(x)\frac{\partial^2 y'}{\partial t^2}}, \qquad \mathfrak{P}' = -\frac{\partial^2\bar{y}}{\partial t^2}\cdot\overline{\rho'(x)\frac{\partial y'}{\partial t}}, \tag{86.24}$$

which characterize time variations in the respective energies (per unit length) of the coupled (mean/random) displacements, turn out to possess equal and opposite values when suitably averaged over position and time, viz.,

$$\langle\bar{\mathfrak{P}}\rangle = -\langle\mathfrak{P}'\rangle < 0, \tag{86.25}$$

signifies an irreversible transfer of energy from mean to random states of excitation. To corroborate and also fix (86.25) in clearer perspective, we first rewrite the expression for $\mathfrak{P}$, employing an approximate version of the correlation product, $\overline{\rho'(x)\partial^2 y'/\partial t^2}$, secured by means of the equation

$$\rho_0 \frac{\partial^2 y'}{\partial t^2} - P \frac{\partial^2 y'}{\partial x^2} = -\rho'(x) \frac{\partial^2 \bar{y}}{\partial t^2} \tag{86.26}$$

that results on neglecting the terms

$$\overline{\rho'(x) \frac{\partial^2 y'}{\partial t^2}} - \rho'(x) \frac{\partial^2 y'}{\partial t^2}$$

which appear in the difference between (86.21) and its ensemble average; the aptness of (86.26) for estimating the correlation product, rather than y' individually, reflects an exclusive dependence of the former on local pairings of y' and ρ'.

Fourier synthesis provides an integral of the inhomogeneous wave equation (86.26), embodying a formal source representation of the random displacement generated by the mean component, namely

$$y'(x, t) = \frac{1}{(2\pi)^2} \int_{-\infty}^{\infty} \frac{\rho'(\xi)\omega^2 \bar{y}(\xi, \tau)}{Pk^2 - \rho_0(\omega + \mathrm{i}\varepsilon)^2} \exp\{\mathrm{i}k(x-\xi) - \mathrm{i}\omega(t-\tau)\}\, \mathrm{d}\xi\, \mathrm{d}\tau\, \mathrm{d}k\, \mathrm{d}\omega; \tag{86.27}$$

and compliance with the outgoing wave (or radiation) condition is assured if ω has an infinitesimal positive imaginary part, say ε. Thus

$$\overline{\rho'(x) \frac{\partial^2 y'}{\partial t^2}} = \frac{1}{(2\pi)^2} \int_{-\infty}^{\infty} \frac{\omega^4 \hat{R}(x-\xi)\bar{y}(\xi, \tau)}{Pk^2 - \rho_0(\omega + \mathrm{i}\varepsilon)^2} \exp\{\mathrm{i}k(x-\xi) - \mathrm{i}\omega(t-\tau)\}\, \mathrm{d}\xi\, \mathrm{d}\tau\, \mathrm{d}k\, \mathrm{d}\omega \tag{86.28}$$

where the correlation

$$\hat{R}(x-\xi) = \overline{\rho'(x)\rho'(\xi)} \tag{86.29}$$

constitutes a symmetrical function of its argument.

The combination of (86.28) with the ensemble average of (86.21),

$$\rho_0 \frac{\partial^2 \bar{y}}{\partial x^2} - P \frac{\partial^2 \bar{y}}{\partial x^2} = -\overline{\rho'(x) \frac{\partial^2 y'}{\partial t^2}},$$

advances an (approximate and self-contained) equation for the mean displacement; and it appears therefrom that, insofar as random density variations are a controlling factor, the overall length and time scales for variations in the mean displacement have an order of magnitude $1/|\overline{\rho'^2}|$. These scales, with a considerable span if $\sqrt{(|\overline{\rho'^2}|)} \ll 1$, may be specifically ascribed to the amplitude factor $A(x, t)$ of the propagating mode form

$$\bar{y}(x, t) = A(x, t)\, \mathrm{e}^{\mathrm{i}(\bar{k}x - \bar{\omega}t)} + A^*(x, t)\, \mathrm{e}^{-\mathrm{i}(\bar{k}x - \bar{\omega}t)}, \tag{86.30}$$

whose local variation is determined by the parameters $\bar{k}, \bar{\omega}$. Pursuant to the utilization of (86.28) and (86.30) we obtain, on averaging $\mathfrak{P}(x, t)$ over coordinate and time intervals of the magnitude $1/\surd(\overline{|\rho'^2|})$ say, which are relatively large in comparison with those characterizing local variations of the mean displacement, though small in relation to the overall scales of the latter, an expression

$$\begin{aligned}\langle\bar{\mathfrak{P}}\rangle &= -\mathrm{i}\int_{-\infty}^{\infty}\frac{\bar{\omega}^5\mathfrak{F}(k+\bar{k})|A|^2\,\mathrm{d}k}{Pk^2-\rho_0(\bar{\omega}-\mathrm{i}\varepsilon)^2}+\mathrm{i}\int_{-\infty}^{\infty}\frac{\bar{\omega}^5\mathfrak{F}(k-\bar{k})|A|^2\,\mathrm{d}k}{Pk^2-\rho_0(\bar{\omega}+\mathrm{i}\varepsilon)^2}\\ &= \mathrm{i}\int_{-\infty}^{\infty}\bar{\omega}^5|A|^2\mathfrak{F}(k-\bar{k})\left\{\frac{1}{Pk^2-\rho_0(\bar{\omega}+\mathrm{i}\varepsilon)^2}-\frac{1}{Pk^2-\rho_0(\bar{\omega}-\mathrm{i}\varepsilon)^2}\right\}\mathrm{d}k\end{aligned}$$

wherein

$$\mathfrak{F}(k)=\frac{1}{2\pi}\int_{-\infty}^{\infty}\hat{R}(x)\,\mathrm{e}^{-ikx}\,\mathrm{d}x=\mathfrak{F}(-k) \tag{86.31}$$

is a real and non-negative function.

Since

$$\operatorname*{Lim}_{\varepsilon\to 0}\left\{\frac{\bar{\omega}^5}{Pk^2-\rho_0(\bar{\omega}+\mathrm{i}\varepsilon)^2}-\frac{\bar{\omega}^5}{Pk^2-\rho_0(\bar{\omega}-\mathrm{i}\varepsilon)^2}\right\}=2\pi\,\mathrm{i}|\bar{\omega}|^5\delta(Pk^2-\rho_0\bar{\omega}^2),$$

it follows that

$$\begin{aligned}\langle\bar{\mathfrak{P}}\rangle &= -2\pi\int_{-\infty}^{\infty}|\bar{\omega}|^5|A|^2\mathfrak{F}(k-\bar{k})\delta(Pk^2-\rho_0\bar{\omega}^2)\,\mathrm{d}k\\ &<0,\end{aligned} \tag{86.32}$$

and a wholly analogous evaluation of $\langle\mathfrak{P}\rangle$ confirms the relationship (86.25).

§87. *The Dispersion Relation and its Role in Stability/Gain Analysis*

Small amplitude excitations (or disturbances) that can propagate freely in homogeneous continua acquire their substantive form through a dispersion relation which connects the frequency and wave number. The totality of such waves or free modes possesses a completeness property; that is, appropriate superposition of modal forms furnishes a representation, to the lowest order of magnitude, for all excitations in different systems (comprising fluids, plasmas, electron beams and mechanical varieties). It may therefore be presumed that general systemic properties, which include stability and amplification, are explicable and classifiable with the help of the dispersion relation.

To elaborate on these matters in a broadly conceptual framework, let $\Psi(x, t)$ describe a physical quantity with a one-dimensional support, and assume the existence of a development for same, viz.

$$\Psi(x, t) = \psi_0(x, t) + \varepsilon\psi(x, t) + \varepsilon^2\psi_2(x, t) + \cdots \tag{87.1}$$

where ε is a small magnitude and the initial term, ψ_0, characterizes an equilibrium or unperturbed state of the related system. We shall suppose, furthermore, that the first order perturbation, $\psi(x, t)$, satisfies a homogeneous linear partial differential equation whose coefficients are independent of x and t; and admits the integral representation

$$\psi(x, t) = \int \mathfrak{A}(k, \omega) \exp\{\mathrm{i}(kx - \omega t)\} \,\mathrm{d}k \,\mathrm{d}\omega \tag{87.2}$$

wherein ω and k are linked by the compatibility condition or so-called dispersion relation

$$D(\omega, k) = 0. \tag{87.3}$$

For each frequency ω or wave number k the dispersion relation prescribes an affiliated parameter, namely the wave number or frequency, that completes a specification of the component wave function $\exp\{\mathrm{i}(kx - \omega t)\}$, in (87.2). When the roots of (87.3) are complex, the significance and realization of said functions invites closer attention, since they incorporate a factor with an exponential order of magnitude; for example, when $\operatorname{Im} \omega > 0$ at a specific (real) value of k, the concomitant wave function evidences an exponential rise with time, which is the mark of an instability in the system.

Among the important prototypes of the representation (87.2) are those which describe the perturbations corresponding to pre-selected initial or boundary conditions. If

$$\psi(x, 0) = \int_{-\infty}^{\infty} f_k(0)\, \mathrm{e}^{\mathrm{i}kx} \,\mathrm{d}k \tag{87.4}$$

specifies a limited initial form of disturbance ($|\psi| \to 0, |x| \to \infty$) and the dispersion equation associates with each (real) value of k the discrete set of values $\omega_j(k)$, $j = 1, \ldots, n$, then

$$f_k(t) = \sum_{j=1}^{n} c_j(k)\, \mathrm{e}^{-\mathrm{i}\omega_j(k)t} \tag{87.5}$$

where the coefficients $c_j(k)$ are determined by $f_k(0)$ and a suitable number of derivatives, $(\mathrm{d}f_k/\mathrm{d}t)_{t=0}, \ldots$, contingent on the order of the perturbation differential equation. Thus, the disturbance $\psi(x, t), t > 0$, which evolves from an initial choice (87.4) is, in accordance with the linear basis for its description and in conformity with the general scheme (87.2), given by

$$\psi(x, t) = \int_{-\infty}^{\infty} \sum_{j=1}^{n} c_j(k) \exp\{\mathrm{i}(kx - \omega_j(k)t\} \,\mathrm{d}k. \tag{87.6}$$

Similarly, if we refer to the local and temporally limited excitation

$$\psi(0, t) = \int_{-\infty}^{\infty} \mathfrak{G}_\omega(0)\, \mathrm{e}^{-\mathrm{i}\omega t}\, \mathrm{d}\omega \quad (\omega \text{ real}) \tag{87.7}$$

and

$$\mathfrak{G}_\omega(x) = \sum_{j=1}^{n} \mathrm{d}_j(\omega)\, \mathrm{e}^{\mathrm{i}k_j(\omega)x}, \tag{87.8}$$

then the appertaining level at any location x is expressed by

$$\psi(x, t) = \int_{-\infty}^{\infty} \sum_{j=1}^{n} \mathrm{d}_j(\omega) \exp\{\mathrm{i}(k_j(\omega)x - \omega t)\}\, \mathrm{d}\omega \tag{87.9}$$

where $k = k_j(\omega)$, $j = 1, \ldots, n$ enumerate the discrete roots of a dispersion relation and the $\mathrm{d}_j(\omega)$ are expressible in terms of $\mathfrak{G}_\omega(0), (\mathrm{d}\,\mathfrak{G}_\omega/\mathrm{d}x)_{x=0}, \ldots$.

A more direct procedure (Convert, 1964) for accomodating the given data (initial or boundary) involves the composition of $\psi(x, t)$ with a pair of complex Fourier integrals, the latter being indicated when it is necessary to represent a function that can have exponential growth at infinity. If we postulate the asymptotic behaviors

$$f(t) \simeq \begin{cases} \exp(\alpha t), & t \to \infty, \quad \alpha > 0 \\ \exp(-\beta t), & t \to -\infty, \quad \beta > 0 \end{cases} \tag{87.10}$$

and define the half-range transforms

$$\begin{aligned} f^+(\omega) &= \frac{1}{2\pi} \int_0^{\infty} f(t)\, \mathrm{e}^{\mathrm{i}\omega t}\, \mathrm{d}t, \quad \operatorname{Im} \omega > \alpha \\ f^-(\omega) &= \frac{1}{2\pi} \int_{-\infty}^{0} f(t)\, \mathrm{e}^{\mathrm{i}\omega t}\, \mathrm{d}t, \quad \operatorname{Im} \omega < -\beta \end{aligned} \tag{87.11}$$

of a function $f(t), -\infty < t < \infty$, its complex Fourier representation may be written as

$$f(t) = \int_{C^+} f^+(\omega)\, \mathrm{e}^{-\mathrm{i}\omega t}\, \mathrm{d}\omega + \int_{C^-} f^-(\omega)\, \mathrm{e}^{-\mathrm{i}\omega t}\, \mathrm{d}\omega \tag{87.12}$$

where the integration contours C^+, C^- pass above and below the singularities of $f^+(\omega)$ and $f^-(\omega)$, respectively; the first and second integrals in (87.12) thus provide characterizations of $f(t)$ in the complementary ranges $t >, < 0$.

Referring again to the excitation (87.7) at a specific place ($x = 0$) and introducing the half-range transforms

$$\mathfrak{G}^{\pm}_{\omega,k} = \pm \frac{1}{2\pi} \int_0^{\pm\infty} \mathfrak{G}_\omega(x)\, \mathrm{e}^{-\mathrm{i}kx}\, \mathrm{d}x \tag{87.13}$$

(whose existence is assured by appropriate stipulations on the magnitude of

Im k in $x \gtrless 0$), it follows that

$$\mathfrak{G}_\omega(x) = \int_{C^+} \mathfrak{G}^+_{\omega,k}\, e^{ikx}\, dk + \int_{C^-} \mathfrak{G}^-_{\omega,k}\, e^{ikx}\, dk \tag{87.14}$$

and thence we obtain the formal representation

$$\psi(x, t) = \int_{-\infty}^{\infty} d\omega \left[\int_{C^+} \mathfrak{G}^+_{\omega,k} \exp\{i(kx - \omega t)\}\, dk + \int_{C^-} \mathfrak{G}^-_{\omega,k} \exp\{i(kx - \omega t)\}\, dk \right] \tag{87.15}$$

of the disturbance at any location. The transforms $\mathfrak{G}^\pm_{\omega,k}$ are individually characterized by equations

$$D(\omega, k)\mathfrak{G}^\pm_{\omega,k} = \pm E(\omega, k) \tag{87.16}$$

that contain the dispersion function $D(\omega, k)$ and another, $E(\omega, k)$, which depends linearly on the values of $\mathfrak{G}_\omega(x)$ and its derivatives at the reference point $x = 0$. After eliminating $\mathfrak{G}^\pm_{\omega,k}$ from (87.15) on the basis of (87.16) there obtains, in keeping with the disposition of the contours C^+, C^-, a representation of more practical nature, viz.

$$\psi(x, t) = \oint_{-\infty}^{\infty} d\omega \int \frac{E(\omega, k)}{D(\omega, k)} \exp\{i(kx - \omega t)\}\, dk \tag{87.17}$$

wherein the k-integral extends over a complete curve that encloses all singularities of the integrand; and it is apparent how, following an evaluation of the latter integral, we recover the prior representation (87.9). The counterpart of (87.17), with interchanged places for ω and k, namely

$$\psi(x, t) = \oint_{-\infty}^{\infty} dk \int \frac{E(\omega, k)}{D(\omega, k)} \exp\{i(kx - \omega t)\}\, d\omega \tag{87.18}$$

is suited to a specification of initial values for ψ and its t-derivatives at $t = 0$ (which enter via the function $E(\omega, k)$).

With the establishment of various (first order) expressions for the response of linear systems to small perturbations (at a definite instant of time or location), we may commence an examination of stability and amplification properties. A system is termed stable if, consequent to an initial ($t = 0$) disturbance of finite extent therein, the response $\psi(x, t), t > 0$, remains bounded everywhere as time goes on; and unstable if $\psi(x, t)$ increases without limit at particular location(s) x when $t \to \infty$. Since the Fourier integral representation for a limited initial disturbance involves real wave numbers, it follows from (87.6) that the system is unstable when, corresponding to a real value of k, one (or more) of the solutions (or branches) $\omega_j(k)$ of the dispersion relation has a positive imaginary part, and the system is stable if, for every real value of k, all the solutions $\omega_j(k)$ have negative imaginary parts. A classification of instabilities, enunciated by Landau and Lifschitz (1958) and also by Sturrock (1958), reflects two contrasting types of evolutionary behavior for a perturbation that initially surrounds the point $x = 0$, say; an essential instability occurs in the circumstance that a growing perturbation is continually registered at and near the point $x = 0$, while a convective instability is manifest by mobile perturbations of increasing magnitude that leave the neighborhood of the origin, where the level of excitation subsequently diminishes.

To formulate instability criteria, let us consider (vide (87.6)) the particular measure, at the fixed point $x = 0$, of a perturbation which is originally confined to a small enveloping range, viz.

$$\psi(0, t) = \int_{-\infty}^{\infty} \sum_{j=1}^{n} c_j(k)\, \mathrm{e}^{-\mathrm{i}\omega_j(k)t}\, \mathrm{d}k. \tag{87.19}$$

Since the coefficient functions $c_j(k)$ possess a bounded variation, the (gross) behavior of $\psi(0, t)$ in the limit $t \to \infty$ may be adduced from that of the integrals

$$I_j = \int_{-\infty}^{\infty} \exp\{-\mathrm{i}\omega_j(k)t\}\, \mathrm{d}k, \quad j = 1, \ldots, n \tag{87.20}$$

conducted along the real axis of the k-plane; if $\omega_j(k)$ has a complex nature, a direct estimate for the latter is impeded by the joint occurrence of rapidly varying and exponentially large integrand factors. A transformation of the integrals (87.20) is thus indicated, assuming that the locus for $\omega_j(k)$, mapped out when k traverses the real axis, is the curve C_j in Fig. 48 which lies partly in the upper half of the complex ω-plane. If it is feasible to displace the contour C_j onto or below the real ω-axis, in conjunction with a shift of the integration contour in the k-plane, we infer that the representative integral I_j approaches zero, or remains bounded, as $t \to \infty$; and hence the system is a stable one if such transformations can be effected in a collective manner.

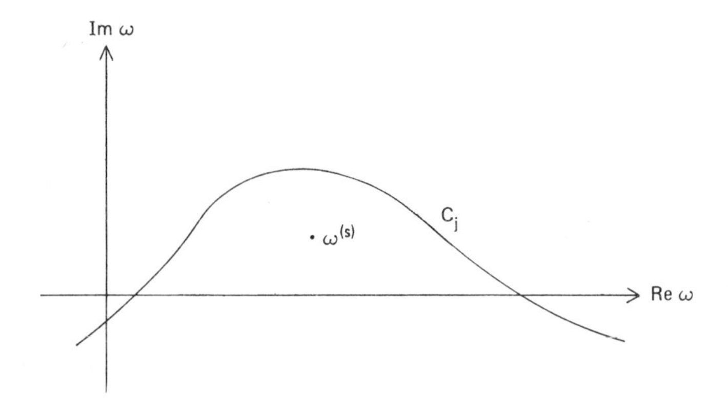

Fig. 48. Dispersion curve and a saddle point.

Analyticity of the function $\mathrm{d}k/\mathrm{d}\omega$ in the region of the ω-plane passed over on displacement of the curve C_j justifies this process and, contrariwise, the way to the intended disposition of said curve is blocked by the presence of any singular point of the derivative above the real axis.

The conditions

$$D(\omega^{(s)}, k^{(s)}) = 0, \quad \left(\frac{\partial D}{\partial k}\right)_{\substack{k=k^{(s)} \\ \omega=\omega^{(s)}}} = 0$$

characterize, for example, a branch type singularity at the point $\omega^{(s)}$ of the function $k(\omega)$ obtained by explicit resolution of the dispersion relation; and the local behavior

$$k - k^{(s)} \doteq A(\omega - \omega^{(s)})^{1/2}, \quad \omega \to \omega^{(s)}$$

makes evident that the inverse function $\omega(k)$ possesses a saddle point at $k^{(s)}$. When the integration contour for I_j is converted to a steepest descent path and reference be made to the (asymptotic) estimate arising from the vicinity of the saddle point for $\omega_j(k)$ at $k^{(s)}$, viz.

$$I_j \sim \left\{ \left\{ -\frac{\mathrm{i}}{2\pi} \left(\frac{\mathrm{d}^2\omega}{\mathrm{d}k^2}\right)_{k=k^{(s)}} t \right\}^{-1/2} \exp\{-\mathrm{i}\omega^{(s)} t\} \right\}, \quad t \to \infty,$$

we deduce that, if the latter corresponds to a point $\omega^{(s)}$ situated below the curve C_j and above the real axis (see figure), the magnitude of I_j increases without limit as $t \to \infty$; and hence the system manifests an essential instability. In practical and general terms, therefore, the criterion advanced for distinguishing between essential and convective instabilities rests on explicit solution of the dispersion relation and the subsequent location of singular points of the function $\mathrm{d}k/\mathrm{d}\omega$ relative to the curve(s) $\omega_j(k)$ traced out as k travels along the real axis.

A different criterion for contrasting instabilities is based on the expression [cf. (87.18)]

$$\psi(0, t) = \int_{-\infty}^{\infty} \mathrm{d}k \int_{C_\omega^+} \frac{E(\omega, k)}{D(\omega, k)} \mathrm{e}^{-\mathrm{i}\omega t}\, \mathrm{d}\omega \qquad (87.21)$$

which, assuming that the contour C_ω^+ passes above all singularities of the integrand, characterizes a vanishing perturbation at $x = 0$ prior to the instant of time $t = 0$; and this is couched in terms of the conditions enabling the contour C_ω^+ to be brought into coincidence with the real axis. The instability is, evidently, a convective one when such a deformation of the contour can be accomplished and otherwise it has an essential nature. Inasmuch as displacements of the singularities of the integrand in (87.21), located at the zeros $k = k_j(\omega)$ of the dispersion function $D(\omega, k)$, are compelled by varying ω, the intended repositioning of C_ω^+ is conditional on the possibility to shift the k-contour and thereby avoid these mobile singularities. The latter option is foreclosed when a pair of singularities tend (as $\omega \to \omega^{(s)}$, $\operatorname{Im} \omega^{(s)} > 0$, say) towards a point on the k-integration contour opposite sides thereof and thus block its freedom of movement.

If the dispersion relation associates complex values of the wave number, $k_j(\omega)$, with a range of real values for ω, positional rather than temporal changes bring out the salient behavior of the affiliated wave functions; evanescent waves, specifically, diminish in amplitude with increasing separation from their source, which may correspond to the site, $x = 0$, of a given perturbation, $\psi(0, t)$, that has a limited time span. A distinction between evanescent and amplifying waves, which are also tied in with complex values of the wave number, stems from an observation (of Sturrock) that the former cannot be 'launched' by the source; thus, amplification is ruled out unless a perturbation of the aforesaid type has the characteristics of a limited (one-dimensional) wave packet, signifying that $\psi(x, t) \to 0$ as $|x| \to \infty$, t arbitrary. Otherwise stated, and in analogy with the criteria bearing on stability, a system supports evanescent waves when the dispersion relation admits complex solutions $k = k_j(\omega)$ for real frequencies and there is a singularity of the function $(\mathrm{d}k_j/\mathrm{d}\omega)^{-1}$ lying between the real axis of the k-plane and a curve C_j traced out by $k = k_j(\omega)$ as ω ranges over the real spectrum.

To illustrate the application of general principles and appreciate the systematization achieved by their use, let us consider the particular dispersion relation of algebraic nature,

$$D(\omega, k) = (\omega - kv)^2 - (kc)^2 - \alpha = 0 \tag{87.22}$$

that is linked with the second order differential equation

$$\left(\frac{\partial}{\partial t} + v\frac{\partial}{\partial x}\right)^2 \psi = c^2 \frac{\partial^2 \psi}{\partial x^2} - \alpha\psi. \tag{87.23}$$

For interpretative purposes, we associate the latter equation with a mechanical system, namely a taut string which undergoes translation along its length at the uniform speed v and experiences a transverse force (per unit length) proportional to the local displacement $\psi(x, t)$, whose sense depends on the sign of α. If $\alpha = \omega_0^2 > 1$, as befits a restoring or stabilizing force, complex values of k are implied by the dispersion relation (87.22) when $c > |v|$ and $|\omega| < (1 - (v^2/c^2))^{1/2}\omega_0$, and are characterized by the pair of equations

$$\operatorname{Re} k = -\frac{v\omega}{c^2 - v^2}, \quad \operatorname{Im} k = \pm \frac{[(c^2 - v^2)\omega_0^2 - c^2\omega^2]^{1/2}}{c^2 - v^2}.$$

On solving the dispersion relation for ω, we obtain the respective branch functions,

$$\omega = \omega_\pm(k) = vk \pm (\omega_0^2 + c^2k^2)^{1/2}, \tag{87.24}$$

whose real values throughout the range $-\infty < k < \infty$ (see Fig. 49) indicate a

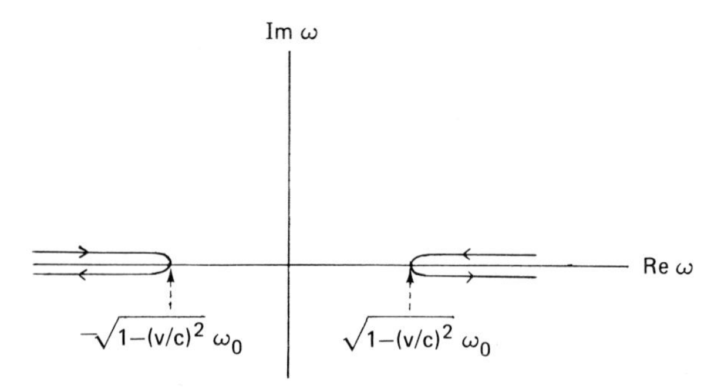

Fig. 49. Branches of a dispersion relation.

stable nature of the system. Differentiation of (87.22) or (87.24) reveals that

$$\left(\frac{\mathrm{d}k}{\mathrm{d}\omega}\right)^{-1} = \frac{\mathrm{d}\omega}{\mathrm{d}k} = v + \frac{c^2 k}{\omega - vk}$$

with a (pole) singularity in the right-hand member (qua function of k) at $k = \omega/v$ on the real axis if ω is real. To infer that the system supports evanescent waves only, we observe that the requisite localization in both time and position for a growing wave packet therein, requiring the interchangeability or transformation of (single) Fourier integrals with respect to the separate variables ω, k, is negated by virtue of the non-overlapping aspect of the branch curves $\omega_\pm(k)$.

If $\alpha = -\mu^2 > 0$ and $v > c$, the determinations which follow from (87.22), viz.

$$\operatorname{Re} k = \frac{v\omega}{v^2 - c^2}, \quad \operatorname{Im} k = \frac{\pm[(v^2 - c^2)\mu^2 - c^2\omega^2]^{1/2}}{v^2 - c^2}, \quad |\omega| < \left(\frac{v^2}{c^2} - 1\right)^{1/2} \mu$$

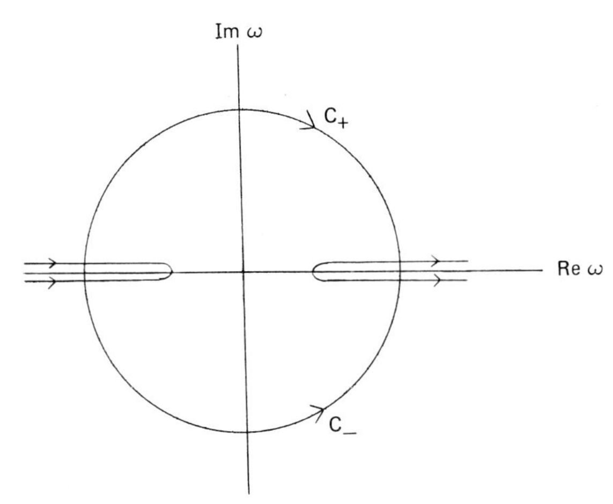

Fig. 50. Interconnection of branch curves.

and

$$\mathrm{Re}\ k = \frac{v\omega \pm [c^2\omega^2 - (v^2 - c^2)\mu^2]^{1/2}}{v^2 - c^2}, \quad \mathrm{Im}\ k = 0, \quad |\omega| > \left(\frac{v^2}{c^2} - 1\right)\mu$$

restrict the possibility of wave amplification to the frequency band $|\omega| < ((v^2/c^2) - 1)^{1/2}\mu$. The inverse function determinations

$$\mathrm{Re}\ \omega = vk, \quad \mathrm{Im}\ \omega = \pm(\mu^2 - c^2k^2)^{1/2}, \quad |k| > \mu/c,$$
$$\mathrm{Re}\ \omega = vk \pm (c^2k^2 - \mu^2)^{1/2}, \quad \mathrm{Im}\ \omega = 0, \quad |k| > \mu/c$$

confirm the existence of curves (namely, C_+, C_- in Fig. 50), bridging the cited frequency band, which are correlated with displacement of k along the real line; and the permissible arrangement of integrals,

$$\int_{-\infty}^{\infty} \mathfrak{G}(\omega) \exp\{\mathrm{i}[k(\omega)x - \omega t]\}\,\mathrm{d}\omega = \int_{-\infty}^{\infty} \mathfrak{G}(\omega(k)) \exp\{\mathrm{i}[kx - \omega(k)t]\}\frac{\mathrm{d}\omega}{\mathrm{d}k}\,\mathrm{d}k,$$

extended over real values for ω and k, respectively, bespeaks the fact that the perturbations defined thereby are of the amplifying type.

Problems III

1. Given the differential equation

$$\frac{d^2\varphi}{dx^2}+k^2\varepsilon(x)\varphi=0, \tag{*}$$

with a symmetric coefficient profile

$$\varepsilon(x)=1-\frac{4p^2-1}{4k^2(L+|x|)^2}, \qquad -\infty<x<\infty$$

that varies continuously between the common limits $\varepsilon(\pm\infty)=1$ and assumes a (positive, negative or zero) value

$$\varepsilon_0=\varepsilon(0)=1-\frac{p^2-1/4}{(kL)^2}$$

at $x=0$, where its first derivative is discontinuous. Determine the explicit form of solutions to (*) upon the intervals $x \gtrless 0$ which have the asymptotic behaviors indicative of primary (incident) wave stimulus from the left, viz.

$$\varphi(x)\simeq \begin{cases}[e^{ikx}+r\,e^{-ikx+2i\delta_p}]\,e^{-i\delta_p}, & x\to-\infty\\ t\,e^{ikx+i\delta_p}, & x\to+\infty\end{cases}$$

with

$$\delta_p=kL-\frac{2p+1}{4}\pi;$$

and, by imposing the continuity of φ and $d\varphi/dx$ at $x=0$, obtain the representations

$$r=\frac{1}{2}\left[\frac{H_p^{(2)}(\alpha)}{H_p^{(1)}(\alpha)}+\frac{H_p^{(2)}(\alpha)+2\alpha H_p^{(2)\prime}(\alpha)}{H_p^{(1)}(\alpha)+2\alpha H_p^{(1)\prime}(\alpha)}\right]$$

$$t=\frac{1}{2}\left[\frac{H_p^{(2)}(\alpha)}{H_p^{(1)}(\alpha)}-\frac{H_p^{(2)}(\alpha)+2\alpha H_p^{(2)\prime}(\alpha)}{H_p^{(1)}(\alpha)+2\alpha H_p^{(2)\prime}(\alpha)}\right]$$

where $H_p^{(1),(2)\prime}(\alpha)=(d/d\alpha)H_p^{(1),(2)}(\alpha)$ and

$$\alpha=kL.$$

Confirm the scattering coefficient relationship

$$|r|^2+|t|^2=1$$

and the individual determination

$$|r|=\left\{1+\left|\frac{4}{\pi\frac{d}{d\alpha}(\alpha H_p^{(1)}(\alpha)H_p^{(2)}(\alpha))}\right|^2\right\}^{-1/2} \tag{**}$$

after employing the Wronskian formula

$$H_p^{(1)}(\alpha)H_p^{(2)'}(\alpha) - H_p^{(1)'}(\alpha)H_p^{(2)}(\alpha) = -\frac{4\,\mathrm{i}}{\pi\alpha}.$$

Examine, with reference to (**), the dependence of $|r|$ on the parameters α, ε_0 and p which characterize the coefficient function $\varepsilon(x)$; and show that if α increases, or the wave length $\lambda = 2\pi/k$ diminishes in comparison with L, the transition between partial and almost total reflection narrows in the ε_0-scale and is effected at very small values of ε_0. Verify that a jump in $|r|$, from zero to unity, occurs at the parametric value $\varepsilon_0 = 0$ for vanishingly small wave length, thereby linking the existence of reflection with the circumstance of a null point for the wave number in a slowly changing environment. Apply suitable approximations for the Hankel functions of equal argument and order to establish the limiting magnitude $|r| \simeq 1/2$ when $\varepsilon_0 = 0$ and $\alpha = p(\gg 1)$ increases.

2. The free surface elevation $\zeta(x, t)$ of unidirectional waves in a canal whose depth, h, is large/small compared to ζ and the wave length, respectively, satisfies a linear equation

$$\frac{\partial^2\zeta}{\partial t^2} = g\frac{\partial}{\partial x}\left(h\frac{\partial\zeta}{\partial x}\right)$$

where g denotes the gravitational constant; and the progressive wave solutions

$$\zeta = f(x - ct) + g(x + ct), \quad c^2 = gh$$

are realized when h is constant. Assume, for purposes of devising an approximate solution in the circumstance of varying depth $h(x)$, that

$$\zeta(x, t) = A(x)\cos\omega(t - \varphi(x))$$

and obtain a pair of coupled ordinary differential equations for the amplitude and phase functions A, φ. Show that the characterizations

$$\frac{\mathrm{d}\varphi}{\mathrm{d}x} = (gh)^{-1/2}, \quad A^2 = \text{const.}\,(g/h)^{1/2}$$

are consistent with these equations if $\mathrm{d}/\mathrm{d}x(h(\mathrm{d}A/\mathrm{d}x)) \ll \omega^2 A/g$ or $\mathrm{d}h/\mathrm{d}x$, $\mathrm{d}^2h/\mathrm{d}x^2 \ll h/\lambda$, where $\lambda = 2\pi/\omega(gh)^{1/2}$; and interpret the results.

3. An infinite string has uniform outer portions $x < 0, x > L$, wherein the phase velocity of progressive wave forms equals the (same) constant c, and a non-uniform intervening portion $0 < x < L$, wherein slow variation of the local phase velocity $c(x)$ is implied by the inequality $(L/c)(\mathrm{d}c/\mathrm{d}x) \ll 1$. If $y = f(t - (x/c))$ represents the assigned and invariant profile of a wave incident from $x = -\infty$, show that its perturbed form after encountering the central portion is

$$y_1 = f(t - \varphi(x)), \qquad \varphi(x) = \int_0^x \frac{d\xi}{c(\xi)}$$
$$y_2 = f(t - \varphi(x))\left\{1 + \frac{1}{2}\log\frac{c(x)}{c(0)}\right\},$$

in first and second orders of approximation. Determine the initial, namely second order, approximation to the wave which progresses in the reverse direction and, on the hypothesis of time periodicity, make comparison with independently achieved results for the same arrangement.

4. Suppose the complex displacement factor

$$s(x) = e^{ikx} + Cs(-l/2)\, e^{ik|x+l/2|} + C's(l/2)\, e^{ik|x-l/2|}$$

pertains to time-periodic excitations on a string with mass loads M, M' at the respective points $x = -l/2, l/2$, and with a primary wave of unit amplitude incident from the left ($x = -\infty$). Verify that the reflection coefficient (based on the reference point $x = 0$),

$$r = Cs(-l/2)\, e^{-ikl/2} + C's(l/2)\, e^{ikl/2},$$

has an alternative representation

$$\begin{aligned} r = {} & 2Cs(-l/2)\, e^{-ikl/2} + 2C's(l/2)\, e^{ikl/2} \\ & - C(1-C)s^2(-l/2) - C'(1-C')s^2(l/2) + 2CC's(-l/2)s(l/2)\, e^{ikl} \end{aligned}$$

which is stationary for arbitrary and independent first order variations of $s(-l/2)$ and $s(l/2)$ relative to their proper values; and obtain the corresponding representation for the transmission coefficient. Employ the stationary forms to derive expressions for r and t in terms of C and C'; and compare the latter with exact determinations. Can a variational characterization be devised if the loads are elastically coupled to the string?

5. The pair of conservation relations

$$\frac{\partial}{\partial t}\left\{\frac{1}{2}\left(\frac{\partial\phi}{\partial t}\right)^2 + \frac{1}{2}\left(\frac{\partial\phi}{\partial x}\right)^2 + \frac{1}{2}\phi^2\right\} + \frac{\partial}{\partial x}\left\{-\frac{\partial\phi}{\partial x}\frac{\partial\phi}{\partial t}\right\} = 0,$$
$$\frac{\partial}{\partial t}\left\{-\frac{\partial\phi}{\partial x}\frac{\partial\phi}{\partial t}\right\} + \frac{\partial}{\partial x}\left\{\frac{1}{2}\left(\frac{\partial\phi}{\partial t}\right)^2 + \frac{1}{2}\left(\frac{\partial\phi}{\partial x}\right)^2 - \frac{1}{2}\phi^2\right\} = 0$$

are associated with the linear partial differential equation

$$\frac{\partial^2\phi}{\partial t^2} - \frac{\partial^2\phi}{\partial x^2} + \phi = 0$$

cast in a non-dimensional form. Consider a progressive wave solution

$$\phi(x, t) - A(x, t)\cos(k(x, t)x - \omega t)$$

whose amplitude $A(x, t)$ and wave number $k(x, t)$ are slowly varying functions; and show that, subsequent to an averaging of the conservation relations, the amplitude and wave number satisfy a non-linear first order system

$$\frac{\partial k}{\partial t}+\frac{d\omega}{dk}\frac{\partial k}{\partial x}=0$$

$$\frac{\partial A}{\partial t}+\frac{d\omega}{dk}\frac{\partial A}{\partial x}+\frac{1}{2}A\frac{d^2\omega}{dk^2}\frac{\partial k}{\partial x}=0$$

with the versions

$$\frac{dk}{dt}=0, \quad \frac{dA}{dt}=-\frac{1}{2}A\frac{d^2\omega}{dk^2}\frac{\partial k}{\partial x}$$

on the characteristic curves $dx/dt = d\omega/dk = k/\sqrt{(k^2+1)}$. Interpret the latter generally and give their particular realizations if the initial disturbance $y(x, 0)$ has a localized aspect relative to $x = 0$.

Note. For additional problems, see Problems IV on p. 493.

*Problems IV**

Problems II(b) (cf. p. 267)

11. The sudden introduction of some additional fluid into a previously steady or quiescent system affords a means of generating non-periodic and solitary surface waves therein; such excitations, first observed in a canal over a century ago, are naturally manifest by large scale river movements. Consider, for definiteness and simplicity, an infinite straight canal with horizontal bottom and vertical side walls at fixed separation, in which a source of additional fluid at one extremity gives rise to fluid motions parallel to the side walls and also a progressive free surface undulation that moves away, along the canal, from the launching site. Let the x, y directions be chosen so that the former is aligned with the principal axis of the canal while the latter points upward from the bottom; and assume that the system of equations

$$\mathrm{d}v_x/\mathrm{d}t = \partial v_x/\partial t + v_x \partial v_x/\partial x + v_y \partial v_x/\partial y = -\frac{1}{\rho}\frac{\partial p}{\partial x}$$

$$\mathrm{d}v_y/\mathrm{d}t = \partial v_y/\partial t + v_x \partial v_y/\partial x + v_y \partial v_y/\partial y = -g - \frac{1}{\rho}\frac{\partial p}{\partial y}$$

$$\partial v_x/\partial x + \partial v_y/\partial y = 0 \qquad (*)$$

governs the stipulated type of fluid motion, in conjunction with a source condition at $x = -\infty$. If H represents the uniform equilibrium depth of the fluid, $\zeta(x, t)$ the variable magnitude of vertical surface displacement, and $U(x, t) = \int_0^{H+\zeta} v_x \,\mathrm{d}y/(H + \zeta)$ the mean horizontal velocity, show that an alternative version of the equation of continuity (*) has the form

$$\frac{\partial \zeta}{\partial t} + \frac{\partial}{\partial x}[(H + \zeta)U] = 0.$$

On the premise of negligible vertical acceleration, $\mathrm{d}v_y/\mathrm{d}t = 0$, which underlies the consequent equality $v_x = U$, and also of a comparative scale ratio $\zeta/H \ll 1$, establish the pair of first order equations

$$\frac{\partial U}{\partial t} + g\frac{\partial \zeta}{\partial x} = 0, \qquad \frac{\partial \zeta}{\partial t} + H\frac{\partial U}{\partial x} = 0$$

that link both ζ and U with the same form invariant progressive wave character typical of small amplitude string displacements, and note the particular wave speed determination $c = (gH)^{1/2}$.

*Additional problems added in proof.

12. There are two significant measures for the configuration in the above problem, namely $Q = \int_{-\infty}^{\infty} \zeta \, \mathrm{d}x$ and $q(x, t) = \int_x^{\infty} \zeta \, \mathrm{d}x$, which specify the entire amount of fluid introduced (per unit width) at the end $x = -\infty$ of the canal and the net amount of displaced surface volume on the far side of the observation point, x. If c^* designates the speed of propagation for a displaced surface volume element, whose magnitude is found through the twin requirements $\mathrm{d}q = 0$ and $\mathrm{d}x = c^* \, \mathrm{d}t$, confirm the representation $c^*\zeta = \int_x^{\infty} (\partial\zeta/\partial t) \, \mathrm{d}x$ and thence obtain a third form of the continuity or mass conservation equation, $(H + \zeta)U = \zeta c^*$. Show that, in the first order approximation previously cited, $c^* = c = (gH)^{1/2}$, and, given the succeeding or second order approximation (Boussinesq, 1877)

$$c^* = (gH)^{1/2}\left\{1 + \frac{3\zeta}{4H} + \frac{H^2}{6\zeta}\frac{\partial^2\zeta}{\partial x^2}\right\}$$

deduce the concomitant estimate

$$\hat{c} = (gH)^{1/2}\left\{1 + \frac{\zeta}{H} + \frac{1}{6}H^2\left[\frac{2}{\zeta}\frac{\partial^2\zeta}{\partial x^2} - \frac{1}{\zeta^2}\left(\frac{\partial\zeta}{\partial x}\right)^2\right]\right\}$$

for the propagation speed of an element of surface energy, implicit in the relations $\mathrm{d}w = 0$ and $\mathrm{d}x = \hat{c} \, \mathrm{d}t$, where $w = \int_x^{\infty} \zeta^2 \, \mathrm{d}x$.

Problems II(c) (cf. p. 320)

11. Study the inhomogeneous linear system (54.16), whose coefficient matrix has a difference form as regards the indices, and apply the results gained thereby to the appertaining initial value problem for a loaded string.

12. Let it be supposed that a single deviation of mass, or impurity, occurs in an otherwise regular array of equal loads on a dense string; and, specifically, locate the former, with mass $M' = \rho\beta/k^2$ at $x = 0$, while the lattice points $x = \pm l, \pm 2l, \ldots$ are occupied by a different, though common magnitude, load $M = \rho\alpha/k^2$. Introduce the transverse displacement factor representations

$$s(x) = \alpha l s(0) \sum_{n=-\infty}^{\infty} \frac{\exp\{\mathrm{i}(k + 2n\pi/l)x\}}{(kl + 2n\pi)^2 - (kl)^2}, \qquad x > 0$$

and

$$s(x) = l^2 \sum_{n=-\infty}^{\infty} [\exp\{\mathrm{i}(k + 2n\pi/l)x\}$$

$$+ R\exp\{-\mathrm{i}(k + 2n\pi/l)\}] \frac{1}{(kl + 2n\pi)^2 - (kl)^2} \qquad x < 0$$

for purposes of calculating the reflection coefficient R appropriate to an incident wave on the left (i.e., from $x = -\infty$), and, taking note of the sums

$$S_1(a, b, \varepsilon) = \sum_{n=-\infty}^{\infty} \frac{e^{2in\pi\varepsilon}}{(2n\pi + a)^2 + b^2} = S_1(-a, b, -\varepsilon)$$

$$= \frac{1}{2b} e^{(b-ia)\varepsilon} \left\{ \frac{1}{1 - e^{-(b-ia)}} - 1 \right\} + \frac{1}{2b} \frac{e^{-(b+ia)\varepsilon}}{1 - e^{-(b+ia)}}, \quad 0 < \varepsilon < 1,$$

and

$$S_2(a, b, \varepsilon) = \sum_{n=-\infty}^{\infty} \frac{(2n\pi + a)e^{2in\pi\varepsilon}}{(2n\pi + a)^2 + b^2} = -S_2(-a, b, -\varepsilon)$$

$$= \frac{1}{2i} e^{(b-ia)\varepsilon} \left\{ \frac{1}{1 - e^{-(b-ia)}} - 1 \right\} - \frac{1}{2i} \frac{e^{-(b+ia)\varepsilon}}{1 - e^{-(b+ia)}}, \quad 0 < \varepsilon < 1$$

confirm the expression

$$R = -\frac{1 - (M'/M)}{1 - (M'/M) - \dfrac{i \sin kl}{\cos kl - \cos kl}}.$$

Study the corresponding problem wherein the impurity is not located at a lattice point, say $x = \delta l$, $0 < \delta < 1$.

Problems III (cf. p. 492)

6. The density of a stretched string increases uniformly but slightly from one end to the other. Show that, if ρ be the mean density, $\Delta\rho$ the difference of the densities at the ends, the longest period is greater than that of a uniform string of the same tension and length in the ratio

$$1 + \frac{1}{96}\left(\frac{15}{\pi^2} - 1\right)\left(\frac{\Delta\rho}{\rho}\right)^2 : 1$$

and that the proper functions are approximately of the form

$$\sin\frac{n\pi x}{l} - \frac{1}{4}\frac{\Delta\rho}{\rho}\left\{\frac{x - l/2}{l}\sin\frac{n\pi x}{l} + \frac{n\pi x(l - x)}{l^2}\cos\frac{n\pi x}{l}\right\}$$

7. Consider the free oscillations of water in a narrow elongated basin, neglecting all save the longitudinal or x component of velocity, u, and relying on the coupled equations

$$\frac{\partial u}{\partial t} + g\frac{\partial \zeta}{\partial x} = 0, \qquad b(x)\frac{\partial \zeta}{\partial t} + \frac{\partial}{\partial x}(A(x)\zeta) = 0,$$

where $\zeta(x, t)$ designates the surface elevation $b(x)$ the breadth of the basin at the (undisturbed) surface and $A(x)$ the area of a transverse section. Assuming a state of time-periodic oscillation, with

$$\zeta(x, t) = s(x)e^{-i\omega t},$$

and the particular depth specification

$$h = h_0\left(1 - \frac{x^2}{a^2} - \frac{y^2}{b^2}\right)$$

which implies

$$b(x) = 2b\left(1 - \frac{x^2}{a^2}\right)^{1/2}, \qquad A(x) = \frac{4}{3}h_0 b\left(1 - \frac{x^2}{a^2}\right)^{3/2},$$

show that the ordinary differential equation for $s(x)$ admits polynomial solutions with regular behavior along the full extent of the basin, $-a \leqslant x \leqslant a$. Determine thereby the proper frequencies of a few modes and contrast these with estimates obtained from a perturbation analysis.

8. Let the equation

$$\frac{d^2\psi}{dx^2} + k^2(1 + \eta(x))\,\psi = 0, \qquad -\infty < x < \infty$$

pertain to excitations in a setting whose measure of inhomogeneity has a finite span, viz. $\eta(x) \neq 0$, x in $\Delta(-l/2, l/2)$. Establish the relationships

$$A_{++} = A_{--} = 2\frac{\alpha^2 - \beta^2 + i\alpha}{1 - \alpha^2 + \beta^2 - 2i\alpha}, \qquad A_{-+} = \frac{a_{-+}}{1 - \alpha^2 + \beta^2 - 2i\alpha}$$

$$A_{+-}/A_{-+} = a_{+-}/a_{-+}$$

where the quantities $A_{++}, \ldots$ enter into the scattering matrix representation

$$S = \begin{pmatrix} A_{-+} & 1 + A_{++} \\ 1 + A_{++} & A_{+-} \end{pmatrix}$$

and

$$a_{++} = a_{--} = \frac{ik}{2}\int_\Delta e^{-ikx}\,\eta(x)\,\varphi_+(x)\,\mathrm{d}x = 2i\alpha, \quad \alpha \text{ real}$$

$$a_{+-} = -a_{-+}^* = \frac{ik}{2}\int_\Delta e^{-ikx}\,\eta(x)\,\varphi_-(x)\,\mathrm{d}x, \quad |a_{-+}| = 2\beta$$

with functions $\varphi_\pm(x)$ that are complex-valued solutions of the integral equations

$$\varphi_\pm(x) + \frac{k}{2}\int_\Delta \sin\{k\,|x - x'|\}\,\eta(x')\,\varphi_\pm(x')\,\mathrm{d}x' = e^{\pm ikx}, \quad x \text{ in } \Delta.$$

Indicate how the foregoing characterizations bear on properties of the scattering matrix.

9. A linear theory of straight-crested waves in shallow water is derivable from the system of equations

$$\varphi_{xx} + \varphi_{yy} = 0, \quad -\varepsilon h(x) < y < 0, \quad -\infty < x < \infty$$

$$g\varphi_y + \varphi_{tt} = 0, \quad y = 0$$

and

$$\varphi_y + \varepsilon h_x \varphi_x = 0, \quad y = -\varepsilon h, \tag{*}$$

where the latter pair express boundary conditions at the free surface and rigid bottom, respectively. Employing an overbar to identify quantities at the free surface, viz.

$$\bar{\varphi}(x, t) = \varphi(x, 0, t)$$

and postulating that the velocity potential φ remains harmonic thereat, confirm the representations

$$\varphi_x = \bar{\varphi}_x - \int_y^0 \varphi_{x\xi}\, \mathrm{d}\xi$$

$$\varphi_y = -\frac{1}{g}\bar{\varphi}_{tt} - y\bar{\varphi}_{xx} - \int_y^0 (\xi - y)\varphi_{xx\xi}\, \mathrm{d}\xi \qquad -\varepsilon h < y < 0;$$

and thence, combining with (*), obtain the homogeneous integro-differential equation (Shinbrot, 1961)

$$\varepsilon(h\bar{\varphi}_x)_x - \frac{1}{g}\bar{\varphi}_{tt} = \frac{\partial}{\partial x}\int_{-\varepsilon h}^0 (y + \varepsilon h)\varphi_{xy}\, \mathrm{d}y$$

where left-hand member alone figures in the elementary analysis of shallow water wave excitation. Justify an estimate for the free surface elevation (exclusive of a time periodic factor),

$$\mathrm{i}\omega\bar{\varphi}/g = \varepsilon h^{-1/4}\left\{A \exp\left(\mathrm{i}\omega \int \frac{\mathrm{d}x}{\sqrt{(\varepsilon g h)}}\right) + B \exp\left(-\mathrm{i}\omega \int \frac{\mathrm{d}x}{\sqrt{(\varepsilon g h)}}\right)\right\} + \mathrm{O}(\varepsilon^{3/2}),$$

and compare the relative orders of magnitude, insofar as the small depth parameter ε is concerned, with the corresponding pair of terms that apply in the case of a fixed depth fluid, viz. $h = h_0$.

10. Adapt the equation of motion for small transverse displacements $y(x, t)$ of a string with variable tension, namely

$$\frac{\partial^2 y}{\partial t^2} - \frac{P(x)}{\rho}\frac{\partial^2 y}{\partial x^2} - \frac{1}{\rho}\frac{\mathrm{d}P}{\mathrm{d}x}\frac{\partial y}{\partial x} = 0,$$

to the circumstances wherein the string has a vertical equilibrium profile and a large magnitude of tension thoughout; and construct an approximate solution involving the tension at the midpoint.

11. Employ the approach and result given in §80 for a series development of wave-like solutions, when the scale of inhomogeneity is small in comparison with the wave length, to the specific cases of a point load on an otherwise uniform string and of a truncated oscillator potential (§81); and relate the outcome with prior deductions.

12. A graphical method of one-dimensional wave analysis developed by Schnyder and Bergeron (1937), with applications to mechanical, electrical and hydraulic systems, is based on theoretical connections of the form

$$\mathcal{P}(x, t) - \mathcal{P}(x_0, y_0) = \pm\frac{P}{c}(v(x, t) - v(x_0, t_0))$$

between the respective coordinate and time derivatives

$$\mathcal{P} = P\frac{\partial y}{\partial x}, \qquad v = \frac{\partial y}{\partial t}$$

of a function $y(x, t)$ which satisfies the homogeneous wave equation

$$\frac{\partial^2 y}{\partial x^2} = \frac{1}{c^2}\frac{\partial^2 y}{\partial t^2}, \qquad c^2 = P/\rho = \text{constant}$$

and may be identified with the displacement of a uniform string, under given tension; here x_0, t_0 specify any initial pair of coordinate and time variables, while the other pair refer to arbitrarily sited points x ($\gtrless x_0$) that are reached at a later time t ($>t_0$) after moving away from x_0 with the speed c. How are such connections altered when inhomogeneity is present and the wave equation for $y(x, t)$ contains a variable coefficient?

References

Airy, G. B. [1838]. On the intensity of light in the neighborhood of a caustic. *Camb. Phil. Trans.* **6**, 379–402.

Atkinson, F. V. [1960]. Wave propagation and the Bremmer series. *J. Math. Anal. Appl.* **1**, 255–276.

Baerwald, H. G. [1930/1]. Über die fortpflanzung von signalen in dispergierenden systeme. *Ann. Physik* **6**, 295–368 and subsequent parts.

Becker, E. [1965]. Die pulsierende quelle unter der freien oberfläche eines stromes endlicher tiefe. *Ingenieur-Archiv.* **24** (2), 69–76.

Beddoe, B. [1966]. Vibration of a sectionally uniform string from an initial state. *J. Sound Vib.* **4** (2), 215–223.

Bellman, R. and Kalabi, R. [1959]. Functional equations, wave propagation and invariant imbedding. *J. Math. and Mech.* **8**, 683–704.

Boulanger, A. [1913]. Étude sur la propagation des ondes liquides dans les tuyaux élastiques. Travaux et Mémoires de l'Université de Lille, Nouvelle série II. *Médicine-Sciences*, vol. 8.

Bremmer, H. [1949]. The propagation of electromagnetic waves through a stratified medium and its WKB approximation for oblique incidence. *Physica* **XV** (7), 593–608.

Brillouin, L. [1932]. Propagation des ondes éléctromagnétiques dans les milieux matériels. Congrès International d'Électricité, Paris. Rapport 18.

Broer, L. J. F. [1970]. On the dynamics of strings. *J. Engrg. Math.* **4**, 195–202.

Burke, W. L. [1971]. Gravitational radiation damping of slowly moving systems calculated using matched asymptotic expansions. *J. Mathematical Phys.* **12** (3), 401–418.

Cantor, M. [1907]. Zür bestimmung der lichtgeschwindigkeit nach Fizeau und akustische analogien. *Ann. Physik* **24**, 439–449.

Convert, G. [1964]. Interprétation des équations de dispersion. *Instabilité. Amplification. Amortissement. Ann. Radioélectricité* **19**, 299–320.

Copson, E. T. [1965]. *Asymptotic Expansions.* Cambridge University Press.

d'Alembert, J. le Rond [1747] *Mémoires de l'Académie de Berlin*, 214.

Debye, P. J. W. [1909] Näherungsformeln für die zylinderfunktionen für grosse werte des arguments und unbeschränkt veränderliche werte des index. *Math. Ann.* **67**, 535–558.

de Hoop, A. [1965] A note on the propagation of waves in a continuously layered medium. *Appl. Sci. Res.* B **12**, 74–80.

Drazin, P. G. [1963]. On one-dimensional propagation of long waves. *Proc. Roy. Soc. London*, Series A, **273**, 400–411.

Eckart, C. [1948]. The approximate solution of one-dimensional wave equations. *Rev. Modern Phys.* **20** (2), 399–417.

Euler, L. [1775]. Principia pro moto sanguinis per arterias determinando. Opera postuma.

Faddeev, L. D. [1967]. Properties of the *S*-matrix of the one-dimensional Schrödinger equation. *Amer. Math. Soc. Transl.*, Series 2, **65**, 139–166.

Filippi, P. [1975]. Wave phenomena in inhomogeneous media. *Quart. Appl. Math.* **33**, 337–350.

Fock, V. A. [1965]. Electromagnetic Diffraction and Propagation Problems. Pergamon Press.

Fröman, N. and Fröman, P. O. [1965]. JWKB Approximation: Contributions to the Theory. North-Holland Pub. Co.

Greenspan, H. P. [1963]. A string problem. *J. Math. Anal. Appl.* **6**, 339–348.

Hankel, H. [1869]. Die cylinderfunctionen erster und zweiter art. *Math. Ann.* **1**, 467–501.

Heading, J. [1962]. *An Introduction to Phase-integral Methods.* Methuen and Co.

Heading, J. [1972]. Further exact and approximate considerations of the barrier problem. *J. Inst. Math. Appl.* **10**, 213–324.

Helmholtz, H. von. [1898]. Vorlesungen über Theoretische Physik III. Mathematischen Principien der Akustik. Leipzig.

Herglotz, G. [1903]. Zür elektronentheorie. Göttingen Nach. **6**, 357–382.

Howe, M. [1971]. Wave propagation in random media. *J. Fluid Mech.* **45**, 769–783.

Jeffreys, H. [1931]. Damping in bodily seismic waves. Monthly Notices of the Royal Astronomical Soc., *Geophysical Suppl. II* (7), 318–323.

Kock, W. [1937]. The vibrating string considered as an electrical transmission line. *J. Acoust. Soc. Amer.* **8**, 227–233.

Korteweg, D. J. [1878]. Über die fortpflanzungsgeschwindigkeit des schalles in elastischen röhren. *Ann. Physik* **5**, 525–542.
Lamb, H. [1898]. On the velocity of sound in a tube, as affected by the elasticity of the walls. Memoirs and Proceedings of the Manchester Literary and Philosophical Society **9**, 1–16.
Lamb, H. [1900]. On a peculiarity of the wave-system due to the free vibrations of a nucleus in an extended medium. *Proc. London Math. Soc.* **32**, 208–211.
Landau, L. D. and Lifschitz, E. M. [1958]. *Fluid Mechanics.* Pergamon Press.
Langer, R. [1934]. The asymptotic solutions of certain linear ordinary differential equations of the second order. *Trans. Amer. Math. Soc.* **36**, 90–106.
Lauwerier, H. A. [1966]. *Asymptotic Expansions.* Mathematisch Centrum, Amsterdam Tract 13.
Levinson, N. [1949]. On the uniqueness of the potential in a Schrödinger equation for a given asymptotic phase. Kgl. Danske Videnskab Selskab. *Mat. fys. Medd* **25**, No. 9, 1–29.
Lewis, R. [1964]. Asymptotic theory of transients. *Electromagnetic Wave Theory Symposium* (Delft). Pergamon Press, 845–869.
Ludwig, D. [1966]. Wave propagation near a smooth caustic. *Bull. Amer. Math. Soc.* **71** (5), 776–779.
Noether, E. [1918]. Invariante variationsprobleme. *Göttingen Nach.*, 235–257.
Radakovič, M. [1899]. Über die bewegung einer saite unter der einwirkung einer kräft mit wanderndem angriffspunkt. *Sitz. der Wiener Akademie* **198**, 577–612.
Riemann, B. [1860]. Über die fortpflanzung ebener luftwellen von endlicher schwingungswerte. *Abh. Königliche Ges. Wiss. Göttingen* **8**, 43–62.
Roseau, M. [1975]. *Asymptotic Wave Theory.* North-Holland Pub. Co. Series in Applied Mathematics and Mechanics.
Roy, L. [1914]. Sur le mouvement des milieux visqueux et les quasi-ondes. *Mémoires présentées par divers savants à l'Académie des Sciences de l'Institut de France* **35**, 1–64.
Saxon, D. [1959]. Modified WKB methods for the propagation and scattering of electromagnetic waves. Symposium, University of Toronto.
Schmidt, A. C. and Wu, T. T. [1974]. Theory of the One-dimensional Inverse Scattering Problem with Dissipation. Thesis (of Schmidt) presented to the Division of Engineering and Applied Physics, Harvard University.
Schrödinger, E. [1914]. Zür dynamik elastisch gekoppelter punktsysteme. *Ann. Physik* **44**, 916–934.
Sturrock, P. [1958]. Kinematics of growing waves. *Phys. Rev.* **112** (5), 1488–1503.
van Kampen, N. G. [1967]. Higher corrections to the WKB approximation. *Physica XXXV*, 70–79.
Watson, G. N. [1944]. A Treatise on the Theory of Bessel Functions, 2nd ed., Cambridge/Macmillan.
Whitham, G. B. [1965]. A general approach to linear and nonlinear dispersive waves using a Lagrangian. *J. Fluid Mech.* **22**, 273–283.
Wu, T. T. [1966]. Born approximation and large-momentum transfer processes in potential scattering. *Phys. Rev.* **143** (4), 1110–1116.
Young, T. [1808]. Croonian lecture on the functions of the heart and arteries. *Philos. Trans. Roy. Soc. London*, Series II, 164–186.

Other books

Achenbach, J. D. [1973]. *Wave Propagation in Elastic Solids.* North-Holland Pub. Co.
Macke, W. [1962]. *Wellen.* Akademische Verlagsgesellschaft Leipzig.
Tolstoy, I. [1973]. *Wave Propagation.* McGraw-Hill.
Whitham, G. B. [1973]. *Linear and Nonlinear Waves.* John Wiley.

Index